Hydrological Processes of the Danube River Basin

Mitja Brilly
Editor

Hydrological Processes of the Danube River Basin

Perspectives from the Danubian Countries

Springer

Editor
Mitja Brilly
University of Ljubljana
Fac. Civil & Geodetic Engineering
Jamova Cesta 2
SI-1000 Ljubljana
Slovenia
mbrilly@fgg.uni-lj.si

ISBN 978-90-481-3422-9 e-ISBN 978-90-481-3423-6
DOI 10.1007/978-90-481-3423-6
Springer Dordrecht Heidelberg London New York

Library of Congress Control Number: 2010930853

Cover images: photo of the Iron Gate provided by Dr. Ulrich Schroeder from the German IHP Committee; photo of the Danube near Belgrade, Serbia, provided by Mr. Borivoje Jovanović.

Printed on acid-free paper

Springer is part of Springer Science+Business Media (www.springer.com)

Preface

The Danube River Basin is shared by 19 countries and there is no river basin in the world shared by so many nations. Europe's second largest river basin with a total area of about 800,000 km^2 is also home to 83 million people of different cultures, languages and historical backgrounds.

Management of common water sources and overcoming difficulties caused by droughts and floods requires co-operation between the countries. In 1971 these common interests stimulated the hydrologists of – at that time – eight Danube countries to begin regional co- operation in the framework of the International Hydrological Decade of UNESCO. The result of this research was The Hydrological Monograph of the Danube and its Catchment, which was published in 1986. Since 1975 this co-operation has continued under the umbrella of the International Hydrological Programme (IHP) of UNESCO. In the past 20 years political turbulence has caused an increase in the number of countries, making the co-operation difficult at times. Nevertheless several projects were launched, and a number of reports were produced. At the 22nd Working Meeting of the Regional Hydrological Co-operation of the Danube Countries in the framework of IHP/UNESCO which took place at the XXIVth Conference of the Danubian Countries, in Bled, Slovenia, in 2008, it was decided that those reports would be summarised in one special publication that would include a few additional reports on the common research theme. The editorial board was also nominated. This present publication is the result of a major collaboration and our efforts to bring together the reports and research papers related to the regional co-operation.

Acknowledgements

The book is the result of a long co-operation in research, which included the efforts of many hydrologists and technical staff from different Universities and Agencies from all countries of the Danube River Basin. The co-operation was undertaken under the umbrella of UNESCO and we are very grateful for its financial support. The research was funded from national sources of individual countries. The IHP National committee and the UNESCO Commission of Slovenia supported the final editing.

Contents

Contributors

Mitja Brilly Faculty of Civil and Geodetic Engineering, University of Ljubljana, Ljubljana 1000, Slovenia, mbrilly@fgg.uni-lj.si

Uwe Ehret Technical University Munich, Department of Hydrology and River Basin Management, Munich, Germany

Lidija Globevnik Faculty of Civil and Geodetic Engineering, University of Ljubljana, Ljubljana, Slovenia, Lidija.Globevnik@guest.arnes.si

László Goda sr. Research Institute for Environment and Water (VITUKI), Budapest H-1453, Hungary, godalaszlo@gmail.hu

Christine Hangen-Brodersen Flood Forecast River Main, Bavarian Environment Agency, Hof, Germany

Ester Heath Department of Environmental Sciences, "Jožef Stefan" Institute, Ljubljana, Slovenia

Ladislav Holko Institute of Hydrology of Slovak, Academy of Sciences, Bratislava, Slovakia

Franz-Klemens Holle Flood Information Centre, Bavarian Environment Agency, Munich, Germany

Aleksandra Ilic Jaroslav Cerni Institute for the Development of Water Resources, Pinosava 11226, Belgrade, Serbia, stevan.prohaska@jcerni.co.rs

Zdeněk Kostka Institute of Hydrology of Slovak, Academy of Sciences, Bratislava, Slovakia

Stefan Laurent State Office for Water Management, Kempten, Germany, stefan.laurent@wwa-ke.bayern.de

Inke Meyer Flood Information Centre, Bavarian Environment Agency, Munich, Germany

Pavol Miklánek Institute of Hydrology of Slovak, Academy of Sciences, Bratislava, Slovakia

Domokos Miklós Surface Waters, Research Institute for Environment and Water (VITUKI), Budapest H-1453, Hungary, domokosm@vituki.hu

Matjaž Mikoš Faculty of Civil and Geodetic Engineering, University of Ljubljana, Ljubljana, Slovenia, Lidija.Globevnik@guest.arnes.si

Radmila Milačič Department of Environmental Sciences, "Jožef Stefan" Institute, Ljubljana, Slovenia

Katja Moritz Flood Information Centre, Bavarian Environment Agency, Munich, Germany

Katarína Mravcová Water Research Institute, Bratislava, Slovakia

Ferenc Neppel Groundwaters, Research Institute for Environment and Water (VITUKI), Budapest H-1453, Hungary

Nives Ogrinc Department of Environmental Sciences, "Jožef Stefan" Institute, Ljubljana, Slovenia

Matej Padežnik Faculty of Civil and Geodetic Engineering, University of Ljubljana, Ljubljana, Slovenia

Sašo Petan Faculty of Civil and Geodetic Engineering, University of Ljubljana, Ljubljana, Slovenia

Péter Kovács Middle-Danube-Valley Environmental and Water Management Directorate, Budapest H-1088, Rákóczi út 41, Hungary, kovacs.peter@kdvvizig.hu

Ana Petkovšek Faculty of Civil and Geodetic Engineering, University of Ljubljana, Ljubljana, Slovenia

Pavel Petrovič Water Research Institute, Bratislava, Slovakia

Stevan Prohaska Jaroslav Cerni Institute for the Development of Water Resources, Pinosava 11226, Belgrade, Serbia, stevan.prohaska@jcerni.co.rs

László Rákóczi Research Centre for Environment and Water (VITUKI), Budapest, Hungary, lrakoczi@t-online.hu

Jenő Sass Surface Waters, Research Institute for Water and Environment (VITUKI), Budapest H-1453, Hungary, sass@vituki.hu

Janez Ščančar Department of Environmental Sciences, "Jožef Stefan" Institute, Ljubljana, Slovenia

Heinz Schiller Bavarian Environment Agency, Augsburg D-86177, Germany, hz-schiller@alice-dsl.net

Alžbeta Stančíková Water Research Institute, Bratislava, Slovakia, alzbeta.stancik@stonline.sk

Andrej Vidmar Faculty of Civil and Geodetic Engineering, University of Ljubljana, Ljubljana, Slovenia

Alfons Vogelbacher Flood Information Centre, Bavarian Environment Agency, Munich, Germany

Editorial Board

Ognjen Bonacci
Split University
Faculty of Civil Engineering and Architecture
Matice hrvatske Str. 15
21000 Split
Croatia
obonacci@gradst.hr

Hans Peter Nachtnebel
Univ. of Natural Resources and Applied Life Sciences
Gregor Mendel Straße 33
A-1180 Wien
Österreich
hans_peter.nachtnebel@boku.ac.at

Ulrich Schroeder
IHP-HWRP Sekretariat, Bundesanstalt für Gewasserkunde
Am Mainzer Tor 1,
D-56068 Koblenz,
Germany
schroeder@bafg.de
tel: 0049 261 1306 5440
fax: 0049 261 1306 5422

Domokos Miklós
Research Institute for Environmental Protection and Water Management
(VITUKI Kht.)
Pf. 27.,
Kvassay Jeni Utc 1
H-1453, Budapest
Hungary
domokosm@vituki.hu
tel: 0036 1 215 6140
fax: 0036 1 216 1514

Chapter 1
History and Results of the Hydrological Co-operation of the Countries Sharing the Danube Catchment (1971–2008)

Domokos Miklós

Abstract In 1971, the hydrologists of the (at that time) eight Danube Countries launched, on a voluntary basis, a regional co-operative endeavor with the aim of producing consistent hydrological information for the whole Danube Catchment, which has an area of 817,000 km^2. Since 1987, this co-operation has been carried out within the framework of the International Hydrological Programme (IHP) of UNESCO, until 2008 co-ordinated by the consecutive IHP National Committees of Germany, Austria, Slovakia, Hungary, and Serbia. Herein I briefly describe the history of this co-operation, focusing on its results published in the Hydrological Monograph of the Danube and its Catchment (1986), as well as in the 13 thematic follow-up volumes issued to date. The information included in these publications provides a sound and consistent informational basis for integrated water management, according to the EU Water Framework Directive, embracing the whole Danube Catchment. Finally, a short comparison is also offered with other international regional collaborative hydrological studies (Rhine Monograph, FRIEND).

Keywords Danube Countries · Danube Catchment · IHP/UNESCO · Regional hydrological co-operation · Hydrological monograph

1.1 The Danube River and Its Catchment

The Danube River, crossing Central and South-eastern Europe over a length of 2,857 km, is with its multi-annual mean discharge of 6,855 m^3 s^{-1} into the Black Sea, the 21st largest river in the World and – after the Volga River – the 2nd largest river in Europe. Its river basin is situated – between the headwater regions of the Rhine and the Dniepr River – on the southern side of the European main continental divide that runs from Gibraltar to the Northern Ural Mountains (Fig. 1.1). The

D. Miklós (✉)
Surface Waters, Research Institute for Environment and Water (VITUKI),
Budapest H-1453, Hungary
e-mail: domokosm@vituki.hu

M. Brilly (ed.), *Hydrological Processes of the Danube River Basin*,
DOI 10.1007/978-90-481-3423-6_1,

Fig. 1.1 Geographical situation of the Danube Catchment (RCDC 1999a)

bee-line distance between the springs of the Danube in the Black Forest Mountains and its embouchure into the Black Sea is 1,630 km. The highest points of its – orographically strongly articulated – watershed divide are: on its southern reach, Piz Bernina (altitude 4,052 m above sea level), and on the northern one, Peak Kriváň (2,496 m). The average altitude of the Catchment is 475 m above sea level. It can be subdivided into three main parts, each of them characterized by different hydro- and geographical features: the Upper Danube Region (between the springs and the Devín Gate near to the mouth of the tributary March/Morava), the Central Danube Region, and the Lower Danube Region (between the Iron Gate and the Danube's embouchure into the Black Sea).

1.2 The Countries Sharing the Danube Catchment

Out of the area of 10,508 million km^2 of Europe, situated between the western coast of Ireland and the Ural Mountains, the Danube Catchment's share is 0.817 million km^2 (7.8%). Out of the 783 million inhabitants of the continent, 82.7 million

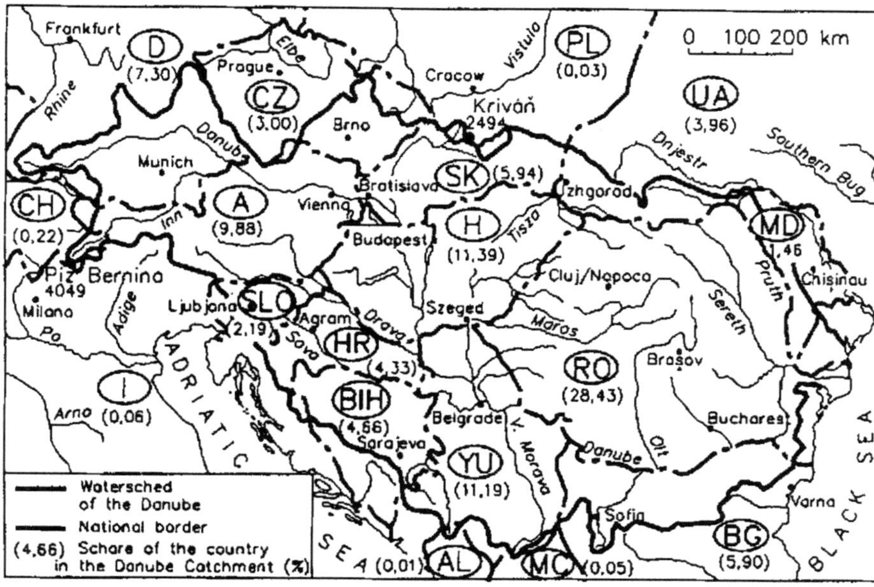

Fig. 1.2 The countries sharing the Danube Catchment in 2000

people (10.55%) live in the Danube Catchment. In the year 2008, there are 18 countries with various superficial shares in the Catchment (Fig. 1.2); among them are 13 countries, with superficial shares between 1.46 and 28.43%, which are considered and hereinafter called "Danube countries" (in a narrower sense), the largest shares in the Catchment belonging to: Romania (28.43%), Hungary (11.39%), Serbia (11.19%), and Austria (9.88%). These four countries in addition to Slovakia (5.94%) and Slovenia (2.19%) lie with 90–100% of their respective national areas within the Danube Catchment. The sum of the superficial percentages of the five countries with minor shares (0.01–0.22%) in the Catchment, which may also be called "peripheral" countries (as related to the Danube Catchment), is merely about 0.4% (Table 1.1).

Table 1.1 contains not only a more detailed characterization of the political subdivision of the Danube Catchment in 2008, as described above, but also its constant status during the – almost half-a-century long – period between the end of Second World War (1945) and the political changes that took place around 1990 which led to further subdivision of various Danube countries (thus the latter status was valid also in 1971, which marked the beginning of the multi-lateral hydrological co-operation we are dealing with in this chapter). It can be seen from Table 1.1 that in 1971, compared to the present 13 + 5 countries, there were only eight "Danube countries" and four "peripheral" countries.

Table 1.1 The countries sharing the Danube Catchment in 2008 and 1971

Status of 2008										Status of 1971			
	Country		Area (1,000 km^2)		Share of		Inhabitans			Country			
No.	Symbol	Name	Total	In the catchment	country in the catchment	catchment in the country	Total (10^6)	in the catchment (10^6)	in the catchment %	Symbol	Name	Share in catchment (%)	No.
					%	%							
"Danube contries" (with major areal shares), ca. in hydrographic order													
1.	D	Germany	357.0	56.2	7.30	16.8	82.1	9.1	11.1	D	Germany	7.30	1.
2.	A	Austria	83.9	80.4	9.88	96.4	8.1	7.7	9.4	A	Austria	9.88	2.
3.	CZ	Czech Rep.	78.9	21.7	3.00	31.1	10.3	2.8	2.5	CS	Czechoslovakia	8.94	3.
4.	SK	Slovakia	49.0	47.1	5.94	99.0	5.4	5.2	6.4				
5.	H	Hungary	93.0	93.0	11.39	100.0	10.0	10.0	12.2	H	Hungary	11.39	4.
6.	SLO	Slovenia	20.3	16.4	2.19	88.8	2.0	1.7	2.1	YU[a]	Yugoslavia	22.42	5.
7.	HR	Croatia	56.6	35.0	4.33	62.5	4.8	3.2	3.8				
8.	BIH	Bosnia and Herzegovina[a]	51.1	36.6	4.66	74.9	3.8	2.9	3.5				
9.	SR	Serbia	88.4	81.6	9.9	92.3	9.8	8.7	10.5				
10.	CG	Montenegro	13.8	7.1	0.9	51.4	0.6	0.3	0.5				
11.	RO	Romania	237.5	232.2	28.43	97.6	22.6	22.6	27.6	RO	Romania	28.43	6.
12.	BG	Bulgaria	110.9	47.4	5.90	43.6	8.3	3.9	4.8	BG	Bulgaria	5.90	7.
13.	MD	Moldova	33.7	12.8	1.46	35.6	4.3	1.1	1.2	SU	Soviet Union	5.42	8.
14.	UA	Ukraine	603.7	30.5	3.96	5.4	50.9	3.1	3.7				
"Peripheric countries" (with minor areal shares)													
15.	CH	Switzerland	41.3	1.8	0.22	4.4	6.7	0.3	0.4	CH	Switzerland	0.22	9.
16.	I	Italy	301.3	0.6	0.06	0.2	57.5	0.1	0.1	I	Italy	0.06	10.
17.	PL	Poland	312.7	0.4	0.03	0.1	37.8	0.04		PL	Poland	0.03	11.
18.	AL	Albania	28.7	0.1	0.01	0.01	3.2			AL	Albania	0.01	12.
19.	MC[a]	Macedonia	25.7	0.1	0.01	0.2	2.1			–	–	–	–
1–19.	Danube Catchment total			817.0	100.00			82.74	100.0	Danube Catchment total		100.00	1–12.

[a]Macedonia, in 2008 independent, was in 1971 a part of Yugoslavia.

1.3 Beginnings of Hydrological Co-operation in the Danube Catchment

As is generally known, the activities necessary for sustainable development in a major geographic region – e.g., in a river basin – can efficiently be planned only on the basis of deep and reliable knowledge about the natural and social conditions and expectations characterizing the whole region, independently from the political borders crossing it.

Hence, indispensable for the planning of optimal measures to be taken for the welfare of all inhabitants of the Danube Catchment – dispersed in a number of countries – would also be the availability of knowledge characterizing the physiographic conditions of the whole Catchment, emphatically including its hydrological features. Such knowledge (data and relationships), independent of national borders, can be collected, systemized and made accessible for utilization (first of all for the planning of catchment-wide water resource development) only through the co-operation of competent experts from all countries existing in the Catchment.

It was this perception that prompted the development of the Working Group for Scientific Hydrology, founded under the chairmanship of K. Stelczer (Hungary) in 1967, of the international Danube Commission (DC) (already residing at that time in Budapest), established, under the regulations of the Belgrade Convention of 1948 in order to regulate the navigation conditions of the Danube, as an international waterway. A suggestion by the Soviet delegate Prof. Korzun in 1971 led to the addition of the new topic of compilation of the Danube Catchment water balance to the already numerous ongoing themes of the working plan, i.e., evapotranspiration, river training, storage reservoir dimensioning. The main purpose of the inclusion of this theme was to collect, harmonize, and publish information characterizing the multi-annual mean water cycle of the Danube Catchment.

Unfortunately, from the at that time eight Danube countries (Table 1.1) only four: Czechoslovakia, Hungary, Bulgaria, and the Soviet Union delegated experts to the Working Group of Scientific Hydrology of the Danube Commission, while the remaining four countries: the Federal Republic of Germany (FRG), Austria, Yugoslavia, and Romania abstained from participation (due to political considerations contesting the competency of the DC for the Danube River Basin beyond the Danube River itself). The FRG, however, although not being a member at that time of the DC, consequently delegated observers to the meetings of the Working Group.

The research institutes of the four countries participating in the activities of the Working Group, compiled, according to their working plan, under the co-ordination of the Water Resources Research Institute (VÚVH), Bratislava, Czechoslovakia, from the contributions of the co-operating countries (Kardos 1975) already in 1972–1973 the multi-annual average water balances for their respective national shares in the Danube Catchment, including isoline maps of the three main components of the water balance (precipitation, evapotranspiration, and runoff). The professional-informative value of this output was of course rather limited, due to the "white patches" of the non-co-operating countries as well as to the impossibility of harmonizing national information with the non-co-operating countries, along the state borders. For example, it is of considerably less value than the former works of two outstanding Danube hydrologists (Kresser and Lászlóffy 1964, Lászlóffy 1965), which – although containing consistent information – also are devoid of any co-operation of the community of hydrologists from all Danube countries. At the same time, the – at that time for most experts baffling – publication of the significantly incomplete water balance by the VÚVH, Bratislava had an important scientific-diplomatic importance. Namely, this publication was a strong, if not compelling, incentive for the hydrologists of the four formerly non-co-operating Danube countries to join the common undertaking since it became obvious that no reliable work of this kind would be possible without the co-operation of all Danube countries.

The framework of the thus eight Danube countries was established in 1974 – after a number of conciliatory meetings, resulting in a politically influenced "hybrid" solution which nowadays may seem rather anachronistic – in such a way that while the first group of countries (BG, CS, H, SU) continued its activity under the auspices of the Working Group of the Danube Commission, under the co-ordination of the VÚVH, Bratislava (Dr A. Sikora and Dr A. Stančík), the hydrologists of the newly attached countries (A, D, YU, RO) participated in the common work through their respective National Committees for the International Hydrological Programme (IHP) of UNESCO, under the co-ordination of a Technical Secretariat established for this purpose at Belgrade (Prof. S. Jovanovič and Dr G. Bozič). The co-ordinators of Bratislava and Belgrade spontaneously synchronized their activities. There were two common languages of this co-operation: German and Russian.

1.4 First Phase of the Danubian Co-operation (1974–1986): Publication of the Danube Monograph

Together with the number of the co-operating countries, the subject-matter of the collaborative work – which in 1971 began as the water balance of the Danube Catchment – was also expanded. The goal of the hydrologists of the by this time eight Danube countries was to compile their work under the title Hydrological Monograph of the Danube Catchment, containing the following three chapters (in brackets, the name of the country responsible for the given chapter):

1. Physical, geographical and water management characteristics of the Danube Catchment (Yugoslavia)
2. Characteristics of the runoff regime (Soviet Union) (Figs. 1.3, 1.4 and 1.5)
3. Hydrological balance (Hungary) (Table 1.2)

It should be noted that originally six chapters were planned for the Monograph. Later, three of them – dealing with water quality, sediment regime, and ice conditions – were, for the time being, cancelled with the idea that they should form follow-up volumes to the Monograph.

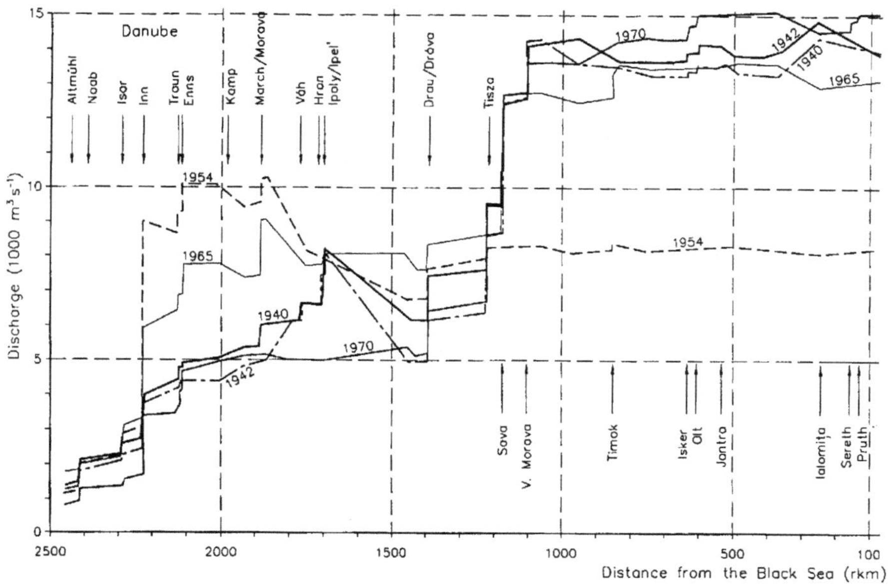

Fig. 1.3 Longitudinal profiles of the greatest floods along the Danube (RZD 1986)

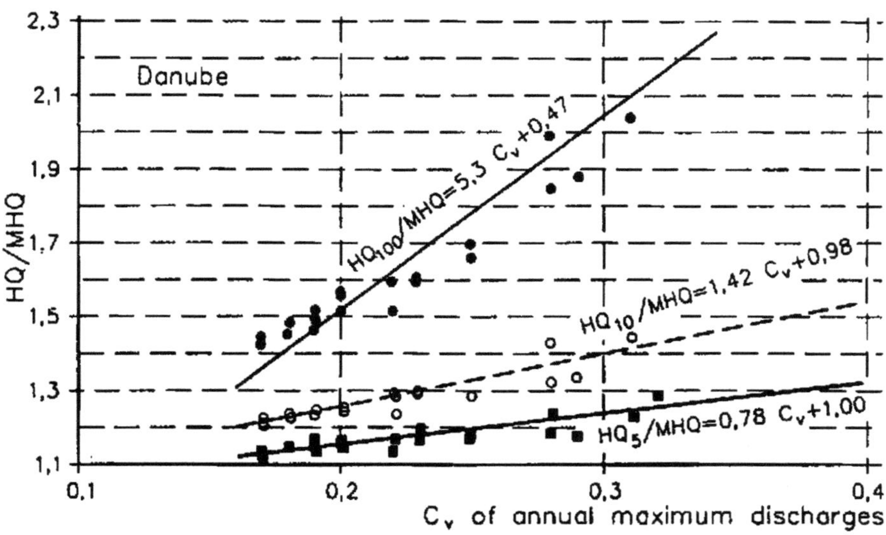

Fig. 1.4 Relationship between the average flood discharge ($Q_{g,m}$), the flood discharges of selected probabilities of nonexceedance ($Q_{g,p\%}$) and the variation coefficient of the annual maximum discharges (C_V) at the gauging stations of the Danube (RZD 1986)

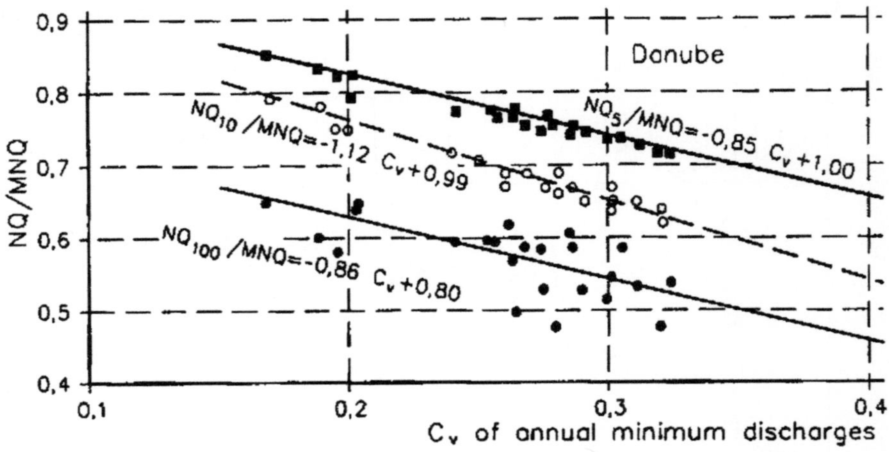

Fig. 1.5 Relationship between the average value of the annual low discharges ($Q_{s,m}$), the low discharges of selected probabilities of exceedance ($Q_{s,p\%}$) and the variation coefficient of the annual minimum discharges (C_v) at the gauging stations of the Danube (RZD 1986)

All eight co-operating Danube countries supplied information for all three chapters. The three isoline maps (scale 1:2,000,000), displaying the regional distribution of the for the period 1931–1970 calculated average values of the three main water balance components (precipitation, evapotranspiration, and runoff), were compiled and issued by the Water Resources Research Center (VITUKI) Budapest, Hungary.

Because of various difficulties of organizational character, the work for the Danube Monograph became rather protracted: it lasted, instead of the originally foreseen 3–4 years, altogether 12 years. Its final editing took place, under the leadership of H. Schiller, in Germany.

The German language version of the Danube Monograph was published in 1986 in Munich, Germany (RZD 1986, Schiller 1989) and included three volumes: I. Text, II. Tables, and III. Maps. The Russian version was issued in 1989 in Leningrad (RSPS 1989). Typical results of the Danube Monograph are displayed in Figs. 1.3, 1.4 and 1.5. Prompted by UNESCO, also a quadrilingual (English, French, German, and Russian) so-called representative (abridged) version was compiled and published in 1988 in Bratislava, Czechoslovakia (Stančík and Jovanovič 1988).

The volumes of the Monograph were disseminated among a wide circle of interested institutions of all Danube countries. Further copies can still be requested from the IHP National Committees of these countries.

From Chapter 3 of the Monograph, prepared in Hungary, a condensed technical paper was also compiled and published not only in an international journal (Domokos and Sass 1990), but also in the national languages of all the eight Danube countries of that time (Altmann 1999).

An interesting feature of Chapter 3 of the Monograph is that it contains not only the multi-annual water balances of the 47 subcatchments of the Danube Basin, but also those of the national partial areas of the 12 countries concerned in 1986 (eight Danube and four peripheral countries) (Table 1.2, RZD 1986).

Table 1.2 Average hydrological balance of the national partial areas in the Danube Catchment for the period 1941–1970

Country's			Mean multi-annual value					Own surface water resources				
No. n	Symbol	Name	Precipitation, P_j	Evapotranspiration, E_j	Runoff, R_j	Balance error $d_j = \frac{P_j-(E_j+R_j)}{P_j}100$	Runoff coefficient $\alpha_j=\frac{R_j}{P_j}$	$W_j=10^{-6}R_jA_{jD}$	$Q_j=\frac{W_j}{0,031536}$	Share in water resources of the Catchment $\gamma_j=\frac{Q_j}{Q_D}100$	Outflowing (not returning) water resources $Q_{j,out}$	Share of own w.r. in outflowing water resources $\gamma_j=\frac{Q_j}{Q_{j,out}}100$
			$mm\ a^{-1}$			%	–	$km^3\ a^{-1}$	$m^3\ s^{-1}$	%	$m^3\ s^{-1}$	%
1.	D	Germany	962	528	415	+1.98	0.43	4.75	785	11.48	867	90.54
2.	I	Italy	1,425	391	1,139	−7.37	0.80	0.54	17	0.25	17	100.00
3.	CH	Switzerland	1,136	358	768	+0.88	0.68	1.40	44	0.64	51	86.27
4.	A	Austria	1,098	508	600	−0.91	0.55	48.44	1,536	22.45	2,392	62.21
5.	PL	Poland	959	525	372	+6.46	0.39	0.10	3	0.04	3	100.0
6.	CS	Czech Rep.	719	513	215	−1.25	0.30	15.70	498	7.28	1,531	32.53
7.	H	Hungary	609	539	60	+1.64	0.10	5.58	176	2.57	3,525	4.99
8.	AL	Albania	1,875	494	1,292	+4.75	0.69	0.13	4	0.06	4	100.00
9.	YU	Yugoslavia	928	606	351	−3.12	0.38	64.31	2,039	29.81	5,851	34.85
10.	BG	Bulgaria	661	487	152	+3.33	0.23	7.32	232	3.39	3,149	7.37
11.	R	Romania	752	557	160	+4.65	0.21	37.16	1,177	17.21	6,765	17.40
12.	SU	Soviet Union	748	509	235	+0.53	0.31	10.40	330	4.82	3,458	9.54
1-12	Danube Catchment		P_D=816	E_D=547	R_D=264	d_D=+0.60	α_D=0.32	W_D=215.83	Q_D=6,841	100.00		

Table 1.3 Major regional hydrological co-operations in Europe, before 2003

Name of co-operation	Areal extension (10^6 km^2)	Number of countries	Phases	Results
Rhine Monograph	0.16	5	I: 1967–1977	Rhine Monograph
			II: 1978–2002	Follow-up volumes
Danube Monograph	0.82	8	I: 1972(75)–1986	Danube Monograph
		13	II: 1987–1992	Two follow-up volumes
			III: 1993–1998	One follow-up volume
		13	IV: 1999–2002	Four follow-up volumes
FREND/FRIEND/ /NE-FRIEND	≈2.5 (≈7.5)	13	I: 1985–1989	FREND report (three volumes)
		17	II: 1990–1993	FRIEND report (three volumes)
		≈22	III: 1994–1997	FRIEND report (one volume)
		≈22	IV: 1998–2002	FRIEND report (one volume)

It is rather enlightening to compare the co-operation of the Danube hydrologists, resulting first in the publication of the Danube Monograph, with the two other most important international hydrological co-operations, i.e., with the activities of the five countries of the Rhine Catchment, yielding also a monograph (CIHBR 1977) and a series of follow-up volumes (thus serving, in a way, as a pattern for the Danubian co-operation) and the hydrological co-operation of the – at its beginning – 13 countries of Northern and Western Europe, launched in 1985 and symbolized, subsequently, with the acronyms FREND, FRIEND, and

Table 1.4 Comparison of the FREND-report of 1989 and the Danube Monograph of 1986 (Domokos 1994)

		FREND	Danube monograph
Data basis	Discharges of great catchments	–	50 stations (average length; 40 years)
	Discharges of small catchments (<500 km^2)	2,100 stations (average length: 17 years)	–
	Catchment characteristics	+	–
	Climatic information	+	–
Empirical relationships	For ungauged small catchments	Many (cluster principle)	–
	For major catchments	–	Few (isoline maps, longitudinal profiles, Figs. 1.3, 1.4, and 1.5, etc.)
Water balances		–	(e.g., Table 1.2)
Experimental/representative basins		+	–

NE-FRIEND, resulting each fourth year in a general conference and the publication of report(s) (FREND 1989, FRIEND 1993, 1997, 2002).

A survey of these three co-operative endeavors is offered in Table 1.3, while the basic features of the first FREND-report (FREND 1989) and of the Danube Monograph (RZD 1986) are compared in Table 1.4. A detailed, analytical comparison of the latter two works was also published (Domokos 1994). According to one of the basic statements of the latter analysis, the expert community compiling the Danube Monograph concentrated its activity – as contrasted with that of the authors of the first FREND-report – indeed on problems that cannot be solved without international co-operation, while its methods are more transparent and its results meet more directly practical requirements (Table 1.4).

1.5 The "Constituent" Expert Meeting of the Danubian Co-operation

After the publication of the Danube Monograph in 1986, the representatives of the IHP National Committees of the Danube Countries held in April 1987 a meeting in Budapest, Hungary. The participants of this meeting declared that an institutional continuation of the co-operation was necessary, following the pattern of the hydrological co-operation of the Rhine countries, but it should be carried out henceforth completely under the aegis of IHP/UNESCO. At the meeting, also the Principles (constitution) of the co-operation were formulated and accepted.

The Principles declare, among others, the following:

- The goal of the co-operation is to jointly elaborate scientific themes of common interest, concerning the whole Danube Catchment (i.e., first of all the topics provisionally left out from the Monograph) as well as to periodically update selected parts of the Monograph itself, to publish the results of these efforts – as with the Rhine Monograph (CIBHR 1977) – in the form of follow-up volumes to the Danube Monograph of 1986. The works on each such theme should be co-ordinated and financed by one of the Danube countries or institutions, on a voluntary basis.
- In order to promote the realization of the goal of the co-operation, every year (at least) one expert meeting must be held, alternating in the various Danube countries.
- The Chief Co-ordinator of the co-operation should be, for a determined period, on a voluntary basis, one institution or one expert from a Danube country, recommended by the IHP National Committee of that country and elected/accepted by the other participants of the co-operation.
- The working languages of communication and publication are henceforward German and Russian (as in the past).

The Principles, as one of the annexes of the official report of the expert meeting of April 1987, Budapest, were signed by the representatives of the IHP National Committees of all the (at that time) eight Danube countries. Today, although there are already as many as 13 Danube countries (Table 1.1), the Principles are practically still considered as basic guidelines of the regional co-operation, although there have been some – so far unsuccessful – attempts to update them. It is mostly the introduction of English as the third (or the only) working language that is being suggested by some countries. Although no such – or any other – formal modification of the Principles has taken place so far, in practice not only the introduction of English as the (third, and increasingly predominant) working language of communication has been adopted, but also some of the follow-up volumes of the Danube Monograph have been published in English (RCDC 1999a, b, 2000, 2004a, b, c, d, 2006, 2008), despite of the formally still effective rules of the Principles.

1.6 Further Phases of Co-operation After 1987

According to a proposal of the Soviet delegation during the "constituting" expert meeting of 1987, the order of succession of Chief Coordinators of the co-operation should possibly follow, in hydrographic order, the Danube itself from its springs to its Delta. This succession, although not fixed in a written form, has prevailed until now. Accordingly, the Chief Co-ordinators have been:

- During the IInd phase of the co-operation (1987–1992): the German IHP/OHP National Committee (Prof. K. Hofius)
- during its IIIrd phase (1993–1998): the Austrian IHP National Committee (Prof. F. Nobilis, Dr O. Behr, and Prof. D. Gutknecht)
- during its IVth phase (1999–2002): the Slovak IHP National Committee (Dr P. Miklánek and Dr P. Petrovič)
- during its Vth phase (2003–2005): the Hungarian IHP/OHP National Committee (Dr M. Domokos)
- during its VIth phase (2006–2008): the Serbian IHP National Committee (Prof. M. Miloradov)

As prescribed in the Principles', the main topic both of the regular annual expert meetings and the short extraordinary meetings – attached to the biennial scientific Danube Conferences, organized, since 1961, independently from the regional hydrological co-operation – is a survey of the progress of the works foreseen in the working plan of the co-operation and the promotion of publication of follow-up volumes to the Danube Monograph.

The regular annual expert meetings took place in the following towns (Hofius and Schröder 1997): 1987: Budapest, Hungary; 1988: Vienna, Austria; 1989: Sofia, Bulgaria; 1990: Kiyv, Soviet Union; 1991: Straubing, Germany; 1992: Kelheim, Germany; 1993: Tulcea, Romania; 1994: Malé Vozokany, Slovakia and Vienna, Austria; 1995: Eckartsau, Austria and Smolenice, Slovakia; 1996: Lednice, Czech Republic; 1997 (realized in January 1998): Vladaia, Bulgaria; 1998: Budapest, Hungary; 1999: Bratislava, Slovakia; 2000: Postojna, Slovenia; 2001: Deggendorf, Germany; 2002: Zagreb, Croatia; 2003: Sofia, Bulgaria; 2004: Brno, Czech Republic; 2005: Passau, Germany; 2006: Beograd, Serbia; 2007: Novi Sad, Serbia; 2008: Bled, Slovenia.

The expert meetings are chaired by the representative of the IHP National Committee of the country actually serving as Chief Co-ordinator.

The joint work of the period 1987–2008 resulted in 13 follow-up volumes to the Danube Monograph of 1986, as listed in Table 1.5 (each of format 210 × 300 mm), from which selected typical illustrations are presented in Figs. 1.6, 1.7, 1.8, 1.9, 1.10, 1.11, 1.12 and 1.13. Note that the three subprojects, No. 5.1, 5.2, and 5.3 (i.e., the corresponding follow-up volumes No. VIII/1, VIII/2, and VIII/3) have the aim of updating respective chapters of the Danube Monograph of 1986. The 13 follow-up volumes were duly disseminated among the Danube countries and also discussed in various scientific journals (Behr 1995, Belz 2000, Domokos 2000, Schiller 1989, etc.).

Note that four of the 13 volumes were compiled under Hungarian co-ordination; the abridged versions of three of them have also been published so far in Hungarian (Rákóczi 1993, Goda 1995, Neppel et al. 1999) and German scientific journals (Domokos et al., 2000) and it is the intention of the authors to publish them also in other national languages in order to make their contents accessible to a wider circle of interested experts (Domokos et al., 2001, Domokos and Neppel 2002). In the case of two more follow-up volumes (No. VIII and X), Hungary was the co-coordinator with Germany and in the case of No. VII, it co-operated with Romania (Domokos et al., 1999).

1.7 The Future of the Hydrological Co-operation of the Danube Countries

Further projects to be initiated or continued after 2008 under the co-operative endeavor, have been listed in Table 1.5. The list includes one project, No. 3, which commenced as early as during the IInd phase of the co-operation, which has not yet been completed due to unforeseen difficulties.

The practically spontaneously started hydrological co-operation of the Danube countries, today with a history of nearly four decades, first of all based on the generosity of these countries and the erudition of their experts, has brought (Table 1.5) and is hopefully to bring soon further important results that could not have been obtained without this co-operation: notably the Danube Monograph and its follow-up volumes. These works provide – in accordance with the Principles of the co-operation, signed in 1987 in Budapest – an indispensable, mutually agreed and accepted informational basis for integrated water resources management, water damage prevention, and environment protection measures, all of which are increasingly urgent in the Danube Catchment.

As described, at the beginning the co-operation had to cope with certain political-diplomatic difficulties. Today, there are other kinds of difficulties, mostly arising from

- limited professional experience in some "new" Danube countries, created by further subdivision of the "old" countries as well as
- (hopefully only provisional) uncertainties of the financial background of the co-operation in the countries in political/economical transition, and finally

Table 1.5 History and results of the regional hydrological co-operation of the Danube Countries in the frame of IHP/UNESCO, including proposed projects, as of September 2008

Phase of co-operation	Period	Number and list of Danube Countries (with major areal shares)	Chief co-ordinating		Original symbol of project in Working Plan	Serial number of volume/ follow-up volume	Title of publication[a]	Language of text (and summaries)	Co-ordinating country of the project[b]	Co-ordinator(s) of project/ /leading author(s) of volume	Extension of volume format (A4) pp: pages tt: tables ff: figures and maps	Year and place of publication	External financial supporter[b]	Regular annual (I., II., etc.) and extraordinary (#) expert meetings
			Institution(s)	Expert(s)										
I	1971–1986	8 (D, A, CS, H, YU, RO, BG, SU)	(a) Water resources Institute VÚVH, Bratislava (CS)[c] (b) Technical Secretariat, Beograd (YU)[d]	(a) A. Sikora (CS), A. Stančík (CS) (b) S. Jovanovič (YU), M. Andjelič (YU), S. Prohaska (YU), M. Miloradov (YU)	–	(0)	Die Donau und ihr Einzugsgebiet – Eine hydrologische Monographic Teil 1, 2 und 3	G (–)		H. Schiller (D), K. Toduta (D), 0. Behr (A), M. Domokos (H), P. Petrovic (CS)	pp: 377+228+0 tt: 25+7+0 ff: 33+0+8	1986 München	German IHP/OHP NC	During Phase I of the co-operation various expert meetings were held alternatively in Bratislava and Beograd, further one meeting in München
					–	(0)[e]	Dunay i ego basseyn. Gidrologicheskaya Monografiya. Chast' 1 : Tekst	R (–)			pp: 379 tt: 25 ff: 33	1989 Leningrad	IHP NC of the USSR	
					–	(0)[f]	Hydrology of the River Danube/ Gidrologiya reki Dunaya/Hydrologie du Danube/Hydrologie der Donau	E, R, F, G		A. Stančík (CS), S. Jovanovič (YU), S. Sikora (CS), L. Ürge (CS), M. Domokos (H)	pp: 271 tt: 25 ff: 33	1988 Bratislava	UNESCO	
II	1987[g] –1992	8/13	German IHP/OHP National Committee	K. Hofius (D)										I. Budapest 1987[g] ; II. Wien 1988; III. Sofia 1989; IV. Kiev 1990; V. Straubing 1991; VI. Kelheim 1992
III	1993–1998	13	Austrian IHP National Committee	O. Behr (A), D. Gutknecht (A), F. Nobilis (A)	1	I	Sediment regime of the Danube	G,R (E)	H	L. Rákóczi (H)	pp: 83, tt: 8 ff: 27	1993 Budapest		VII. Tulcea 1993; VIII. M. Vozokany/Wien 1994; # Budapest 1994; IX. Eckartsau/Smolenice 1995; X. Lednice 1996; # Graz 1996; XI. Sofia, 1998, # Osijek 1998; XII. Budapest 1998
					2	II	Thermal and ice conditions of the Danube and its major tributaries	G, R (E)	CS	A. Stančíková (CS)	pp: 116, tt: 5 ff: 40	1993 Bratislava		
					6.4	V/1	The fords of the Danube	G, R (E)	H	L. Goda sr (H)	pp: 187, tt: 5, ff: 63	1996 Budapest		
IV	1999–2002	13	Slovak IHP National Committee	P. Miklánek (SK), P. Petrovič (SK)	4	IV[h]	Coincidence of flood flow of the Danube River and its tributaries	E (–)	YU	S. Prohaska (YU)	pp: 184, tt: 45 ff: 153	1999 Bratislava	UNESCO/ ROSTE	XIII. Bratislava 1999; XIV. Postojna 2000; # Bratislava 2000; XV.Deggendorf 2001; XVI. Zagreb 2002; # Bucureşti 2002
					6.1	V/2	Palaeogography of the Danube and its Catchment	E (G, R)	H	F. Neppel (H), S. Somogyi (H), M. Domokos (H)	pp: 62, tt: 21 ff: 30	1999 Budapest	UNESCO/ ROSTE	
					6.2	V/3	The Danube River channel training	G, R (E)	SK	A. Stančíková (SK)	pp: 171, tt: 24 ff: 84	1999 Bratislava	UNESCO/ ROSTE	
					8.2	VI/1	Danube River Basin coding	E, R.G	SLO	M. Brilly (SLO)	pp: 99, tt: 2 ff: 12	1999 Ljubljana	UNESCO/ ROSTE	

Table 1.5 (continued)

Phase of co-operation	Period	Number and list of Danube Countries (with major areal shares)	Chief co-ordinating: Institution(s)	Chief co-ordinating: Expert(s)	Original symbol of project in Working Plan	Serial number of volume/ follow-up volume	Title of publication[a]	Language of text (and summaries)	Co-ordinating country of the project[b]	Co-ordinator(s) of project/ /leading author(s) of volume	Extension of volume format (A4) pp: pages tt: tables ff: figures and maps	Year and place of publication	External financial supporter[b]	Regular annual (I., II., etc.) and extraordinary (#) expert meetings
V	2003–2005	13	Hungarian IHP/OHP National Committee	M. Domokos (H)	7	VII	Regional analysis of annual peak discharges in the Danube Catchment	E (G, R)	RO	V.-Al. Stănescu (RO), V. Ungureanu (RO), M. Matreata (RO), M. Domokos (H)	pp: 64, tt: 11 ff: 16	2004 București		XVII. Sofia 2003; XVIII Brno 2004; XIX. Passau 2005
					5.1	VIII/1[i]	Inventory of the main hydraulic structures in the Danube Basin	E (G,R)	RO	I. Pașoi (RO)	pp: 62, tt: 5 ff: 14	2004 București		
					5.2	VIII/2[i]	Flow regime of River Danube and its Catchment	G, E, R	D&H	J.-U. Belz (D), L. Goda jr (H), Zs. Buzas (H), M. Domokos (H), H. Engel (D), J. Weber (D)	pp: 152, tt: 9 ff: 43	2004 Koblenz/Baja	UNESCO/ ROSTE	
					5.3	VIII/3[i,j]	Basin-wide water balance in the Danube River Basin	E(–)	SK	P. Petrovič (SK) & Steering Committee	pp: 139, tt: 100 ff: 24	2006 Bratislava	UNESCO/ ROSTE	
VI	2006–2008	13	IHP National Committee of Serbia	M. Miloradov (SR)	10	X	The hydrological meta-database of the countries sharing the Danube Catchment	E, G, R	D&H	J.-U. Belz (D), L. Goda jr (H), M. Hils (D)	pp: 7, tt: 5 app.: 1, cd: 1	2008 Baja/Koblenz		XX. Beograd 2006 XXI. Novi Sad 2007 XXII. Bled 2008
					11	XI	Characterization of the runoff regime and its stability in the Danube Catchment	E(–)	H	P. Kovács (H)	pp: 40, tt: 7 annex: 17	2006 Budapest	UNESCO/ ROSTE	
VII	2009–2011	13 (D, A, CZ, SK, H, SLO, HR, BIH, SR, RO, BG, MD, UA)	IHP National Committee of Croatia	...	3	III	Long-term fluctuations of precipitation in the Danube Basin		A	O. Behr (A) ...		... Wien		...
					9	IX	Flood regime of rivers in the Danube Basin		SK	P. Miklánek (SK), P. Pekarova (SK)		Bratislava		
					12	XII	Sediment balance in the Danube Basin[k]		A	H.-P. Nachtnebel (A) & Steering Committee		Wien		
					13	XIII	Low flow and hydrological drought in the Danube Basin[k]		BG	C. Dakova (BG)		Sofia		
						...	...							

[a]All titles of the published volumes are translated (whenever necessary) in this table into English, except the three versions of the Danube Monograph itself (see in the first three rows of the table), identified with the original titles.
[b]The expenses of editing and printing each project report, issued as a follow-up volume to the Danube Monograph, were regularly met, as requested by the "Principles" of 1987, by the respective IHP National Committee, delegating the co-ordinator(s) of the works on that project. Additional external financial supports are listed in a separate column of this table.
[c]The Institute VÚVH, Bratislava was commissioned by the Working Group for Scientific Hydrology (chairman: K. Stelczer) of the Danube Commission (member states of the Group: CS, H, BG, SU) to act as one of the two co-ordinators.
[d]The Technical Secretariat in Beograd was installed for acting as one of the two co-ordinators on behalf of the IHP National Committees of D, A, YU and RO.
[e]Volume I of the Danube Monograph (issued in 1986), translated from German to Russian.
[f]Abridged, illustrated quadrilingual version of the Danube Monograph (issued in 1986).
[g]In 1987, the 1st s ession of a new series of regular annual expert meetings was held in Budapest, establishing the "Principles" for the further co-operation.
[h]It is planned to issue after 2008 also a quadrilingual version (E, G, R, Serbian) of follow-up volume No. IV, whose monolingual (English) version was issued in 1999.
[i]The follow-up volumes No. VIII/1, VIII/2 resp. VIII/3 are the updated versions of the subchapters I.5, II resp. III of the Danube Monograph (issued in 1986).
[j]The implementation of Project No. 5.3 was prompted by ICPDR.
[k]The work on Project No. 12 will be a contribution to the International Sedimentation Initiative (I.S.I.) of UNESCO.

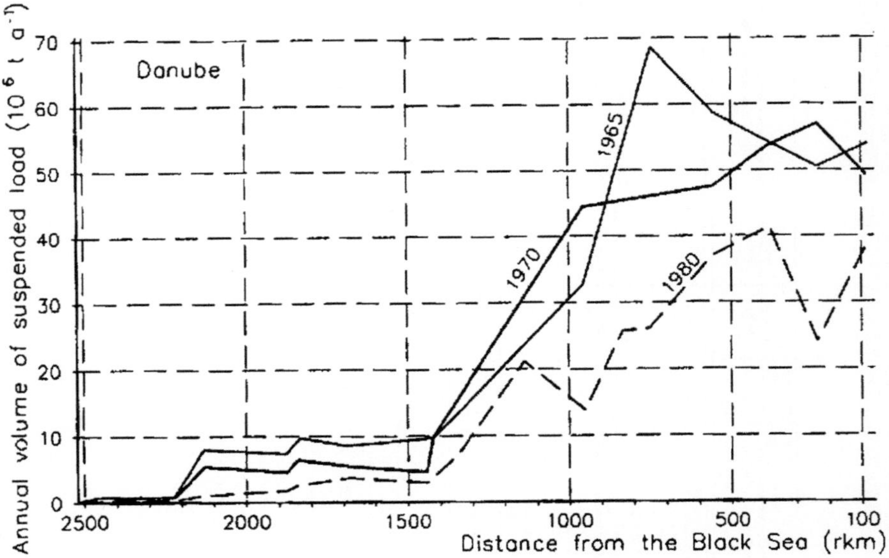

Fig. 1.6 Distribution of the annual volumes of suspended load along the Danube in three particularly watery years (RZD 1993a)

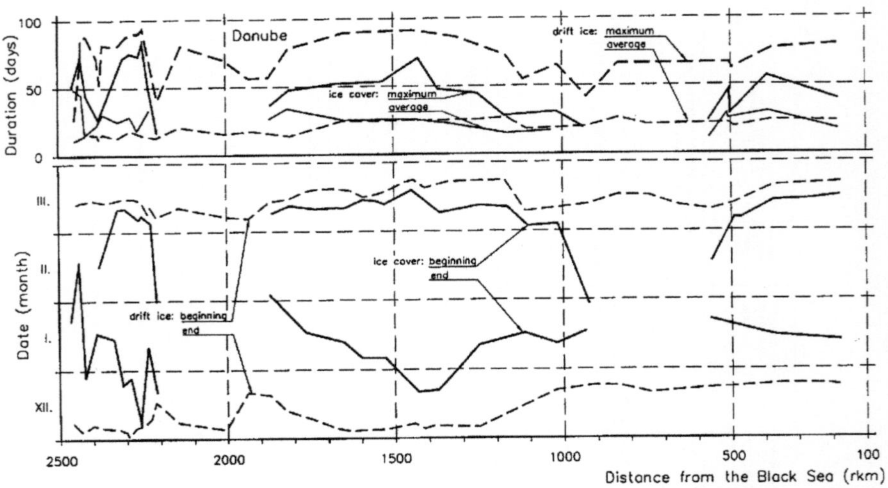

Fig. 1.7 Characteristics of the ice regime of the Danube during the period 1955–1985 (RZD 1993b)

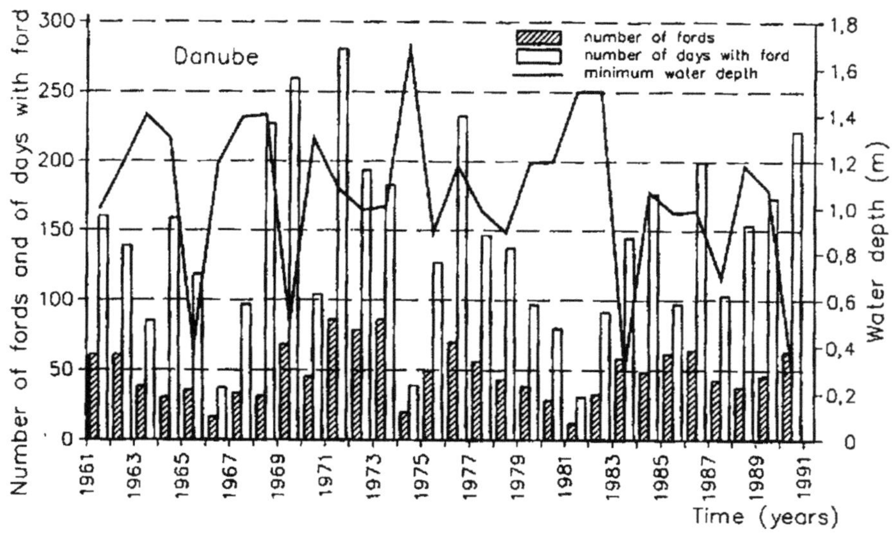

Fig. 1.8 Characteristics of the fords of the Danube reach between Regensburg and Sulina, during the period 1961–1990 (RZD 1996)

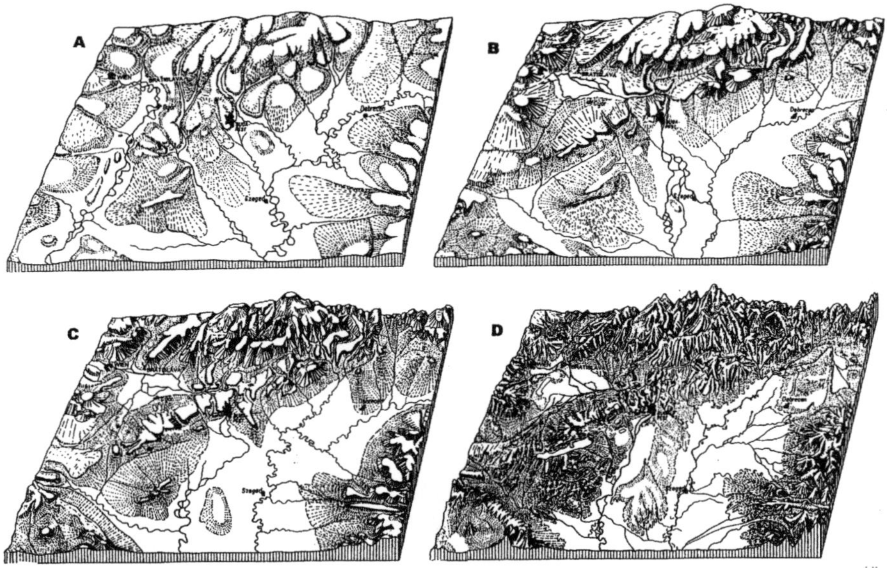

Fig. 1.9 Reconstruction of the (**a**) Eopleistocene, (**b**) Pleistocene, (**c**) Neopleistocene and (**d**) Holocene orography and hydrography of the Carpathian Basin (RCDC 1999a, after Mike 1970)

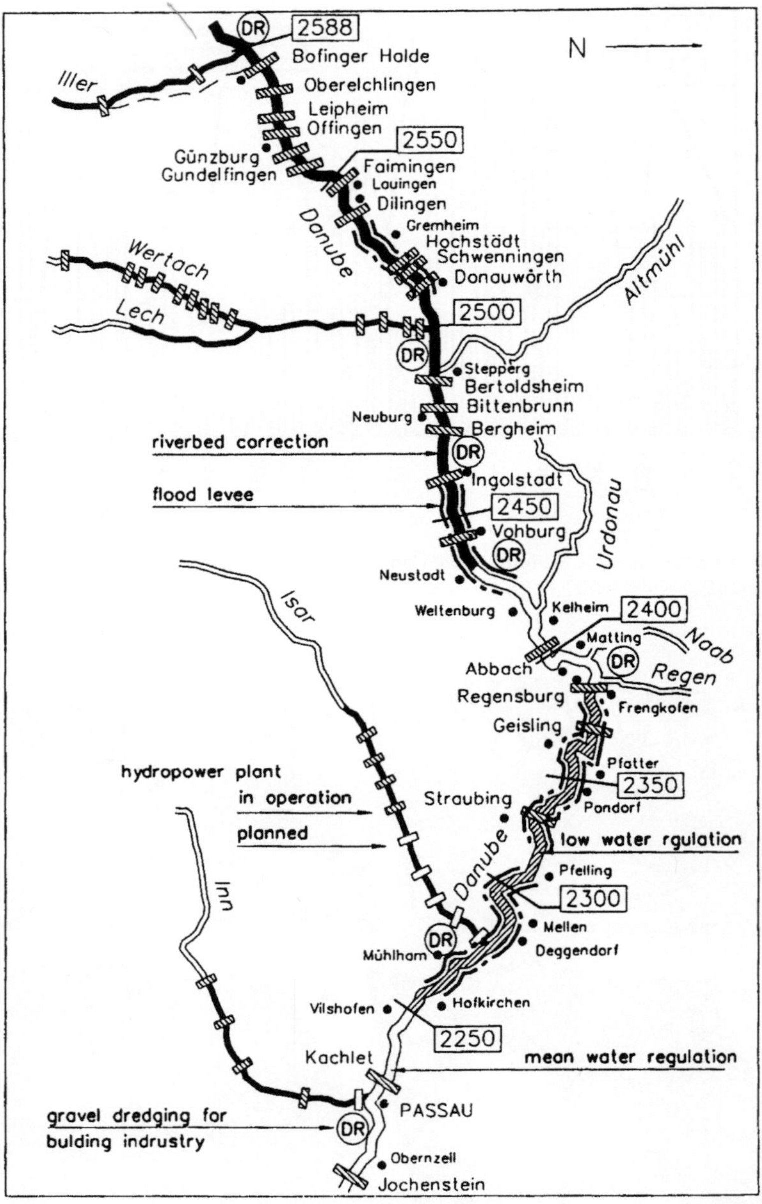

Fig. 1.10 River training and flood control works on the Bavarian Danube reach in 1995 (RZD 1999)

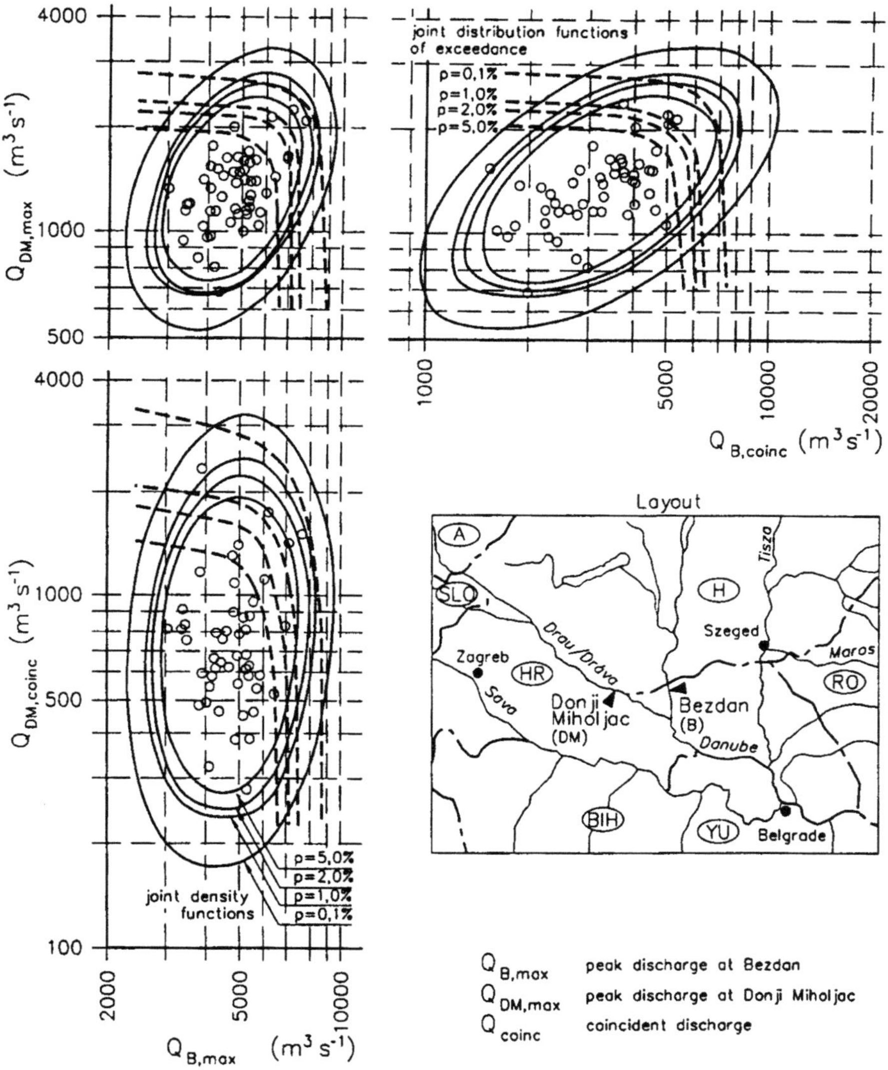

Fig. 1.11 Characterization of the coincidence of flood discharges observed at the gauging stations Bezdan/Danube and Donji Miholjac/Drava with joint density and distribution functions (RCDC 1999b)

- the increasing average age of the international staff of the co-operation, and the lack of a sufficient number of younger experts prepared to join the community.

It is hoped, however, that the governments of the Danube countries recognizing the importance and necessity of further results, will do their best to revitalize

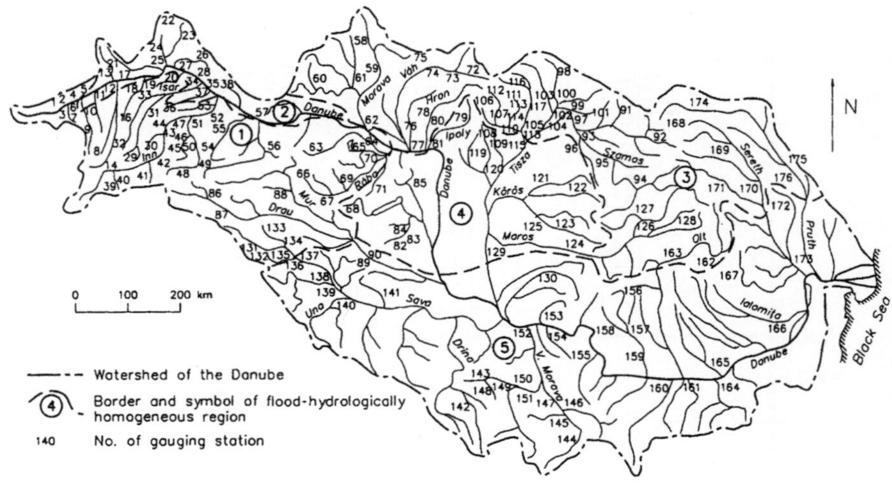

Fig. 1.12 The flood-hydrologically homogeneous regions of the Danube Catchment (Domokos et al., 1999)

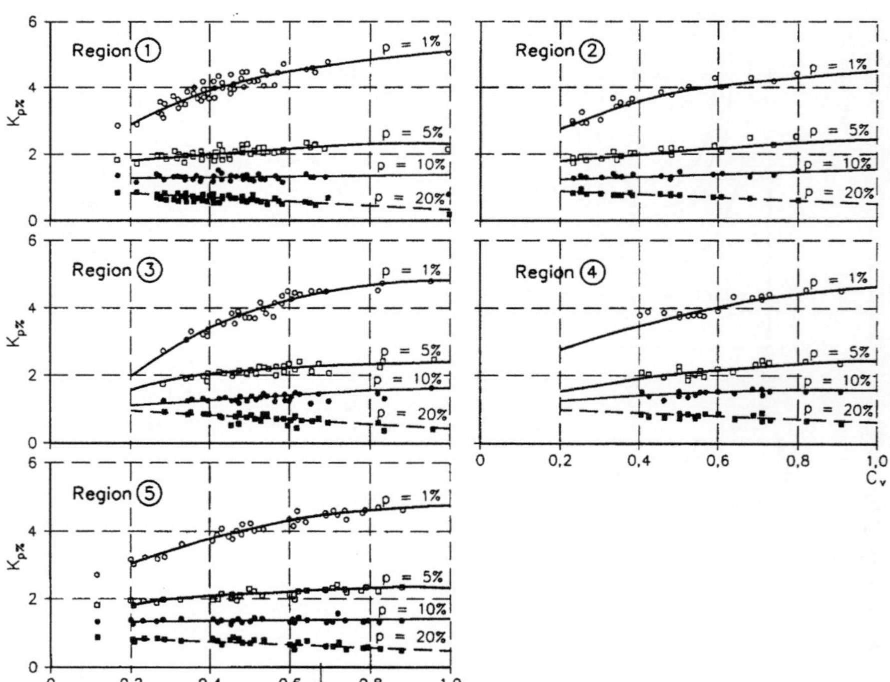

Fig. 1.13 Nomograms for estimating annual maximum discharges in the five regions of the Danube Catchment, displayed in Fig. 1.12 (Domokos et al., 1999)

the hydrological co-operation of the Danube Countries – which seems to be one of the most effective activities of this kind in the Catchment (Pfündl 1994) – for the common benefit of their inhabitants (Liška 2001), with special regard to its usefulness for implementing the goals of the EU Water Framework Directive in the whole Danube Catchment (Jekel 2003).

References

Altmann K (1999) Die Verbreitung der hydrologischen Bilanz des Donaueinzugsgebietes in sieben und mehr Sprachen. Hydrologie und Wasserbewirtschaftung 43:H.2

Behr O (1995) Langfristiges Verhalten des Niederschlags im Donaueinzugsgebiet. Stand der Abschlussarbeiten. Manuskript. IX. Expertensitzung der hydrologischen Zusammenarbeit der Donauländer, Smolenice, Slowakei

Belz J-U (2000) Aktualisierung der hydrologischen Monographie des Donaueinzugsgebietes. Hydrologie und Wasserbewirtschaftung 44:H.3

CIHBR (Comm. Int. Hydrol. Bassin di Rhin) (1977) Le Bassin du Rhin (Das Rheingebiet). Monographie Hydrologique

Domokos M (1990) Bestrebungen für eine koordinierte Nutzung der Wasservorräte und deren Schutz im Donauraum. In: Schriftenreihe des Österreichischen Wasserwirtschaftsverbandes, Heft 82: Internationale Wasserwirtschaft. Wien

Domokos M (1994) The Danube Monograph and FREND: comparison of two approaches. IAHS Publication No. 221

Domokos M, Neppel F (2002) Paleogeografska poviest sliva Dunava. Hrvat Vode 10 (2002):40

Domokos M, Neppel F, G Gergov (2001) Palaeogeographic history of the Danube Catchment Basin. Problems of Geography, 2001/3-4. Sofia, Bulgaria

Domokos M, Neppel F, Somogyi S (2000) Paläogeographische Geschichte der Donau und ihres Einzugsgebietes. Hydrologie und Wasserbewirtschaftung 44:H.4

Domokos M Sass J (1990) Long-term water balances for subcatchments and partial national areas in the Danube Basin. J Hydrol 112:267–292

Domokos M, Stănescu V-A, Ungureanu V (1999) Analyse der jährlichen Hochwasser im Donauraum als Beitrag zum Hochwasserschutz und der Regionalisierung hydrologischer Daten. Hydrologie und Wasserbewirtschaftung 43:H.5

FREND (Flow Regimes from Experimental and Network Data) (1989) Report. vol. I: Hydrological studies, vol. II: Hydrological data. Institute of Hydrology, Wallingford, UK

FRIEND (Flow Regimes for International Experimental and Network Data) (2002) Regional hydrology: bridging the gap between research and practice. IAHS Publication No. 274. Cape Town, South Africa

FRIEND (Flow Regimes from International Experimental and Network Data) (1993) Report. Vol. I.: Hydrological studies, vol. II: Hydrological data, vol. III: Inventory of streamflow generation studies. Institute of Hydrology, Wallingford, UK

FRIEND (Flow Regimes from International Experimental and Network Data) (1997) Third Report: 1994–1997. Cemagref Éditions, Antony, France

Goda L sr (1995) A Duna gázlói Pozsony és Mohács között [The fords of the Danube between Bratislava and Mohács] Vízügyi Közlemények 1

Hofius K, Schröder U (1997) Die regional Zusammenarbeit der Donauländer im Rahmen des IHP der UNESCO.Deutsche Gewässerkundliche Mitteilungen H.5

Jekel H (2003) Water Framework Directive and public participation. Danube Watch, No. 2/2003

Kardos M (1975) A Duna-medence magyar részének vízmérlege [Water balance of the Hungarian part of the Danube Catchment] In: Beszámoló a VITUKI 1972. évi munkájáról. Budapest, 1975

Kresser W, Lászlóffy W (1964) Hydrologie du Danube. La Houille Blanche 2:133–178
Lászlóffy W (1965) Hydrographie der Donau. In: Liepolt R (ed) Limnologie der Donau. Schweizerbart'sche Verlagsbuchhandlung, Stuttgart, Lfg. 1
Liška I (2001) Co-operation under ICPDR to address the hydrological issues. Manuscript, 15th working meeting of representatives and experts of the regional cooperation of the Danube Countries, Deggendorf, Germany, May
Mike K (1970) Magyarország ősvízrajza és felszíni vizeinek története [Palaeohydrography and the history of surface waters in Hungary]. AQUA Kiadó, Budapest
Neppel F, Somogyi S, Domokos M (1999) A Duna és vízgyűjtőjének ősvízrajza (Papaeogepgraphy of the Danube and its catchment), vol 3/LXXXI. Vízügyi Közlemények, Budapest
Pfündl D (1994) Internationale Zusammenarbeit im Donauraum auf dem Gebiet der Wasserwirtschaft. In: Proceedings of the 17th Konferenz der Donauländer, Sammelband, Band I. Budapest, September
Rákóczi L (1993) A Duna hordalékjárása [Sediment regime of the Danube]. Vízügyi Közlemények, 2
RCDC (Regional Cooperation of the Danube Countries) (1999a) Palaeogeography of the Danube and its catchment (Follow-up volume No. V/2 to the Danube Monograph). Budapest
RCDC (Regional Cooperation of the Danube Countries) (1999b) Coincidence of flood flow of the Danube River and its tributaries (Follow-up volume No. IV to the Danube Monograph). VÚVH, Bratislava
RCDC (Regional Cooperation of the Danube Countries) (2000) Danube River basin coding (Follow-up volume No. VI/1 to the Danube Monograph). Ljubljana
RCDC (Regional Cooperation of the Danube Countries) (2004a) Regional analysis of annual peak discharges in the Danube Catchment (Follow-up volume No. VII to the Danube Monograph). Bucureşti
RCDC (Regional Cooperation of the Danube Countries) (2004b) Inventory of the main hydraulic structures in the Danube basin (Follow-up volume No. VIII/1 to the Danube Monograph). Bucureşti
RCDC (Regional Cooperation of the Danube Countries) (2004c) Flow regime of river Danube and its catchment (Follow-up volume No. VIII/2 to the Danube Monograph). Koblenz/Germany – Baja/Hungary
RCDC (Regional Cooperation of the Danube Countries) (2004d) Basin-wide water balance in the Danube River basin (Follow-up volume No. VIII/3 to the Danube Monograph). Bratislava
RCDC (Regional Cooperation of the Danube Countries) (2006) Characterization of the runoff regime and its stability in the Danube Catchment (Follow-up volume No. XI to the Danube Monograph). Budapest
RCDC (Regional Cooperation of the Danube Countries) (2008) The hydrological metadatabase of the countries sharing the Danube Catchment (Follow-up volume No. X to the Danube Monograph). Baja/Hungary – Koblenz/Germany
RSPS (Regialnoe Sotrudnichestvo Pridunayskih Stran)(1989) Dunay I ego basseyn. Gidrologicheskaya monografiya. Chast I: tekst.[Danube Monograph, Russian version]. Gidrometeoizdat, Leningrad
RZD (Regionale Zusammenarbeit der Donauländer) (1986) Die Donau und ihr Einzugsgebiet. Eine hydrologische Monographie. Teil 1: Texte, Teil 2: Tabellen, Teil 3: Karten [Danube Monograph, German version] Bayerisches Landesamt für Wasserwirtschaft, München
RZD (Regionale Zusammenarbeit der Donauländer) (1993a) Schwebstoff- und Geschieberegime der Donau [Follow-up volume No. I to the Danube Monograph]. VITUKI, Budapest
RZD (Regionale Zusammenarbeit der Donauländer) (1993b) Temperatur- und Eisregime der Donau und ihrer wichtigeren Zubringer [Follow-up volume No. II to the Danube Monograph]. VÚVH, Bratislava

RZD (Regionale Zusammenarbeit der Donauländer) (1996) Die Furten der Donau [Follow-up volume No. V/2 to the Danube Monograph] VITUKI, Budapest

RZD (Regionale Zusammenrabeit der Donauländer) (1999) Regulierung des Donaubettes. Eine Darstellung der flußbaulichen und hochwasserschutztechnischen Maßnahmen an der Donau (Follow-up volume No. V/3 to the Danube Monograph). VÚVH, Bratislava

Schiller H (1989) Die Donau und ihr Einzugsgebiet. Zur Herausgabe einer hydrologischen Monographie. Deutsche Gewässerkundliche Mitteilungen 33:H.2

Stančík A, Jovanovič S et al. (1988) Hydrology of the River Danube [Quadrilingual, abridged version of the Danube Monograph]. Príroda, Bratislava

Chapter 2
The Danube River and its Basin Physical Characteristics, Water Regime and Water Balance

Heinz Schiller, Domokos Miklós, and Jenő Sass

Abstract The Danube Monograph (RZD 1986, Die Donau und ihr Einzugsgebiet. Eine hydrologische Monographie. Band I, II, III (The Danube Monograph, German version). Bayerisches Landesamt für Wasserwirtschaft, München) was the first published work derived from the results of the hydrological co-operation of the Danube countries. The Danube Monograph included three chapters:

(1) Physical, geographical and water management characteristics of the river basin.
(2) Water regime of the River Danube and its most important tributaries for the period 1931–1970.
(3) Hydrological balance for the period 1931–1970.

Since the publication of the Monograph several follow-up volumes (please see Chapter 1 of this book, Table 1.3) have been published. These follow-up volumes provide additional studies, more detailed information and studies with an updated database. Thus, some sections of the 1986 Monograph chapters are out of date or to a greater or lesser extent have been improved upon:

- Chapter 1: Additional studies have been carried out (e.g. palaeogeography of the catchments in addition to geography and geology; river training and hydraulic structures), but they do not change the original content. With regard to the number of countries within the catchment's area it has increased for political reasons.
- Chapter 2 was based on data for the period 1931–1970. However, since its publication new data over a longer period has been collected (with data capture continuing in the future). This led to the publication of an updated study based on the period 1931–1990 in 2004 (RCDC 2004a). However, in spite of a slightly modified methodology some of the former results are still interesting. They will

H. Schiller (✉)
Bavarian Environment Agency, Augsburg D-86177, Germany
e-mail: hz-schiller@alice-dsl.net

M. Brilly (ed.), *Hydrological Processes of the Danube River Basin*,
DOI 10.1007/978-90-481-3423-6_2, © Springer Science+Business Media B.V. 2010

be offered in this chapter together with some results of a few special studies (e.g. flood regime).

- Chapter 3 (Water balance) is not influenced by the updated hydrological database: Between the mean annual discharges of the period 1931–1970 and those of the period 1931–1990 there are only small or minute differences – please see Table 2.1.

An abridged and partly improved version of Chapter 3 of the Danube Monograph (1986) is offered herein (subchapter 2.4). The simplest type of multi-annual water balance, expressing the equilibrium of precipitation P versus evapotranspiration E plus runoff R, has been compiled for the period 1931–1970, first for 47 subcatchments and then for 12 partial national areas, as balance units. For each of these units the regional average values of the three balance components, expressed in mm year^{-1}, have been determined by transforming the hydrological isoline maps (scale: 1:2,000,000), printed in 1984 in Budapest, into planimetric maps. The errors of balance do not exceed, with a few exceptions, the yearly favourable limit of $\pm 5\%$.

As hydrologic characteristics, the runoff coefficients were calculated for each balance unit both individually and as a longitudinal function of the whole Danube. The latter also includes the (cumulative) regional average values of precipitation and runoff for the catchments belonging to any arbitrary Danube section. With the help of this longitudinal profile, it was possible to compare the discharges derived from the runoff isoline map with the corresponding values calculated from observed data series of Danube gauges. The result of this comparison was satisfactory, too.

As water management characteristics for each of the 12 national catchments areas, the following indices have been determined: "own" surface water resources; relative contribution to the total water resources of the Danube Basin; and the ratio of "own" to "transit" water resources, the lowest value of the latter characterising the national area of Hungary.

On the basis of the results obtained, it can be stated that the isoline maps, printed beforehand in 1984 as annexes to the Danube Monograph (1986), reflect the regional distribution of long-term average values of the water balance elements in compliance with reality and in comprehensible interrelation. Both the maps and the data in the balance tables derived from the former may be recommended as internationally co-ordinated basic information, reflecting our present level of knowledge. This information may be useful for all further work aiming at water resources development in the Danube Basin.

Keywords The Danube Monograph · Geography · Water management · Water regime · Water balance · Danube Catchments · Hydrological balance · Isoline maps · Precipitation · Evapotranspiration · Runoff · Runoff coefficient · Hydrological longitudinal profile · Balcerski index · Tributary

Table 2.1 Comparison of main discharges of the periods 1931–1970 and 1931–1990 for the stations of the Monographs 1986 and 2004 (Location of the stations see Figs. 2.3 and 2.4)

Nr.	Station	Area (km^2)	Distance from estuary	Observation period	MNQ ($m^3 s^{-1}$)	MQ ($m^3 s^{-1}$)	Δ MQ:MQ_2 %	Mq ($l s^{-1} km^2$)	MHQ ($m^3 s^{-1}$)	Daily mean discharges ($m^3 s^{-1}$) with selected exceedance 10%	50%	90%
1	Danube/Ingolstadt	20,001	2,613.0	1931–1970	123	308	1.2	15.4	1,084	144	275	506
				1931–1990	129	312			1,079	159	274	502
2	Danube/Regensburg	35,399	2,376.1	1931–1970	193	435	2.0	12.3	1,468	208	382	718
				1931–1990	198	444			1,468	201	393	736
3	Danube/Hofkirchen	47,496	2,256.9	1931–1970	307	645	0.1	13.6	1,864	333	574	1,030
				1931–1990	315	646			1,860	365	556	1,015
4	Danube/Achleiten	76,653	2,223.0	1931–1970	583	1,430	0.7	18.7	4,163	700	1,300	2,320
				1931–1990	599	1,420			4,100	785	1,349	2,317
5	Danube/Linz	79,490	2,135.2	1931–1970	593	1,509	2.4	19.0	4,279	719	1,370	2,420
				1931–1990	619	1,473			4,212	687	1,346	2,309
6	Danube/Stein-Krems	96,045	2,002.7	1931–1970	773	1,864	0.5	19.4	5,426	912	1,670	2,990
				1931–1990	789	1,854			5,530	872	1,747	2,996
7	Danube/Wien-Nussdorf	101,700	1,934.1	1931–1970	804	1,943	1.2	19.1	5,500	951	1,730	3,150
				1931–1990	832	1,920			5,547	1,013	1,703	3,045
8	Danube/Bratislava	131,338	1,868.8	1931–1970	844	2,020	1.2	15.4	5,750	960	1,810	3,380
				1931–1990	896	2,044			5,724	1,062	1,860	3,225
10	Danube/Dunaalmás	171,720	1,751.8	1948–1970	924	2,314	4.2	13.5	5,703	1,096	2,106	3,733
				1948–1990	945	2,220			5,319	1,187	2,058	3,415
11	Danube/Nagymaros	183,534	1,694.6	1931–1970	956	2,379	3.6	13.0	5,556	1,129	2,133	3,862
				1931–1990	976	2,296			5,293	1,170	2,045	3,579

Table 2.1 (continued)

Nr.	Station	Area (km^2)	Distance from estuary	Observation period	MNQ (m^3 s^{-1})	MQ (m^3 s^{-1})	Δ MQ:MQ$_2$ %	Mq (l s^{-1} km^2)	MHQ (m^3 s^{-1})	Daily mean discharges (m^3 s^{-1}) with selected exceedance 10%	50%	90%
12	Danube/Mohács	209,064	1,446.9	1931–1970	1,068	2,389	2.3	11.4	5,095	1,187	2,191	3,963
				1931–1990	1,046	2,335			4,841	1,146	2,205	3,719
13	Danube/Bezdan	210,250	1,425.5	1931–1970	976	2,479	4.5	11.8	4,924	1,350	2,340	4,150
				1931–1990	992	2,372			4,788	1,300	2,159	3,748
14	Danube/Bogojevo	254,593	1,367.3	1931–1970	1,231	3,060	2.2	12.2	5,795	1,700	2,900	4,910
				1931–1990	1,293	2,944			5,533	1,681	2,761	4,519
15	Danube/Pancevo	525,009	1,153.3	1931–1970	2,242	5,490	2.1	10.4	10,072	2,900	5,220	8,930
				1931–1990	2,229	5,375			9,934	2,679	4,818	8,545
16	Danube/Veliko Gradiste	570,375	1,059.8	1931–1970	2,383	5,745	2.3	10.1	10,636	2,970	5,500	9,250
				1931–1990	2,364	5,616			10536	2,976	5,002	8,939
17	Danube/Orsova	576,232	955.0	1931–1970	2,334	5,699	1.6	9.9	10,631	2,930	5,560	9,580
				1931–1990	2,246	5,611			10,604	2,641	5,394	8,912
19	Danube/Lom	588,860	743,3	1941–1970	2,417	5,766	1.3	9.8	10,534	3,250	6,100	10,500
				1941–1990	2,559	5,842			10,632	3,178	5,574	9,084
21	Danube/Zimnicea	658,400	554.0	1931–1970	2,489	6,152	1.8	9.3	11,090	3,090	6,030	10,650
				1931–1990	2,520	6,041			11,166	2,812	5,775	9,440
22	Danube/Ruse	669,900	495.6	1931–1970	2,646	6,264	1.1	9.3	11,037	3,500	6,800	11,200
				1931–1990	2,711	6,195			11,058	3,017	5,662	9,739
23	Danube/Silistra	689,700	375.5	1941–1970	2,632	6,300	0.1	9.1	10,952	3,400	6,800	11,200
				1955–1990	2,206	6,292			11,009	3,317	5,902	9,990

Table 2.1 (continued)

Nr.	Station	Area (km^2)	Distance from estuary	Observation period	MNQ ($m^3 s^{-1}$)	MQ ($m^3 s^{-1}$)	Δ MQ:MQ_2 %	Mq ($l s^{-1} km^2$)	MHQ ($m^3 s^{-1}$)	Daily mean discharges ($m^3 s^{-1}$) with selected exceedance 10%	50%	90%
24	Danube/	709,100	252.3	1931–1970	2,621	6,216	0.4	8.7	10,812	3,110	6,100	10,710
	Vadu-Oii-Hirsova			1931–1990	2,692	6,188			10,861	3,171	5,952	9,718
25	Danube/Ceatal Izmail	807,000	72.0	1931–1970	2,934	6,550	1.0	8.1	10,621	3,340	6,440	11,460
				1931–1990	2,901	6,486			10,889	3,328	6,042	10,148
26	Inn/Passau-Ingling	26,084	3.1	1931–1970	257	743	1.5	28.6	2,983	310	610	1,350
				1931–1990	267	732			2,936	318	659	1,275
×	Salzach/	6,650	11.4	1931–1970	72.6	245	1.2	37.3	1,334	92.5	206	452
(25)	Burghausen			1931–1990	76.0	248			1,336	94.2	204	459
28	Enns/Steyr	5,915	30.9	1951–1970	46.8	200	1.0	33.8	1,297	65.8	150	389
				1951–1990	50.0	198			1,210	60.8	156	390
29	Morava/Moravsky Jan	24,129	67.6	1931–1970	29.0	110	0	4.6	629	31.0	75.0	228
				1931–1990	29.0	110			584	27.6	74.5	222
30	Vah/Sala	10,620	33.9	1931–1970	38.0	152	3.4	14.3	1,048	48.0	115	287
				1931–1990	32.0	147			985	54.4	115	260
31	Hron/Brehy	3,821	102.5	1931–1970	13.0	50.0	3.3	13.1	381	14.0	32.0	104
				1931–1990	13.0	48.4			400	13.7	29.4	102
32	Ipel/Ipelsky Sokolec	4,838	42.9	1931–1970	1.90	21.0	5.5	4.34	208	2.00	8.20	52.0
				1931–1990	2.00	19.9			201	1.78	7.93	48.8
34	Mur/Landscha from	8,340	153.4	1951–1970	46.3	143	1.4	17.1	610	57.9	118	259
	1971 spielfeld	9,480	144.3	1951–1990	48.0	141			681	56.9	115	257

Table 2.1 (continued)

Nr.	Station	Area (km^2)	Distance from estuary	Observation period	MNQ ($m^3 s^{-1}$)	MQ ($m^3 s^{-1}$)	Δ MQ:MQ_2 %	Mq ($l s^{-1} km^2$)	MHQ ($m^3 s^{-1}$)	Daily mean discharges ($m^3 s^{-1}$) with selected exceedance 10%	50%	90%
36	Drava/Donji Miholjac	37,142	74.6	1931–1970	235	554	2.4	14.9	1,341	312	511	910
				1931–1990	234	541			1,359	307	468	852
×	Tisa/Vásárosnamény	29,057	684.5	1931–1970	72.6	357	2.2	12.2	2,134	80	232	762
(35)				1931–1990	74.0	366			2,155	72.5	232	778
×	Tisa/Szolnok	73,113	334.6	1931–1970	117	541	2.3	7.2	1,786	125	369	1,220
(36)				1931–1990	111	529			1,650	118	356	1,180
39	Tisa/Szeged	138,408	173.5	1931–1970	192	813	4.1	5.88	2,298	182	588	1,774
				1931–1990	209	848			2,325	237	634	1,844
40	Tisa/Senta	141,715	46.4	1931–1970	174	766	3.3	5.40	2,119	235	610	1,680
				1931–1990	179	792			2,142	224	571	1,641
41	Szamos/Csenger	15,283	23.7	1931–1970	22.8	127	3.1	8.31	879	25.0	83.0	273
				1931–1990	26.0	131			1,019	22.0	87.6	276
43	Maros/Makó	30,149	23.7	1931–1970	46.0	175	1.7	5.80	710	45.0	125	393
				1931–1990	47.0	178			745	51.6	122	387
44	Sava/Sremska Mitrovica	87,996	136.0	1931–1970	390	1,613	2.6	18.3	4,272	525	1,450	3,150
				1931–1990	401	1,572			4,154	427	1,333	3,032
45	Velika Morava/ Ljubičevski Most	37,320	39.9	1931–1970	52.0	238	14.1	6.38	1,302	80.0	171	550
				1931–1990	55.0	277			1,260	54.7	154	508
47	Olt/Stoenesti	22,683	71.2	1959–1970	41.7	162	5.8	7.14	988	61.1	130	370
				1960–1990	48.0	172			908	63.2	121	340

Table 2.1 (continued)

Nr.	Station	Area (km^2)	Distance from estuary	Observation period	MNQ ($m^3\ s^{-1}$)	MQ ($m^3\ s^{-1}$)	Δ MQ:MQ_2 %	Mq ($l\ s^{-1}\ km^2$)	MHQ ($m^3\ s^{-1}$)	Daily mean discharges ($m^3\ s^{-1}$) with selected exceedance 10%	50%	90%
48	Siret/Storozinec	672	448.0	1953–1970	0.66	5.50	12.4	8.18	186	1.00	2.40	11.5
				1962–1990	0.89	6.28			174	1.30	2.61	12.6
49	Siret/Lungoci	36,036	74.0	1950–1970	43.4	172	18.1	4.77	1,091	57.5	125	383
				1960–1990	52.0	210			1,294	67.3	141	400
50	Prut/Cernovci	6,890	772.0	1945–1970	7.00	61.0	9.0	8.85	1,213	14.3	37.3	116
				1945–1990	10.0	67.0			1,200	13.5	40.1	136

Station Nr. in Brackets: Station is only involved in the 2004 study.

2.1 Introduction

The Danube River is the only one of the big European waterways that flows across the continent from west to east. On its course of 2,857 km from the heights of the Schwarzwald Mountains to the Black Sea, it crosses 22 geographical longitudes touching 11 countries: Germany, Austria, Czech Republic, Slovakia, Hungary, Croatia, Serbia, Romania, Bulgaria, Moldavia and Ukraine.

The river basin includes glacier-covered mountains, mid-mountains covered with dense forests, karst regions with little vegetation, uplands and lowlands, tablelands with deeply carved river valleys, and wide plains.

Since pre-historic times the fertile riparian territories of the Danube have attracted tribes and nations have been founded and developed into rich cultures.

In 1908, a very old settlement was found 14 km downstream from Belgrade on the right bank of the Danube (the village of Vinča). The oldest excavated cultural layer dates back to about 5500 BC. On the basis of similar artefacts excavated from numerous sites in the south-eastern part of Europe, this regional culture was named the "Vinča-Culture". Some archaeologists named it also the "Old-European Danube-Civilisation". This culture had already created a kind of script. There are special symbols on various objects (like pottery, figurines and amulets) excavated from different sites which repeat themselves. These symbols are now widely interpreted as a writing system – the earliest form of writing ever found, predating ancient Egyptian and Sumerian writing by about 2000 years. However, the script cannot be read as this Old European language has disappeared. Nowadays only the Basque language and a lot of geographic names are related to this "Vasconian" language.

In 1972, a horde of gold was found together with many copper objects in a necropolis near Varna on the Bulgarian coast of the Black Sea. It is the biggest (more than 6 kg) and oldest gold horde excavated to date. The objects are dated between 4500 and 4200 BC, which is the climax of the Copper Age of this region. The more than 3,000 artefacts prove that people at that time had been able to work to a high standard with metals (copper and gold).

The settlement itself consists of houses built of stone while in other places the houses were made of wood, clay and straw (twisted walls). As to the inhabitants it is assumed that they had been part of the Suvorovo culture that existed in the region at about 4500–4000 BC. This culture was created by the first wave of Indo-European migrations. This new culture was a mixture of two components: the traditions of the original Old-European inhabitants and the imported culture of the Indo-European immigrants from the eastern part of Europe, who were politically dominating.

The historic age of the Danube started in the eighth to seventh century BC when Phoenicians and Greeks (in part also Egyptians) penetrated mostly via the estuary and entered into trade contacts with the local population. They also explored the river upstream but knew only the lower area.

The Greeks called this lower reach "Istros" (Herodotus; Virgil) and later the Roman author Cicero mentioned the lower Danube as "Histerus".

In the sixth century BC the Persian king Darius the First tried to conquer the territories adjacent the lower Danube. Alexander the Macedonian tried to do the same in 334 BC. In the first century AD the upper Danube region became a part of the Roman Empire. In the years 101–106 the Roman Emperor Traianus conquered the lower Danube region by defeating the Dacian tribes. He ordered the construct of Traian's Bridge across the Danube at Turnu-Severin in the Iron Gate Gorge and Traian's Road on the right bank of the Danube. The existence of this road which was a prolongation of the military road "Danube South" built about 45 AD from Donaueschingen (near the origin of the Danube) to Regensburg is documented by the "Tabula Traiana".

During the period of Roman rule the Danube achieved important strategic significance and this has continued to the present day. There has been a considerable migration of nations along the Danube throughout the centuries. The river became more and more a strategic waterway and an artery of life influencing the development of European history. At the same time the economic significance of the Danube has been growing.

In order to investigate the origin of the name "Danube" one has to go back to the Celtic tribes who lived around the upper Danube region already before they became a part of the Roman Empire. The word "Danu" means in the Celtic language "swift, rapid, violent". The Roman Emperor Caesar had named the stream Danubius in his book "De Bello Gallico" and later other Roman and Greek authors used this name also. Later other nations migrating along the stream modified and transformed the name according to their languages as Donau, Dunaj, Duna, Dunav and Dunarea.

2.2 Physical, Geographical and Water Management Characteristics of the Danube River Basin

2.2.1 Layout and Subdivision of the Basin

It is commonly accepted that the River Danube begins at the confluence of the source rivulets Breg and Brigach near Donaueschingen in the Schwarzwald Mountains. Nevertheless, it is also partly accepted that the origin of the Danube is at the source of the Brigach (46 km) which is the longer of the headwater streams. From this point, the River Danube crosses Middle and Southeastern Europe in a 2,826-km long course down to the Black Sea. With its multi-annual mean discharge of 6,855 $m^3 s^{-1}$ (Fig. 2.5), it is the 21st greatest river in the world and – after the river Volga – the second greatest of Europe. The Danube catchment, covering an area of 817,000 km^2 (Fig. 2.4), is situated south of the European main continental divide, running from Gibraltar to the Ural Mountains.

The source of the Danube River is in the Schwarzwald Mountains at 8° 09′ longitude east and its mouth into the Black Sea is at longitude 29° 45′ east. Therefore, the length of the axis of the catchment is 1,630 km. The northernmost point of the Danube Basin is at latitude 50° 15′ near the source of the River Morava while the

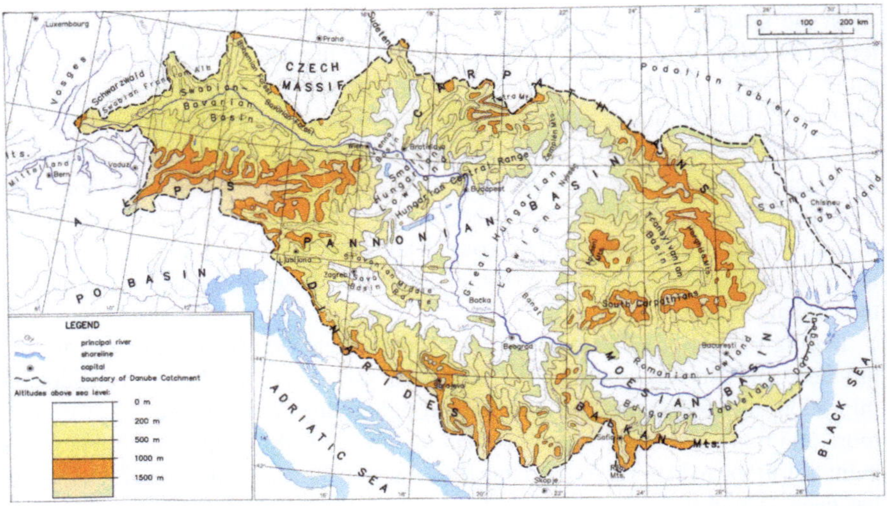

Fig. 2.1 Orography of the Danube Catchment (taken from RCDC 1999a), see also Chapter 3

southernmost point is at latitude 42° 05′ at the source of the river Iskar in the Rila Mountains.

The basin of the Danube and its tributaries adjoin the Rhine basin in the west and northwest, the Weser, Labe, Odra and Visla river basins in the north, the Dnjestr river basin in the northeast and the basins of the rivers flowing into the Adriatic and Aegean Sea in the south.

The elevation of the spring of the main source river Brigach is 1,076 m above sea level (a.s.l.), the elevation of the confluence of the two source rivulets Breg and Brigach is at 676 m a.s.l. The orography of the whole Danube Basin is rather uneven: while it reaches on the southern part the height of 4,052 m a.s.l. (Piz Bernina), the highest point in its northern part is only 2,496 m a.s.l. (Peak Krivan in the High Tatra Mountains). About one third of the basin consists of mountains while the hills and plains fill the other two thirds. The average elevation of the catchment is 475 m.

The geographical layout (Fig. 2.1) is defined by the mountain ridges – mostly the Alps, the Bavarian–Bohemian Forest, the Carpathians, the Dinarides and the Balkan Mountains (Stara Planina). According to this structure, the Danube River itself and the basin are conveniently divided into three regions: the Upper, Middle and Lower Danube.

2.2.1.1 The Upper Danube Region

The Upper Danube Region with an area of 132,000 km^2 covers the catchment from the source in the Schwarzwald Mts. down to the Devin Gate (Porta Hungarica) east of Vienna, which connects the little Carpathians and the last promontories of the

Alps (Leitha Mt.). The basin south of the Danube includes major parts of the Alps up to the watershed in the crystalline Central Alps and then the adjoining Swabian-Bavarian-Austrian foothills belt. The northern part of the catchment is significantly smaller; it is confined by the heights of the Swabian and Franconian Alb, the Upper Falconian Forest, then parts of the Bavarian–Bohemian forests – which all are of medium altitude – down to the Austrian Mühl- and Waldviertel and the Bohemian-Moravian Upland.

2.2.1.2 The Middle Danube Region

The Middle Danube Region presents an imposing and unique geographic unit inside Europe. It extends from the Devin Gate to the mighty fault section between the Southern Carpathians and the Balkan mountains near the Iron Gate Gorge. With an area of 445,000 km^2 the Middle Danube Basin is 3, 4 times as large as the Upper Region and thus the largest one of the three regions. It is confined by the Carpathians and a section of the Balkan Mts. in the north and east, by the Alps (Carnian Alps, Karavanken Mountains and Julian Alps) in the west and the Dinarides in the south. This closed circle of mountains embraces the "Pannonian Basin System", consisting of the South- and East-Slovakian Lowland, the Small and Great Hungarian Lowland, the Transylvanian Basin, the Slavonian Middle Range, the Sava Basin and a part of the Velika Morava Basin.

2.2.1.3 The Lower Danube Region

The Lower Danube Region with a catchment area of 241,000 km^2 is the reach from the Iron Gate to the Delta in the Black Sea. In the western part it is bordered by parallel mountain ridges: the Carpathians in the north and the Balkan Mts. (Stara Planina) in the south. In between there is the "Moesian Basin", consisting of the Romanian-Bulgarian Lowland, and the Bulgarian Tableland. The eastern part of the Lower Danube Region narrows in the south because of the Dobrogea Hills, but the northern extension is considerable: it reaches up to the Eastern Carpathians and includes the rivers Siret (45,500 km^2) and Prut (29,000 km^2). The northern and the eastern watersheds of these rivers are on the western margin of the Moldavian-Ukrainian Hills (also known as the Bessarabian Upland Plateau).

2.2.2 Geological Structure and Geomorphologic Conditions

The Danube has unlike the majority of the European rivers (e.g. Rhone, Rhine, Vistula) its source outside a high mountain range (like the Alps). It comes from an old, middle height mountain massif, and then it flows through hilly terrain, mountainous ranges and lowland valleys changing its course and longitudinal profile several times. This peculiar character of the stream and the various catchment surface features are a result of its geological evolution (from Pliocene to Alluvium).

About one third of the basin consists of mountains, while hills and plains fill the other two thirds. On the basis of the geological evolution (RCDC, 1999a) the following tectonic units can be distinguished:

- Alpine mountainous systems, being the youngest (Mesozoic) tectonic formation in Europe, cover the major part of the Danube basin. To these systems belong the Alps, the Carpathians, the Dinaric system and the Balkan Mts. (Stara planina), as well as the foothills and the inter-mountain plains. The two largest inter-mountain plains are the Pannonian Basin and the Lower Danube Plain (Rumanian and Bulgarian Lowland and the Bulgarian Tableland).
- Variscan (also called Hercynian) units, formed in the ancient (Palaeozoic) era of the earth and mostly consisting of crystalline rocks, are situated mostly in the upper Danube basin: the Schwarzwald Massif, the Upper Falconian Forest, the Bavarian–Bohemian Forest, and the Bohemian-Moravian Uplands are remnants of former much higher mountain ridges. Small units of this type occur in the form of uplands in the lower Danube Basin in Romania and Bulgaria (Dobrogea, Ludogorie). These mountains and hills are of medium and lower altitude.
- The Ante-Palaeozoic Russian Platform covers the north-eastern part of the Danube basin, called the Moldavian, or Podol Plateau.

Below we provide a little more information on the Alpine System , which is the most important in the Danube Catchment:

The alpine folding of mountains has involved all existing units back to the Varistic one. Thus, there is a mighty Variscan core in most parts of the alpine range of mountains: the crystalline rock (granite, gneiss) predominates in the central Alps, the central chain of the Eastern Carpathians, the Southern Carpathians and sections of the central Stara Planina. According to the geological periods there is a general pattern for the structure of the alpine mountain ranges: The Variscan core has on both sides mantles and zones of succeeding (rock-) formations: at first there is a schist mantle, then follows a zone of limestone (partly dolomite), and next comes the flysch region, mostly followed by tertiary molasses. According to local conditions there are some variations within this pattern, for example in the core of the Dinarides there are only parts of the (crystalline) schist layer on the surface or due to erosion in some regions a formation may have widely disappeared. Generally, the features of the present mountainous relief have been considerably influenced by significant glacial activity during the ice ages. The above-described orogenic pattern can for example be fully seen in the Upper Danube Basin: The main ridges (crystalline, calcareous etc.) are separated by longitudinal valleys in which the most southern tributaries originate and proceed as far as the points where cross valleys break through the ridges. There they have to change their original directions northward and can reach the Danube via a shorter path. This model is typical for the rivers Lech, Isar, Inn, Salzach, Traun, Enns and also for the Mur, but not for the rivers Drau/Drava and Sava which are already in the Medium Danube

Basin. The foothills and inter-mountain plains which are filled with (tertiary) sediments belong also to this alpine system. The two largest inter-mountain plains are the Pannonian Basin and the Lower Danube Plain (Rumanian and Bulgarian Lowlands).

2.2.3 Soils

In the Danube Basin many types of soil occur due to different soil-forming factors: Basic rock, relief, climate, precipitation, vegetation and human utilisation caused the evolution of a variety of soils. They can be roughly divided into two different families: the soils of the mountains and the soils consisting of loose sediments and alluvial fills.

- In the mountains the evolution of soils is influenced by the type of bedrock (mostly of crystalline or limestone origin) and with increasing altitude by increasing precipitation and decreasing air and soil temperature. About one-third of the high-mountainous regions are covered with rock-virgin soil (syrozem) and with soil without clayey subsoil (leptosol) and with only a slight humus layer. Brown soils (cambisols) and partly also Gray soils (podzol-cambisols) prevail below these high regions due to the increasing influence of higher temperature and more vegetation, sometimes varied by the influence of water.
- The soils consisting of loose sediments and alluvial fills in the lower altitudes of the basin are usually much deeper. The various soils of the family of the floodplain soils (with uniform and non-uniform grain size) had evolved on the sediments of the river valleys and riverine plains. In the areas free of flooding, depending on the base content, there are cambisols, mostly, however, podzol-cambisols and podzols, and on loess the more fertile luvisols. The plains of the Lower Danube Region consist mainly of fertile black soils (chernozem) because of the increasing continental character; in the more southern regions with a greater Mediterranean influence the soils show a reddish-brown colour (rhodic or chromic cambisol-variants: Terra-Fusca, Terra-Rossa).

There are also many locations in the Danube Basin – mostly in the lowlands – where under the influence of water (high groundwater, slack water, on the margins of surface-water bodies) subhydric soils have developed. These marshy soils are mostly histic gley soils and diverse types of eutric histosols; soil improvement measures have led to significant changes in these soils.

In depressions of the Pannonian Basin and the plains of lower Danube there are saline soils in areas where higher groundwater levels, low precipitation and greater evaporation rates coincide.

2.2.4 *Climate*

2.2.4.1 Rainfall

Because of its large extension – mainly in the west-eastern direction – several climatic zones cover the Danube Basin. The climate extends from the western regions with high Atlantic influence to continental dry and cold in the eastern part. Generally alternating effects of these two climates can be observed within the entire Danube Basin. The southern sections of the upper and middle Danube Basin – especially in the Drava and Sava basin – are rather influenced by the Mediterranean. This rough zoning is greatly influenced by relief features: with increasing elevation above sea level there is an increase of precipitation mostly because of more Atlantic characteristics. On the other hand the low-lying basins are more influenced by continental conditions and are therefore deficient in precipitation. Another climate variation relates to the (windward and leeward) exposition of the mountain slopes. It is also an Atlantic characteristic if the "weatherside", exposed to the west and northwest, receives more precipitation than the leeward slopes of the opposite side. The maximum average precipitation up to about 3,200 mm a^{-1} is in the high mountain range of the Alps. The most arid area is the lowland at the Black Sea with an average of only 350 mm a^{-1}. In between there is a great variety depending on the location in the basin, altitude and exposition. Already in the Upper Basin there are large differences. In the Alps precipitation is generally high, up to 2,000 and 2,500 mm a^{-1}, but the lowest precipitation is in the upper Inn river valley, extending in a west-easterly direction (Lower Engadin), with only 600–700 mm a^{-1}. Similar values occur outside the Alps in some river valleys, including parts of the Danube valley itself (e.g. the Vienna basin), the Naab and Morava valley. In the middle and lower Danube Basin the highest amounts of precipitation also occur in the mountains: in some sections of the Carpathians up to 2,000 mm a^{-1}, in the southern-oriented mountain chains of the Julische Alps and the Dinarides up to even more than 2,000–2,500 mm a^{-1} because of the influence of humid-warm air masses coming from the Mediterranean. In the plains of the middle and lower Danube there is already a dry climate regime. The precipitation falls to 600 mm a^{-1} and partly to 500 mm a^{-1} in the Pannonian Basin and to lower values in the plains near the Black Sea.

The maximum of the long-term mean precipitation falls in the western part of the Danube Basin in the midsummer (July) and the minimum in spring (April). To the East the maximum shifts forward in the season towards June and May due to increasing continentality and the minimum occurs in midwinter (February, sometimes January) when the Asiatic region of high air pressure blocks the transfer of Atlantic air-masses to the east. In the mid-mountains under the influence of the Atlantic climate there are frequently additional precipitation peaks in the winter months of December and January. In some places they can be higher than in the summer as for example in the Bavarian and Bohemian Forest. In southern regions that are influenced by Mediterranean conditions the months October–December show the maximum precipitation and the summer is dry.

Humid and Dry Years

Precipitation of a single year can deviate considerably from the long-term mean value. In a humid year, for example 1965, the long-term values can be exceeded in the Alps by 140–150% and in the Carpathians by 130–150%. In a dry year like 1947 precipitation can decrease to below 400 mm in the German and Austrian Lowlands and in inter-mountain basins. The value of 233 mm recorded at Targovište in the South Carpathians in 1944 is considered to be the minimum annual precipitation in the Danube Basin.

2.2.4.2 Snow

The proportion of snow of the total precipitation and the duration of snow cover is clearly dominated by the elevation above sea level, but it is also influenced from west to east by rising continentality. The medium duration of snow cover in rising order is: Black Sea 9–12 days a^{-1}, plains in the Middle Danube Basin 20–30 days a^{-1}, lower regions of the Upper Danube 40–60 days a^{-1} (for both 10–15% of precipitation a^{-1}), Alpine Foothills and high regions of mid-mountains up to 100 days (20–30% of precipitation a^{-1}), in the Alps at heights above 1,500 m 4–6 months, above 2,000 m 6–8 months and above 2,500 m 8–10 months (80–90% of precipitation a^{-1}). The snowline is at heights between 2,900 and 3,200 m depending on the orientation of the mountain sides. The snow cover stays longer in mountain regions of the eastern catchment compared with the Alps; for instance at heights above 2,000 m the snow cover lasts 7–8 months in the Carpathians.

2.2.4.3 Temperature

The spatial pattern of temperature conditions is also clearly dependant on the altitude and from west to east on the rising continentality. The highest annual average temperatures with values of +12° can be found in the middle and lower Danube lowlands, the coldest regions are at the heights of the mountains, the Carpathians and, above all, the Alps (Sonnblick Observatory in the Austrian Alps, altitude a.s.l. 3,107 m, mean annual temperature: –6.2°). The increase of the fluctuation of mean monthly air temperatures from west to east is regarded as a criterion of the increasing continentality. The value increases from the Upper Danube Basin (Ulm 20°C, Vienna 21°C) to the Middle Danube Basin (Budapest 23°C, Beograd 23°C) down to the lower flatland (Bucharest 26°C). Also the rise of the extremes of single temperatures measured is a characteristic for increasing continentality. The maximum difference observed is 74.3° in the Hungarian plains.

2.2.4.4 Evaporation

Evaporation, being an essential part of the water balance of a catchment, is a function of several factors and therefore too difficult and costly to be measured directly. For this reason it is normally estimated by means of water balance (difference precipitation – runoff). The factors influencing evaporation are: available energy (depending on radiation, air temperature and wind) and available water (depending on precipitation, soil humidity, plant cover and groundwater conditions). In regions with high precipitation such as mid- and high-mountains energy (temperature) is the factor limiting evaporation while in hot plains rich in solar radiation water is the limiting factor. For this reason the mean annual evaporation shows considerably lower variation compared to precipitation and a different spatial distribution. The highest amounts of evaporation are in those regions where high amounts of precipitation are combined with high temperatures, situated in the middle Danube Basin: the Sava valley north of the Dinarides (725 mm a^{-1}) and the slopes of the Carpathians (700 mm a^{-1}). The greatest part of the catchment shows yearly rates of between 500 and 650 mm a^{-1}. Evaporation decreases in mountains with rising altitude because of decreasing temperatures. The lowest evaporation rates are found at the very high central alpine mountains (100 mm a^{-1}), at the heights of the South Carpathians and the Stara Planina and – due to lack of precipitation – in the Danube delta with about 400 mm a^{-1}.

2.2.5 Vegetation and Land Use

The natural vegetation cover of the Danube Basin is wood except in very extreme locations referring to climate, soil and relief, for example above the timberline in the high mountainous regions or in salt steppes. On the basis of dependence on climate, soil, regional and local conditions the natural wood is differentiated into numerous woody-plant populations among which many kinds of trees can be found, the most important ones in order being for example spruce, pine trees, beech trees, oak trees, maple and sycamore, elm trees, birch trees, willow and chestnut. Anthropogenic interventions over a long time period (partly since the Roman age) however have essentially changed the plant cover; the present vegetation and surface cover is mainly a result of deforestation and forestation. The region with comparatively rich woodland (mostly as a result of anthropogenic forestation) is the Upper Danube Basin with a forest cover of more than one third of its area. Extremely scarcely wooded are for example the Hungarian Lowland with 7% and the Moldavian Lowland with 8% tree cover. The main reason for the deforestation was to obtain farm and grazing land. Anthropogenic influences have also rather diminished the formerly large floodplain woods.

2.2.6 Lakes

There are a large number of natural lakes in the Danube Basin, mostly smaller ones but also large ones. They influence the runoff regime according to the local situation – sometimes very significantly.

The Alps and their foothills have large numbers of lakes. Within the German and Austrian Basin there are 29 larger lakes of more than 3 km^2 size (four of them in the upper basin of the Drava). The total surface area is about 435 km^2 and the water volume about 21,300 million m^3. On the other hand, the region north of the Danube is – except for a few tarns – a landscape without lakes.

In the Middle Danube region there are some lowland lakes. Lake Balaton is the largest one with an area of 589 km^2. It is also the largest Central-European lake; mean depth is 3.2 m and maximum depth is 12 m. The second largest is Lake Fertö tó (Neusiedler See), which is situated at the border between Austria and Hungary. It has a maximum depth of only about 2 m and usually covers around 315 km^2 (240 km^2 on the Austrian side). This size can fluctuate considerably due to precipitation, because the lake has a close catchment of only 1,237 km^2. The lake has suffered dry periods causing the lake to dry out in the past, and the maximum documented size is 515 km^2. Today there is a sluice on the Hungarian side to prevent flooding – but there is not yet an installation that protects against drying out. Most of the lake is surrounded by reeds. Within the Pannonian Basin there are six more lowland lakes of a size between 3.1 and 25 km^2. In the mountain range around the Middle Danube region there are several small lakes and tarns and this is also the case in the mountains around the Lower Danube region (also the Carpathians, then Stara Planina and Rila mountains).

The most significant lakes in the Lower Danube region are the riverine lakes along the Danube itself and some tributaries near their mouths. Most of these riverine lakes are connected by – mostly artificial – channels with the main river. Thus, these lakes will be filled during floods and reduce the flood peaks in this way.

Last but not least, a peculiarity of the Black Sea concerning its water quality may be of interest: Below 120–150 m in depth anoxic conditions prevail because of the presence of very toxic hydrogen sulphides. Because of the great depth of the Black Sea (more than 2,000 m) it is the biggest natural resource of hydrogen sulphides in the world. After strong winds it can occur that some of this hydrogen-sulphide rich water rises as bubbles to the surface. People living around the Black Sea know that they are poisonous and dangerous. Wooden vessels that come into contact with such water become black in colour and this may be the origin of its name "Black Sea" (Schiller, 2006).

2.2.7 River Construction and Water Engineering

In the Danube Basin there are very many hydrotechnical structures for example river channel training works, dykes, barrages, reservoirs and diversions. Every one of them has an important purpose for example fixing the river course, navigation, flood protection, drainage, irrigation and electric power (Fig. 2.2: General Map). The most important ones are described and listed in the 1986 Monograph itself and in the follow-up volumes V/3 (RCDC 1999b) and IX/1 (RCDC 2004b). Each of these structures affects more or less the natural flow regime. Therefore, a small summary will be given here:

First, a short review of the history is useful because there is a very close connection to the development of hydrology. Large-scale systematic river training works began approximately at the beginning of the nineteenth century. Reasons for such

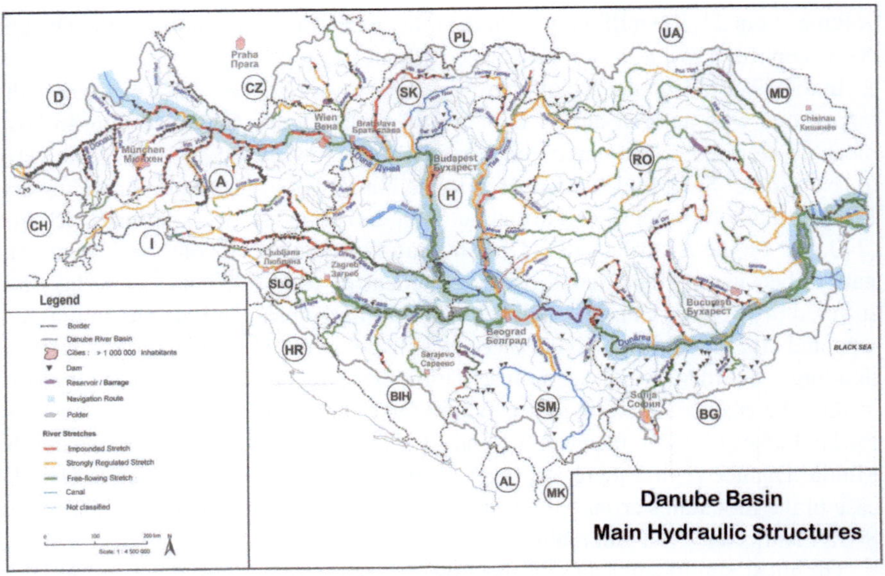

Fig. 2.2 General map of the main hydraulic structures in the Danube Basin (composed from maps in the follow-up volumes VIII/2 and IX/1)

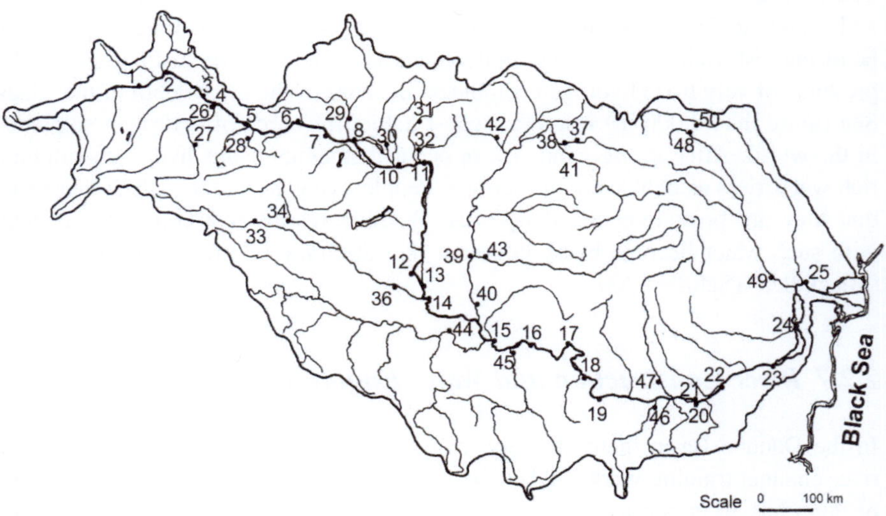

Fig. 2.3 General map of Danube Basin with selected stations (Monograph 1986)

works first included health problems (epidemic fever), and protection of agricultural areas, traffic lines (first mostly railroads) and settlements against destruction by meandering river-courses – and later also against flooding. In order to obtain the necessary fundamentals for planning the first gauges had been installed by the government departments which had to realise these works. One of the first systematic

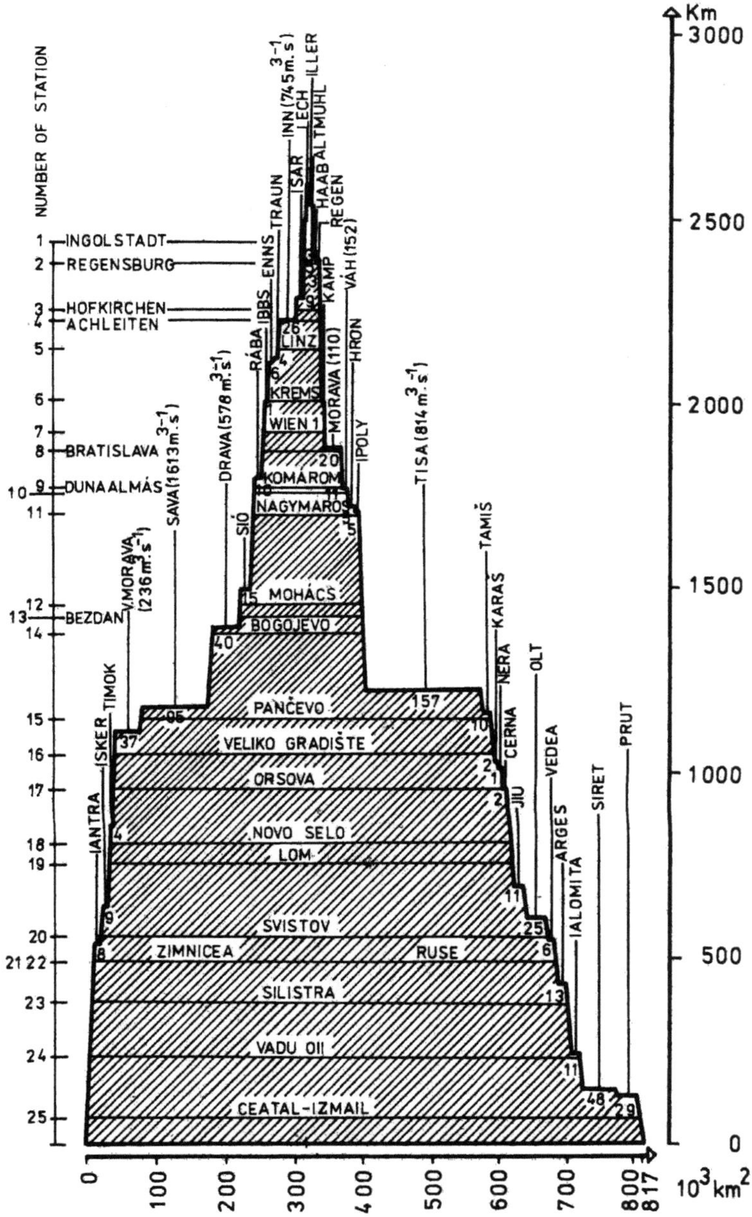

Fig. 2.4 Increase of the catchment area of the Danube River (with selected stations No. 1–25 of the 1986 Monograph on the left axis)

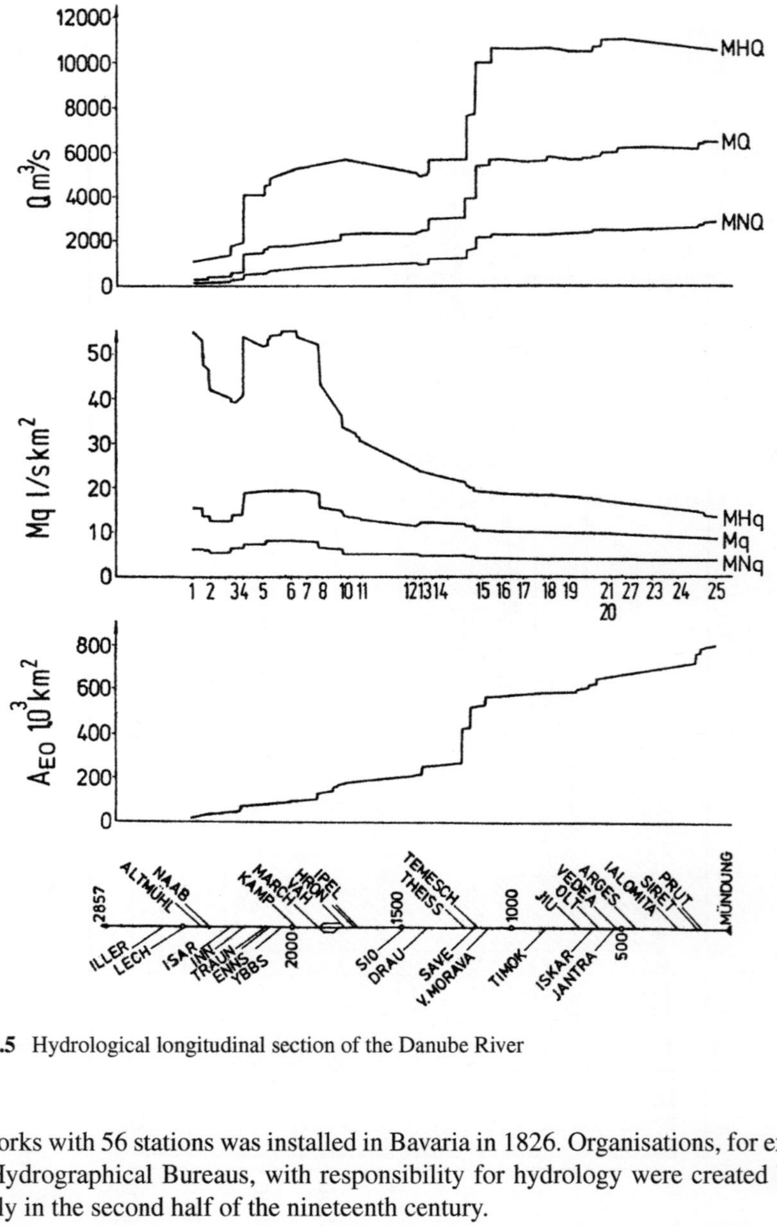

Fig. 2.5 Hydrological longitudinal section of the Danube River

networks with 56 stations was installed in Bavaria in 1826. Organisations, for example Hydrographical Bureaus, with responsibility for hydrology were created later, mostly in the second half of the nineteenth century.

The river bed of the whole length of the Danube itself and of the main tributaries has been already regulated for a long time – perhaps excluding some of the uppermost stretches.

Practically all important potential flood plains in the valleys of the Danube itself and the important tributaries are protected against flooding by dykes – generally against a 100-year flood. The total length of the dykes is about 12,500 km^2; most of them are in the Middle and the Lower Danube region.

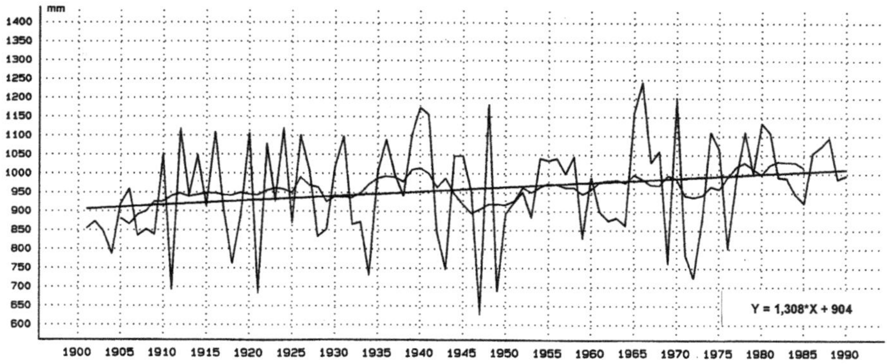

Fig. 2.6 German catchment of the Danube Basin: Trend of rainfall 1901–1990

There are more than 200 barrages, 12 of them are situated at the borders of two countries. Mostly the barrages are built as cascades along river courses for example the upper Danube (33 barrages) and the rivers Iller, Lech, Isar, Inn, Salzach, Enns, Mur/Mura, Drau/Drava, Sava, Váh, Jiu, Iskar, Osam, Olt and Arges. In the Lower Danube (Iron Gate) there are the two big barrages "Djerdap I" and "Djerdap II".

Reservoirs have a major direct impact on the natural flow regime because they change the redistribution of the water over time (for an example please see Fig. 2.11 right graph). In the Danube basin there are 269 permanent reservoirs with a volume of 17,769 million m^3 and 28 non-permanent reservoirs with a volume of 2,632 million m^3.

Additionally, there are a lot of deviations, diversions and intakes. The impact on the flow regime always depends on the purpose and the features of each single plant (Schiller, 1983). They mostly affect only a limited area.

2.3 Water Regime of the River Danube and Its Most Important Tributaries

2.3.1 General

This part of the 1986 Monograph was based on the data of the period 1931–1970, as already mentioned. In spite of the restricted possibilities of using computers at that time nevertheless every important evaluation had been made. And all the results, presented in many tables, graphs and figures had to be printed. Today, the widespread use of Personal Computers (PC) has made evaluation and publishing (adding a CD-ROM with the data and typical evaluations) much easier.

An updated study, based on the period 1931–1990 and using the tools of today, was published in 2004 (RCDC 2004a). But in spite of a slightly modified methodology, some of the former results are still interesting. They will be presented in this subsection together with some results of a few studies on special hydrological topics (e.g. flood regime).

2.3.2 Comparison of the Main Discharges Between the Periods 1931–1970 and 1931–1990

In order to get an idea of how much the results of the two evaluations may differ a comparison was made of the main discharges for both periods for almost every station (please see Table 2.1). The comparison produced the following result:

> The differences between the MQ of the periods are very small at the Danube river itself – mostly ca. 0.5–2.4% (the three higher values in the middle of the chain of stations do not seem to be realistic). The tributaries of the Upper and the Middle Danube show differences between 1.5% and mostly 3.3% and only two values up to 5.5%. The tributaries of the Lower Danube show higher differences – perhaps because of shorter observation periods and the increased continental regime. And it is somewhat surprising that also other values, for example MHQ, show only relatively small differences. This means that the results of the 1986 Monograph still provide very good evidence as far as the reference to the middle range of the discharges (MQ and a little more) is concerned. But the longer periods are always better for the estimation of extreme discharges (e.g. 100-years flood) by means of empirical and mathematical distributions because of their better representativeness.

2.3.3 Trend of Annual and Seasonal Series of Mean Discharges and Precipitation

Trends of the series of mean annual discharges, flood peaks and minimum low-flows were analysed in the 2004 study (RCDC 2004a).

Additional information is given by a study which analysed the trends of the seasonal mean discharges (period 1901–1990) of the station Danube/Achleiten and also the yearly values of aerial precipitation onto the German catchment. The results are presented in Figs. 2.6, 2.7, 2.8 and 2.9: The positive trend of the winter discharges (1 November–30 April) is exactly as great as the negative trend of the summer discharges (1 May–30 October) and, for this reason, the yearly discharges have no trend at all. The station Danube/Achleiten represents roughly the German catchment and therefore it is surprising that the rainfall has a positive trend. But in the water balance this additional amount of water is compensated by more evaporation. All these trends are not large enough to be significant. The two results however may give the impression that more studies on those two topics might be useful in the future.

2.3.4 Mean Monthly Discharges of an Observation Period

Usually the runoff regime of a station will be described by the monthly discharges of an observation period from January to December. If one divides the monthly

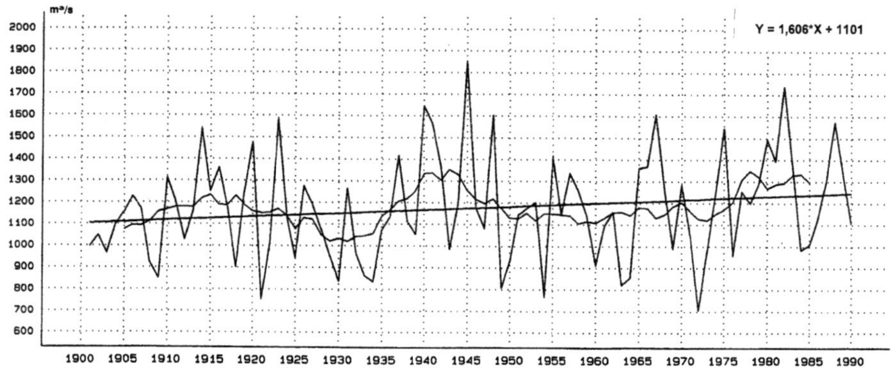

Fig. 2.7 Danube/Achleiten 1901–1990: trend of mean discharges of winter (1 November–30 April)

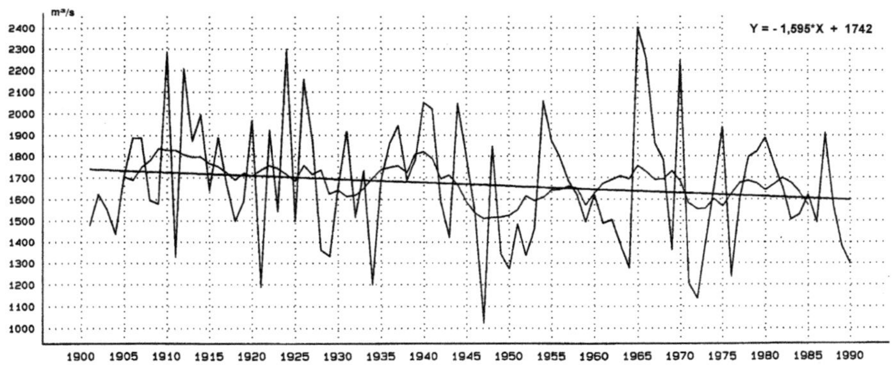

Fig. 2.8 Danube/Achleiten 1901–1990: trend of mean discharges of summer (1 May–30 October)

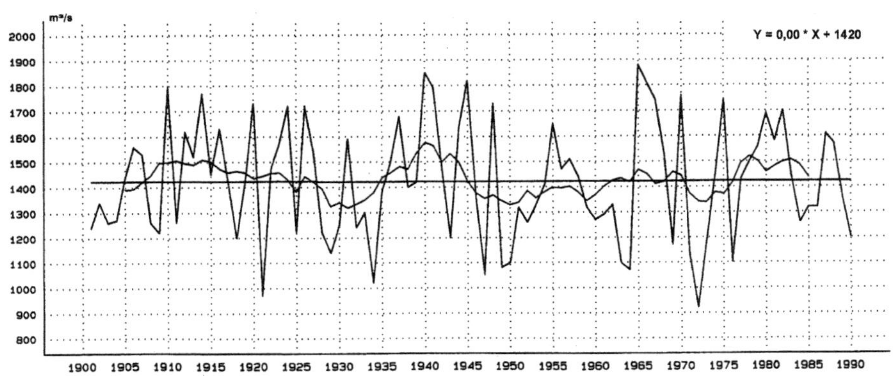

Fig. 2.9 Danube/Achleiten: trend of yearly mean discharges

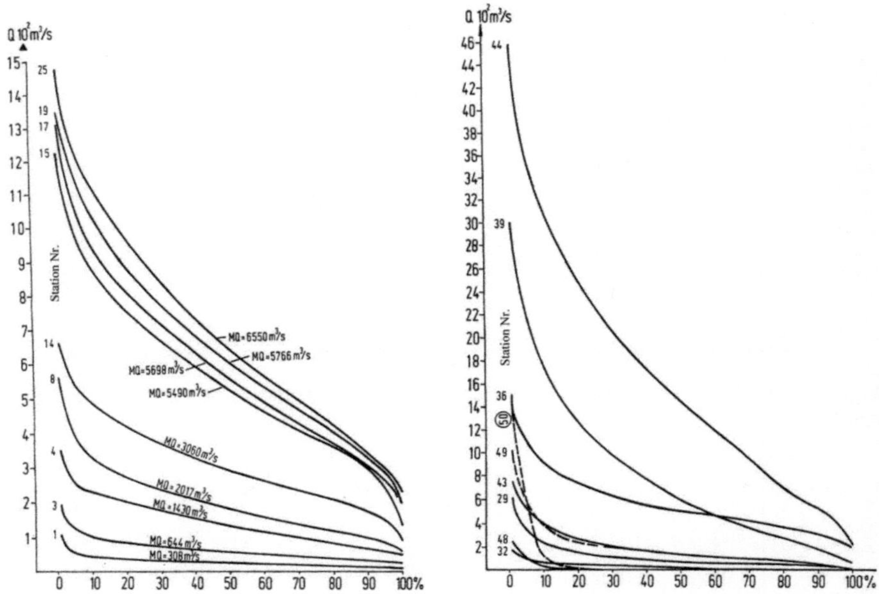

Fig. 2.10 Duration curves (m^3/s) of selected stations at the Danube (*left*) and at the tributaries (*right*)

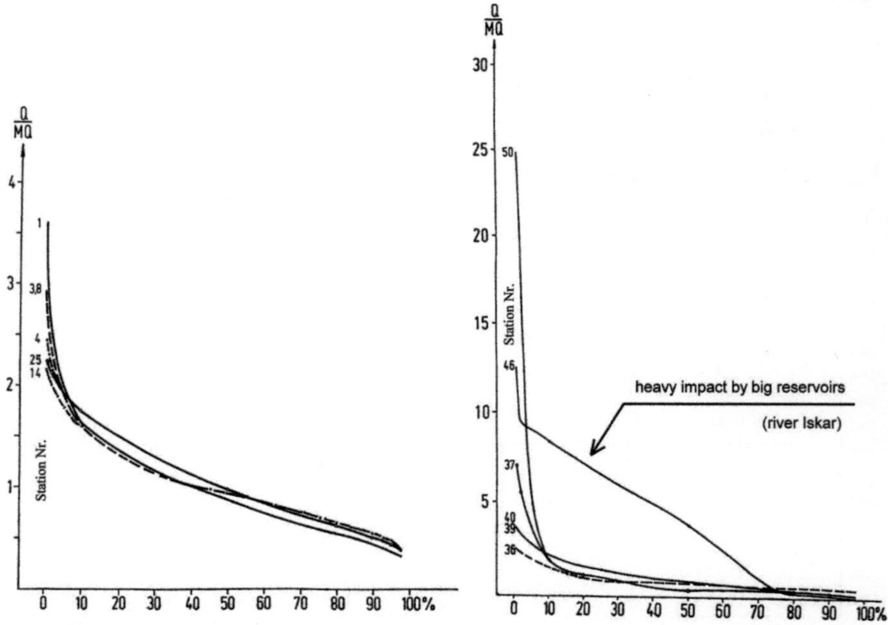

Fig. 2.11 Standardized duration curves (Q:MQ) of selected stations at the Danube (*left*) and at the tributaries (*right*)

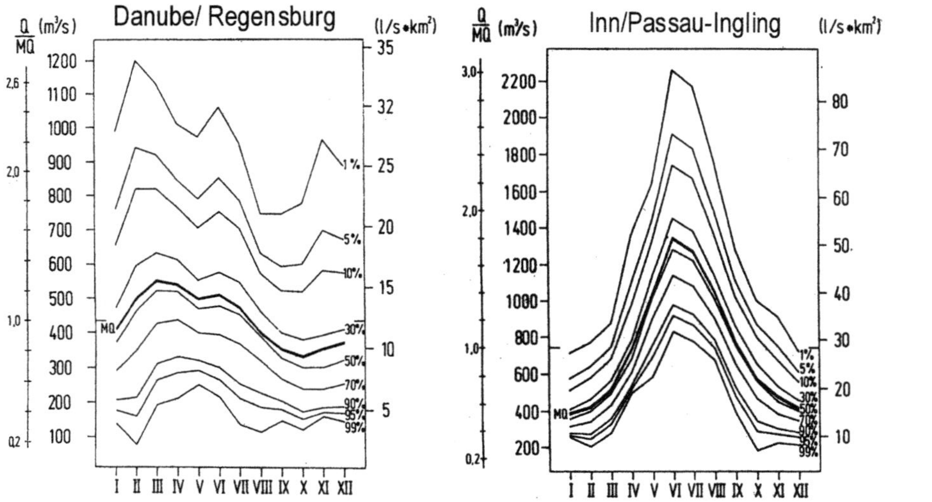

Fig. 2.12 Probability of monthly means, period 1931–1970, of selected stations at the Danube and at the tributaries

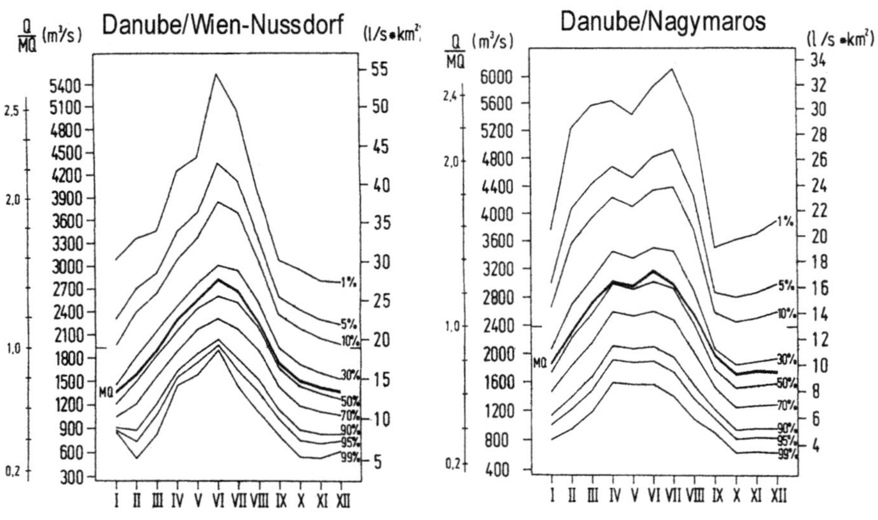

Fig. 2.13 Probability of monthly means, period 1931–1970, of selected stations at the Danube and at the tributaries

discharges by the mean discharge MQ of the period dimensionless values are obtained – the Pardé-coefficients. But it is possible to get more information on the runoff regime by calculating the probability of the monthly means, for example from 1 to 99%. These values indicate also the months of a year in which floods

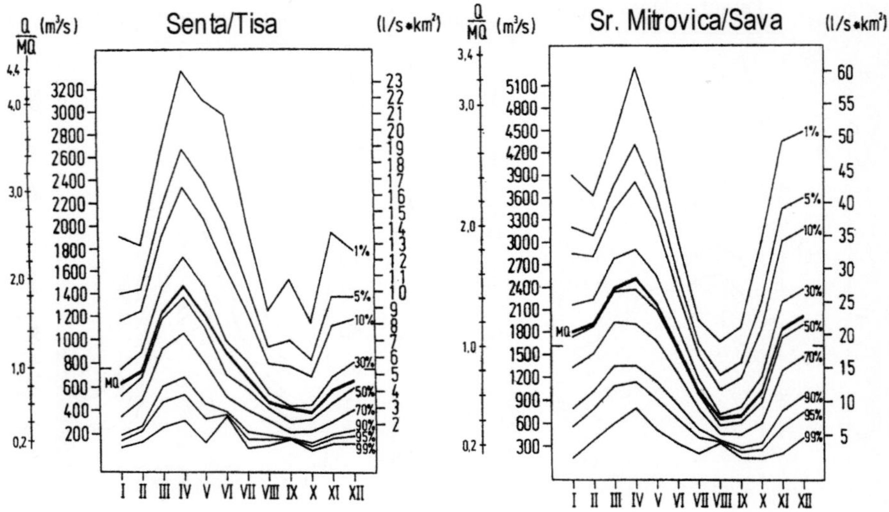

Fig. 2.14 Probability of monthly means, period 1931–1970, of selected stations at the Danube and at the tributaries

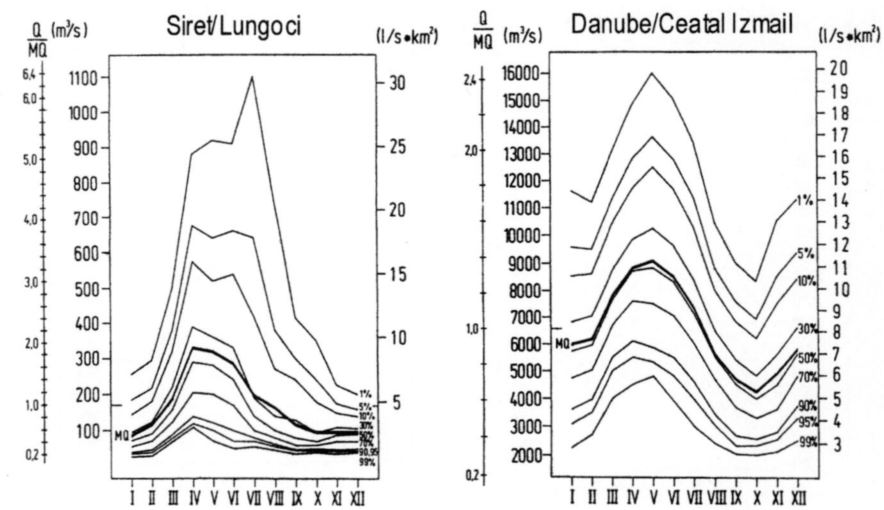

Fig. 2.15 Probability of monthly means, period 1931–1970, of selected stations at the Danube and at the tributaries

and low flows usually occur. Sometimes there are two seasons in which there is good probability for those events (Figs. 2.12, 2.13, 2.14 and 2.15) and then this can have additional influences on the mathematical calculation of flood discharges of different recurrence intervals.

2.3.5 *Estimation of Flood Extremes*

In practice, flood discharges of small probabilities (1–0.1% or 100-year up to 1,000-year flood) are needed as fundamentally important data for planning hydraulic structures like weirs, dykes, dams and others. This task will not be finished if one takes only the yearly flood peaks of the nearest station and makes only a mathematical calculation (best fitting distribution function and confidence intervals). In addition, two essential conditions for the use of these calculations have to be checked: representativeness and homogeneity.

Representativeness means that the recorded sample has to be a true pattern of the whole population. This requires that the longest series available along a river or within an area have to be taken. Sometimes it may also be helpful to use information of very significant historic flood events. But it is necessary for reliable marks of the highest water levels to be present on very old buildings and that it is possible to convert this information into discharges. This requires special studies and, in some places, this can be done with sufficient accuracy.

Homogeneity means that any flood of the sample has to be generated through the same or at least similar physical processes, for example rainfall floods, snowmelt with rain without and those with frozen soil. Probability of mean monthly discharges of the station Danube/Regensburg (Fig. 2.12 – left) indicates that there might be a family of yearly flood peaks which is not homogeneous. But in order to distinguish that there is really a big problem with

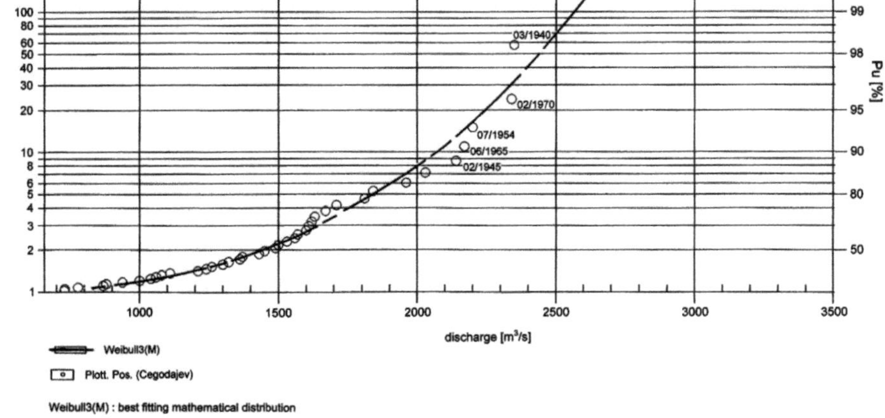

Fig. 2.16 Station Danube/Regensburg: flood probability derived from period 1931–1970

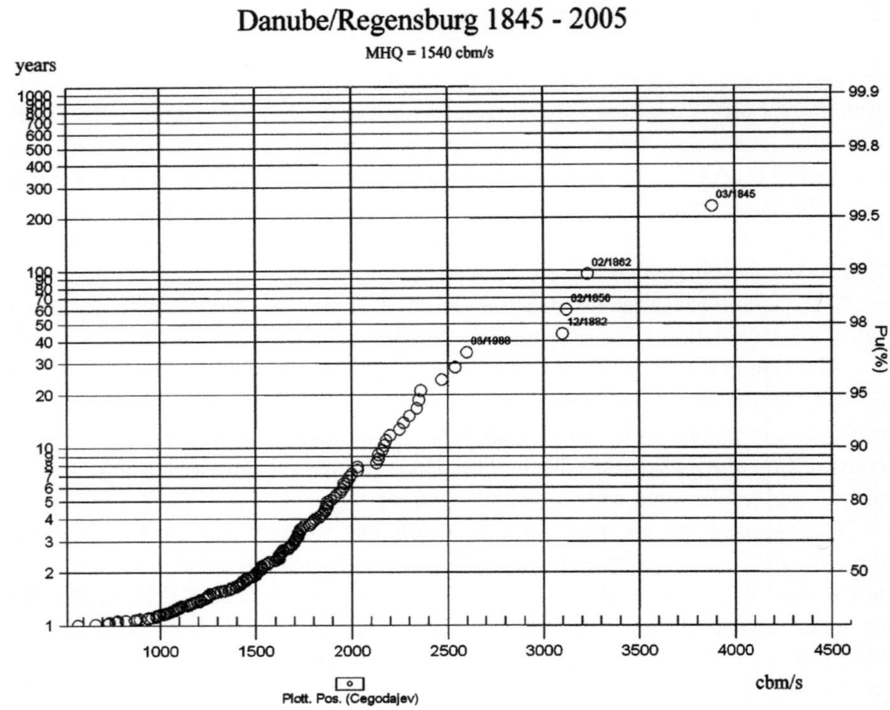

Fig. 2.17 Station Danube/Regensburg: flood probability derived from the total observation period 1845–2005 (plotting position)

homogeneity at the station Danube/Regensburg it is necessary to take the longest available series of annual flood peaks: 1845–2005. This reveals that all the big floods >2,400 $m^3\ s^{-1}$ (Fig. 2.17) had been winter floods and – except for 1988 (Schiller and Deisenhofer, 1989) – they had occurred before the year 1931 (Unbehauen, 1970). These floods are not involved in the studies for the 1986 Monograph or the updated study of 2004 because the series began in 1931.

The longest available series of three stations, which are partly very long, are shown in Figs. 2.18 and 2.19. The results for HQ_{100} for these stations are compared in the following list:

River/station	HQ_{100} (1931–1970) $m^3\ s^{-1}$	HQ_{100} (1931–1990) $m^3\ s^{-1}$	HQ_{100} (long series) $m^3\ s^{-1}$
Danube/Regensburg	2,600[a]	2,610[a]	3,250 (1845–1990)
Danube/Achleiten	7,900[a]	7,830[a]	8,300 (1861–1990)
Inn/Passau-Ingling	6,530[a]	6,220[a]	6,700 (1501–1990)
Salzach/Burghausen	3,000[a]	2,950[a]	3,400 (1598–1990)

[a]Best-fitting mathematical distribution function.

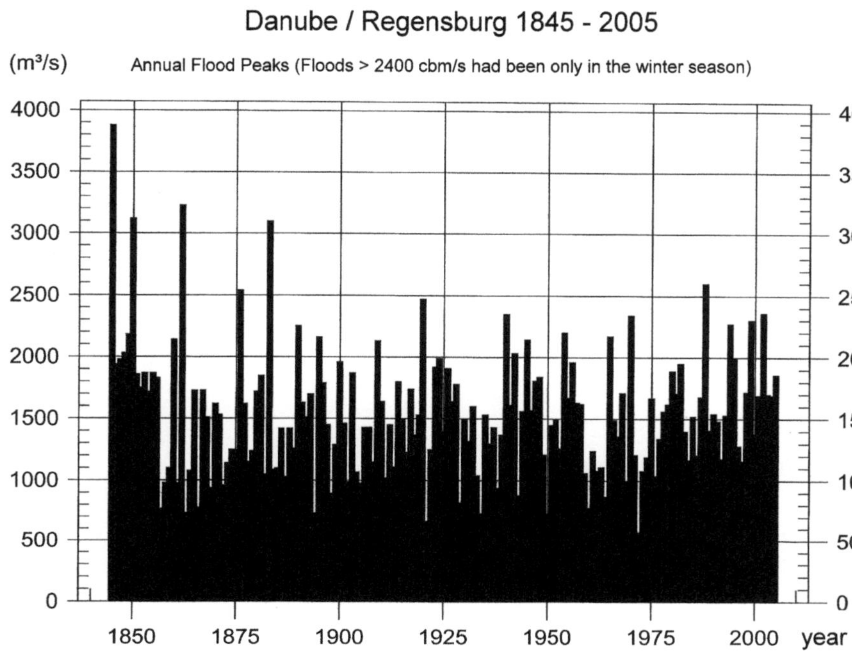

Fig. 2.18 Station Danube/Regensburg: annual flood peaks 1845–2005

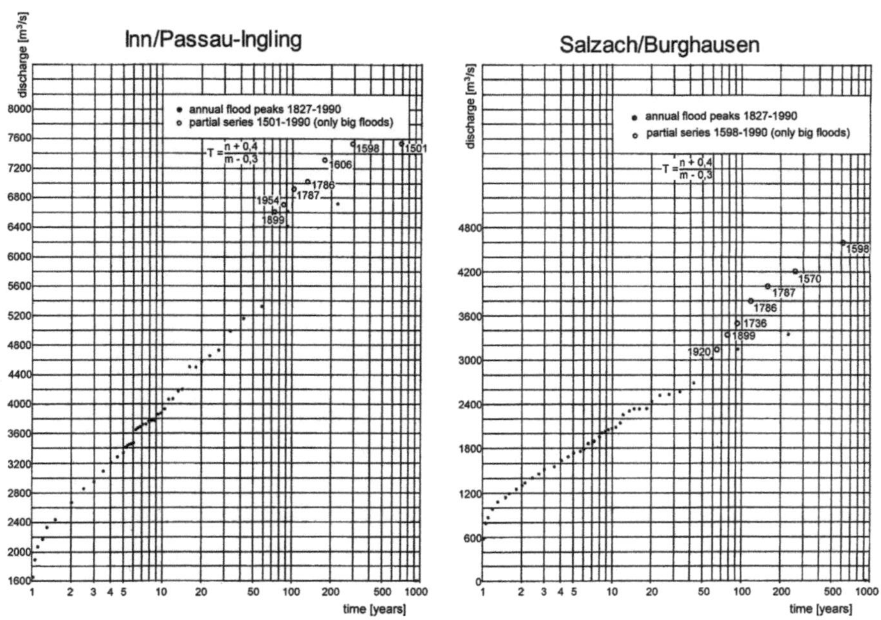

Fig. 2.19 Plotting positions of the longest available flood series of the stations Inn/Passau and Salzach/Burghausen

These four stations are very good examples demonstrating that it is necessary to always take the best available information. The next step, in order to obtain the best results along a river or within a region, is to transfer the information from the best stations with very long series to those with shorter records. There are some good tools for such a regional flood analysis for example regressions to improve the reliability of MHQ, the index flood method (Dalrymple, 1969) and longitudinal sections of indexes and specific discharges in l s^{-1} km^2 (Schiller 1993).

2.4 Long-Term Hydrological Balances of Subcatchments and Partial National Areas in the Danube Catchment for the Period 1931–1970

2.4.1 Introductory Remarks

This is an abridged and partly improved version of Chapter 3, Water Balance, of the Danube Monograph (RZD 1986), the first result of the regional hydrological co-operation of the (at that time) eight Danube Countries, whose history is briefly reviewed in the foregoing chapter, while the general content of the Monograph itself is outlined in the first part of the present chapter.

Chapter 3 of the Monograph, to be dealt with in detail herein, essentially summarises numerical information concerning the hydrological cycle from the preceding two chapters of the Monograph (Chapter 1 Physical Geography and Chapter 2 Runoff Regime) and its aim is to show, at least in the case of the long-term means, an agreement of the data presented. This is a rather significant aspect, since – after important previous individual efforts (e.g., Kresser and Lászlóffy 1964) – for the first time in the history of hydrology of the (at that time) eight Danubian countries an experiment was carried out to include internationally unified hydrological information on the whole Danube Catchment, collected and processed necessarily by means of different methods in the different Danube Countries. The agreement of data arranged in this way will be demonstrated without suppressing deviations, but by their numerical expression and evaluation.

Since the hydrological balance to be described here, was compiled more than two decades before the publication of the present work, it is appropriate to mention the following facts at the outset:

- The work herein is confined to the simplest variant of the hydrological balance: that of the long-term mean values of the three basic balance elements. It does not divide them into further components (i.e. precipitation into rainfall and snow; areal evaporation into soil evaporation and free water surface evaporation and transpiration; runoff into direct runoff and base flow) nor does it deal with their time or probability distribution. Genesis, including a description of the hydrological processes and data collection is dealt with in Chapter 1 of the Monograph (RZD 1986); runoff, the balance element of foremost importance in water resources development and water engineering works, is dealt with, including its time and probability distribution, in Chapter 2 of the Monograph and in its follow-up volume no. VIII/2 (RCDC 2004a).

- In the year 1984, when this water balance was compiled for 47 subcatchments and 12 partial national areas, the Danube Catchment was shared by 12 countries (among them the eight so-called "Danube Countries" with areal shares of more than 5%, see Table 2.3), while by the year 2009, when the present manuscript is being compiled, the number of countries interested in the Catchment has grown, because of political changes that have occurred in the meantime, to 20 (including 13 "Danube Countries"). The bases for compiling in 1984 these balances were the isoline maps of the three main elements of the balance, prepared previously by each of the eight co-operating Danube Countries separately.
- In the course of the hydrological co-operation of the Countries with an interest in the Danube catchment, the water balance of the Danube Catchment was re-compiled once more, this time for its 109 subcatchments (but for no national partial areas) by using hydrometeorological data series of the period 1961–1990 and advanced modelling methods (as compared with the more traditional approach of 1984). In this case, the starting inputs were not the isoline maps prepared previously, but hydrometeorological data series elaborated in the framework of modelling. The results of this second water balance of the Danube Catchment are included in one of the follow-up volumes of the Danube Monograph (RCDC 2006).

2.4.2 A Concise Description of the Working Procedure Adopted

The assessment of water balances and the computation of related indices included the following working steps.

(a) As a first step, isoline maps were constructed to the scale 1:2,000,000 for the regional distribution of long-term mean values of precipitation P, evaporation E and runoff R within the Danube Catchment.
(b) The Danube Catchment was divided into hydrological and national balance units.
(c) For each balance unit the regional average values of the three balance elements P, E, and R were assessed by means of planimetering the areas occurring between the isolines developed in Step (a).
(d) Substitution of the aforementioned average values into the balance equation enabled assessment of the final errors in the hydrological balances as well as the long-term runoff coefficients for the various balance units.
(e) By evaluating the results obtained for the subcatchments in Step (c) and by cumulating them along the Danube River, the regional variability of balance elements within the Catchment could be evaluated and the hydrological longitudinal profile of the River could be assessed.

(f) For selected gauging sections of the Danube, long-term mean discharges computed from the runoff isoline map (item a) or from balance results (item c) – and appearing in the longitudinal profile of the River (item e) – could be compared with discharges obtained from observation series.
(g) Finally, two indices for water resources availability in the Danube countries were calculated: their respective contribution to the whole water resources of the Danube basin and the ratios for "proper" to "transit" water resources (called also Balcerski indices).

It is obvious that the procedure described under steps (a)–(g) anticipates certain iteration steps so as to obtain acceptable final results. If, for instance, the final errors stated in steps (d) and (f) exceeded certain admissible limiting values, the need would arise to amend the isoline maps of (a), while maintaining the determined values of observation stations, or eventually to change the division of the Catchment under (b). However, it is also obvious that since the isoline maps are the result of a previous stage of co-operation between eight countries, the possibility to modify them was rather limited. All the same it was necessary to carry out some minor iteration steps, especially in connection with the regional distribution under (b). A more detailed description of the respective working steps follows.

2.4.3 Assessment of Isoline Maps of Balance Elements

As the basic map for compiling isoline charts of the regional distribution of mean annual values of the three hydrological balance elements, the map issued in 1972 by the Danube Commission to the scale 1:1,000,000 was used, reduced to the scale 1:2,000,000. The major rivers of the Danube Catchment, the main watershed divide, national frontiers, more important towns and the main stream gauging stations are presented on the basic map. Names of the countries are quoted according to the international abbreviations used in vehicular traffic (Table 2.3).

The isoline maps for the three elements of the hydrological balance were compiled on working maps by the respective Danube Countries, according to a methodology agreed in advance. For this purpose, data originating from measurements performed over the period 1931–1970 within the observation network for discharges and precipitation monitoring were adopted as well as empirical formulae for evaporation estimation used in respective countries, hydrological maps from previous periods (with isolines), and topographical and thematic (geological, vegetation, etc.) maps. Data on precipitation were converted to ground level prior to the construction of isohyets.

When constructing isolines, attention was paid to meeting the requirement of the envisaged subcatchments as well as the requirement for agreement among discharge and runoff data, at least in the case of the stream gauging stations entered in the basic map.

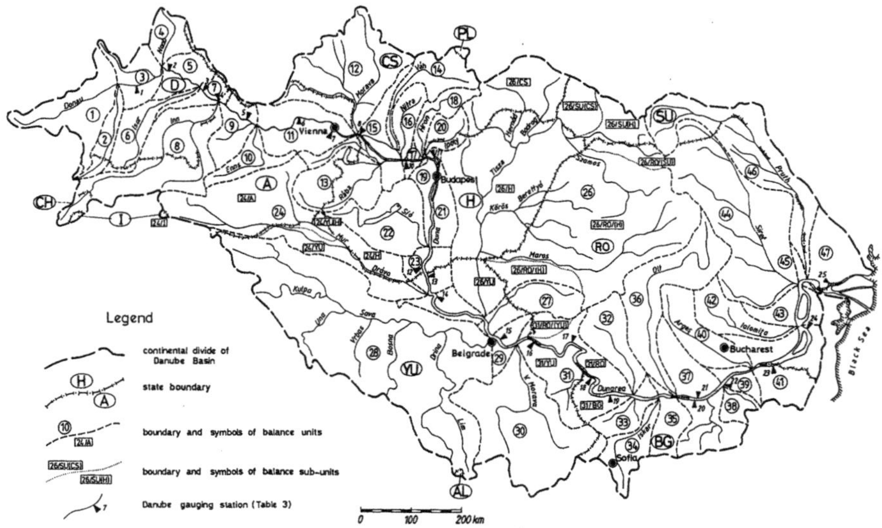

Fig. 2.20 General map of the Danube Catchment with subcatchments, as the balance units

Isolines drawn independently for the regions by the respective Danube Countries were fitted along the national borders. This work was performed by the experts of the countries in the course of working sessions or bi-lateral negotiations.

An exemplary fragment of the isoline map of mean annual specific runoff is shown in Fig. 2.20.

2.4.4 Determination of Balance Areas

Water balances were to be compiled on the basis of the isoline maps described, both for subcatchments and the partial national areas within the Danube Basin.

2.4.4.1 Subcatchments

The subcatchments to be considered as balance units were identified according to the following criteria:

(a) Generally catchments of more important Danube tributaries (Lech, Naab, Isar, etc.) were considered as independent units, and even if they were as large as for example the Sava, Drava, and Tisza catchments, they were not further divided.
(b) The second group of regional units: immediate catchment areas of the Danube sections between tributaries noted in item (a) (for instance from the Danube's

springs to the mouth of the Lech, between the Lech and the mouth of the Naab, between the Naab and the mouth of the Isar, etc.).

Finally, in such a way the Danube Catchment was divided into 47 subcatchments (Fig. 2.20). They were numbered strictly according to their hydrographical order. Sequence numbers, names and surface areas of regional units are identified in the first four columns of Table 2.2.

As a consequence of having adopted the jointly accepted rules (a) and (b) noted above, the areal extensions of the subcatchments thus obtained vary within a quite wide range (from 131 to 158,182 km^2), which may be considered as a drawback of the methodology.

2.4.4.2 Partial National Areas

When compiling water balances for the partial national areas, the main territorial units are the portions of the 12 country areas sharing the Danube Catchment in 1984 (Fig. 2.20). Eight of them – having portions more than 5% in the Danube Basin – are called "Danube Countries". The 12 countries, denoted by symbols used in international vehicular traffic, are listed in columns 2 and 3 of Table 2.3.

Depending on the relative situation of primary and secondary catchment divides and state boundaries, various sub-units appear, necessarily considered when compiling the balances, but not to be explained further in this paper. Some of the symbols of such partial areas appear in Fig. 2.20 as well; for instance, the symbol 26/SU(CS) denotes the Soviet part of the Tisza River catchment area, whose surface waters flow from the Soviet Union into Czechoslovakia.

2.4.4.3 Determination of Surface Extents of Balance Units

For determining the surface areas of the above identified natural and political water balance units, the values obtained by planimetering on the map to scale 1:2,000,000 were used as starting values.

These values were then adjusted with respect to probable variations in paper dimensions, in such a way that: (a) both the sum of the areas of the 47 subcatchments and of the 12 partial national areas should give the internationally accepted drainage area of the Danube Catchment, namely 817,000 km^2; and (b) the surface areas of the three countries falling totally (H) or almost totally (A, RO) in the Danube Basin, should be in accordance with their official values (Europa Year Book 1985).

The thus necessary corrections of "rough" extension values obtained by the planimeter, were: Austria –0.5%; Hungary –4.2%; Rumania –2.6%; the other nine national areas +2.7%; and for the whole Danube Catchment the average correction was +2.5%.

The surface areas A_i (subcatchments) and A_{jD} (partial national areas within the Danube Basin) are listed in column 5 of Table 2.2 and column 5 of Table 2.3, respectively.

Table 2.2 Mean water balance of subcatchments in the Danube Catchment

Balance unit					Regional mean annual value (mm year^{-1}) of				
No. (Fig. 2.1)	Drainage basin of river	From	To	Area A_i (km^2)	Precipitation P_i	Evaporation E_i	Run-off Ri	Balance error di (Eq. 2.2) (%)	Runoff coefficient $a_i = Ri/Pi$ (–)
1	2	3	4	5	6	7	8	9	10
1	Danube	Spring	Lech	15,654	914	539	370	+0.55	0.40
2	Lech	Spring	Mouth	4,398	1,349	536	769	+3.26	0.57
3	Danube	Lech	Naab	7,614	748	518	231	–0.13	0.31
4	Naab	Spring	Mouth	5,645	767	490	288	–1.56	0.37
5	Danube	Naab	Isar	6,203	850	509	348	–0.82	0.41
6	Isar	Spring	Mouth	8,369	1,174	536	614	+2.04	0.52
7	Danube	Isar	Inn	3,741	947	502	415	+2.85	0.44
8	Inn	Spring	Mouth	26,976	1,315	436	868	+0.84	0.66
9	Danube	Inn	Enns	7,614	1,148	512	677	–0.68	0.59
10	Enns	Spring	Mouth	5,940	1,483	483	1016	–1.08	0.68
11	Danube	Enns	Morava	14,112	829	537	299	–0.84	0.36
12	Morava	Spring	Mouth	27,633	640	483	141	+2.50	0.22
13	Rába	Spring	Mouth	14,702	738	592	130	+2.17	0.18
14	Váh	Spring	Mouth	9,714	906	534	415	–4.75	0.46
15	Danube	Morava	Nitra	10,567	677	542	141	–0.89	0.21
16	Nitra	Spring	Mouth	5,415	694	549	157	–1.73	0.23
17	Danube	Nitra	Hron	1,444	576	516	63	–0.52	0.11
18	Hron	Spring	Mouth	5,251	809	514	290	+0.62	0.36
19	Danube	Hron	Ipoly	131	588	525	81	–3.06	0.14
20	Ipoly	Spring	Mouth	4,594	661	531	139	–1.36	0.21
21	Danube	Ipoly	Sió	10,469	586	534	53	–0.17	0.09
22	Sió	Spring	Mouth	15,129	663	572	81	+1.51	0.12
23	Danube	Sió	Dráva	7,023	644	565	62	+2.64	0.10
24	Dráva	Spring	Mouth	41,810	998	575	435	–1.20	0.44

Table 2.2 (continued)

Balance unit					Regional mean annual value (mm $year^{-1}$) of				
No. (Fig. 2.1)	Drainage basin of river	From	To	Area A_i (km^2)	Precipitation P_i	Evaporation E_i	Run-off *Ri*	Balance error *di* (Eq. 2.2) (%)	Runoff coefficient $a_i = Ri/Pi$ (–)
25	Danube	Dráva	Tisza	5, 382	689	594	59	+5.22	0.09
26	Tisza	Spring	Mouth	158, 182	744	560	177	+0.94	0.24
27	Danube	Tisza	Sava	10, 666	764	582	136	+6.02	0.18
28	Sava	Spring	Mouth	94, 778	1, 091	634	514	–5.22	0.47
29	Danube	Sava	V. Morava	4, 332	689	644	51	–0.87	0.07
30	V. Morava	Spring	Mouth	38, 233	756	540	216	–1.34	0.29
31	Danube	V. Morava	Jiul	27, 239	746	571	162	+1.74	0.22
32	Jiul	Spring	Mouth	10, 731	831	576	278	–2.77	0.33
33	Danube	Jiul	Iskar	6, 498	658	494	135	+4.41	0.21
34	Iskar	Spring	Mouth	7, 811	725	455	230	+5.52	0.32
35	Danube	Iskar	Olt	6, 826	690	489	192	+1.30	0.28
36	Olt	Spring	Mouth	24, 810	873	592	234	+5.38	0.27
37	Danube	Olt	R. Lom	17, 984	676	520	130	+3.85	0.19
38	R. Lom	Spring	Mouth	3, 380	599	512	66	+3.51	0.11
39	Danube	R. Lom	Arges	3, 938	573	500	25	+8.38	0.04
40	Arges	Spring	Mouth	11, 814	80	577	190	+4.13	0.24
41	Danube	Arges	Ialomita	16, 573	537	479	20	+7.08	0.04
42	Ialomita	Spring	Mouth	10, 305	738	556	146	+4.88	0.20
43	Danube	Ialomita	Siret	6, 925	508	438	25	+8.86	0.05
44	Siret	Spring	Mouth	45, 420	757	550	157	+6.61	0.21
45	Danube	Siret	Prut	1, 116	462	411	8	+9.31	0.02
46	Prut	Spring	Mouth	28, 945	606	470	96	+6.60	0.16
47	Danube	Prut	Mouth	14, 964	420	412	7	+0.24	0.02
1-47	Danube	Spring	Mouth	817, 000	816	547	264	+0.60	0.32

Table 2.3 Long-term water balances for the partial national catchment areas in the Danube Catchment, with water management indices

Countries			Long-term mean values													
			Area (km^2)		Share of A_{jD} (%)		Regional average (mm $year^{-1}$) of					"Proper" surface water resources		"Transit" water resources		
Serial number j	Symbol	Name	Entire (Europa Year Book 1985) A_j	In the Danube Basin A_{jD}	In country's area	In Danube Basin	Precipitation P_j	Evaporation E_j	Runoff Rj	Balance error (%) (Eq. 2.2) d_j	Runoff coefficient (–) $a_j = R_j/P_j$	Discharge (m^3 s^{-1}) Q_j	Share (%) in resources of Danube Basin (Eq. 2.13)	Total Qj,out	Non-returning $Q^*_{j,out}$	Ratio (%) of "proper" to "transit" water resources (Eq. 2.14) λ_j
1	2	3	4	5	6	7	8	9	10	11	12	13	14	15	16	17
1	D	F.R.G.	248, 687	59, 634	23.98	7.30	962	528	415	+1.98	0.43	785	11.48	867	867	90.54
2	I	Italy	301, 278	471	0.16	0.06	1, 425	391	1, 139	–7.37	0.80	17	0.25	17	17	100.00
3	CH	Switzerland	41, 293	1, 819	4.41	0.22	1, 136	358	768	+0.88	0.68	44	0.64	51	51	86.27
4	A	Austria	83, 855	80, 731	96.27	9.88	1, 098	508	600	–0.91	0.55	1, 536	22.45	2, 460	2, 392	64.21
5	PL	Poland	312, 683	270	0.09	0.03	959	525	372	+6.46	0.39	3	0.04	3	3	100.00
6	CS	Czechoslovakia	127, 896	73, 040	57.11	8.94	719	513	215	–1.25	0.30	498	7.28	1, 547	1, 531	32.53
7	H	Hungary	93, 030	93, 030	100.00	11.39	609	539	60	+1.64	0.10	176	2.57	3, 525	3, 525	4.99
8	AL	Albania	28, 748	101	0.35	0.01	1, 875	494	1, 292	+4.75	0.69	4	0.06	4	4	100.00
9	YU	Yugoslavia	255, 804	183, 210	71.62	22.42	928	606	351	–3.12	0.38	2, 039	29.81	6, 074	5, 851	34.85
10	BG	Bulgaria	110, 912	48, 178	43.44	5.90	661	487	152	+3.33	0.23	232	3.39	3, 157	3, 149	7.37
11	RO	Rumania	237, 500	232, 249	97.79	28.43	752	557	160	+4.65	0.21	1, 177	17.21	7, 270	6, 765	17.40
	SU	Soviet Union	22, 402, 200	44267	0.20	5.42	748	509	235	+0.53	0.31	330	4.82	3, 773	3, 458	9.54
	Danube Basin			817,000 (=A_D)		100.00	816 (=P_D)	547 (=E_D)	264 (=R_D)	+0.60 (=d_D)	0.32 (=a_D)	6,841 (=Q_D)	100.00			

2.4.5 Long-Term Water Balances

2.4.5.1 Basic Equation

The long-term mean hydrological balance for the respective regional unit (subcatchment or partial national area) was assessed in the following form:

$$P = E + R \text{ (mm year}^{-1}\text{)}, \tag{2.1}$$

where: P = annual precipitation onto the unit: E = annual evaporation (or more exactly: evapotranspiration) from the unit; and R = annual runoff from the unit.

These three balance elements are presented in millimetres of water of the areal annual mean. So as to provide homogeneous results, the period 1931–1970 was chosen for the computation of long-term annual and monthly hydrological balance. (Some Danube countries, however, were compelled to use data from shorter periods, for example 1931–1960, or 1941–1970. Attempts were however made to convert this information to the selected standard period 1931–1970.)

The simplified hydrological balance according to Eq. (2.1) does not consider the distribution of balance elements into components, as well as other factors of which the more important are: (a) variations of water storage within the subcatchment; (b) water volume, infiltrating into the deeper strata, or emerging from the deep strata to the surface as springs, or by means of pumping; and (c) water losses due to evaporation from the transit water resources crossing the unit.

On the basis of experience, it is possible to neglect water storage under (a) if there is a sufficiently long observation period (in our case 40 years). As far as vertical water motion is concerned according to (b), it is possible to suppose in the case of sufficiently large surveyed regions, that they become mutually balanced. For item (c) it may be anticipated that water losses due to free water surface evaporation from transit discharges, will be considered as a supplemental component of the actual evaporation within a subcatchment.

Thus, it may be presumed that – in comparison with inaccuracies caused by inadequate hydrometeorological observation, station density and other uncertainties of known and unknown origin – the simplifying assumptions mentioned would not substantially influence the accuracy of the hydrological balance computations.

2.4.5.2 Determination of Mean Regional Values of Water Balance Elements

For each respective regional unit (subcatchment or national area), the mean regional values of precipitation P, evapotranspiration E and runoff R, characterising the period 1931–1970, were determined in mm year^{-1} by means of double planimetering of the respective isoline map (e.g., Fig. 2.21) and by averaging the two results. The values of P, E and R thus obtained are listed, for the 47 subcatchments; in columns 6, 7 and 8 of Table 2.2, and for the 12 national areas in columns 8, 9 and 10 of Table 2.3.

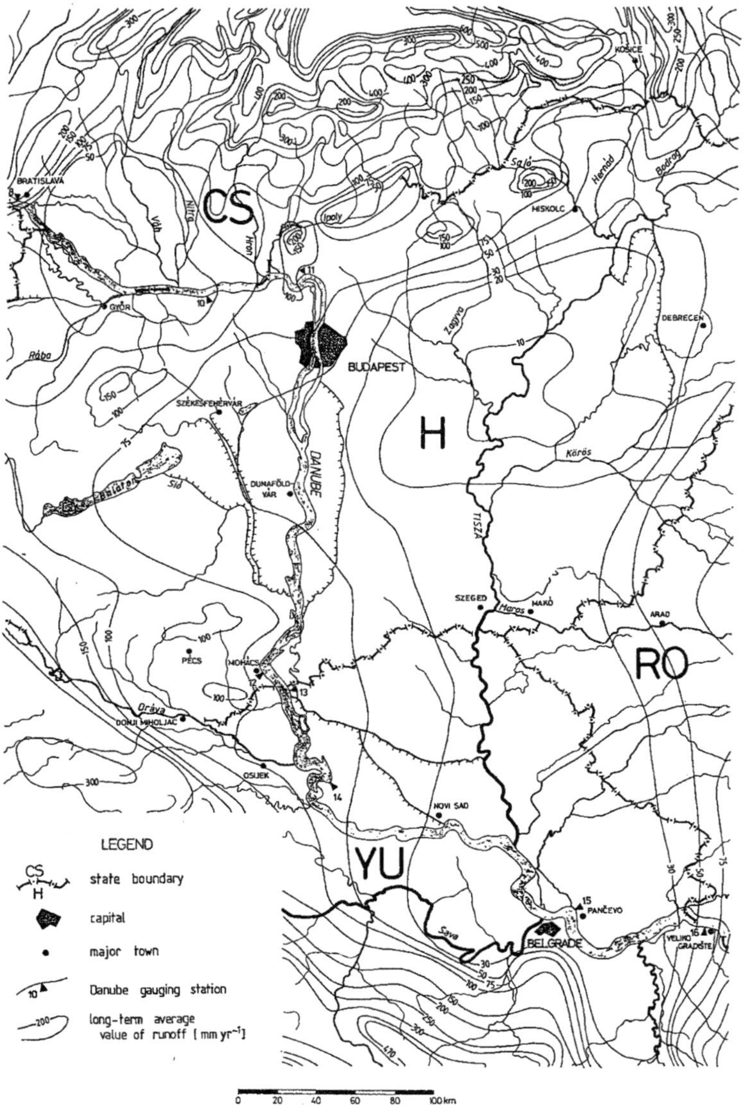

Fig. 2.21 Isoline map of the annual specific runoff (detail)

2.4.5.3 Computation of Final Balance Errors

If for an arbitrary balance unit (subcatchment or national area) the assessed numerical values P_i, E_i and R_i (or P_j, E_j and R_j) are substituted into Eq. (2.1), expressing (with the above-mentioned assumptions) the long-term equilibrium of the water budget, this results in equality when the equilibrium really exists and in an inequality when it does not exist. The difference between the two sides of Eq. (2.1) is the "final error" of the water balance. Expressed as the percentage of precipitation P; [i.e., the supposedly most exactly known term of Eq. (2.1)], the final error d_i of the ith balance unit can be calculated as:

$$d_i = \frac{P_i - (E_i + R_i)}{P_i} 100\ (\%) \tag{2.2}$$

The final errors d_i and d_j, calculated on the basis of Eq. (2.2) for each balance unit, are listed in Table 2.2, column 9 and Table 2.3, column 11, respectively, As for the balance of subcatchments, the maximal absolute error, +9.31% was assessed for unit no. 45 (sub-Danube basin between Siret and Prut). The arithmetic average of the final errors of the 47 subcatchments is:

$$\bar{d}_i = +1.75\% \tag{2.3}$$

the arithmetic average of their absolute errors being:

$$\left|\bar{d}_i\right| = 2.98\% \tag{2.4}$$

As for the balance errors obtained for the partial national areas, Table 2.4 shows that their absolute values lie below the very favourable limit of 5% for each state except the extremely small partial areas of Italy (d = –7.4%) and Poland (d = –f 6.5%), where the planimetration of isoline maps to scale 1:2,000,000 must be very unreliable. The simple unweighted arithmetic average of the final error d_j of the 12 state regions is:

$$\bar{d}_j = +0.96\% \tag{2.5}$$

and the arithmetic average of their absolute values is:

$$\left|\bar{d}_j\right| = 3.07\% \tag{2.6}$$

The mean values presented, as compared with the known uncertainties of hydrographical observations and hydrologic computations may be considered as acceptable, or even satisfactory. The maximum final error noted exceeded 9%, and some others exceeded 5%. However, since within the scope of international co-operation there were no further possibilities for correction, verification, and further approval of the previously printed maps of hydrological balance elements, one has to content oneself with the results obtained and to take into account the most significant balance errors mentioned above whenever making practical use of the isoline maps.

The final error in the hydrological balance of average values of P_D, E_D, and R_D, computed for the entire Danube Catchment is negligible:

$$d_D = \frac{P_D - (E_D + R_D)}{P_D} 100 = \frac{816 - (547 + 264)}{816} 100 = 0.6\%$$

Table 2.4 Hydrologic characteristics, integrated according to hydrographical order along the Danube (on the basis of Table 2.2) with a comparison of long-term mean discharges

Bordering section n							Gauging section k			Discharge ($m^3\ s^{-1}$)		
Symbol $n = \Sigma_i$ (Fig. 2.1)	Location[a]	Catchment area (km^2) $A_n = \Sigma_i{}^n{=}1A_i$	Regional mean value of Precipitation (mm $year^{-1}$) P_n	Regional mean value of Runoff (mm $year^{-1}$) R_n	Regional mean value of Runoff coefficient ($m^3\ s^{-1}$) a_n	Mean discharge ($m^3\ s^{-1}$) Q_n (Eq. 2.8)	Symbol k (Fig. 2.1)	Name	Catchment area (km^2)	Computed Q_k (Eq. 2.9)	Observed (from Chapter 2 of the Monograph) Q_k^*	Relative error (%) Δk (Eq. 2.10)
1	2	3	4	5	6	7	8	9	10	11	12	13
1	Lech	15,654	914	370	0.40	184						
2	Lech	20,052	978	458	0.47	291						
							1	Ingolstadt	21,268	300	307	−3.26
3	Naab	27,666	915	376	0.41	331						
4	Naab	33,311	890	361	0.41	383						
							2	Regensburg	36,415	417	435	−4.14
5	Isar	39,514	883	359	0.41	451						
6	Isar	47,883	934	404	0.43	614						
							3	Hofkirchen	48,139	617	643	−4.04
7	Inn	51,624	935	405	0.43	663						
8	Inn	78,600	1,066	564	0.53	1,405	4	Achleiten	78,600	1,421	1,430	−0.63
							5	Linz	81,480	1,467	1,509	−2.78
9	Enns	86,214	1,073	574	0.53	1,568						
10	Enns	92,154	1,099	602	0.55	1,759						
							6	Krems	97,658	1,811	1,863	−2.79
							7	Vienna	103,034	1,862	1,942	−4.12
11	Morava	106,266	1,063	562	0.53	1,893						
12	Morava	133,899	976	475	0.49	2,017						
							8	Bratislava	133,899	2,017	1,986	+1.56
13	Váh	148,601	952	441	0.46	2,078						
14	Váh	158,315	950	439	0.46	2,206						
15	Nitra	168,882	933	421	0.45	2,253						
16	Nitra	174,297	925	412	0.45	2,280						
							10	Dunaalmás	174,809	2,281	2,314	−1.43

Table 2.4 (continued)

Bordering section n							Gauging section k			Discharge ($m^3 s^{-1}$)		
			Regional mean value of									
Symbol $n = \Sigma_i$ (Fig. 2.1)	Location[a]	Catchment area (km^2) $A_n = \Sigma_i^n = 1 A_i$	Precipitation (mm $year^{-1}$) P_n	Runoff (mm $year^{-1}$) R_n	Runoff coefficient ($m^3 s^{-1}$) a_n	Mean discharge ($m^3 s^{-1}$) Q_n (Eq. 2.8)	Symbol k (Fig. 2.1)	Name	Catchment area (km^2)	Computed Q_k (Eq. 2.9)	Observed (from Chapter 2 of the Monograph) Q_k^*	Relative error (%) Δk (Eq. 2.10)
17	Hron	175,714	922	410	0.44	2,283						
18	Hron	180,992	919	406	0.44	2,331						
19	Ipoly	181,123	919	406	0.44	2,332						
20	Ipoly	185,717	912	399	0.44	2,353						
							11	Nagymaros	186,037	2,353	2,378	−1.05
21	Sió	196,186	895	381	0.43	2,370						
22	Sió	211,315	878	359	0.41	2,409						
							12	Mohács	213,363	2,413	2,389	+1.00
							13	Bezdan	214,579	2,416	2,479	−2.54
23	Dráva	218,338	871	350	0.40	2,423						
24	Dráva	260,148	891	363	0.41	3,000						
							14	Bogojevo	260,948	3,001	3,060	−1.93
25	Tisza	265,530	887	357	0.40	3,010						
26	Tisza	423,712	834	290	0.35	3,898						
27	Sava	434,378	832	286	0.34	3,944						
28	Sava	529,156	878	327	0.37	5,489						
							15	Pancevo	529,476	5,490	5,490	±0.00
29	V.Morava	533,488	877	325	0.37	5,496						
30	V.Morava	571,721	868	318	0.37	5,758						
							16	Gradište	576,969	5,785	5,750	+0.61
							17	Oršova	579,401	5,797	5,700	+1.70
							18	Novo Selo	584,809	5,825	5,842	−0.29
							19	Lom	596,048	5,883	5,766	+2.03
31	Jiul	598,960	863	311	0.36	5,898						
32	Jiul	609,691	862	310	0.36	5,993						
33	Iskar	616,189	860	308	0.36	6,021						
34	Iskar	624,000	858	308	0.36	6,078						

Table 2.4 (continued)

Bordering section n							Gauging section k			Discharge ($m^3\ s^{-1}$)		
			Regional mean value of									
Symbol $n = \Sigma_i$ (Fig. 2.1)	Location[a]	Catchment area (km^2) $A_n = \Sigma_i{}^n{=}1A_i$	Precipitation (mm $year^{-1}$) P_n	Runoff (mm $year^{-1}$) R_n	Runoff coefficient ($m^3\ s^{-1}$) a_n	Mean discharge ($m^3\ s^{-1}$) Q_n (Eq. 2.8)	Symbol k (Fig. 2.1)	Name	Catchment area (km^2)	Computed Q_k (Eq. 2.9)	Observed (from Chapter 2 of the Monograph) Q_k^*	Relative error (%) Δk (Eq. 2.10)
35	Olt	630,826	856	306	0.36	6,120						
36	Olt	655,636	857	303	0.35	6,304						
							20	Svistov	658,228	6,315	6,168	+2.38
							21	Zimnicea	658,228	6,315	6,150	+2.68
37	R.Lom	673,620	852	299	0.35	6,378						
38	R.Lom	677,000	851	298	0.35	6,385						
							22	Ruse	677,000	6,385	6,264	+1.93
39	Argeş	680,938	849	296	0.35	6,388						
	Argeş	692,752	848	294	0.35	6,459						
							23	Silistra	697,328	6,462	6,300	+2.57
							24	Vodu Oii	719,630	6,518	6,220	+4.79
41	Ialomita	709,325	841	288	0.34	6,470						
42	Ialomita	719,630	840	286	0.34	6,518						
43	Siret	726,555	836	283	0.34	6,523						
44	Siret	771,975	832	276	0.33	6,749						
45	Prut	773,091	831	276	0.33	6,750						
46	Prut	802,036	823	269	0.33	6,838	25	Ceatal				
							808,180	Izmail	6,839	6,550	+4.41	
47	Mouth into Black Sea	817,000										

[a]First is upstream, second is downstream.

2.4.6 Hydrological Characteristics

2.4.6.1 Regional Variability of Water Balance Elements

Remarkably high variability in elements P_i, E_i and R_i of the hydrological balance in respective subcatchments, presented in Table 2.2, is ascribed to the exceptional diversity of physical-geographical (topographical, geological, soil, vegetation, climatic, etc.) conditions within the Danube Catchment.

Precipitation P, representing the input side of the hydrological balance, depends on the mutual effect of the distance from the Atlantic Ocean and the altitude above sea level. These alternating effects are responsible for the decreasing of long-term average precipitation values from $P_i > 900$ to $P_i < 400$ mm in the west–east direction. (With increasing continentality the dispersion of extreme annual values also increases.) In comparison with the above general trend, high mountains (the Alps, Carpathians) cause positive anomalies from 200 to 500 mm, lowlands (as for instance the Vienna Basin, and the Large Hungarian Plain) negative anomalies from 100 to 200 mm. A further significant high-mountainous phenomenon is the increasing portion of precipitation falling as snow with increasing altitude, contributing in the case of concentrated melting in spring even up to 80–90% of the runoff.

From the two expenditure elements of the hydrological balance, the average annual values of the actual areal evapotranspiration E_i are mostly influenced by the same physical-geographical factors as is precipitation, though having an opposite effect, the magnitude of E_i being defined by the amount of available precipitation. In accordance with that and due to increasing continentality in the eastward direction, also an increasing precipitation portion generally evaporates. With regard to evaporation, mountains with their lower average temperatures induce negative deviations, and lowlands (due to higher average temperatures and close groundwater level) represent positive deviations in comparison with the general trend.

The most sensitive element, recording all variability in physical-geographical factors is the overland runoff R_i. In addition to the factors already mentioned (effect of precipitation amount and evapotranspiration rate) it depends mainly upon the orographic parameters (slope angle, relief energy, exposure, etc.). In the Enns river catchment (balance unit no. 10), having an extremely mountainous character (steep slopes, relief energy exceeding 2,000 m) the average long-term runoff is more than a hundred times as high as that in the east part of the Danube Catchment (unit nos. 45, 47).

2.4.6.2 Runoff Coefficient

The values of multi-annual mean runoff coefficient, i.e., the ratio of runoff to precipitation:

$$\alpha = \frac{R}{P}(-), \tag{2.7}$$

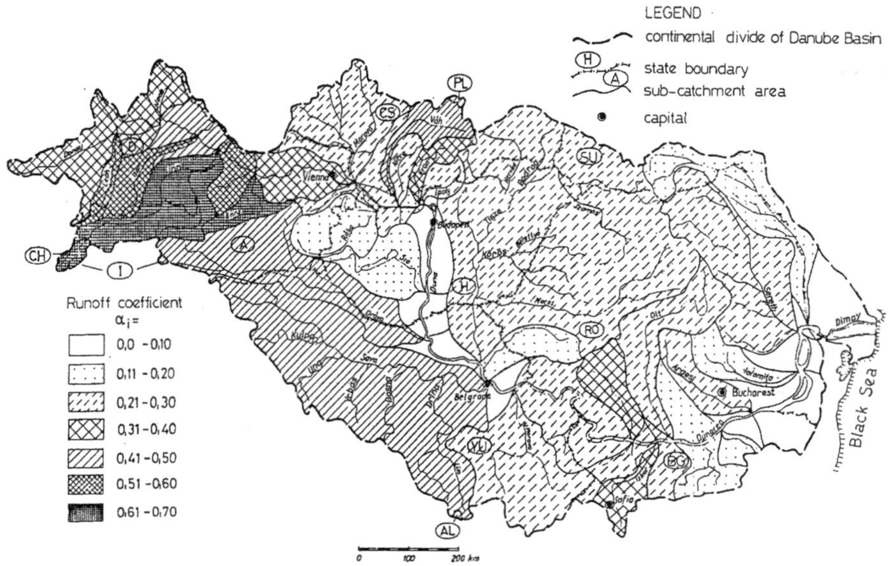

Fig. 2.22 Values of runoff coefficients in the subcatchments of the Danube Catchment (for names see also Fig. 2.20)

are listed for the 47 subcatchments investigated, in Table 2.2, column 10, (α_i, see Fig. 2.22), and for the 12 partial national areas in Table 2.3, column 12 (α_j). As for the latter, except for extraordinary high α_j values in the case of countries which participate in the Danube Basin only with a small portion of high-mountainous regions (I: $\alpha = 0.80$; CH: $\alpha = 0.68$; AL: $\alpha = 0.69$) it may be observed that the highest average values of α_j, valid for the state territories, consistent with orographic conditions, characterise the territory of Austria (0.55) and of Germany (0.43), while the lowest runoff coefficient (0.10) represents the Hungarian region, consisting mainly of lowlands.

2.4.6.3 Hydrological Longitudinal Profile of the Danube River

The aim was to plot selected hydrological indices in the form of a longitudinal profile (Kresser and Lászlóffy 1964, Lászlóffy 1965, Újvári 1967). For each important section of the Danube (called "bordering section" and denoted with n) the catchment area A_n belonging to it as well as the regional mean values P_n, R_n, α_n and the discharge Q_n were computed in columns 3–6 of Table 2.4, from the respective values given in Table 2.2. The areal structure of the Danube Basin – slightly modifying the values so far published by Kresser and Lászlóffy (1964) and by Újvári (1967) – is shown in Fig. 2.23 and the hydrological longitudinal profile of the Danube River in Fig. 2.24.

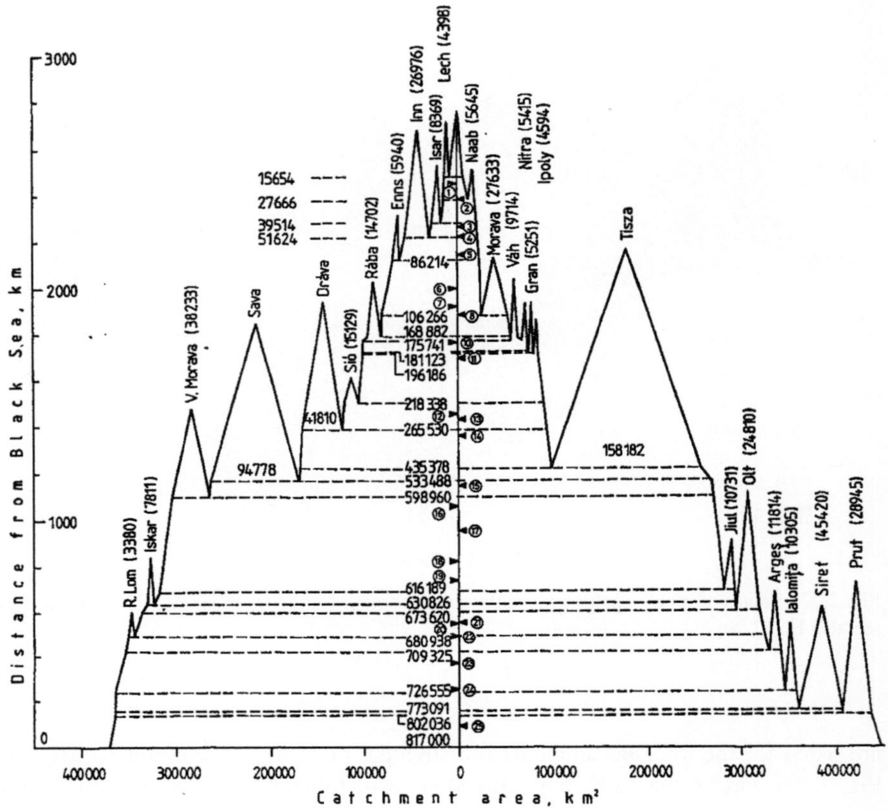

Fig. 2.23 Modified areal structure of the Danube Catchment, after Kresser and Lászlóffy (1964) and Újvári (1967)

2.4.6.4 Checking on Discharge Values

A possibility to check on the Q_n discharge values derived from the runoff isoline map (Fig. 2.21), is a comparison thereof with the long-term mean discharge values computed from the observation series of 1931–1970 in Chapter 2 of the Monograph (RZD 1986).

The comparison is carried out for all Danube gauging stations denoted with $k = 1, 2, \ldots, 25$ in Figs. 2.20 and 2.24 and listed in columns 8 and 9 of Table 2.4.

As a first step, for each "bordering section" of the Danube, the discharge value Q_n resulting from the runoff isoline map (Fig. 2.21) is calculated by cumulative summation of products of A_i (km^2) and R_i (mm year^{-1}) values taken from Table 2.2, for all subcatchments upstream of section n:

$$Q_n = \frac{1}{31{,}536} \sum_{i=1}^{n} A_i R_i \quad (\mathrm{m^3\,s^{-1}}) \tag{2.8}$$

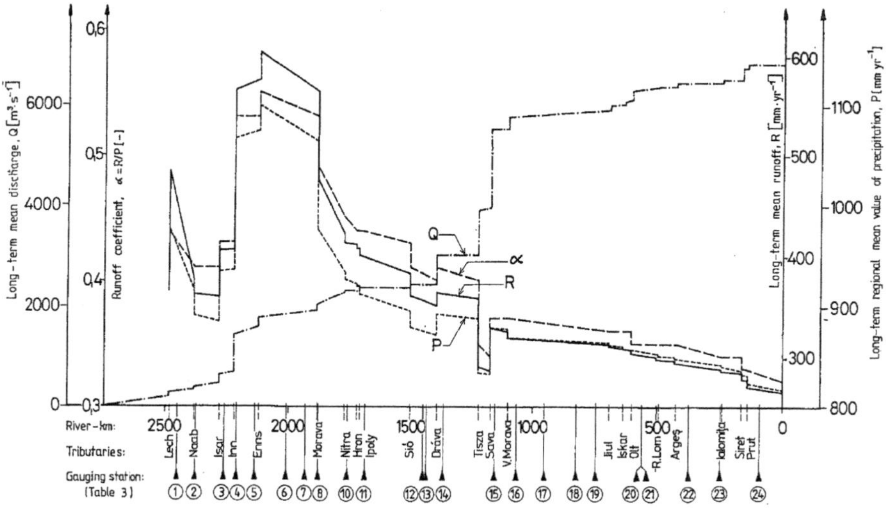

Fig. 2.24 Hydrologic characteristics of the Danube Catchment, represented in a longitudinal profile

As the next step, for each gauging section k the discharge resulting from the runoff isoline map is estimated by linear interpolation between the two adjacent "bordering sections"

$$Q_k = Q_u + \frac{A_k - A_u}{A_d - A_u}(Q_d - Q_u) \quad (\mathrm{m}^3\,\mathrm{s}^{-1}), \tag{2.9}$$

where u is the subscript for the upstream and d the subscript for the downstream bordering section next to the k gauging section.

Finally, the difference between this value Q_k (derived from the runoff isoline map) and the value Q_k^* (computed from long-term observation series and listed in Chapter 2 of the Monograph) is characterised with the relative error:

$$\Delta_k = \frac{Q_k - Q_k^*}{Q_k^*} 100(\%) \tag{2.10}$$

Values Δ_k are shown in column 13, Table 2.4. It can be seen that absolute values of deviations are lower than 5%. This result can be considered in discharge measurement as favourable with respect to the internationally required accuracy of ±5%. It confirms the reliability of the runoff isoline map for the Danube Basin (Fig. 2.21) on one hand, and the reliability of hydrological balance computations (Tables 2.2 and 2.3) carried out, on the other. Thus indirectly also the reliability of the remaining two isoline maps is confirmed.

It is interesting to note, that while the relative deviations, occurring in the upper Danube reach, are mostly negative, the deviations downstream from the gauging station Pančevo are exclusively positive. In addition it can also be seen that the deviation indices Δ_k considered from the spring of the river downstream to the estuary are continuously increasing.

2.4.7 Water Resources Indices for the Partial National Areas

2.4.7.1 "Proper" Surface Water Resources of the Danube Countries

For the partial national areas A_{jD} of the 12 Countries situated in the Danube Catchment, the "proper" surface water resources W_j resulting directly from precipitation, were computed using the formula:

$$W_j = 10 - 6\, R_j\, A_{jD}\ (\text{km}^3\text{year}^{-1}) \tag{2.11}$$

or eventually converting to discharge values Q_j, by means of an equation analogous to Eq. (2.8):

$$Q_j = \frac{1}{31{,}536} R_j\, A_{jD} = \frac{1}{0.031536} W_j \quad (\text{m}^3\,\text{s}^{-1}) \tag{2.12}$$

Q_j values computed in this way are presented in column (13) of Table 2.3.

2.4.7.2 Relative Contribution to the Total Water Resources of the Danube Basin

Relative values, characterising the ratio of the "proper" surface water resources in the total surface water resources of the Danube Catchment:

$$\gamma_j = \frac{Q_j}{Q_D} 100 = \frac{Q_j}{6{,}841} 100(\%) \tag{2.13}$$

are presented in column (14) of Table 2.3.

Out of the eight Danube Countries existing in 1986, Yugoslavia had the highest contribution (30%) to the stream flow rate at the Danube Delta estuary or to the total potential surface water resources of the Danube. Indices γ_j, considerably exceeding the respective proportions A_{jD}/A_D, characterise Germany and Austria; in these countries also occur the highest runoff values R_j. Out of the eight Danube Countries, the lowest contribution of surface water resources is that of Hungary ($\gamma = 2.6\%$) in spite of the fact that Hungary covers 11.4% of the total Danube Basin area. This is ascribed to the lowland character of the country, and thus to an extraordinary low runoff coefficient α, as mentioned above.

2.4.7.3 Ratio of "Proper" to "Transit" Water Resources

Finally, as an important index for water resources development, for each of the 12 national areas the ratio (often referred to also as the Balcerski index)

$$\lambda_j = \frac{Q_j}{Q^*_{j,\mathrm{out}}} 100\,(\%) \tag{2.14}$$

was computed in column 17 of Table 2.3. Q_j is the "proper" water resources as defined by Eq. (2.12) and $Q^*_{j,\mathrm{out}}$ (column 16) is the non-returning part of the total outflowing discharge $Q_{j,\mathrm{out}}$ (column 15) of the given country (also called "transit resources"). The assessment of Q_j and $Q^*_{j,\mathrm{out}}$ values (and thus of λ_j indices) is in some cases necessarily based on working assumptions to be explained in connection with the following examples.

Discharges leaving respective state regions were divided into returning and non-returning portions. The necessity to proceed in this way will be explained on the basis of Fig. 2.25. From the balance unit no. 26/RO occurring in Rumania, the runoff passes onto the regions of neighbouring countries (SU, H, YU), and from the unit nos. 27/RO, 29/RO and 31/RO the whole surface runoff passes to Yugoslavia. All such discharge would later come back into the Danube and would run through its channel again into Rumania, from which they flow as a part of the Danube discharge partly onto the region of the Soviet Union, and partly directly into the Black Sea. Thus, if the outflowing discharges would be mechanically summed along the boundary of Rumania, the sum (7,286 $m^3\ s^{-1}$) would exceed the discharge, which does not return into Rumania (6,781 $m^3\ s^{-1}$). The difference between the two quoted discharge values is the sum of the discharges (505 $m^3\ s^{-1}$) which leaves though later returns to Rumania.

Some reaches of the Danube and its tributaries form the borders between countries. The basic rule of computing Q_j values in such a case is shown by means of the example in Fig. 2.26. Within the boundary section of the Danube A/CS, Austria delivers the whole discharge of the Danube measured immediately upstream of the mouth of the River Morava, as well as half of the River Morava discharge, since the Morava on its lower reach forms the boundary A/CS (i.e. half of the water resources from the whole subcatchment no. 12). From the point of view of water resources development, the distance between the Danube boundary section A/CS and the upper starting point of the Danube reach forming the state boundary between Czechoslovakia and Hungary was neglected. Thus, Czechoslovakia obtains 50% and Hungary 50% of the hypothetical discharge:

$$Q^{(A)}_{D,\mathrm{out}} + \frac{1}{2} Q_M$$

The portion accepted by Czechoslovakia is increased by half of the River Morava discharge, by the whole of the tributaries from the balance units 14,15/CS, 16,17/CS and 18, as well as by half of the River Ipoly discharge within its mouth, which also forms the Czechoslovak–Hungarian border. Czechoslovakia delivers the sum of this discharge at the mouth of the Ipoly to Hungary. From there on, the Danube no longer

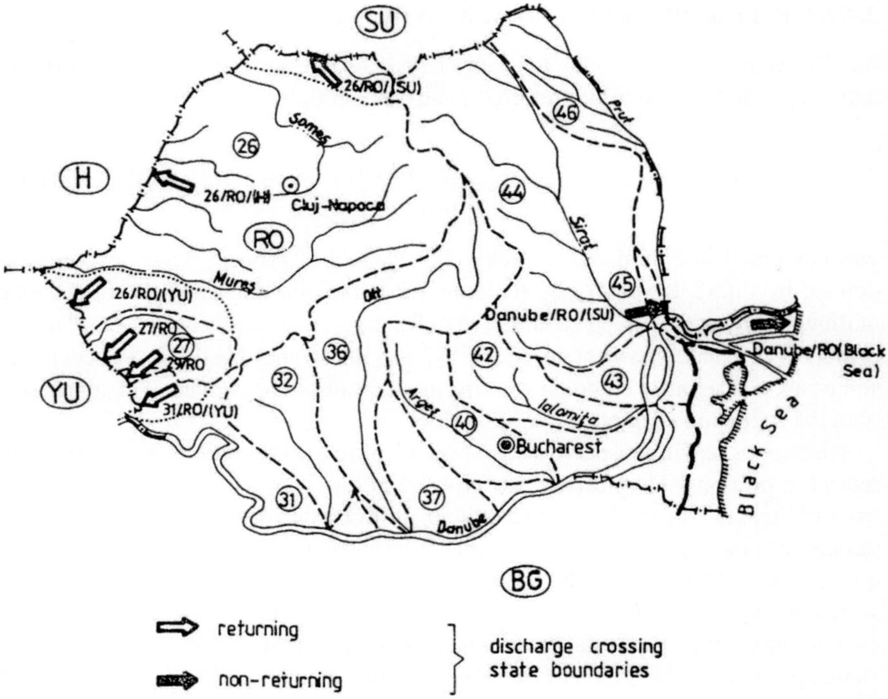

Fig. 2.25 Theoretical sketch for the interpretation of "returning" and "non-returning" outflows, by the example of Rumania

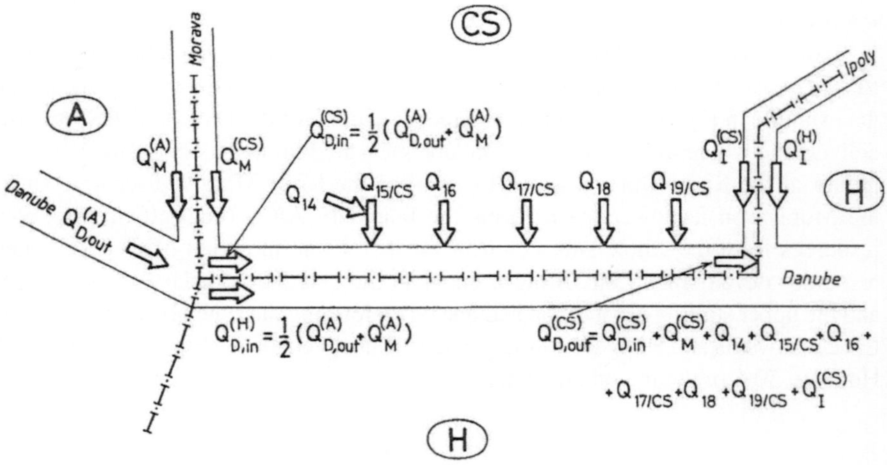

$$Q^{(CS)}_{D,in} = \frac{1}{2}(Q^{(A)}_{D,out} + Q^{(A)}_{M})$$

$$Q^{(H)}_{D,in} = \frac{1}{2}(Q^{(A)}_{D,out} + Q^{(A)}_{M})$$

$$Q^{(CS)}_{D,out} = Q^{(CS)}_{D,in} + Q^{(CS)}_{M} + Q_{14} + Q_{15/CS} + Q_{16} + Q_{17/CS} + Q_{18} + Q_{19/CS} + Q^{(CS)}_{I}$$

Fig. 2.26 Theoretical sketch for distribution of water resources among neighbouring countries, by the example of the frontier Danube reach between Czechoslovakia and Hungary (see Fig. 2.20) forms a state border. The same principle was followed for instance in the case of the River Inn (forming the D/A boundary), of the River Drava (forming the H/YU boundary), as well as the Danube reaches forming state borders RO/YU, RO/SU.

The sequence of eight Danube countries based on the characteristic λ_j, is as follows:

1. Federal Republic of Germany 90.5%
2. Austria 64.2%
3. Yugoslavia 34.8%
4. Czechoslovakia 32.5%
5. Rumania 17.4%
6. Soviet Union 9.5%
7. Bulgaria 7.4%
8. Hungary 5.0%

The portion of the "transit" water resources are naturally expected to increase in the downstream direction. Thus, the above sequence of countries should follow, as a tendency, the hydrographical sequence of the Danube Countries. Conspicuous exceptions from this tendency are two countries: Yugoslavia and Hungary. The reasons for this phenomenon have already been mentioned in connection with the runoff coefficients α_j. Therefore, Hungary being lowland in character and having an extraordinarily low runoff coefficient – and with it, very low "proper" surface water resources – is at the end of the list arranged according to the λ_j values in spite of the fact that geographically she has a central position within the Danube Catchment (Fig. 2.27).

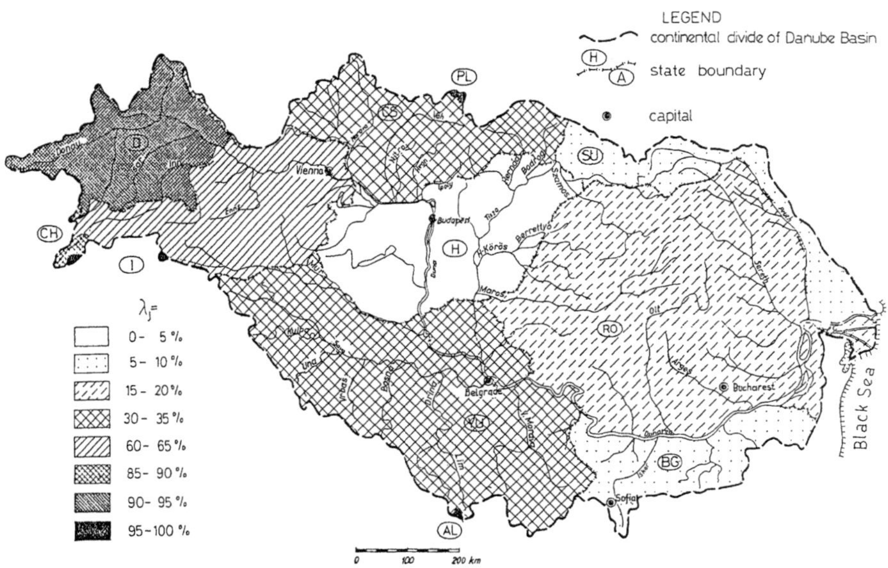

Fig. 2.27 Ratios of "proper" to "transit" water resources (Balcerski indices) for partial national areas in the Danube Catchment (for names see also Fig. 2.20)

2.4.8 Conclusions

On the basis of the final errors in the hydrological balances of subcatchments (Table 2.2, column 9) and the deviations between mean discharges computed from the observations originating from stream gauging stations and discharges assessed using the isoline map of runoff (Table 2.4, column 13), the following conclusions may be drawn:

1. The isoline maps (e.g., Fig. 2.21) according to available information give a true picture of the regional distribution of the long-term average values of the hydrological balance elements within the Danube Basin during the period 1931–1970.
2. Data, available in the isoline maps and Tables 2.2, 2.3 and 2.4 show the present state of knowledge and represent internationally approved basic hydrological information. They can be recommended for application in all future works focused on water resources development within the Danube Catchment, especially for planning of internationally co-ordinated resources utilisation.

It is hoped, that this publication represents a further approach towards the ever-increasing intensive co-operation and better information transfer among the Danube Countries as far as the basic hydrological and water-management data are concerned on the one hand, and existing conditions on the other.

Acknowledgments The author of the subchapters 2.1 to 2.3, H. Schiller heartily thanks his former colleagues for supplying information and data. Special thanks to Mr. Hans Weber and to Mr. Franz-Clemens Holle.

The authors of subchapter 2.4, M. Domokos and J. Sass, heartily thank all hydrologists of the Danube Countries who both by supplying information and by commenting on the tentative results obtained, substantially contributed to the assessment of the water balance of the Danube Catchment. Special thanks to Mr. Heinz Schiller (Germany), Dr. Oskar Behr (Austria), Dr. Pavel Petrovič (Slovakia), Dr. Mária Kardos (Hungary) and Dr. Károly Stelczer (Hungary).

References

Dalrymple T (1969) Flood frequency analysis. Geol Survey, Water-Supply Paper 1543-A

Europa Year Book (1985), Europa Publisher Ltd. London

Kresser W, Lászlóffy W (1964) Hydrologie du Danube. La Houille Blanche 2:133–178

Lászlóffy W (1965) Hydrographie der Donau. In: Liepolt R (ed) Limnologie der Donau. E. Schweizerbartsche Verlagsbuchhandlung, Stuttgart, Lfg. 1

RCDC (Regional Cooperation of the Danube Countries) (1999a) Palaeogeography of the Danube and its catchment (Follow-up volume No.V/2 to the Danube Monograph). Budapest

RCDC (Regional Cooperation of the Danube Countries) (1999b) The Danube river channel training (Follow-up volume No. V/3 to the Danube Monograph). Bratislava

RCDC (Regional Cooperation of the Danube Countries) (2004a) Flow regime of the River Danube and its catchment (Follow-up volume No. VIII/2 to the Danube Monograph). Koblenz

RCDC (Regional Cooperation of the Danube Countries) (2004b) Inventory of the main hydraulic structures in the Danube Basin (Follow-up volume No. IX/1 to the Danube Monograph). Bucharest

RCDC (Regional Co-operation of the Danube Countries) (2006) Basin-Wide water balance in the Danube River Basin (Follow-up volume No. VIII/3 to the Danube Monograph). Bratislava

RZD (Regionale Zusammenarbeit der Donauländer) (1986) Die Donau und ihr Einzugsgebiet. Eine hydrologische Monographie. Band I, II, III (The Danube Monograph, German version). Bayerisches Landesamt für Wasserwirtschaft, München

Schiller H (1983) River canalization with preservation of natural retention areas within the flood plains. Proceedings of the international conference on the hydraulic aspects of floods & flood control, London, England, 13–15 September, Paper A3

Schiller H (1993) Flood frequency analysis with consideration of representativeness and homogeneity of the flood series. Proceedings of the 2nd international conference on FRIEND, Braunschweig, 11–15 October

Schiller H (2006) The culture and the development of the Danube Area and Eastern Europe in Prehistoric Times – from the Black Sea Deluge until the end of the copper age. Proceedings of the 23rd conference of Danubian countries on hydrological forecasting and hydrological bases of water management, Belgrade, 28–31 August 2006

Schiller H, Deisenhofer J (1989) Das Märzhochwasser 1988 in Bayern. Informationsbericht des Bayerischen Landesamtes für Wasserwirtschaft Nr. 3/89, München

Újvári J (1967) Mittlere monatliche Abflüsse im Donaugebiet. Dtsch Gewässerk Mitt Sonderh 11:14–22

Unbehauen W (1970) Die Hochwasserabflußverhältnisse der Bayerischen Donau. Besondere Mitteilungen zum Deutschen Gewässerkundlichen Jahrbuch Nr. 30, München

Chapter 3
Palaeogeography of the Danube and Its Catchment

Domokos Miklós and Ferenc Neppel

Abstract The paper describes, as one of the possible interpretations, the present knowledge about the development of the Danube Catchment and its associated river system over some 25 million years. Both the fragmentary information scattered over a great number of pertinent publications in the palaeogeographic literature and hints given in the correspondence by the participating experts from 13 Danube Countries of the period 1993–1999, were pooled in one coherent concept after some necessary co-ordination and filling of gaps. The information presented in the paper is mostly fact-based, although necessarily some hypotheses had to be included as well. Consequently, one can expect that the picture drawn here will undergo further refinements in the future when new insights will be gained. A compact overview of the palaeogeographic history of the Danube Catchment is given in Table 3.1, listing the major events taking place during the various eras of the Earth's history and referring to the pertinent illustrations. The major findings regarding the development of the network of water bodies in the Catchment are illustrated in a series of nine palaeogeographic maps, each of which displays the state of the water system during a certain geological epoch (Figs. 3.9–3.17).

Keywords Danube Catchment · Danube River · Synoptic palaeogeographic map · Palaeogeographic development · Orogeny · Erosion · Accretion · Paratethys Sea · Pannonian Sea · Glaciation · Tributary · Tectonic uplift · Transgression · Alps · Carpathians · Great Hungarian Lowland · Romanian Lowland

3.1 Introduction

The present paper is an abridged version of the homonymous volume (RCDC 1999), issued in 1999 in Budapest as one of the results of the regional hydrological co-operation of the IHP National Committees of the Danube Countries in the

D. Miklós (✉)
Surface Waters, Research Institute for Environment and Water (VITUKI),
Budapest H-1453, Hungary
e-mail: domokosm@vituki.hu

M. Brilly (ed.), *Hydrological Processes of the Danube River Basin*,
DOI 10.1007/978-90-481-3423-6_3, © Springer Science+Business Media B.V. 2010

framework of the International Hydrological Programme (IHP) of UNESCO. The project leading to the publication of that volume was started in 1993, directed by the Authors of this paper and Prof. S. Somogyi (Budapest), under the overall co-ordination of the Austrian IHP National Committee from 1993 to 1998, and that of the Slovak IHP National Committee in the closing year of the project, 1999.

With a catchment of 817,000 km^2, the Danube is the second largest river in Europe and the 21st of the Earth. Its Catchment, with its wide West–East extension has experienced since the Miocene, i.e., over the past 25 million years, manifold changes, which have formed in their complexity the present orography and thus influenced today's natural and cultural landscape.

To give a well-balanced, up-to-date overview of this development, all relevant information on this topic – from scientifically substantiated facts to theories and hypotheses – had to be collected, selected, and arranged for presentation. The review given in this paper is based on material supplied by experts in the co-operating Danube countries (see Acknowledgement) as well as on a wide spectrum of relevant literature and is – of course – only one of several possible concepts . Before going to press, the manuscript of the project report taken as a basis for preparing the present paper, had been reviewed by the experts of the 13 co-operating Danube Countries, so that the statements included therein can be considered as a kind of collective opinion of all those who contribute.

The hydrological co-operation between the Danube Countries had begun as early as in 1971. As its first major result, in 1986 the Danube Monograph was published (RZD 1986), and it set the frame for a series of follow-up volumes, including that on the palaeogeography of the Danube and its Catchment (RCDC 1999). This follow-up volume of the Monograph is confined in its scope in so far as some topics, which are treated in other follow-up volumes, like questions of solids transport (RZD 1993) or impacts of anthropogenic activities (RZD 1999), are omitted. The "backbone" of the volume consists of nine synoptic palaeogeographic maps (Figs. 3.9–3.17), which each illustrate the general hydrographic situation for a certain geologic period. With regard to the diversity of ethnic groups and languages in the Danube basin, it seemed inevitable to use unified principles for the spelling of geographic names (giving priority to English denominations, if available, or else to those used in the country(-ies) concerned) and the geologic eras (by preferring the IUGS International Terminology).

The knowledge of the present situation in the Danube basin in physicogeographic, hydrographic, hydrologic terms (Figs. 3.1 and 3.2) gives, on the one hand, a useful orientation in the study of the palaeogeographic development, where regular reference to today's conditions is made, and on the other hand, it defines (Fig. 3.2a) and characterises the three main regions (Upper, Central, and Lower Danube region) and the two outer boundary areas (Uppermost Danube region and Danube Delta).

A historical overview on research in the palaeogeography of the Danube Catchment was found necessary to pay due tribute to the scientists to whom we owe our recent palaeogeographic compilation. In the project report (RCDC 1999) a number of selected works are cited, from Marsigli (1726), Cvijič (1908), Sümeghy

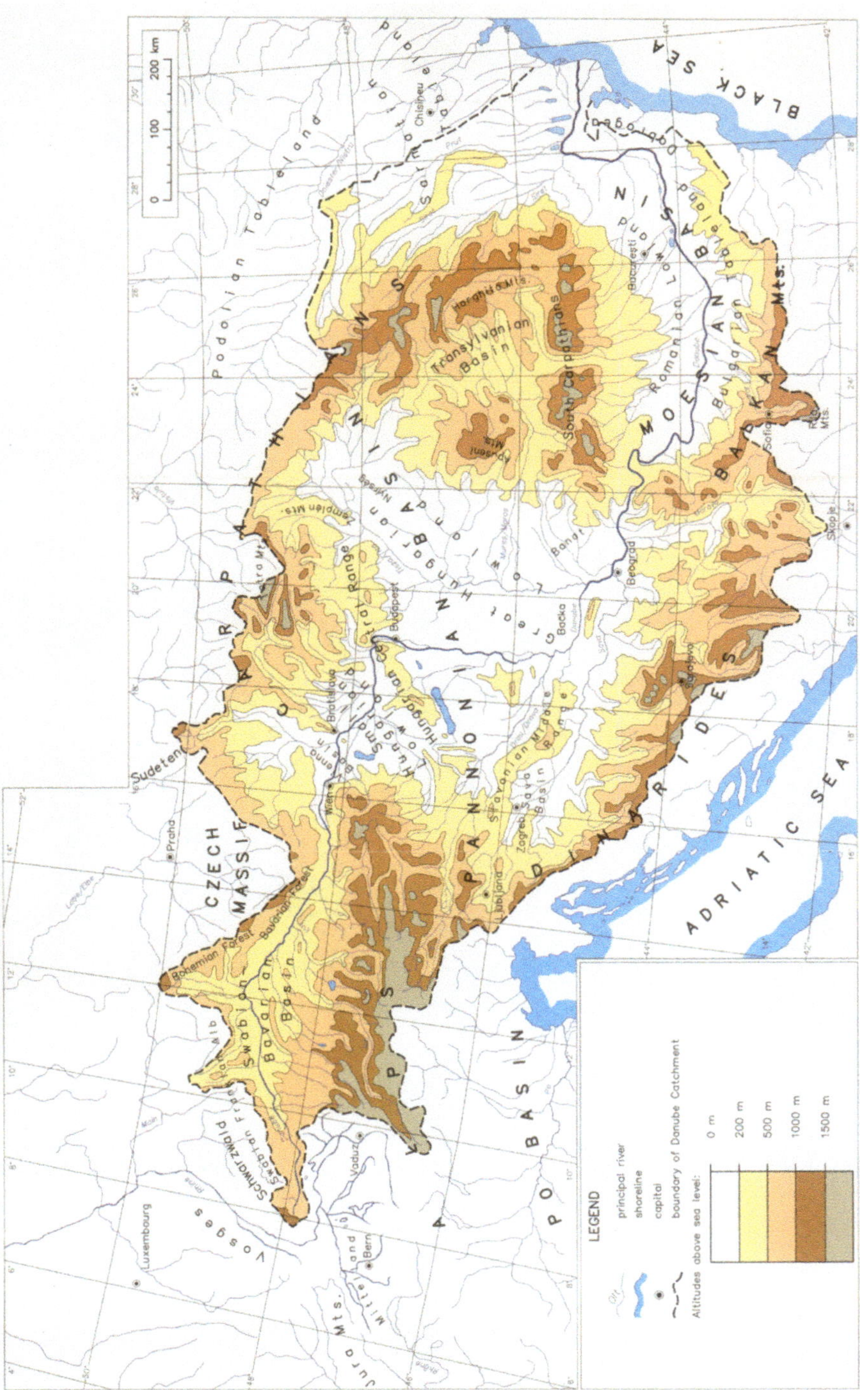

Fig. 3.1 Present-day orography of the Danube Catchment

Fig. 3.2 Present-day river system of the Danube Catchment

(1955), Fink (1966), Niculescu (1977) to Mike (1970) and other contemporary authors (Figs. 3.3, 3.4 and 3.5). A first series of modern palaeogeographic maps comprising three sheets had been designed already by Fink (1966) (Fig. 3.6) and gave the inspiration for the development of nine similar maps in the present paper.

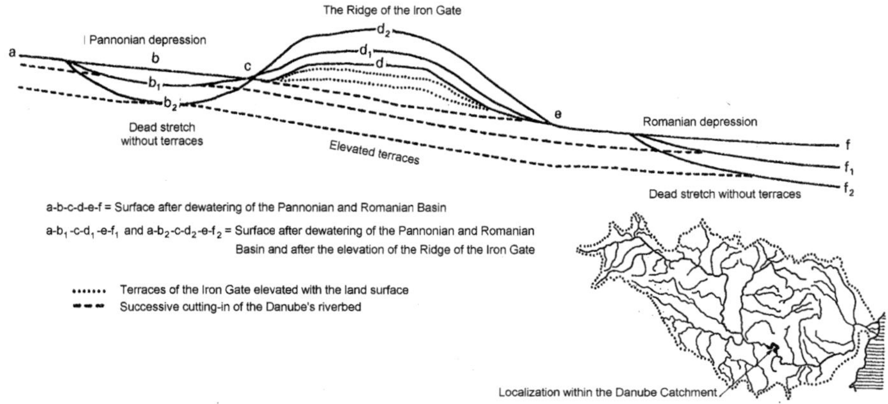

Fig. 3.3 Terraces of the Lower Danube, according to Cvijič (1908)

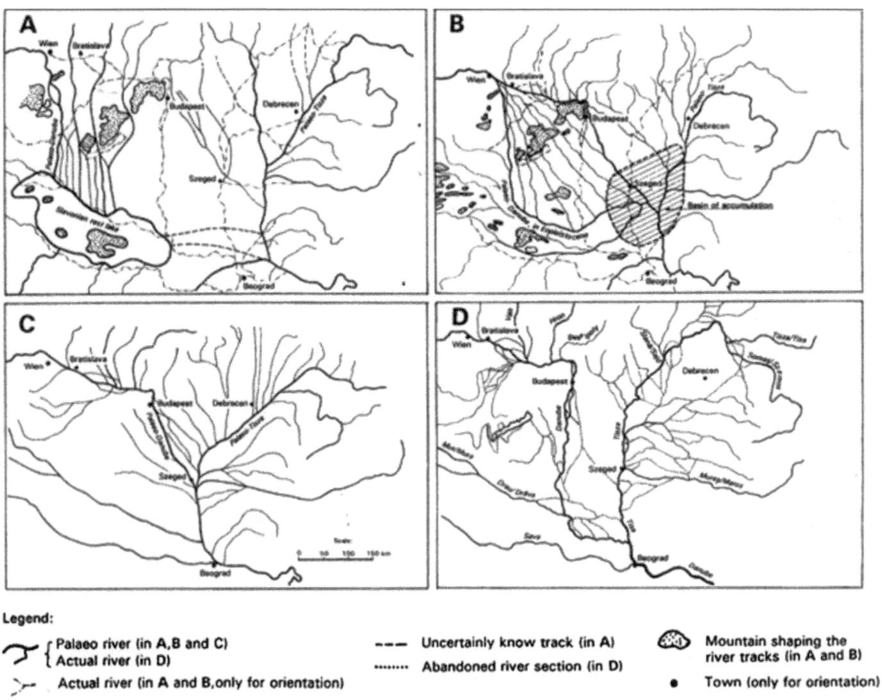

Fig. 3.4 Reconstruction of the hydrography of the Carpathian Basin, by various Hungarian authors. (**a**) End of the Pliocene, according to Sümeghy (1955), completed by Somogyi (1960); (**b**) Pleistocene, according to Sümeghy (1955); (**c**) Middle Pleistocene, according to Borsy and Félegyházi (1983); (**d**) Holocene, according to Sümeghy (1955), completed by Somogyi (1960)

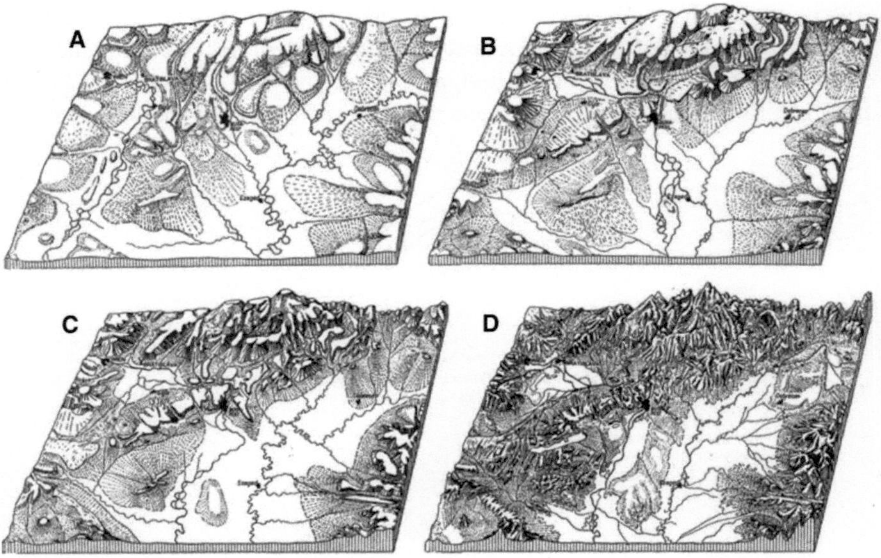

Fig. 3.5 Reconstruction of the orography and hydrography of the Carpathian Basin in various ages by Mike (1970) (**a**) at the end of the Eopleistocene; (**b**) in the middle of Pleistocene; (**c**) at the beginning of Neopleistocene; and (**d**) in the Holocene

Scientific development after 1966 has been directed mainly to continue Fink's work. A number of palaeogeographic results about various regions – including different parts of the Danube Catchment – have been published (sometimes only as chapters of studies on more general topics). Particularly valuable information is included in the National Atlantes of the various Danube Countries as well as in the "Atlas of the Danube Countries", published in Austria. The Danube Monograph (RZD 1986) was also an important contribution to synthesising the knowledge on the Danube Catchment, including also some aspects of its palaeogeography. Among the most recent works, the maps by Hámor (1995) have to be highlighted, which synthesise the Miocene palaeogeography of the Carpathian Basin.

The present paper is intended to become one more item in the series of works aiming at synthesising the results of joint scientific efforts of the experts in the Danube Countries. First, it will be focussed on palaeogeographic events due to basin accretion. The conditions of the mountainous areas of the catchment could be only sketchily described, since in this field a synthesising work has still to be done.

After the foregoing introductory comments, a more detailed treatment of the palaeogeographic development of the Danube Catchment follows, commenting the palaeogeographic maps designed by F. Neppel (Figs. 3.9–3.17). A general overview of the palaeogeographic events that have taken place over the last ca. 25 million years in the three main regions of the Danube Catchment is offered in Table 3.1, based on a quasi-logarithmic time-scale.

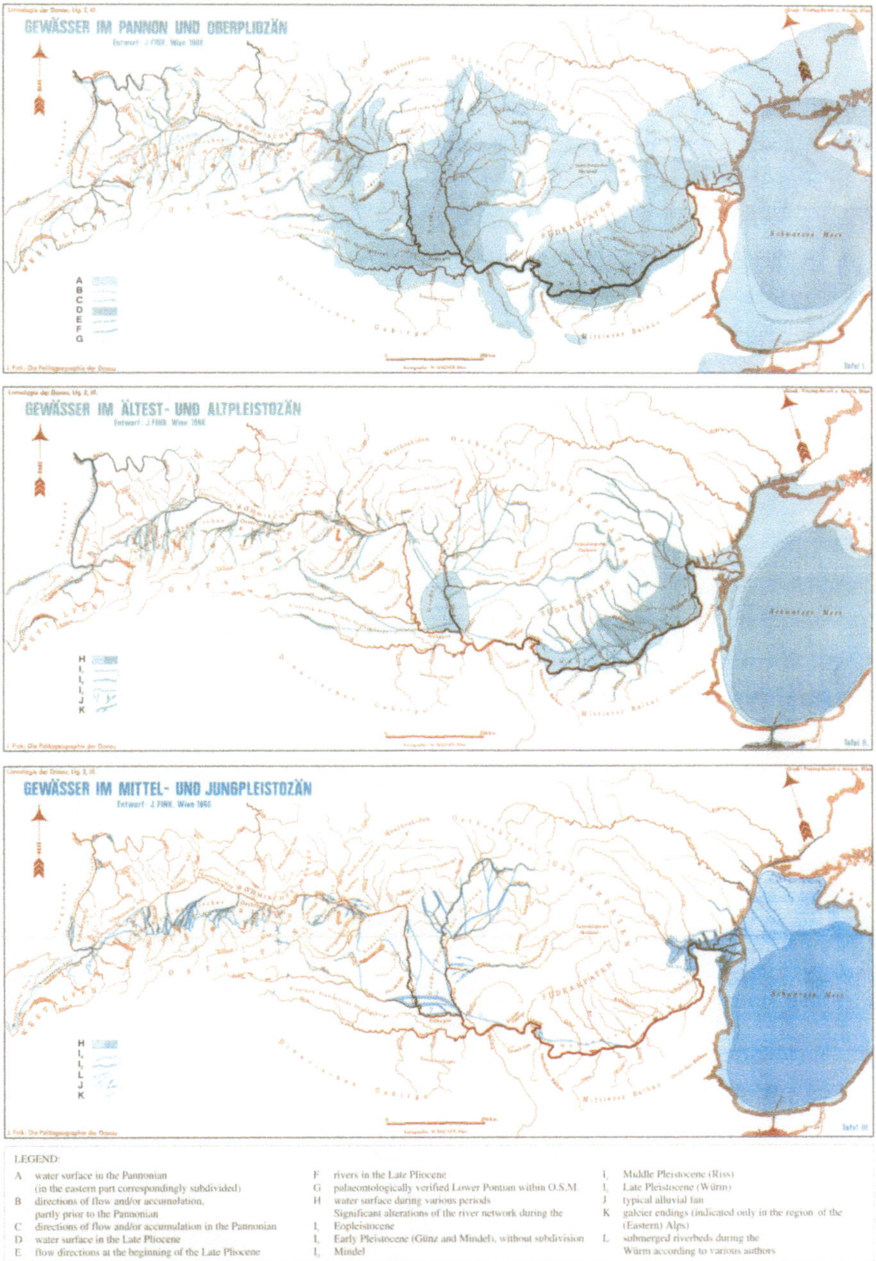

Fig. 3.6 Palaeogeographic evolution of the Danube Catchment according to Fink (1966)

Table 3.1 The palaeogeographic development of the Danube Catchment

Million years BC	Geological period				Palaeogeographic events in the Upper Danube region	Palaeogeographic events in the Central Danube region	Palaeogeographic events in the Lower Danube region	Ref. to Figures
0.001	Quaternary	Holocene	Younger Holocene		Due to continued karst development of the Schwäbische Alb, increasing seepage from the Danube's spring region to the Rhine	The Tisza River is forced to make a northern detour by the uplift of the Nyírség Region	Beginning of successive accretion of the present Delta area	17.
					Formation of the actual river network			16.
	Quaternary	Holocene	Older Holocene		Formation of lakes in the valleys generated by glacial erosion	Formation of lakes and swamps in local depressions (Neusiedlersee/ Fertő, Balaton)	Due to transgression of the Black Sea, inundation of the actual Delta area up to Brăila	16.
0.01	Quaternary	Pleistocene	Upper Pleistocene	Würm-Glacial		Due to tectonic processes, the Danube abandons its talus and forms its actual route along the Transdanubian depressions	Due to the uplift of the Dobroge, the lowest stretch of the Danube turns northwards	15.
0.1	Quaternary	Pleistocene	Upper Pleistocene	Riss-Würm Intergl.	Due to the vigorous uplift of the Alps and to the glacial erosion, strong deepening of valleys and sediment production. Genesis of terraced valleys	Terraced valleys in the mountains along the brims of the basin	Starting from the Iron Gate the Danube forms a terraced valley through the Romanian Lowland up to Călăraşi	14.
	Quaternary	Pleistocene	Upper Pleistocene	Riss-Galacial	Separation of the Alp-Rhine from the Danube Catchment	The riverbed of Tisza gradually migrates westwards		14.
0.5	Quaternary	Pleistocene	Middle Pleist.		A repeated glacification of higher mountainous regions supplies debris for filling up the basins. In the middle heights and along the basin perimeters terraced valleys are developing. Within the basins, significant migrations of riverbeds take place			13.
1	Quaternary	Pleistocene	Lower Pleist.		The catchment of the River Proto-Main joins the River Rhine	The Danube turns towards the Pass of Visegrád and fills up, jointly with the rivers originating in the Carpathians (Proto-Tisza, Proto-Maros, etc.) the Great Hungarian Plain.		13.
2	Quaternary	Pleistocene	Eo-Pleist.		Vigorous uplift of the Alps	Within the Small Hungarian Plain, the Danube is directed southward, along the perimeter of the Pannonian Basin, down until the Iron Gate	After filling up of the rest lakes Petroşani and Geta, the whole Rumanian Plain is becoming a continent. The main branch of the Danube reaches the Black Sea at Constanţa	12.
	Neogene/Younger tertiary	Pliocene	Upper Pliocene		The catchment of the Upper Neckar is incorporated by that of the Rhine	The subsidence of the Pannonian Basin stops, it is gradually filled up. Rest lakes in the Pannonian Basin	The Moesian Sea joins the Black Sea	11.
5	Neogene/Younger tertiary	Pliocene	Upper	Pannonian	The upper catchments of the Alpine-Rhine and of the Aare are incorporated by that of the Danube. The southwestern rivers of the Bohemian Massif join the River Elbe	Filling up of the Vienna Basin. The Paleao-Danube reaches the edge of the Small Hungarian Plain	On the Sarmatian Plate, gradual progression of the continent at the expense of the Pontian Sea	10.
10	Neogene/Younger tertiary	Pliocene	Lower	Pannonian	Capture of the Proto-Alpine-Rhine by the Proto-Aare	The gradually sinking Pannonian Basin is being filled up by voluminous maritime layers originating from the erosion along the perimeters		9.
	Neogene/Younger tertiary	Miocene	Upper	Miocene	Filling up of the Swabian-Bavarian Basin. The upper reach of the Palaeo-Danube comes into being	Vigorous volcanic activity on the perimeter of the Pannonia Basin	The southern coastal region of the Moesian Sea is being filled up by the rivers of the Balkan Mountains. The submountain region of the Southern Carpathians is gradually subsiding	9.
	Neogene/Younger tertiary	Miocene	Middle	Miocene		Salt production in the Transylvanian Basin, provisionally cut-off		8.
20	Neogene/Younger tertiary	Miocene	Lower	Miocene	There is no coherent catchment area yet. Characteristics of an archipelago prevail. Significant erosion takes place on the basin perimeters and in the low flatlands			7.

3.2 Palaeogeographic History of the Danube Catchment

3.2.1 Continentalisation of the Area

Casting a glance at Fig. 3.7, one can see that the major part of the Danube's course is situated in the so-called Neo-Europe, the youngest part of Europe, which reached its final form only during the Tertiary. The overwhelming part of its catchment area is also included in Neo-Europe. Even where some smaller parts of the catchment cover older continental fragments (Meso-Europe), these occur under thick young alluvia (e.g., in the Romanian Lowland). The Danube Catchment came into being within the frame of the Eurasian Mountain System, i.e., in a region having the most eventful geological history. Consequently, also the evolutionary history of its river system is rather complex. Its present structure is very recent and is closely connected with Alpine tectonism.

The mountain system of the Alpides is situated in a great sediment-collecting zone of the Earth, called Tethys (Fig. 3.8) where since the Palaeozoic the predominance of continents and seas – and thus the areal share of the continental river system – was alternating rhythmically. According to geological research, the deeply subsided fragments of the European proto-continent meet here with the masses of the southern (African) one. Consequently, the sediments – both continental and marine – have been shaped repeatedly into extended mountain chains by three successive phases of orogenic processes. While, however, the ranges of the Older Palaeozoic Caledonian and the Palaeozoic Variscan orogeny can be found nowadays mostly in great depths, forming the basements, the major part of the mountain

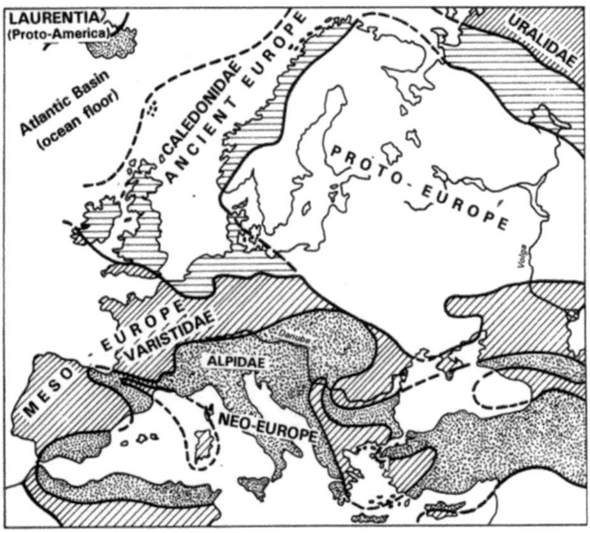

Fig. 3.7 Genetic structure of Europe, after Borsy and Félegyházi (1988)

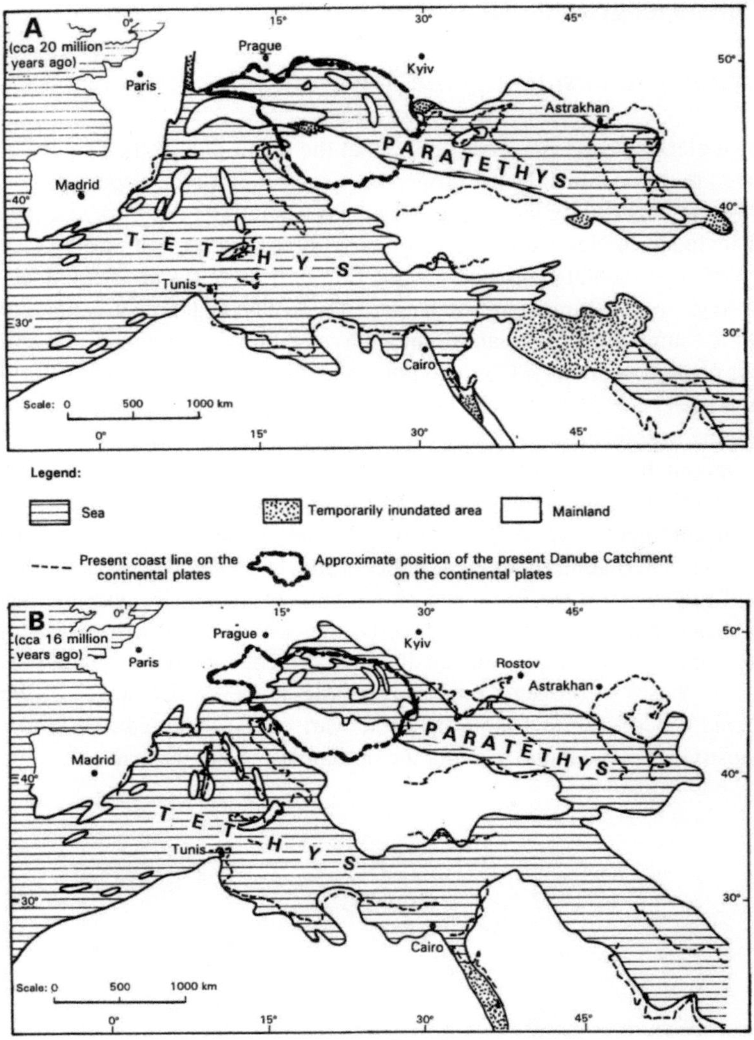

Fig. 3.8 Situation of the Tethys and Paratethys within the Mediterranean region (**a**) in the Early Miocene, and (**b**) in the Middle Miocene, according to Rögl and Steininger (1983)

chains originating from the Mesozoic-Cainozoic Alpine orogeny can still be found on the surface as top formations.

During the first phases of the rather long-lasting Alpine orogeny, Mesozoic sediments were pressed into the folds of mountain chains, a part of which also have subsided since then – during the Tertiary – under the surface. The younger part of the basement consists of these mountain chains containing subsided Mesozoic sediments. The mountain ranges pressed around and in-between the blocks of the mountain chains remaining on the surface are referred to as top formations.

However, there are large areas where even the Tertiary sediments subsided and became covered by a very thick Quaternary sequence of terrigenous origin. This latter is of course also a part of the top formations.

In the course of orogeny, also great masses of volcanic materials reached the surface along the great structural lines of the Earth's crust, considerably contributing to the diversity of the top formations.

According to the presently prevailing theory of plate tectonics, the ancient ranges once existing in the basins of the Alpine mountain system – thus also in the area of the Danube Catchment – were never able to form a consistent continent. On the contrary, they were always fragmented by shallow and deep, narrow and wide sea-troughs and gulfs, permanently changing in space and time. Both during and after the genesis of the various blocks and sediment belts they had their individual dynamics of movement, although being controlled also by the movements of the great lithospheric plates. Thus, they participated in the subsequent uplifting and sinking movements with different rates. Consequently, not only the orography of the surface, but also the surface of the basin sockets filled with fluvial, lacustrine and marine sediments is rather varied. Deep drilling has shown great differences in depth, within short distances, between the fragments of originally continuous formations.

Subsidence and horizontal movements took place along the lines defined by the dynamics of the German block mountains bordering the Tethys to the Northwest, the Czech Massif bordering the latter to the North, with the Podolian Massif to the Northeast and the Thrakian-Macedonian Massif (Rhodope) to the South. The proof of this statement is the frequent sharp turns of the surface watercourses.

The Proterozoic and Palaeozoic blocks of the Danube Catchment and its surroundings (Vogeses, Black Forest, Czech Massif, Sudetes, Podolian Massif, Rhodope and Dobrogea) are today situated far from each other. The sediments of Mesozoic and Tertiary seas cover the subsided basins between them. Although various minor blocks reached (or approached) again the surface during the Alpine orogeny, their young cover in most cases persisted, so that the younger rocks were exhumed only in a few cases.

The Cretaceous Austrian phase of the Alpine orogeny was particularly intensive, not only in the Alps but also in the Carpathians. This was the time of the genesis of the outer (Tatra) and the interior (Fatra) crystalline core zone of the Western Carpathian Mountains. At the same time, the large masses of the "flysch", mostly consisting of sandstone and originating from the eroded materials of the uplifted ranges, were also accumulated, to be later pressed and folded around and in-between the blocks.

It was also during the various phases of the Alpine orogeny that the Swiss Jura, the Swabian and Frank Alb, the Northern and Southern Limestone Alps, the Mesozoic ranges of the Dinarides as well as the limestone Klippen Belt of the Carpathians (between Devín and the Lower Danube) were formed. These mountain chains containing mostly Mesozoic materials consist of rocks more or less inclined to karstification, although subordinate sandstones, marls and shale were also produced at the same time. Tectonism was accompanied by significant volcanism.

Its products (lava rocks and tuffs) can be seen at numerous sites in the Danube Catchment.

From the tectonic and sedimentation conditions described above as well as from the products of continental erosion (e.g., Permian Red sandstone) still existing on the fragments appearing on the surface, one can conclude that the first consistent river system extending itself on a major area may have been formed during the Carboniferous-Permian as a follow-up to the birth of the Variscan Mountains on the present catchment area of the Danube.

In the frame of the described palaeogeographic setting, however, it seems to be only natural that the original elements of the present-day river system came into being on the surface of the crystalline high mountains, several geological ages before the genesis of the Danube. Such are suspected to exist near the springs of the Rivers Inn, Mur/Mura, Drau/Drava/Dráva, Velika Morava and Isker, areas that presently belong, due to epirogenetic uplift, to the highest parts of the Danube Catchment. These proto-rivers became later, following the changes in the extension of the mainland, gradually longer or shorter during the subsequent geological ages. This was due to the filling-up of basins and gulfs situated between the mountain blocks, governed also by the trans- and regressions of the seas.

In the Bulgarian part of the Danube Catchment water courses directed to the North appeared already in Palaeozoic times. The erosion products originating from the surface of today's Rhodope Mountains reached the deeper lying Palaeozoic and Mesozoic basins. In such a way, the area of the Moesian Table was filled up in the northern part of Bulgaria. These palaeogeographic conditions prevailed also during the Late Tertiary, when the Balkan orogeny produced a great change. In the final, Alpine phase, the shaping of the present-day river system started with the genesis of the valley of the river Isker, analogously to the break-through of the Iron Gate.

3.2.2 Evolution of the River System in the Neogene

3.2.2.1 Miocene

Figure 3.7 shows the localisation of the Danube Catchment within Europe's Miocene geography. During the Early and Middle Miocene, the area of the catchment was part of a side-branch called Paratethys of the ancient sea Tethys. The Paratethys covered more or less the areas of the actual basins in an extension varying due to the trans- and regressions, caused by the uplifting and sinking movements of the surrounding mountain blocks. The plate fragments forming the basement were shifted and rotated to their present sites practically during the Miocene. Since then their movements have become considerably slower, exerting only a very slight effect on the development of the river system of the Danube Catchment. During the Miocene, however, significant movements occurred, as indicated by the dotted cast lines in Fig. 3.8.

Thus, the palaeogeographic setting of the Danube Catchment was quite different from the present-day one (Fig. 3.1). The Paratethys, including a number of islands and continuously changing its depth, inundated the major part of its area. It was bordered partly by old block fragments and partly by mountain ranges in the phase of uplift. From the mainland, huge masses of denudation products were removed into the various branches and gulfs of the Paratethys, settling down in very thick layers. This sedimentary was later folded and lifted, thus becoming a major component of the mountains. The visible proofs of the Miocene denudation and accumulation are the remainders of the Miocene gravel blankets on the tops of the present-day middle ranges.

The renewed orogenic movements changed this process of surface development during the second half of the Miocene, i.e. the so-called Styrian phase of the Alpine orogeny. Because of these movements, the Palaeozoic and Mesozoic mountain ranges formerly existing on the sites of today's interior basins, within the intensively rising frame of the Alps, the Carpathians, the Dinarides and the Balkan Mountains – started to subside, with the exception of some blocks and other remains of mountains surviving until now. These tectonic movements, however, were far from being uniform. Interior seas and lakes inundated some of the sinking areas, for longer periods. Others – particularly those situated on the borders, such as the Northern foreland of the Alps – were filled-up in a relatively short time by the huge bed load amounts of the rivers originating from the uplifted areas.

Within the frame of rising mountain ranges continuous connection existed between the subsiding basins of the Inland Sea, such as the Vienna Basin, the Small and the Great Hungarian Lowland, the Transylvanian and the Moesian Basins. The precipitation income of the mountains surrounding the subsiding basins generally surpassed the amount of local evaporation. Because of this fact, currents started rather early, through the basin connections and narrows, toward the Eastern parts of the Paratethys, whose largest unit at that time was extended, through what is now the Black Sea, to the Caspian Sea and Lake Aral. Simultaneously with the currents originating from the interior areas also the desalination process of the seawater of the in-between basins started. It was more and more intensive from the West towards the eastern part of the sea, according to the increasing intensity of the current.

The Transylvanian Basin had a very different development. Its great salt deposits indicate that this area probably was land-locked for a certain time. In the rain-shadow of its bordering mountains, under rather dry climatic conditions, a concentration of the salty seawater took place in the Basin (Fig. 3.9). At the same time, the area of the Great Hungarian Lowland did not become drainless within the same rising mountainous frame, due to the intensive accretion activity of the rivers.

During the Miocene, an important period of volcanic activity occurred in several phases. This culminated along the tectonic lines separating the rising areas from the subsiding ones in the second half of the Miocene, involving particularly the area extending from the Eastern Alps, through Slovakia and Hungary, to the eastern margin of Transylvania. Beginning in the Mid-Miocene, the shaping of the surface by synorogenic volcanism took place in a regional extension. It proceeded in time from West to East, producing the volcanic range referred to also as the most interior independent zone of the Carpathian Mountain System. It was of intermediate to acidic

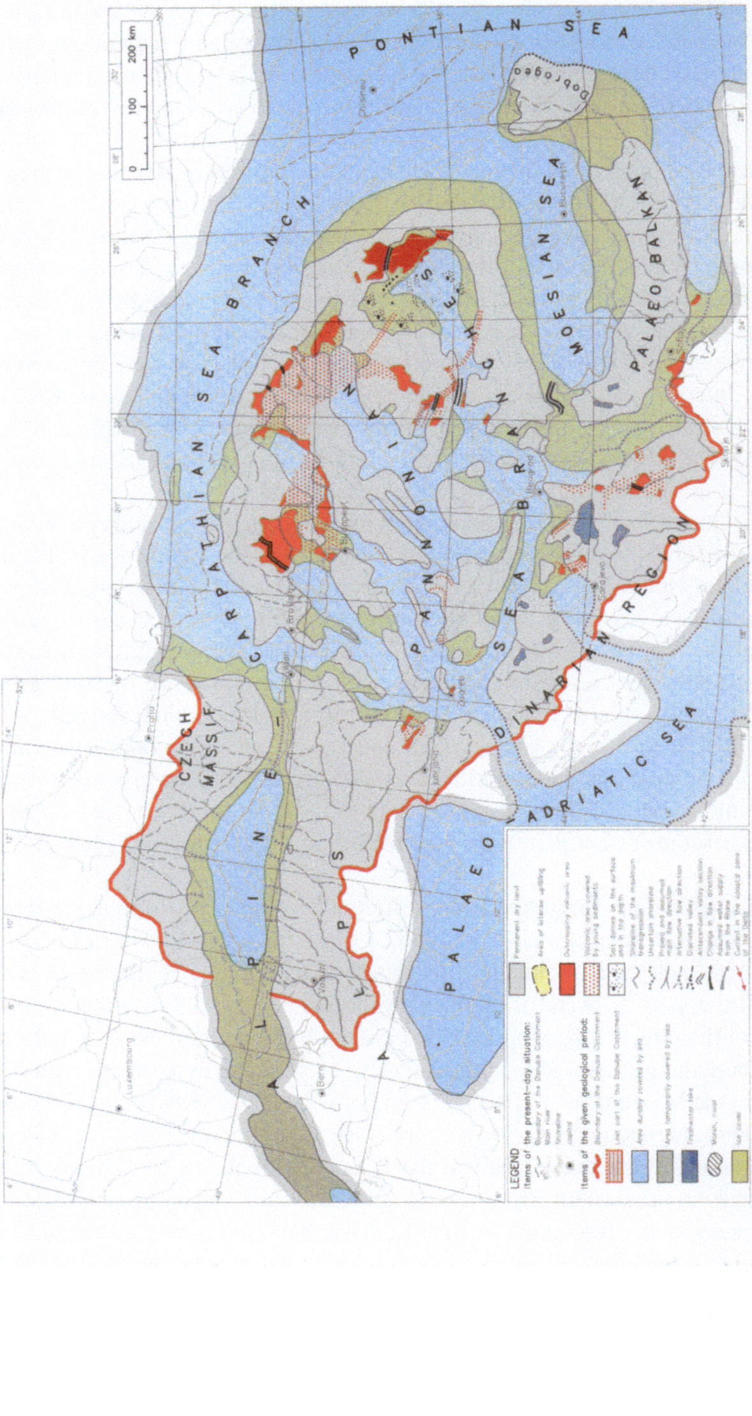

Fig. 3.9 Beginnings of development of the river system of the Danube Catchment in the Middle Miocene, compiled by Neppel (RCDC 1999)

(andesitic-dacitic-rhyolitic) composition. A similarly significant volcanic activity took place on the northern slopes of the Dinarides and in the Apuseni Mountains (Fig. 3.9).

One of the consequences of the Middle Miocene great orogenic surface-shaping movements was also the genesis of the basic lines of the present river system of the Danube Catchment. It could develop only after the major structural elements of the present orography had come into being.

The palaeogeographic situation at the end of the Middle Miocene (i.e., the Badenian Age) is shown in Fig. 3.9. At that time, the sea branches of the Paratethys still covered considerable areas, but there was already also considerable mainland with a well-developed river system. These ancient rivers are the predecessors of the recent ones. Their traces can be followed, with some certainty, even in the catchment areas of the present rivers.

Until the 1970s, the theory – based on the investigation of terrace morphology – prevailed that the Alpine-Carpathian sea branch was gradually filled up by the rivers, starting in its western part, at the Bern Alps and proceeding gradually to the East. This theory, clearly displayed on the maps of Fink (1966) (Fig. 3.6), has not fully been supported by the investigations carried out after the 1970s, using more advanced methods. According to the latter results, the maximum western extension of the proto-rivers of the future Danube Catchment reached the catchment of the Alpine Rhine, which in turn had reached, following the slopes of the Alps, the sea branch in the surroundings of Augsburg, at that time already strongly accreted (Tillmanns 1984).

The rivers of the Eastern Alps, although they were longer, transported a considerably lower amount of water and sediments (Ehlers 1994, Tillmanns 1984).

Further ancient rivers existed on the northern side of the sea branch, where the extension of the catchment was considerably greater than today. The Proto-Neckar (the headwater region of the present river) as well as the rivers of the Proto-Main-Naab System and the western and southern rivers of the Czech Massif were also the components of this hydrographic region, changing frequently their flow courses. The micaceous sand (Glimmer sand) derived from the Bavarian Woods can be traced over an extensive area in the eastern part of the Swabian-Bavarian Basin (Hantke 1994).

Similarly, the more important rivers of the northern side of the Dinarides between the Sava and the Velika Morava, as well as the northern rivers of the Balkan Mountains have appeared by the Middle Miocene. The Dinaric region, at that time only slightly emerging from the sea, had contacts – through two proved and several supposed straits – with the Proto-Adriatic Sea. The area itself was characterised by freshwater lakes and brackish-water lagoons.

In the region of the Carpathians – at one time uniform and archipelagic at some other time – the rivers were short. Their directions depended on the (changing) distance from the shoreline; most of them were flowing to the South. In the volcanic areas, certain rivers (Hron, Ipel'/Ipoly, Zagyva, Laborec, Upper Tisza/Tisa and Mureş/Maros) developed at the same time as the volcanic accumulations.

During the Miocene seawater level fluctuation – in the Sarmatian with certainty – the cut-in process of the valley of the Lower Danube started. However, a great part of this period was characterised by a narrow strait like the Dardanelles (Fig. 3.9).

The evolution of the river system of the Upper Danube region was terminated in the Sarmatian. A typical situation of this age is displayed in Fig. 3.10. At that time, the ancestors of the present Alps, Carpathians, Dinarides and Balkan Mountains emerged from the Paratethys partly as consistent chains and partly as a series of island-like blocks. The waters running down from the mountains were directed toward the sea branches as erosion bases, accreting them gradually. The extended fluvial-erosive activity is witnessed by the great sediment masses deposited at the foothills and filling up the subsiding basins.

At the beginning of the Sarmatian age, one part of the catchment area of the Proto-Rhine was conquered by the River Proto-Aare, providing at that time a more favourable route toward the western sea branch. This means that due to the uplift of the Black Forest and the Alps a connection between the western (Rhône Valley) and the eastern part of the Alpine sea branch was opened. The Swabian-Bavarian Basin continued its evolution independently. At the end of the age, however, the Alpine Rhine turned to the South, supposedly due to the rise of the Jura Mountains.

The mountain rivers originating from the rising ranges of the Alps filled the Swabian-Bavarian Basin with enormous debris fans. A major river came into being only on the eastern margin of the basin, due to the frequent changes of flow directions of the rivers building their debris fans. In the meantime, the area became definitively dry land. At the same time, the rivers of the northwestern part of the Czech Massif turned to the northeast – partly because of the area's rising and partly because of the more favourable runoff route provided by the Labe/Elbe – disconnecting themselves from the Danube system. This event is proved by terraces sloping opposite to the present flow directions.

At the eastern outlet of the Swabian-Bavarian Basin the meandering riverbed of the Proto-Danube cut itself into the southern margin of the Czech Massif. It has cut through the loose gravel-sand sediments of the marine-lacustrine-fluvial alluvium of the area. This occurred first between Vilshofen-Passau-Aschbach, then between Ottensheim and Linz as well as between Ardagger and Persenbeug, and finally between Melk and Krems, sculpturing into the crystalline basement rock beautiful steep-walled canyons.

East of the Czech Massif, the archipelago character persisted also after the Sarmatian. Although the sea became shallower – as evidenced by the widely extended carbonic shelves –, its area increased. The rising Dinarides cut off the connection with the Proto-Adriatic Sea. The brackish-water lagoons of the Transylvanian Basin were inundated by deep water depositing a thick clay sequence. Palaeontological data indicate that a connection may have existed between the Transylvanian Basin and the Pontian Sea, the latter reaching at that time as far as the foothills of the Carpathians.

A similar connection can be assumed in the surroundings of the middle reach of the River Timok, between the Pannonian Basin and the Sarmatian Sea of the Moesian Basin. On the western margin the latter, the Lom-Petroşani depression

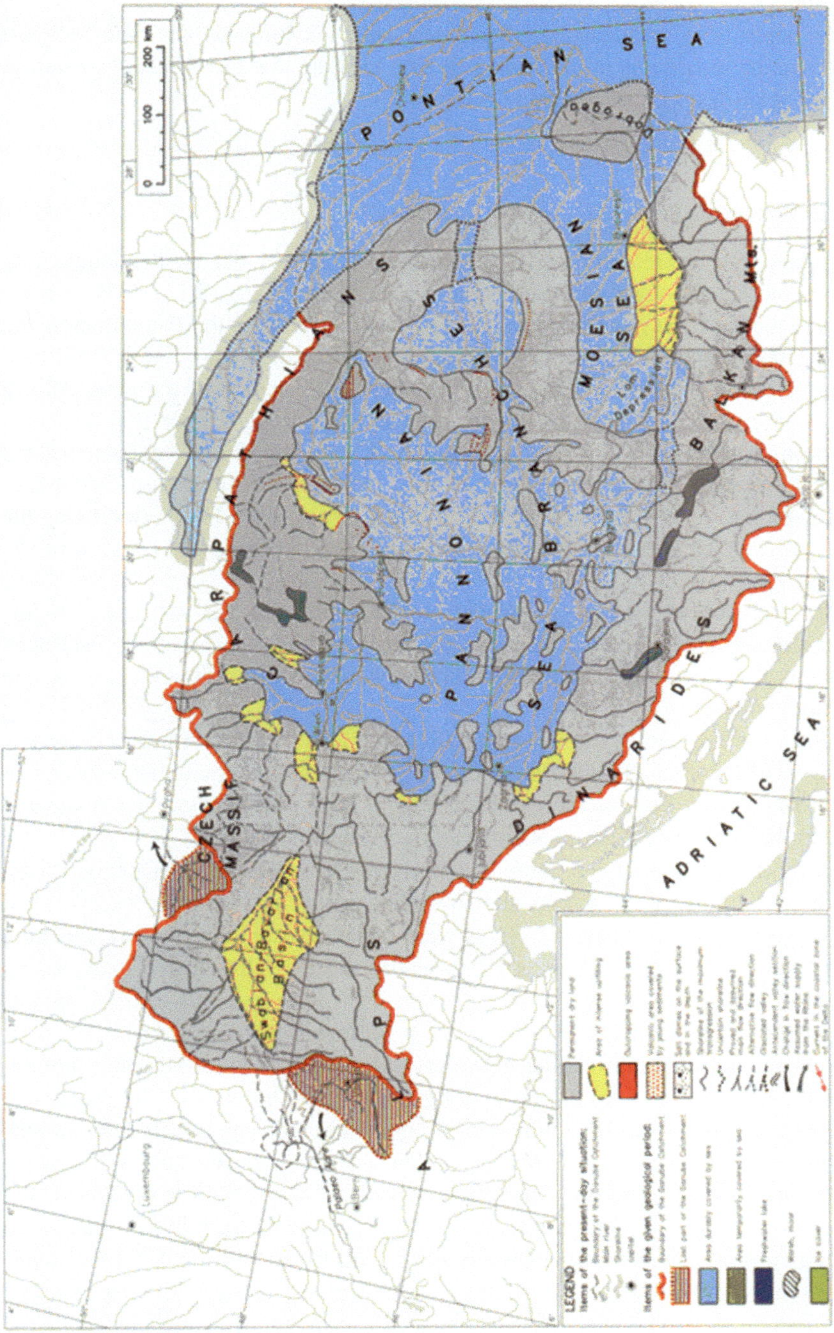

Fig. 3.10 Hydrography of the Danube Catchment in the Sarmatian, compiled by Neppel (RCDC 1999)

came into being, keeping this area covered by sea practically until the end of the Pliocene.

The deposits of various large freshwater-lakes existing at that time between quickly rising mountains in the Dinaric region indicate that the climate was humid (and not semi-desertic) also during the Sarmatian.

3.2.2.2 Pliocene s.l.

The palaeogeographic setting of the Danube Catchment during the Pannonian, Pontian and Dacian Ages of the Pliocene s.l. is shown in Fig. 3.11. (N.B.: we are aware of the fact that most stratigraphers nowadays consider the Pannonian as the terminal stage of the Miocene.) Accordingly, the watershed was slightly modified, as compared with the Sarmatian age, in the headwater region of the River Vltava and – within the Tatra region – in the surroundings of the Upper Poprad River. This brought about areal losses to the Danube Catchment. On the other hand, the Alpine Rhine re-joined it, increased even by a major part of the Reuss Catchment, along with the eastern slopes of the then strongly rising Black Forest (including the present spring region of the Danube itself) and with the headwater region of the Neckar.

The extension of the area inundated by the Pannonian Sea surpassed all former measures. Nevertheless, the sea was very shallow, its shorelines oscillated frequently. Lagoons and marshes were common, as evidenced by extended lignite deposits.

During the lifetime of the Pannonian Sea, the Danube was already a considerable river, filling up the depression of the Tulln Basin. Other young deposits can be found from this age only in a few peripheral gulfs. One of the major ones is situated between the Hungarian Central Range and the Carpathians.

During the Pannonian age, there was powerful tectonism in the Danube Basin, including the uplift of mountains and subsidence of basins. With the exception of the Swabian-Bavarian Basin, all basins were inundated again by the sea, reaching to the western edge of the Vienna Basin and to the foothills of the rapidly rising mountains.

The Pannonian Sea, however, was not a uniformly subsiding basin at all. It included alternating shallow and deep waters and dry lands. The continental-marine-lacustrine deposits vary accordingly along the former shorelines. The process of surface development was determined by progradations (during which delta mouths were built by the rivers) and by retrogradations (when the debris fans followed the retirement of the sea). Thus the marine and continental deposits, instead of following each other regularly, are interfingering. The fine sediments (turbidites) transported by deep currents, the advancing (prograding) delta-fronts and the different finger-like wedged sediment bodies of the delta can be well recognised. The basins of the Pannonian Sea are filled everywhere with similar sedimentary sequences.

During the Pliocene, the Alpine orogeny had various strong impacts on the evolution of the river system of the Danube Catchment.

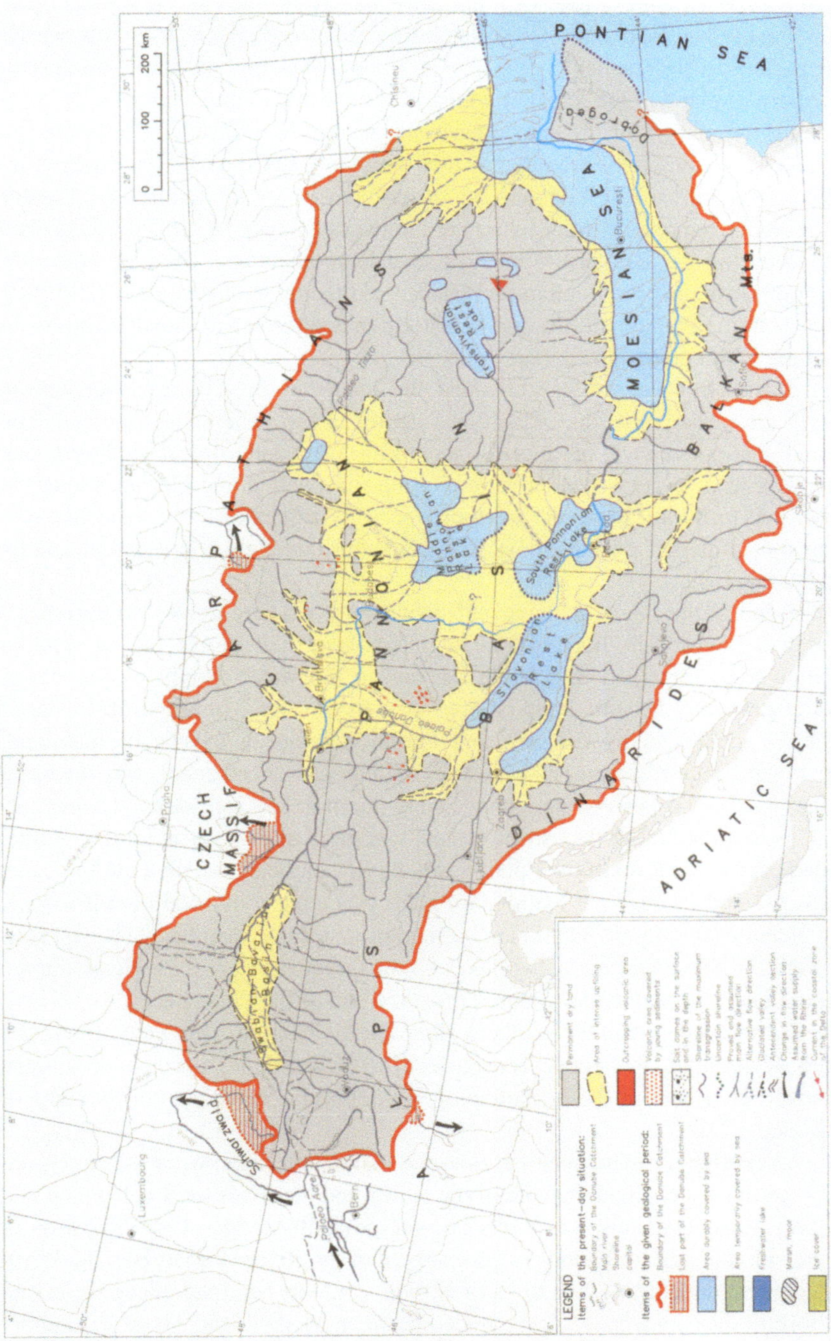

Fig. 3.11 Hydrography of the Danube Catchment in the Early and Middle Pliocene, compiled by Neppel (RCDC 1999)

Simultaneously with the intensive subsidence of basins, the mountains of the periphery were rising very fast. Consequently, the rivers started to produce coarse or even gravely bed load and to transport it into the erosion base. While during the former periods of surface development such sediments occur only rarely (mainly in the northern foreland of the Alps), fluvial gravel deposits became increasingly common along the margin of the subsiding basin during the Pliocene. Therefore, the extended basin depressions are filled mostly with sand. Because of their high porosity, these sediments of the Pontian and Dacian ages are very important for water resources management, contrary to the predominantly clayey i.e., less porous layers deposited during the tectonically much quieter Pannonian age. The predominantly sandy composition of the Pontian and Dacian sediments indicates also that during that age both the Pannonian Sea and the Moesian Sea remained mostly shallow, in spite of the high rate of subsidence.

The crustal movements also changed the character of orographic evolution. Formerly, the characteristic surface-types had been the peneplains articulated by wide and shallow valleys, produced – according to the subtropical-mediterranean climate of that time – by areal erosion. During the Pontian and Dacian, the rivers became, due to the differing altitudes produced by fast rises and subsidence, the starting sites of deeply cut-in valleys. In other terms, areal erosion was replaced by linear-lateral erosion.

At the end of the Pannonian age, subsidence was slowing down and during the Pontian and Dacian only minor basin fragments kept their former rate of sinking. The latter areas remained inundated, but due to the water supply by the rivers, the water gradually became brackish and then freshwater. The filled-up lowlands became, as evidenced by marshes and lignite deposits, a mosaic of mainland, brackish-water and freshwater lakes, while the deltas advanced deeply into the basin at its margin. The lakes were interconnected by more or less well-defined water currents. This so-called fluvio-lacustrine system was particularly typical for the middle, so-called oscillation level of the Pontian and the Levantine age of the Late Pliocene. Towards the end of the Late Pannonian, there was a short episode of subsidence, after which the completely Pannonian basin system became mainland. The last rest-lakes were the Slavonian Lake, two lakes on the Great Hungarian Lowland and the Moesian Lake, completely drying up only at the end of the Pliocene.

The rivers coming from the southern slopes of the Eastern Alps were principally directed, according to the slopes, consequently to SSE and they reached directly the Pannonian Sea. Later, in the Levantine, they became the tributaries of the delta of the advancing Danube.

At the beginning, the mainland around the Pannonian Basin widened only gradually, since the sea reached the gate of the valley and the rivers built deltas first and then debris fans. Thus, also the rivers became gradually longer. One indication of these considerable prolongations is the example of the River Hornád/Hernád. It had its first delta in the North of the town of Košice, while its bed load materials, mixed with those of the Slaná/Sajó, can be traced between Miskolc and Tiszafüred, under the Pleistocene sediments.

The water release from the Pliocene Lake of the Transylvanian Basin was through the Someş/Szamos- and/or the Mureş/Maros valley. The first version is contested by a number of experts, while the second one is certainly proved: some stretches of the Mureş/Maros had been river valleys already in the Sarmatian Age, while its huge debris fan could have come into being at the edge of the Great Hungarian Lowland only after the accretion of the Pliocene rest-lake of the Transylvanian Basin, i.e., at the end of the Pliocene, when also the headwater reach of the River Olt was a part of the Mureş/Maros Catchment.

Note that at that time there were several freshwater lakes in the inter-mountain basins of the Eastern Carpathians, drained by the Proto-Olt into the Transylvanian Lake and later into the River Mureş/Maros.

An important event of the Pliocene evolution was the development of the heavily sloped canyon on the Lower Danube from the former Dardanelles-like strait, because of terrain rise in its surroundings. It is highly improbable that the catchment area of more than 500,000 km^2 belonging to this Danube reach was drainless, even under dry climatic conditions. The basin of the Great Hungarian Lowland has not become drainless within the rising mountainous frame, because its gradual subsidence was compensated by fluvial accretion and thus the water level of the interior lake always exceeded the also gradually rising threshold level of the Lower Danube. This statement is proved by the absence of salt deposits in the Great Hungarian Lowland, in contrast to the existence of such deposits in the Transylvanian Basin, which was drainless during the Miocene.

The development degree of the river system downstream the Iron Gate was rather limited, like that of the Pannonian Basin. The rivers originating in the Southern and Eastern Carpathians and the Balkan Mountains were not able to compensate, even through their increased replenishing activity (caused by terrain rise), for the subsidence of the central part of the Moesian Basin, i.e., the Romanian Lowland. At the beginning of the Early Pannonian the uplift of the Balkan Mountains was particularly intensive, due to which the ratio of coarse sediments (gravel, sand) considerably increased. This phenomenon was particularly typical for the Isker delta on the intensively subsiding periphery of the Lom Subbasin.

As compared with the Early Pannonian, no considerable prolongations of the river courses took place even until the end of the Pliocene. For example, the River Seret was prolonged from the site of the town of Pascani until Bacău, and the River Prut from Iaşi only to Birlad.

The palaeogeographic setting of the latest period of the Pliocene, the Levantine/Romanian (or "fluvio-lacustrine") age is shown in Fig. 3.12.

During this period, only four minor changes occurred in the divide of the Danube Catchment. The Proto-Neckar joined the Rhine Catchment, the Proto-Vltava the Labe/Elbe Catchment, the Proto-Poprad the Vistula Catchment. There was, according to Heim (1921–1922) also a minor areal loss in the headwater region of the River Inn. Upstream of Tulln there were no considerable changes on the Danube. The changes occurring within the Pannonian Basin, on the other hand, were considerable. The basin was filled up by the rivers, leaving only at a few sites

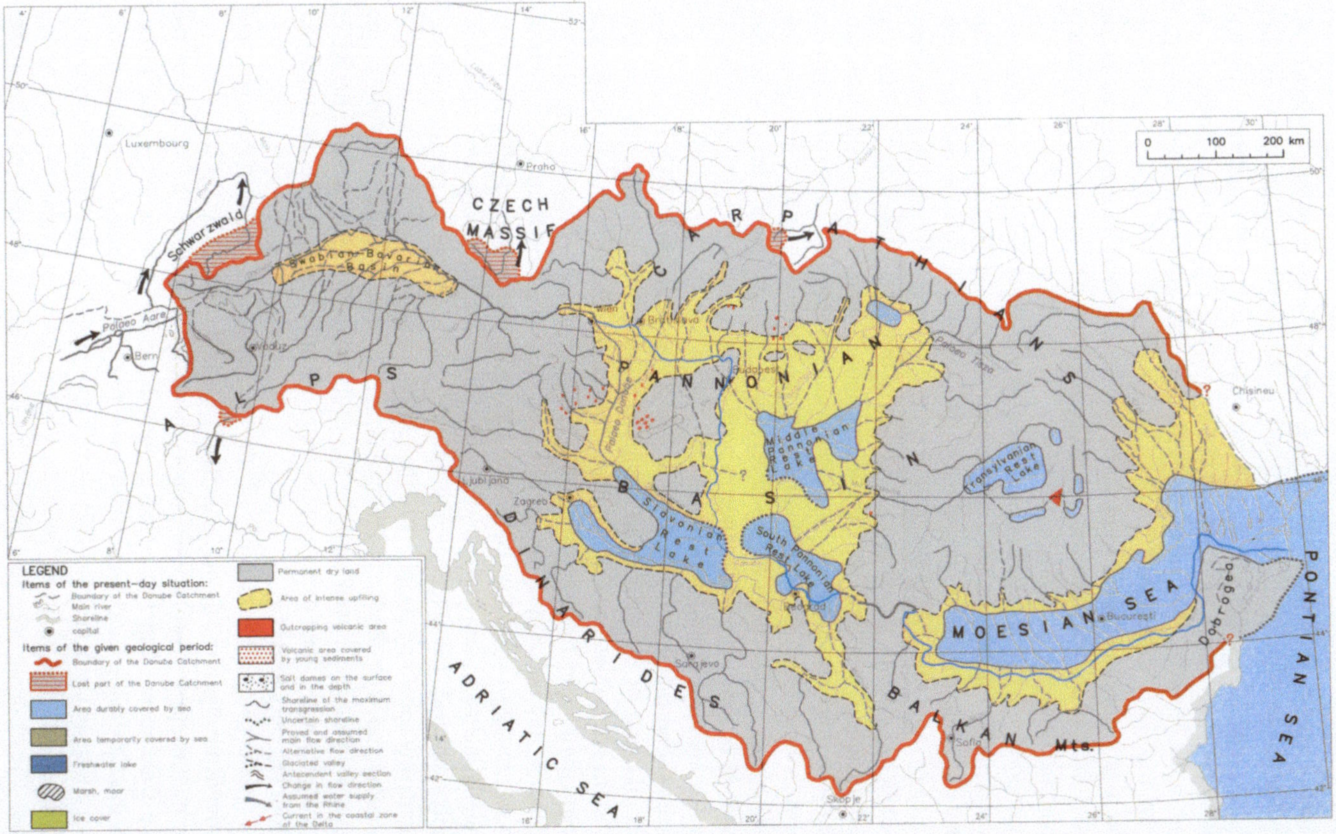

Fig. 3.12 Hydrography of the Danube Catchment at the end of the Pliocene (the "fluvio-lacustrine" time), compiled by Neppel (RCDC 1999)

brackish and freshwater rest-lakes. The increased filling-up activity of the rivers was due to the high-rate rise of the mountains surrounding the basin.

At the end of the Dacian age, the Danube filled up the Vienna Basin in a very short time. It made a sharp turn to the South, reaching the Hungarian Small Lowland – as proved by the gravel layers of the Kőhida Basin – through the so-called Ebenfurt Gate, near the present town of Sopron. The sediments of the Alpine rivers (particularly the River Leitha/Lajta) soon closed this gate, compelling the Danube to find its way through the so-called Bruck Gate, between the Leitha and Hainburg Mountains. After this breakthrough, it kept its southern direction, flowing, until the end of the Pliocene, into the Slavonian rest-lake. From its bed load, only huge masses of sand reached and settled down in the Small Hungarian Lowland. The material of the prograding delta of the Danube, the so-called "Levantian cross-bedded sand" can be found down to the southern part of the Hungarian county Somogy (Fig. 3.12).

At the same time, the River Morava/March had not yet reached the Danube in the Vienna Basin. Cutting its way at that time between the Small Carpathians and the Hainburg Mountains and forming the so-called Devín Gate, it joined the Danube only on the Small Hungarian Lowland.

Since the rough bed load of the Danube was caught by the depressions of the Small Hungarian Lowland and the Slavonian Lake, the filling up of the at that time even greater depression of the Great Hungarian Lowland was mostly up to the River Tisza/Tisa and its tributaries originating from the Carpathians. Their joint water and sediment yield being considerably lower than those of the Danube, also their accretion activity was less intensive. Consequently, the situation of inland sea and the transitional (fluvio-lacustrine) system lasted longer there, before it was transformed into a linear river system, probably as late as at the beginning of the Pleistocene. On the Great Hungarian Lowland, the mostly sandy Upper Pannonian sequence is covered by extensive clayey sediments (originating from marshes and other stagnant waters), called also "Levantian speckled clay".

The rivers originating in the Western Carpathians probably crossed the Hungarian Central Range through transversal valleys, reaching either individually or jointly the Pliocene rest-lakes of the Great Hungarian Lowland. Clayey sediments settled down on the dried-up marshy basin areas. They are covered today by Pleistocene sediments and occur in great depths, so that the courses of the Levantian Rivers can hardly be traced. Their directions can be deduced only from morphological features and a few drilling data.

The Transylvanian rest-lake was completely filled up by the end of the Pliocene. Only a number of freshwater lakes persisted between the blocks of the Transylvanian Carpathians and in the foreland of the Zemplín/Zemplén Mountains until the end of that age.

During this age, still no significant changes occurred in the Moesian Basin and in the eastern foreland of the Carpathians. The area of the Romanian Lowland continued to subside during the Late Pliocene, so that it remained covered by the Moesian Sea. The Bulgarian Table went on rising and the rivers leaving the Balkan Mountains started to cut their valleys into the surface of the Table. The inundation of the Sarmatian Table decreased considerably, due to the regression of the Pontian Sea. On the dried-up land, fine sediments were deposited by the rivers.

One of the accompanying phenomena of the orogenic movements, strongly affecting the Danube Catchment, was the volcanic activity renewed during the Pliocene. It produced various types of basalt within a wide band between the Styrian Basin and Lake Balaton (along the so-called Gleichenberg-Keszthely line), in the present-day Slovak-Hungarian border region and in the southern part of the Harghita Mountains (Romania). Only the surface-shaping effects of this volcanism are conspicuous. They had only a limited impact on the shaping of the surface river system as well as on the chemistry of subsurface waters affected by the post-volcanic activity.

3.2.3 *Evolution of the River System in the Quaternary*

The last main age of the Earth's history, the Quaternary, lasted about 2.4 million years. Its major first part is the Pleistocene, and the last 15–16,000 years is the Holocene.

3.2.3.1 Pleistocene

The Quaternary was characterised by extreme climate changes, including repeated extremely cold periods, called glacials.

The Mediterranean and sub-Atlantic warm temperate climate of the Levantine age cooled considerably in the Pleistocene. Europe belonged to the cold-Atlantic, cold-continental and arctic climatic zones. There were a number of alternations of glacial and mild periods. The latter were similar to the present temperate climate, having a more continental character in the East. According to our present knowledge, six major glacial phases can be distinguished within the Pleistocene. The first one, called Biber (Beaver) Boreal did not bring yet – partly due to the limited altitudes of the mountains – a real glaciation. Only the flora and the fauna were considerably transformed (as compared with that of the Levantian age). During the warmer periods – i.e., the interglacials and the short-lasting so-called interstadials – the ice retired more or less behind its present borderline, so that under the temperate conditions erosion and accumulation processes could continue.

During the glacials, the extent of the arctic ice fields was considerably increased. Their border in the German Lowland almost reached the neighbourhood of the Danube Catchment. In addition, the mountains – although far from being as high as today – were covered by ice. Also on the highest blocks of the Carpathians and the Dinarides, ice caps of limited extension commenced to develop. The Black Forest and the Jura Mountains belonged during the Riss-Glacial to the Alpine ice cover, forming independent ice islands during the other, less intensive glacials.

In contrast to the previous geological ages, a particular feature is that most Pleistocene formations are to great extent components of the orography of the Danube Catchment even today.

The phase of the Alpine orogeny at the Pliocene/Pleistocene boundary, the so-called Romanian phase, brought a renewed strong subsidence of the Small and Great Hungarian Lowlands and the Romanian Lowland. Simultaneously an increased rise of the ranges of the Alps, the Carpathians, the Balkan and Dinarides took place. The extension of the latter process is characterised by the fact that the last marine deposits were found on the eastern edge of the Alps in an altitude of 560 m, and on the slopes of the Carpathians above 800–900 m. The order of magnitude of these contrasting movements considerably exceeded 1,000 m over great areas.

The increasing difference of altitudes brought about increased fluvial erosion and accumulation. This was further intensified by strong climate fluctuations. During the dry climatic periods of the glacials such great masses of coarse bed load reached the rivers that the reduced discharges of the latter were not able to carry them completely away, but filled up their valley bottoms with gravel. During the mild and more humid interglacials, however, the bed load was decreased and the discharge increased. The rivers cut themselves into their filled-up valley bottoms and carried away a great part of the bed load previously accumulated therein, forming terraces from the remaining part of the former bed load. Obviously, such terraces could develop only in the rising areas. Their number and relative altitudes indicate not only the intensity of past crustal movements, but also the succession and the magnitude of the alternating glacial and milder phases creating the accumulations and the incisions of the valleys. In the subsiding areas, on the contrary, continuous accretion was going on, independently from climatic changes. Here the reconstruction of the palaeogeography is facilitated by the changes in the composition of the accumulated materials and the differences between the thicknesses of sequences.

During the Pleistocene, the development of the Danube river system occurred in several areal sections, depending on altitudes. The highest parts were covered repeatedly by ice, making the glacial deepening erosion prevail (e.g., U-shaped valleys, carr-niches, etc.), while on the perimeter of the ice cover glacial accumulation took place (moraines). In the ice-free mountainous and hilly regions, the terrace-producing processes described above prevailed. In the subsiding basins accumulation continued, but the changes in the sedimentation (gravel-sand-silt-clay), induced by climate changes, occurred also in the interior of the basins. In the rising mountainous areas, the flow directions of the rivers hardly changed, due to the prevailing cut-in tendencies. At the very most, only some neighbouring valley stretches became connected through captures, but only with a minor areal effect. The greatest of such captures was the turning of the Proto-Olt to the South. On the subsiding lowland areas, however, very significant changes of flow directions took place, partly due to the differences in the accumulation processes (talus formation) and partly to the differential subsidence of deep-seated blocks (Figs. 3.13, 3.14, 3.15 and 3.16).

The last significant volcanism in the Danube Catchment occurred in connection with the tectonic movements (Romanian orogenic phase) at the end of the Levantian and the beginning of the Pleistocene. In the West-Pannonian depression, along the Gleichenbert-Keszthely line, a strong uplift occurred in a wide belt, accompanied by the birth of a series of small basalt volcanoes. A more limited volcanic activity

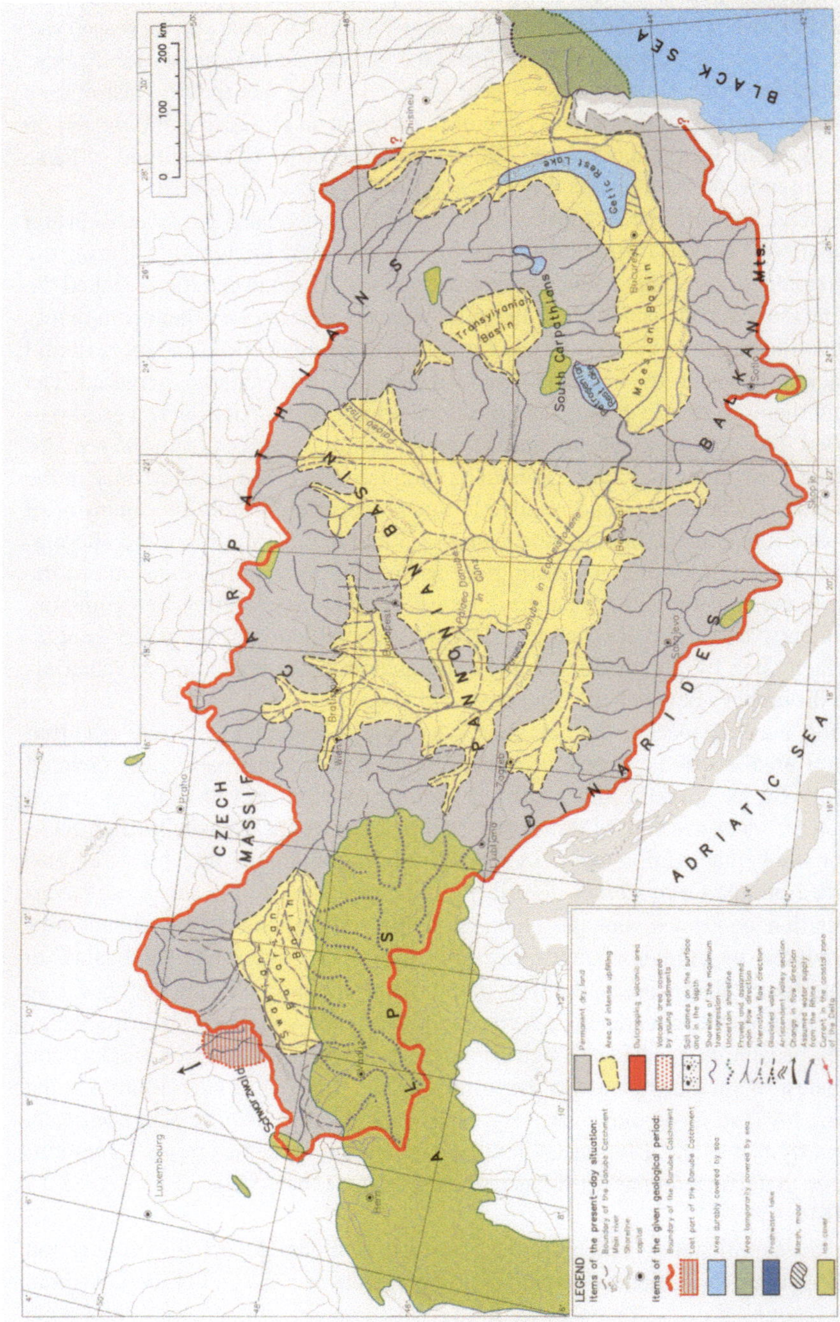

Fig. 3.13 Hydrography of the Danube Catchment at the beginning of the Pleistocene, compiled by Neppel (RCDC 1999)

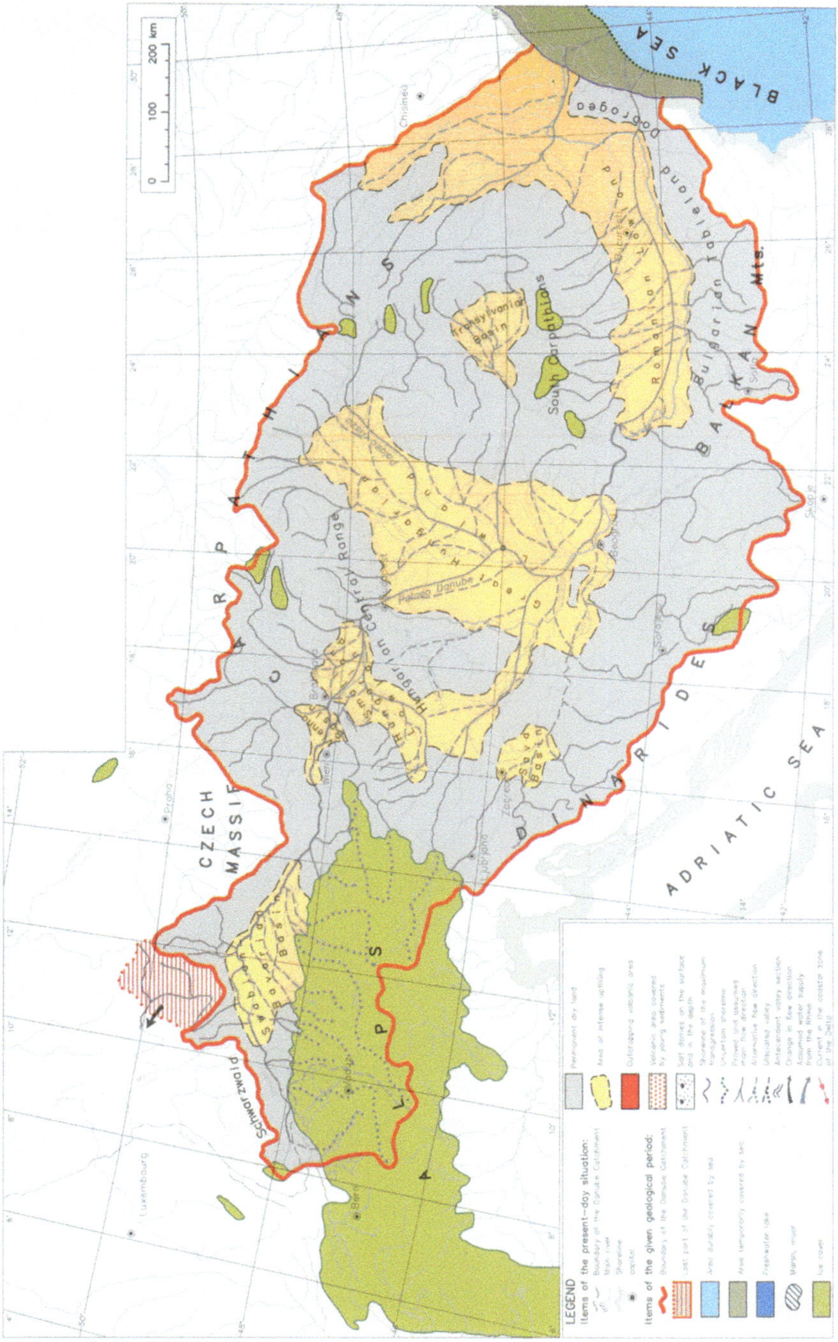

Fig. 3.14 Hydrography of the Danube Catchment in the Mid-Pleistocene, compiled by Neppel (RCDC 1999)

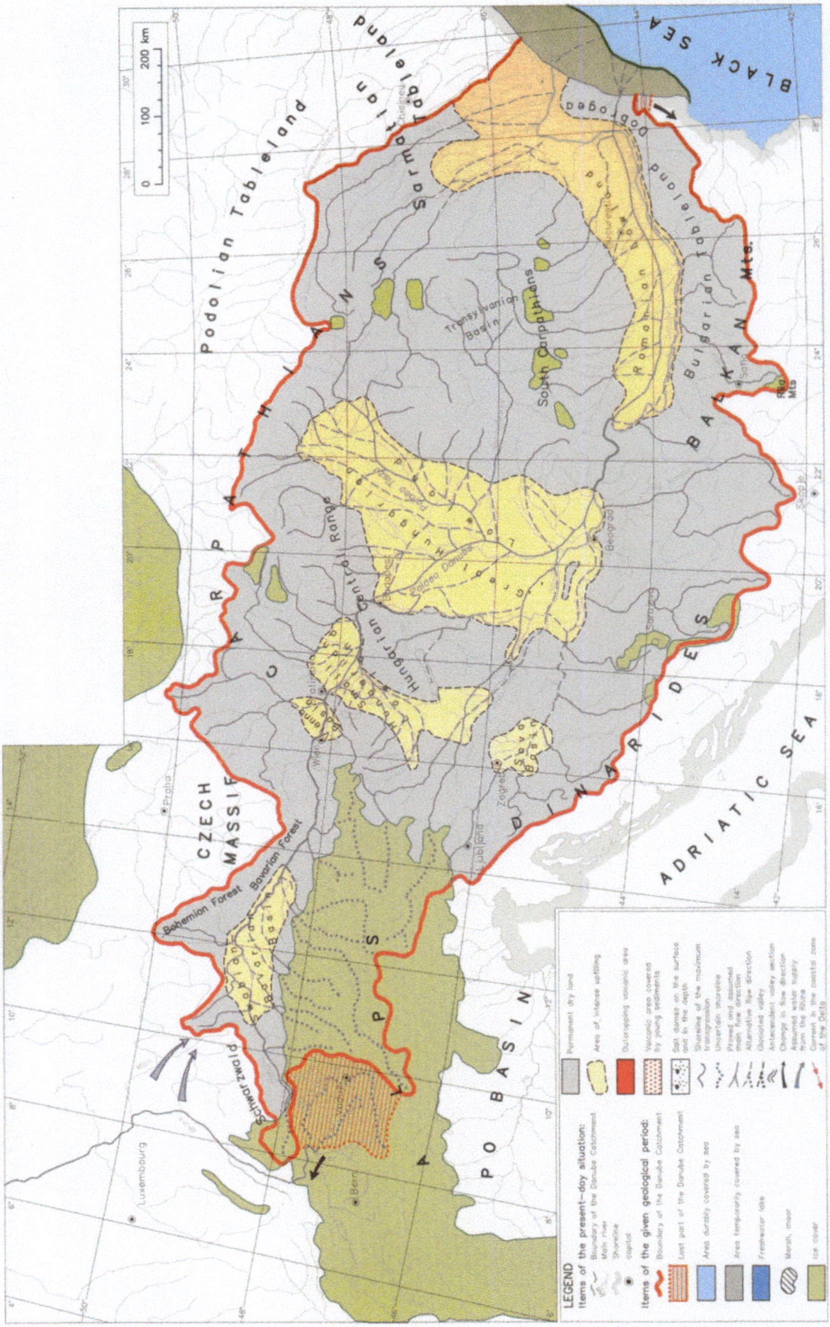

Fig. 3.15 Hydrography of the Danube Catchment in the first half of the Neopleistocene, compiled by Neppel (RCDC 1999)

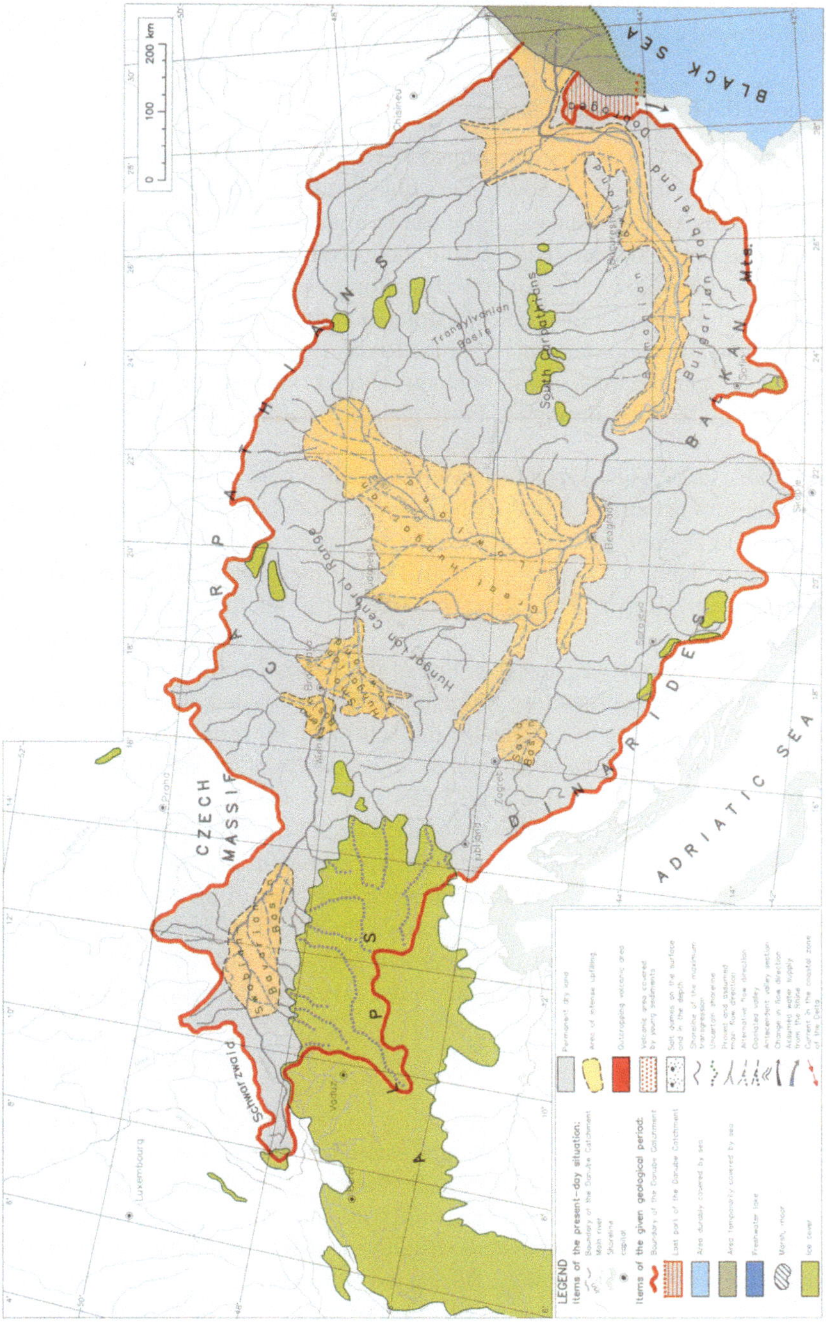

Fig. 3.16 Hydrography of the Danube Catchment in the second half of the Neopleistocene and at the end of the Pleistocene, compiled by Neppel (RCDC 1999)

took place in the surroundings of Salgótarján (along the Slovak-Hungarian border) and in the southern part of the Harghita Mountains. Volcanic activity reached its last maximum during the Eopleistocene (Fig. 3.12).

Pleistocene in the Upper Danube Region

Surface development was mostly determined by the erosive and accumulative effect of the periodically returning ice cover, particularly in the Swabian-Bavarian Basin. During the melting periods, causing significantly higher discharges, the rivers were not always able to re-excavate their beds in their original valleys, which had been in the meantime filled-up with sediments. Therefore, they repeatedly modified their routes, governed by the moraine trenches situated at the downstream outlets of the glaciers. This phenomenon was typical for the Rivers Iller, Lech, Isar, Inn, and Enns as well. For example, the Inn had a significantly shorter route leading to the Danube, in the foreland of the Alps, during the Early Pleistocene, than today (Figs. 3.13 and 3.14).

Considerable changes occurred on the left-side catchment of the Upper Danube. Because of the strong cutting-in of the Rhine, some of the Danube's tributaries were decapitated. The Danube lost, in such a way, two of its minor tributaries already at the end of the Günz Glacial (Fig. 3.13). During the Mindel-Riss Interglacial, the whole (present) Upper Main River became a part of the Rhine Catchment (Fig. 3.14).

The greatest areal loss was, however, that of the Reuss Catchment of the Alpine Rhine. The energically back-cutting Rhine preventing the waters from establishing a stable river valley leading to the Danube broke through the moraines accumulated during the Riss Glacial. On the other side, the karst valleys of the spring region kept their original eastern flow direction (Fig. 3.15).

The cataract reaches on the periphery of the Czech Massif continued to deepen also during the Pleistocene. At the same time, the subsidence of the Vienna Basin gradually stopped and the periodically increasing discharge of the river built a terrace system on its periphery. This complex system resulting from an interaction between erosion and local tectonics is visualised by Fink (1966) (Fig. 3.6). More or less the same terrace system can be found in the valleys of the Morava/March and its tributaries.

Pleistocene in the Central Danube Region

In the first half of the Pleistocene, the Danube entered the Small Hungarian Lowland through the Bruck Gate. From here, it kept flowing to the South, although frequently changing its route, due to the accretion of the Slavonian Lake and to tectonic movements during the Early Pleistocene. Its ancient riverbeds can be traced in the Hungarian counties Zala and Somogy. During the Günz Glacial the Danube reached the depression of the Great Hungarian Lowland in the surroundings of the present town of Paks after flowing through the foreland of the Hungarian Central Range (i.e., through the depression of what is now Lake Balaton) and turning at Siófok to the East, meanwhile pushing forward its debris fan (Fig. 3.13).

During the Eopleistocene, the River Morava/March preserved its original route and emptied itself, after fusing with the Váh, into the Danube in the region of the present Wasen/Hanság Marsh. The River Hron decapitated the headwater reach of the Proto-Nitra, reaching the Great Hungarian Lowland through the Valley of Visegrád. Various sedimentological data indicate that different branches of the ramificated Danube provisionally joined the flow of the Proto-Nitra in the area of the Little Hungarian Lowland. This theory is supported by fine-grained sediments of Alpine origin found from the very beginning of the Pleistocene in the surroundings of the town of Vác (Fig. 3.13).

At the end of the Middle Pleistocene the Danube abandoned, due to tectonic movements, the Bruck Gate, reaching the Little Hungarian Lowland through the Devín Gate, previously well cut-in by the Morava/March. The Parndorf Table (in the foreland of the Bruck Gate), along with the neighbouring Leitha Mountains, started rising also during this age, soon reaching a height of 30–40 m above the surface of the subsiding basin.

Of course, the described shift of the route of the main river Danube had a significant impact on all tributaries within the area concerned. The Morava/March lost its lowest reach downstream the Devín Gate, its sediments being buried under the young Danube-sediments. After the definitive switching of the Danube to the Bruck Gate, the Ebenfurt Gate became the break through of the river Leitha/Lajta (Fig. 3.13). Later on, it turned to the Bruck Gate, due to the accretion of the periphery of the Vienna Basin and to local tectonic movements. During the Neopleistocene, the River Leitha/Lajta turned for a short time North and reached the Danube, but its final path to the Small Hungarian Lowland remained, after all, the Bruck Gate (Figs. 3.14, 3.15 and 3.16).

The Mur/Mura kept consequently its SE direction, although it was no longer a tributary of the Danube, but one of the Drau/Drava/Dráva. The latter occupied in the meantime, after emerging from its terraced Alpine valley, the Early Pleistocene valley reach of the Danube situated along the northern periphery of the Croatian Middle Range. Downstream Barcs, the Drau/Drava/Dráva shifted its riverbed several times during the whole Pleistocene, depending on the actual degree of accretion (talus-creation) and subsidence. Its proto-beds can be traced in the North up the River Karasica and in the South down to the River Vuka. Its sediments can be found even East of its present mouth into the Danube, i.e., in the area of Northern Başka, under the youngest sediments of the Danube. This supports the theory that during the Middle Pleistocene the Drau/Drava/Dráva emptied itself into the Danube in the surroundings of Titel.

Those rivers which – like the Váh-Nitra and the Hron – previously had crossed the Hungarian Central Range East of, and parallel to, the Proto-Danube (at that time flowing in a southerly direction from the Small Hungarian Lowland), lost their reaches to the South of today's Danube Valley. After that, the minor water courses originating from the Hungarian Central Range, have sharply been separated by the divide through the Bakony and Vértes Mountains, into those flowing to the North and those flowing to the South (Figs. 3.14, 3.15 and 3.16).

At the beginning of the Mindel Glacial, the tectonic uplift taking place along the Gleichenberg-Keszthely line blocked the previous southward flow direction of the

Danube. Consequently, the river definitively turned to the Visegrád Canyon at the beginning of the Middle Pleistocene. This transposition is supposed to have been a long-lasting process, during which there were periods when certain branches of the Danube were directed to the South and others to the East, meeting each other only on the Great Hungarian Lowland (Fig. 3.14).

It was also during the Middle Pleistocene that the area of the Small Hungarian Lowland that was extending itself between Bratislava and Győr started to subside, thus fixing the main river Danube in its present position. Its alluvial fan had a certain effect also on the lower stretches of the tributaries reaching the Danube in this region. Because of the subsidence and continuous talus production, the main riverbed of the Danube had not – and could not have – a uniform character. Its branches compiled with the areal changes in the rates of subsidence and accretion, by producing frequently alternating ramifications and fusions. Because of the great amount of its bed load the Danube was continuously able to keep the crest of the talus above the surface level of its surroundings. Thus, both the northern and southern tributaries were able to reach the main riverbed only after making a turn around that talus, at its eastern end as a classic case of mouth-dragging. A similar history characterises the lower reach of the River Hron, the talus of which forced its tributaries to long parallel courses to the main river.

The already-mentioned rise along the Gleichenberg-Keszthely line, the subsidence of the interior basin of the Small Hungarian Lowland and the forced transposition of the Danube's riverbed into the Visegrád Canyon compelled a part of the rivers of the Eastern Alps to change again their flow directions. Since the former slope direction, directed from North to South, was inverted into its opposite, also these rivers turned, describing wide curves, to the North. The previously unimportant River Raab/Rába collected their waters and sediments. Its course ran considerably farther East than the present one, along the margin of the depression situated in the foreland of the Hungarian Central Range, creating a large talus during the Pleistocene (Figs. 3.14 and 3.15), today called "Rábaköz".

In the southeastern part of the Middle Pleistocene gravel mantle of the River Raab/Rába, sloping from South to North, the river complex Proto-Zala-Marcal came into being, draining also waters of the southwestern part of Hungarian Transdanubia into the Danube. Later on, however, the rise of the Zala hill-line caused a cutback of a watercourse from the mouth region of the River Mur/Mura, reaching the Proto-Zala-Marcal system in the West of the Bakony Mountains, and turning its headwaters towards the Drau/Drava/Dráva. Since then only the waters of the Bakony Mountains (Figs. 3.14, 3.15 and 3.16) have fed the thus truncated Marcal.

Between the taluses of the Danube and the Raab/Rába, a weakly drained indentation came into being, creating in the West, jointly with a slight tectonic subsidence, the Neusiedlersee-Wasen/Fertő-Hanság Basin. This basin had already existed since the Middle Pleistocene, but Danube branches periodically eroded its surface. Accordingly, only very young sediments can be found in this area.

Like in the western part of the Pannonian Basin, also in its greater, eastern part, the tectonic movements – of changing intensity in time and space – also very considerably affected the rivers, frequently compelling them to change their flow directions on the Great Hungarian Lowland. Various local depressions attracted the rivers to themselves. Such a local depression existed during the end of the Pliocene and the Early Pleistocene along the line Gödöllő-Cegléd-Szolnok, attracting to itself the northern rivers and those breaking through the Visegrád Valley, along with their first sandy and later coarse gravel sediments. This depression was later shifted to the South, fitting the axis Budapest-Kecskemét-Szeged.

After getting out from the Visegrád Valley, the Danube developed an enormous alluvial cone within the triangle Budapest-Szeged-Szolnok. In the East, it reached even the edge of the depression of the River Criş/Körös, which was an areally reduced remainder of the Middle Pannonian rest-lake. The rate of subsidence of the latter being more or less constant during the whole Pleistocene, the Danube attracted to itself various rivers leaving the mountains (Figs. 3.14, 3.15 and 3.16).

While the Danube carried during the Early Pleistocene mostly fine-grained sediments (occasionally depositing also coarse gravel of a thickness of not more than 5–10 m), the share of the latter type increased after the Middle Pleistocene, reaching even the area of the Great Hungarian Lowland. The Danube crossed with various branches the talus fan between Budapest and Szeged.

Apparently, towards the end of the Pleistocene the subsidence of the southern areas slowed. A younger series of N–S directed depressions came into being South of Budapest, bordered on its western side by the rising table of the Mezőföld. It was this young depression, which gradually turned the Danube to its present, N–S flow direction downstream from Budapest. The latter transposition is evidenced by the fact that the terrace sediments deposited at the beginning of the Würm Glacial as well as the layers forming their continuation are directed southeast from the Pest Plain, while the terrace floor of the end of the Würm already accompanies the present valley of the Danube (Fig. 3.16).

The present, N–S-directed Danube Valley between Vác and Sombor cut through the flat talus of the ancient rivers coming from the West (among them its own Eopleistocene talus). Downstream Sombor it turned to the East, participating in the erosion already started by the Drau/Drava/Dráva of the Telečka Ridge. Then the united Danube and Drava reached its present position at the northern margin of the Fruška Gora Mountains (Figs. 3.15 and 3.16).

The rivers of the Dinarides have deepened their valleys balancing the continuous rise of the mountains. Because of climate fluctuations, terraces are also present there. The small glacial patches of the top regions did not influence the river system. So far, there is no reliable knowledge concerning the position of the lower reach of the River Sava during that age. Anyhow, its talus fan reached the surroundings of Brod in the Middle Pleistocene. Downstream Brod, the river was rather marshy until the Holocene, so that its riverbed became fixed only during historical times.

The characteristic feature of the Carpathian Basin – and within it, of the Great Hungarian Lowland – namely its being separated into two major sub-watersheds, whose main rivers Danube and Tisza/Tisa are parallel to each other on a long stretch, is, geologically speaking, a very young phenomenon. It is mostly due to the tectonic movements, which helped the River Tisza/Tisa to conquer a considerable part of the catchment originally draining directly into the Danube, becoming the main river of the eastern, larger part of the Great Hungarian Lowland.

In the Pliocene, the Tisza/Tisa was no more than one of the proto-rivers originating from the chains of the Carpathians. It produced gravely talus fans and then filled up with fine sediments and thus transformed into mainland the rest-lakes of the Late Pliocene.

The Tisza/Tisa can be considered as an established river in the eastern part of the Great Hungarian Lowland from the very beginning of the Pleistocene. However, at that time it was far from being situated at the same place and of the same magnitude as today. It emerged in the Late Pliocene from the Khust Gate. The effect of its accreting activity was balanced by the continuous subsidence of the Maramureş Basin. The rise of the mountain ridges enframing the valley, however, destroyed the talus fans of the proto-rivers, lifting their remainders.

Continuing its way along the eastern border of the Great Hungarian Lowland, the Tisza/Tisa reached its subsidence centre, i.e., the present Criş/Körös-region. This was the erosion base of the northern and eastern part of that Lowland, for a longer time until the end of the Pleistocene. The Proto-Tisza/Tisa conducted the surplus water collected by that depression in the direction Szeghalom-Csongrád, into the Proto-Danube meandering in these surroundings at that time. At the end of the Pleistocene, the subsidence rate of the surroundings of today's Mureş/Maros-mouth already decreased, so that the latter river reached the Tisza/Tisa also through the Criş/Körös region.

The right-side (northern) tributaries of the Tisza/Tisa arrived from their terraced valleys between the mountains and reached the accreting lowland through a talus zone of varying width. They were directed, through periodically migrating branches, to the Criş/Körös Depression.

In the surroundings of the town of Miskolc, the common valley gate of the Hornád/Hernád and the Slaná/Sajó is a wide entrance to the Great Hungarian Lowland. It was there that a gulf of the Pannonian Sea reached northward as far as to the surroundings of the town of Košice. The reminders of the talus fans of these rivers are now between the altitudes of 200 and 300 m above sea level. This indicates that at that time the river terraces must have reached into the Lowland considerably deeper than presently.

East of the Zemplín/Zempléni Mountains, the talus zone disappears. Instead, there are a number of depressions (the Bodrogland area between the Rivers Bodrog and Tisza/Tisa, the Ecsedi Marsh, etc.) as well as the rising area of the Nyirség region. The elevated surface of the latter originates from a Holocene geomorphologic inversion.

Until the end of the Pleistocene, the rivers of the northern part of the Lowland crossed the Nyírség region and reached the northern margin of the Criş/Körös Depression, joining the river system of the Proto-Tisza/Tisa. When they were decapitated – because of the change in course of the Tisza/Tisa at the end of the Pleistocene – to their present localisation (Fig. 3.17), their abandoned riverbeds were partly filled by wind-blown sand and partly continued their existence as periodically inundated pools between dunes. A fine example is the protected marsh at Bátorliget (relict Ice-Age plants).

The Criş/Körös Depression was filled up only by the end of the Holocene, becoming a marshy region dotted with abandoned riverbeds (Sárrét). Durable flow-directions were created later only in historical times by human interference, such as river training.

The various members of this river family filled up the eastern part of the Criş/Körös Depression. On the foothills, they built a short series of talus fans dropping then suddenly onto the intensively subsiding lowland where they did not create established riverbeds before the Holocene.

The Mureş/Maros is one of the major rivers originating in the Transylvanian Basin. It reaches the Great Hungarian Lowland along the southern periphery of the Criş/Körös Depression. By the end of the Late Pliocene, the Transylvanian Basin was filled up by rivers of its mountain frame, which were short, but carrying much bed load. The Mureş/Maros became the main river of the Basin, when the Gheorgheni Basin, situated between the Eastern Carpathians and the interior volcanic belt, was dewatered partly by back cutting and partly by antecedence during the Eopleistocene. The river occupied its present position at the end of the Pleistocene, due to tectonic movements. Its tributaries mostly follow its abandoned former riverbeds.

In the first half of the Pleistocene, the main tributary of the Mureş/Maros was the river Olt. It could fill up the Pliocene lakes, impounded by the volcanic activity of the Harghita Mountains, only by the middle of the Eopleistocene. It remained one of the tributaries to the Mureş/Maros until the Riss-Würm Interglacial when it was decapitated by a river breaking through the Southern Carpathians by back cutting, thus turning the Olt to the South.

After having collected the waters of the Transylvanian Basin, the River Mureş/Maros proceeded on its way, already prepared in the Pliocene, and crossed the canyon between Deva and Lipova. Reaching the margin of the Great Hungarian Lowland, it built up an enormous talus fan, whose radius is up to 80–100 km.

South of the Mureş/Maros, the river system started taking shape at the end of the Pliocene and at the beginning of the Pleistocene. At that time, the subsidence of the foreland of the Banat Mountains could hardly be balanced by the accretion activity of the Bega, Tamiş and other rivers of the region. The Bega and Tamiş often changed their routes on their talus fans, emptying alternately into the Tisza/Tisa and the Danube. In addition, the Danube itself modified several times its route between Titel and the Banat Hills, establishing its riverbed only as late as during the Würm Glacial.

Also the break-through stretch of the Danube (the Iron Gate), connecting the Pannonian and Moesian Basins was transformed during the Pleistocene. It reached its present form simultaneously with the rise of the Southern Carpathians and the Balkan Mountains, producing a rock threshold in the riverbed. The evolutionary history of this Danube stretch – although dealt with by a great number of outstanding scientists, reaching contradictory results – has not yet been unambiguously determined. This is because even the same phases of evolution have – due to the

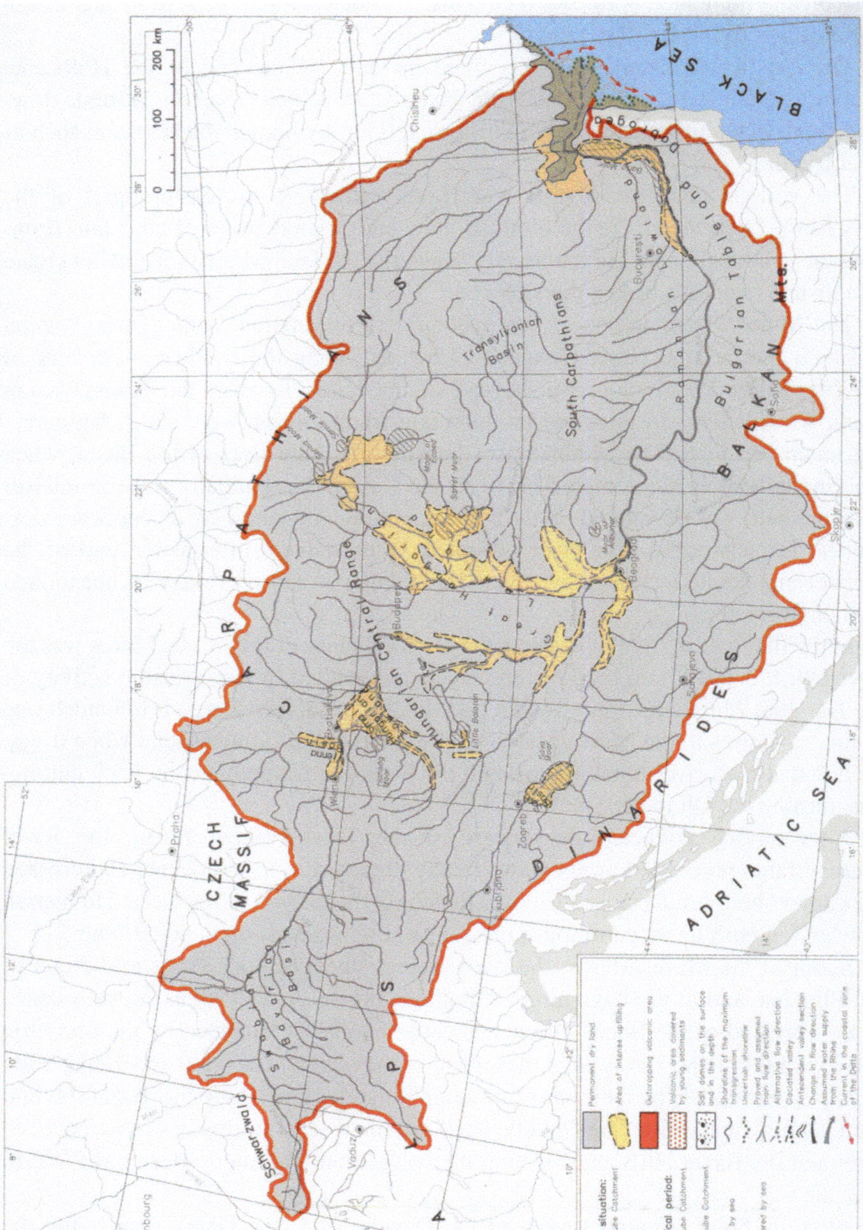

Fig. 3.17 Hydrography of the Danube Catchment during the initial stages of the Holocene, compiled by Neppel (RCDC 1999)

fragmentary tectonic character of the region – dispersed and the available data are not sufficient for their correlation. For example, the number of terraces is decreasing towards the East (Niculescu 1977).

The Pleistocene in the Lower Danube Region

Leaving the Iron Gate, the Danube crosses the Moesian Basin in an easterly direction, following its slope. The surface area of that basin is the Romanian Lowland. It had been filled up by the end of the Pliocene and transformed into a fluvio-lacustrine region. This accretion was a rather slow process, since, first, only fine-grained (mostly suspended) sediment could pass the threshold of the Iron Gate and, second, the rivers of limited discharges coming from the Southern Carpathians and the Balkan Mountains had dropped their bed load already along the debris slopes of the foothills. The Romanian Lowland, however, kept on subsiding during the entire Pleistocene.

The northern foreland of the Balkan Mountains, i.e., the Bulgarian Table, however, stopped sinking during the Pleistocene, maintaining a higher position than the subsiding Romanian Lowland. Consequently, the rivers coming from the Balkan Mountains cut their beds in the carbonic surface of the Table, forming very characteristic valleys.

In the North the rivers originating in the Southern (and partly in the Eastern) Carpathians created wider debris slopes, since the intensively rising blocks of these mountains, heavily glaciated during the various Glacials, supplied a far greater amount of bed load than the mountains on the southern side of the Basin. It was this abundant debris yield that during the Eopleistocene filled up the rest-lakes Petroşani and Geta still existing in the Carpathian foreland. It supplied also the material for terrace building of the Danube Valley following the line of the depression called the Romanian Lowland.

The Danube itself produced a great talus fan during the Eopleistocene on the margin of the Moesian Basin, reaching down to Călăfat and Craiova. On its northern side the rest-lake Petroşani was filled up as early as in the first half of the Eopleistocene, its surface melting into the debris slope of the Southern Carpathians.

After leaving this alluvial fan, the Danube first established its riverbed somewhat North of its present course, at the foot of the Carpathian debris slope, although its side branches can be traced down to the rim of the Bulgarian Table. It occupied its present riverbed – according to some opinions because of minor tectonic movements – during the Pleistocene, leaving behind a well-marked terrace system.

During the Eopleistocene, the Danube emptied itself into the last rest-lake of the Moesian Sea, i.e., the Geta Lake, situated in the deepest depression between the Carpathians and Dobrogea. It collected also the rivers Seret and Prut, coming from the foreland of the Eastern Carpathians, on the western margin of the Sarmatian Table. The Geta Lake was soon filled up.

During the Mindel Glacial, there was a minor river flowing on the southern side of the Dobrogea block. Supposedly due to the considerable water level oscillations of the Black Sea it cut itself strongly back, decapitating the Danube stretch directed to the Geta Lake, thus establishing the Danube's mouth in the surroundings of Constanţa by the Würm Glacial (Figs. 3.14 and 3.15). At the same time Dobrogea was slightly elevated, blocking the river's path to the East, compelling it to turn North and reach the Black Sea through the present-day Delta region.

Here a new factor joined the previous ones of riverbed development: namely, the glacial-ecstatic changes of the sea level. Its magnitude may have been during the glacials of the Pleistocene between –40 and –80 m on the Black Sea. During the interglacials, on the other hand, the water level of the Black Sea may even have surpassed the present one (i.e., in the Riss-Würm Interglacial, by +12 to 15 m), due to the melting of the ice cover. The actual shaping of the Delta Mouth of the Danube was of course strongly influenced by the changes to the shoreline. The fall of the seawater level caused an advancing of the shoreline while its rise resulted in over-flooding of certain parts of the mainland (Banu 1964, Újvári 1973).

The present contour lines of the Danube Delta became established at the beginning of the Holocene. During the previous transgressions, the sea intruded deeply into the mainland through the Danube Valley, and even along the Rivers Seret and Prut. The oscillations of the sea level formed the so-called lagoon estuaries of the next rivers emptying into the Black Sea (Dnester, Southern Bug) as well as the so-called "wing-lakes" of the Danube Delta. During regressions, the river intruded into the sea, so that its routes can still be traced on the sea bottom. For example, an ancient riverbed supposedly having come repeatedly into life emerges from under Holocene sediments at the depth of 30 m just in front of the mouth of the Danube branch Sfintu Gheorghe. It turns to the South and can be traced over a length of 80 km – down to Constanţa – on the sea bottom.

Of course, whenever the Danube advanced during regressions, its slope increased. This had a back-effect also on the rivers of the Romanian Lowland and the Bulgarian Table, increasing their erosive activity. During transgressions, on the contrary, the, slope diminished and sediment deposition and accretion increased. On the Romanian side, this effect concerned mostly the Rivers Prut, Seret and Ialomiţa. It had only minor influence on the Bulgarian side, due to its higher position. This sea level oscillation created also the so-called "balta" formations on the western side of the Dobrogea block. There are marshy areas interlaced with a number of accreted side branches practically without any surface slope.

3.2.3.2 Holocene

The Holocene is the last period of the Quaternary. Climatically it is considered one of the interglacials. The climate change, of course, has not been uniform: there were slight fluctuations even within the Holocene, facilitating its finer subdivision.

During the Holocene, only minor changes took place in the Danube Catchment, the river system practically was already established during the Pleistocene. The ulterior changes worth mentioning mostly took place in the spring region of the Danube, in the lowland part of the Tisza/Tisa Catchment and in the Danube Delta.

The factors prompting changes are the same as in previous times. However, their effects during the relatively short duration of the Holocene could not reach those of the previous ages, which were longer by several orders of magnitude. For example, the Holocene climate changes had fewer extremes, due to which valley accretions and river cut-ins of the magnitude that occurred during the foregoing glacials and interglacials did not take place. As a consequence, the Holocene terraces of the river valleys – probably produced by the continental warm-dry climate of the Hasel Stage of the Older Holocene – cannot be called real terraces as yet, since they are still occasionally overflowed. Their height, referred to the relative zero level, is generally between 4 and 6 m along the Danube and between 2 and 3 m along its tributaries. There are stretches where these pseudo-terraces have even doubled in size through tectonic movements. Because of their exposition to floods, they had not been suitable sites for human settlement before river training and the construction of flood levees were started.

One of the major changes that occurred during the Holocene is the subsurface water loss of the Danube in the Swabian-Bavarian Basin. A considerable part (presently about half) of the river's discharge is seeping from its limestone bed between Immendingen and Tuttlingen through karst passages to the Aach-springs (situated to the South, at a considerably lower level), to feed the Boden Lake, i.e. the discharge of the River Rhine. This phenomenon can be considered as the initial stage of a capture, possibly leading to the total loss, after some 100,000 years, of the spring region of the Danube.

In most parts of the Danube Catchment, and particularly on the northern slopes of the Alps, the position of the riverbeds is generally definitive with a prevailing cutting-in character, due to the reduced bed load yields and the continued rise of the mountains. In the foothills, there is also still build-up of talus fans to some extent, due to the breaking slopes, particularly in the case of the Alpine tributaries of the Danube, from the Iller to the Traisen.

The break-through character of the German and Austrian cataract stretches of the Danube, carved in crystalline rocks, has become more intense, due to the rise of the surrounding mountains, necessarily compensated by the deepening of the riverbed.

In the smaller basins breaking the slope of the Danube accretion stretches have become into being, creating shorter or longer talus fans, ramifications and occasionally wide, marshy valley bottoms. Evidence for this rapid accretion includes finds of tree trunks that fell thousands of years ago into the riverbed and are now covered by over 10 m of sediment. At some sites also lateral riverbed shifting can be observed, such as between Gönyű and Komárno/Komárom, where most Roman fortifications, once built on the Danube embankment, have been washed away by the river. The young depression character of the about 50-km long Danube stretch downstream Vienna clearly demonstrates, with its ramificated side branches and islands, the impact of tectonism on the shaping of the riverbed and the valley.

In addition, the recent changes in flow direction of the Danube and its river system within the Pannonian Basin are mostly due to neotectonism. The slight climatic changes of the last periods have generally had only minor consequences. One exception was the further development of the depression of the Small Hungarian Lowland, reducing strongly the Danube's slope – between Bratislava and Komárno/Komárom from 30 to 7 cm km^{-1} – thus compelling it to deposit there almost its entire coarse bed load. Consequently, during the Holocene only a negligible amount of coarse bed load has passed the Visegrád Valley. Downstream Bratislava, just at the upper end of the great accreting talus fan, two meandering wild branches fork off from the Danube – the Mal Dunaj in the North and the Mosoni Duna in the South –, collecting a number of tributaries and then re-entering the Danube only at the lower end of this talus fan. The main riverbed, positioned on the crest of the fan, changed its location frequently and has ramificated into a number of side branches. It reached its present position only because of river training.

In addition, the present riverbeds of the tributaries on the left side of the River Raab/Rába are rather young, due to continuous shifting on the surface of their respective talus fans toward the depression of the Small Hungarian Lowland. Similarly, the valley of the Raab/Rába itself is fixed by young tectonic lines, the western prolongation of which is the Styrian Basin. The wide valley of the Raab/Rába is filled with a flat talus accompanied by wild branches along its edges (Lahn, Herpenyő).

The Holocene natural hydrographical status of the Neusiedlersee-Wasen/Fertő-Hanság Basin can be described as a steppe-lake. Water regulation put an end to this status. The Wasen/Hanság Moor has been dewatered and only a very careful water management can now secure the permanent water level of the Neusiedlersee/Fertő.

In addition, the major northern tributaries of the Danube reach bordering the Small Hungarian Lowland were compelled by the great talus fan to shift their mouths eastwards and to empty in the Danube united with each other and the Mal Dunaj only at the lower end of the talus.

Already at the end of the Pleistocene, a series of depressions came into being on the southern margin of the Hungarian Central Range, heavily influencing the formation of today's river system. In the depression situated in the southern foreland of the Bakony Mountains, the basin of Lake Balaton came into being. It includes also some remainders of the Proto-Danube existing here at the beginning of the Pleistocene. Further elements of this series of depressions are Lake Velence and various other marshy areas. At the time of its genesis, the basin of Lake Balaton was considerably greater than today. Climate changes during the Holocene brought about also here significant surface changes.

During the Holocene, the most significant change in flow direction was that of the River Tisza/Tisa. As already mentioned, a series of depressions came into being along the southern margin of the Hungarian Central Range and other mountains, bordering the Great Hungarian Lowland. These depressions could not be filled up even by the middle of the nineteenth century, continuing their existence as famous marshes (Fig. 3.7). The various talus fans also hampered their drainage chances.

Contrary to the peripheral depressions of the Great Hungarian Lowland, the talus fan of the Nyírség region (i.e., the NE part of present-day Hungary), which grew during the Pleistocene, was uplifted at the end of that age. A reversal occurred of the original slope direction, completely transforming the river system of the region. Since that time the River Tisza/Tisa, after having passed the valley gate of Khust, has been turning – instead of SW, as previously, to the NW, making a turn, around the Nyírség region, taking advantage of the depressions of Bereg and Bodrogland. Thereafter it intercepted the valleys of the Rivers Slaná/Sajó-Hornád/Hernád, Eger and Zagyva, and reached its former valley downstream Csongrád. All these transformations took place, of course, only gradually. For some time the discharge of the Tisza/Tisa may have been shared between its old and new riverbeds. Later on, the tributary Someş/Szamos – which previously had left riverbed remains on the northern margin of the Criş/Körös region – also joined the Tisza/Tisa. The ancient riverbed of the latter was used for a long period by its tributary Krasna/Kraszna, whose floods used the ancient bed as a flood spillway even until the river regulations of the nineteenth century.

The rivers that had created the Nyírség region – i.e., the source rivers of today's River Bodrog – were captured by the Bodrogland Depression, situated at the foothills of the Zemplín/Zempléni Mountains and collecting a number of tributaries coming from the Carpathians. What is more, this depression conducted even a part of the discharges of the Tisza/Tisa during various stages of the Holocene.

The riverbed of the Tisza/Tisa downstream Tokaj could develop only gradually. The downstream stretches of the riverbeds of the Slaná/Sajó, Hornád/Hernád, Eger Tarna and Zagyva, cut through by the new route of the Tisza/Tisa, served for a certain time as the flood spillways of the latter conducting the water in the previous erosion base, i.e., the Criş/Körös Depression.

The latter depression, after having lost almost all its northern tributaries, captured by the Tisza/Tisa by the end of the Pleistocene, had a reduced rate of subsidence during the Holocene. Thus, it remained a marshland (called Sárrét), interlaced by the source branches of the Criş/Körös and their side channels, originating in the Apuseni Mountains. They were, however, no longer able to fill up with their diminished bed load discharges the widely meandering riverbeds of the ancient rivers, and only filled their embankments.

During the Holocene, also the river Mureş/Maros abandoned the Criş/Körös region, turning near the town of Arad directly westwards, cutting through the loess-covered surface of its own old talus fan and emptying itself at the town of Szeged into the Tisza/Tisa. Its previous channel, directed towards the Criş/Körös region (called Száraz-ér), served as a flood spillway until the river regulations in the nineteenth century.

As for the rivers originating from the Transylvanian mountains ranging between the Mureş/Maros and the Danube and then reaching the Banat Lowland, they also changed their flow routes during the Holocene (Bega, Timiş/Tamiš, Karas, Nera). This is confirmed, among other pieces of evidence, by the young depressions (e.g., the Alibunar Marsh) situated along the periphery of the Pleistocene sand talus called Deliblat, resulting both from fencing-off by sediments and recent tectonic movements. Even the emptying directly into the Danube of the river Timiş/Tamiš has a rather short history. In general, the frequent migrations of riverbeds in the Great Hungarian Lowland are evidenced not only by morphological, lithological, botanical and archaeological finds, but also by the maps of Ptolemaios, the famous Greek cartographer of the second century AD.

Numerous similar but minor riverbed changes occurred, governed by the same processes of sedimentation and tectonism, in the valley of the River Drau/Drava/Dráva downstream Barcs and that of the Sava downstream Zagreb. At the same time, neither in the Danube-drained part of the Balkan Peninsula nor in the Transylvanian Basin (both in a continued stage of rising) did any changes of river routes occur during the Holocene. What prevailed in these locations was a deepening of the cutting-in and fixation of the river valleys that had already developed.

In the Lower Danube region, between the Iron Gate and Călăraşi, the route of the riverbed had no chance of migrating during the Holocene, either. In the mountains, in their forelands and on the Bulgarian Table, the cutting-in character of the rivers prevailed. The Danube itself was making its way in a wide, terraced valley, producing at most some cut-off meanders.

Downstream Călăraşi, the Danube continued its way in a series of depressions on the western side of Dobrogea, along the line Cernavoda-Hîrsova-Brăila-Galaţi. These depressions situated partly in the foreland of the Carpathians, have altitudes of only a few metres above the Black Sea level, thus representing the lowest part of the whole Danube Catchment. To the South, the side branches of the Danube interlace the marshy depressions. This area is called the "balta" region. From the North, the united rivers Buzău and Siret coming from the Eastern Carpathians, empty into the Danube.

At Galaţi, the Danube makes a sharp turn to the East, breaks through Dobrogea and the Sarmatian Table and reaches its Delta estuary, created in its present form during the second half of the Holocene. The to-date last great transgression of the Black Sea, with its water table rise of 5 m, inundated the river's mouth stretch up to Brăila (Fig. 3.17). Both in the present delta area and offshore, only a number of small islands remained, made up of Pliocene and Pleistocene sediments.

By the middle of the Holocene, the Danube had filled up the area lying to the West of Tulcea and started to build its present Delta. Herodotus mentions in the fifth century BC three branches in the foreland of the Dobrogea hills (Fig. 3.18). By the first to second centuries AD, the narrow shore lane was accreted and Roman sources mention already five (possibly navigable) delta branches. In the Middle Ages, the Sulina Branch carried most of the bed load and the middle part of the Delta significantly protruded into the sea. The latter phenomenon may have been promoted by the increased bed load yield during the so-called "Little Ice Age" (Fourteenth to Seventeenth centuries). From the beginning of the eighteenth century, the development of the Delta has become more balanced. The major part of bed load transport has shifted to the Kilia Branch, but there is also a significant accretion of the interior marshland. Part "E" of Fig. 3.18 illustrates the most recent development of the Kilia Branch, showing a significant area gradually conquered by this Delta branch from the Black Sea over 130 years. It indicates, however, also some abrasion activity by the sea. (Another evidence of the latter activity is Pence Island, situated in front of the Delta in Roman times, traces of which can now hardly be found.) Because of the interaction between freshwater and seawater, sea currents exist parallel to the shoreline, creating so-called fencing sandbanks. The latter promote a gradual advancing of the shoreline, considerably hampering navigation. When accretion reaches these

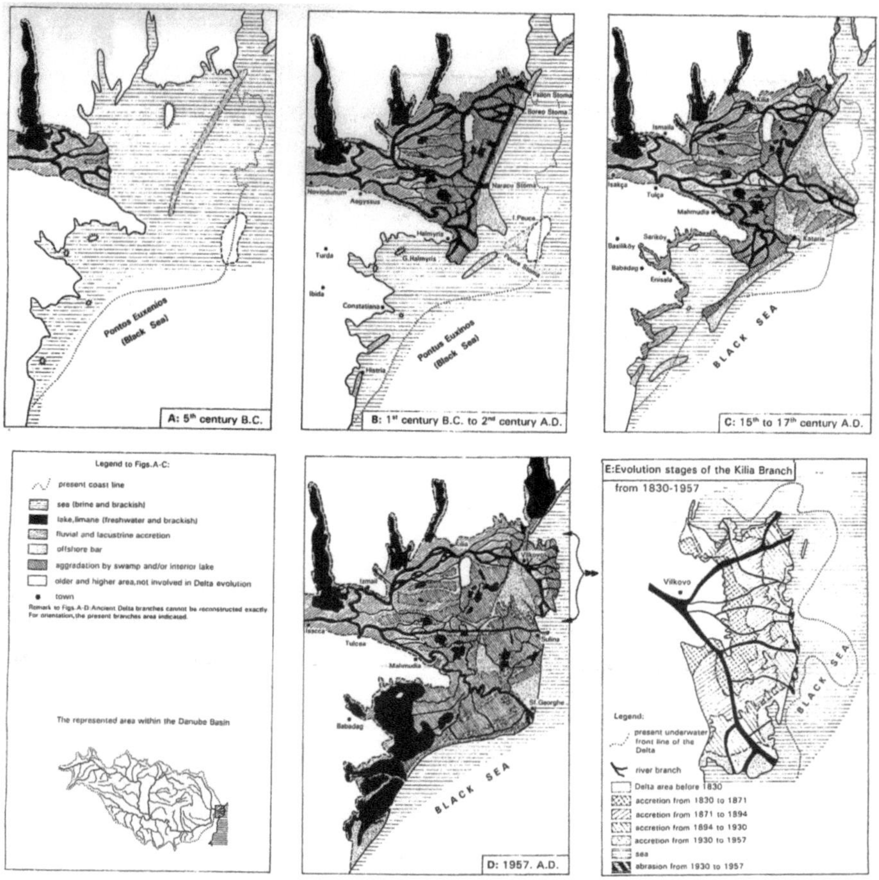

Fig. 3.18 Evolution stages of the Danube Delta in the Holocene, after Banu (1964), Diaconu et al. (1963) and Újvári (1973)

sandbanks, they become beach drifting, facilitating a theoretical reconstruction of delta progradations (Fig. 3.18).

3.2.3.3 Natural Evolution and Recent AnthropogenicImpacts

At the end of the Holocene, i.e., during the last centuries, the process of natural evolution of the Danube Catchment was changed. In addition to natural forces, human activity has gradually grown significant. The changes occurring in the catchment can mostly be detected in riverbed dynamics.

The group of natural factors can be characterised by the following basic phenomena.

In the areas of tectonic uplift (i.e., predominantly in the mountains), both the Danube and its tributaries go on cutting-in their riverbeds. On the intensively rising areas riverbed deepening is intensive, producing great amounts of mostly coarse bed-load material. In the moderately rising areas (including hilly areas) moderate valley widening (meandering) takes place along with deposition of a part of the bed load on the valley bottom (talweg).

In the subsiding areas (first of all, in the basins) the accumulative action of the rivers is still prevails. The transported bed load is being deposited, depending on slope conditions, in an assorted way. On the peripheries of the basins – and of the wide great valleys – talus fans continue to be built from coarse bed load. In a great part of the subsiding areas, the meandering rivers create filled-up plains, occasionally dispersed with marshlands and/or minor lakes (e.g., the Neusiedler-see/Fertő). Only a certain part of its suspended and dissolved load reaches the Danube Delta.

All processes described above are due both to tectonic and climatic factors. Various combinations of natural processes may arise over the catchments, producing typical landscapes, only slowly changing naturally.

In the past 300 years, however, natural processes have increasingly been influenced by human activity. Increasingly effective technical means were adopted, substantially modifying the course of natural evolution.

Remarkable works of water regulation took place as early as in the Roman times of history (e.g., dewatering in the surroundings of Lake Balaton, the bridge and road of Emperor Traianus near the Iron Gate, the construction of harbours at Russe and Vidin). Such works, although far from being interrupted, had minor extensions and effects during the Middle Ages (e.g., Charlemagne's Ditch in 793 AD), as compared with the natural factors. The anthropogenic measures taken to some extent already during the eighteenth, but particularly in the nineteenth and twentieth centuries – first drainage, river training and flood protection, then the facilities for agricultural irrigation and water supply, finally those for hydropower utilisation – have modified the hydrographic and hydrologic conditions basically and irreversibly, the riverbed morphology of the Danube and its tributaries. However, they have practically left unchanged the morphology of the catchment area.

Changes in riverbed dynamics will be dealt with in a future follow-up volume of the Danube Monograph.

From among these effects, only a brief reference should be made in the present Chapter 3, to the sediment regime of the Danube and its tributaries, heavily changed by the construction of barrages during the last decades. Consequently, the accumulation processes compensating natural subsidence have been considerably reduced. The majority of the bed load is intercepted in the impoundments of the barrages that generally do not coincide with the subsiding areas. Therefore, cutting-in tendencies began over long river stretches. Future consequences generally cannot be predicted, only that they may lead occasionally to severe problems. For example, in the so far accreting valleys of the Lower Danube region, deepening processes have started recently. Because of the reduced bed load yield, the sea is washing away from the mainland of the Danube Delta a strip of several metres per year (Bondar and Cernea 1998, Gergov 1998).

Acknowledgments The Authors heartily thank the experts co-operating in the compilation of the project report serving as a basis for preparing the present paper: C. Bondar, V.-A. Stănescu and D. Cernea (Romania), D. Gavrilović (Croatia), S.G. Gergov (Bulgaria), E. Gölz and K. Hofius (Germany), I.V. Grepachevsky (Moldavia), M. Koscal, L. Menković and M. Marković (former Yugoslavia), V. Manukalo (Ukraine) and A. Stančík (Slovakia). Without their valuable contributions, the compilation of the follow-up volume "Palaeogeography of the Danube and its Catchment" to the Danube Monograph (RCDC 1999) would not have been possible. Special thanks are due to the reviewers of that follow-up volume: E. Dudich (Budapest) and E. Gölz (Koblenz, Germany), for their valuable suggestions and corrections. The technical production of the series of coloured palaeogeographic maps (Figs. 3.9–3.17) was carried out by K. Stelczer.

References

Banu AC (1964) Date asupra unei transgresiuni de vîrstă istorică in bazinul Mării Negre şi al Deltei inferioare [Data about a transgression occurring in historical times in the basin of the Black Sea and along the Lower Delta], vol 5. Hidro-biologia, Bucureşti

Bondar C, Cernea D (1996 and 1998) Comments on the preliminary variants of the project report of follow-up volume No. V/2. Manuscripts, Bucureşti

Borsy Z, Félegyházi E (1988) Evolution of the network of water courses in the North-eastern part of the Great Hungarian Plain, from the end of the Pleistocene to our days. Quaternary Studies in Poland, vol 4. Warszawa

Cvijič J (1908) Entwicklungsgeschichte des Eisernen Tors [Evolution history of the Iron Gate] Petermanns Mitteilung, vol 106. Ergänzungsheft 160, Leipzig

Diaconu C, Nichiforov J, Iacov D (eds) (1963) Zona de vărsare a Dunarii – Monografia Hidrologică [Region of the Danube Delta – A hydrologic monograph]. In-stitutul de Studii şi Cercetări Hidrotehnice, Bucureşti

Ehlers J (1994) Allgemeine und historische Quartärgeologie [General and historical geology of the Quaternary]. Enke Verlag, Stuttgart

Fink J (1966) Die Paläogeographie der Donau [Palaeogeography of the Danube]. In: Liepold R (ed) Limnologie der Donau, Lieferung 2. Schwetzerbart'scher Verlag, Stuttgart

Gergov SG (1996 and 1998) Comments on the preliminary variants of the project report of follow-up volume No. V/2. Manuscripts, Sofia

Hámor G (1995) A Kárpát-medence miocén ősföldrajzi és fáciestérképei [Miocene palaeogeographic and facies maps of the Carpathian Basin]. ELTE Térképtudományi Tanszék, Budapest

Hantke R (1994) Fluβgeschichte Mitteleuropas [River history of Middle Europe]. Enke Verlag, Stuttgart

Heim A (1921–1922) Geologie der Schweiz [Geology of Switzerland]. Chr. Herm. Tauchnitz, Leipzig

Marsigli LF (1726) Danubius Pannonico-mysicus … [The Pannonian-Moesian Danube …], vol I–VI. Apud Herm. Uttwwerrf & Franç· Changuion, Amsterdam

Mike K (1970) Magyarország ősvízrajza és felszíni vizeinek története [Palaeohydrography and the history of the surface waters of Hungary]. AQUA Kiadó, Budapest

Niculescu Gh (1977) Hartă teraselor Dunarii, 1: 1,000,000 [Map of the Danube terraces]. Acad. R.S. România, Inst. Geol. şi Geografic, Bucureşti

RCDC (Regional co-operation of the Danube Countries) (1999) Palaeogeography of the Danube and its Catchment (Follow-up volume No. V/2 to the Danube Monograph). VITUKI, Budapest

Rögl F, Steininger FF (1983) Vom Zerfall der Tethys zu Mediterran und Paratethys – Neogene Paläogeographie [The break-up of the Tethys into Mediterranean and Paratethys – Neogene Palaeogeography] Ann. Naturhist. Mus., Wien

RZD (Regionale Zusammenarbeit der Donauländer) (1986) Die Donau und ihr Einzugsgebiet Eine hydrologische Monographie [The Danube and its Catchment A hydrologic monograph]. München

RZD (Regionale Zusammenarbeit der Donauländer) (1993) Schwebstoff- und Geschieberegime der Donau und ihrer wichtigeren Zubringer [Suspended matter and bedload transport of the Danube and its major tributaries]. Folgeband Nr.1 zur Donaumonographie, Budapest

RZD (Regionale Zusammenarbeit der Donauländer) (1999) Die Furten der Donau [The fords of the Danube]. Folgeband Nr. V/3 zur Donaumonographie, Budapest

Somogyi S (1960) Hazánk folyóhálózatának kialakulása az árterek szabályozás előtti természeti földrrajzi viszonyinak vázlatával [Genesis of the river network of Hungary with a sketch of the physico-geographical conditions prior to the regulation of flood plains]. Dissertation, Manuscript, Budapest

Sümeghy J (1955) Medencéink pliocén és pleisztocén ré-tegtani kérdései [Stratigraphic aspects of the Pliocene and Pleistocene of the Hungarian basins]. MÁFI, Budapest

Tillmanns W (1984) Die Flußgeschichte der Oberen Donau [River history of the Upper Danube], vol 26. Jahreshefte des Geologischen Landesamtes Baden-Württemberg, Friburg im Breisgau

Újvári J (1973) A Duna-delta fejlődéstörténet vázlata és szerkezeti egységei [Evolutional sketch and structural units of the Danube Delta], vol 103, Nr. 2–3. Földtani Közlöny, Budapest

Chapter 4
Danube River Basin Coding

Mitja Brilly

Abstract Water database management requires structured hydrological data determined on the watershed. Different systems were derived in the past for this purpose. An analysis of the systems used in the Danube basin countries is presented and a system developed in Germany is adopted for coding of Danube River Basin. The coding system is derived up to the third level of the watershed for which all major tributaries of the Danube River are given a unique code. Use of the system for integrated water management is discussed.

Keywords Danube River Basin · Tributaries · Integrated water management · Watershed coding levels

4.1 Introduction

Both hydrological modelling and environmental database management require a methodology for the determination of watershed separation into subwatersheds and for the topology of a stream network. Many different methods of watershed and stream network numbering have been developed. A classical approach is the European system, in which the main stream is always classified as being of the first order and the tributaries are numbered beginning with the downstream part (from the mouth of the main river) in the upstream direction. In the USA, the Horton approach is widely used (Strahler 1964). The stream order is related to the degree of branching or bifurcation within the basin. The first-order streams are small finger-like tributaries of the second-order streams, which are the tributaries of the third-order streams and so on. The main stream has the highest stream order. The numbering of the streams starts from the upstream part of the watershed and goes downstream.

M. Brilly (✉)
Faculty of Civil and Geodetic Engineering, University of Ljubljana, Ljubljana 1000, Slovenia
e-mail: mbrilly@fgg.uni-lj.si

M. Brilly (ed.), *Hydrological Processes of the Danube River Basin*,
DOI 10.1007/978-90-481-3423-6_4,

The main disadvantage of both systems is that they greatly depend on the map scale or the accuracy of river and creek network. Different types of information give us completely different coding for a given system. The coding gives us only some information about the position of a stream but nothing about the size of the watershed or about the position of any sub-watersheds. In the past few years, different systems of stream network numbering using GIS technology have been developed (Smart 1970, Chorowicz et al. 1992 and Brilly et al. 1993).

Watershed-oriented coding systems have been developed for water resources management purposes at the national level, but have not been published in international publications. The methodologies of watershed-oriented coding systems are presented in national legislation acts (Brilly et al. 1993). Development of watershed-oriented coding systems began more than 30 years ago and since then, many different kinds of systems, depending on the natural conditions in place, have been developed in various European countries. The question of how to develop and implement a single coding system has only become an issue over the past few years due to international co-operation and water resources management of large transnational river basins like the Rhine and the Danube River Basin.

The Proposal was made as part of project No. 8: Hydrological Bibliography of The Danube River Basin related to the Protocol of the VIII working meeting in Male Vozokony, Slovakia, April 11–15, 1995. The project is to be carried out using the German coding system (LAWA 1993), the so-called Pfafstetter system (Verdin and Verdin1999), although parts of the river basin have been numbered up to the tertiary level using a different coding system.

4.2 Concept for Watershed Coding

It is very difficult to develop a simple coding scheme for different hydrological conditions, which would be easily accessible and useful for different purposes including hydrological modelling. The question is how to design an informative, concentrated, easily accessible, and user-friendly system of river basin coding.

Informative means that a code that is the identifier of a specific river basin should give additional information about the relationship between different basins and sub-basins, and the descriptions of the size of the basin etc. The identification code should not just be a group of randomly assembled numbers and characters that uniquely distinguish one basin from another. The code should also incorporate the relationships between a basin and its sub-basins, upstream and downstream basins, basins of different size, etc.

Concentrated means that a code should be as short as possible. A large string of numbers and characters is not practical and some parts of the string would rarely be used. Long codes are also not user-friendly, take up a lot of computer memory, etc.

An accessible system should be easily adaptable to any international system from the uppermost level to the smallest watersheds of only a few square kilometres. The code should be able to be used on any level from the global to the regional or municipal level. At the global level, watersheds of a few 100 km^2 could be easily identified with only a few digits and we could simply add new digits to the existing code for smaller and smaller sub-basins. A string of seven numbers was enough to identify sub-basins of 10 km^2 at the national level in Slovenia (Brilly and Vidmar 1996).

User-friendly means that a code should be as easy as possible for users to understand and as easily manipulated as possible for different needs, particularly for computation purposes. The use of alphabetical characters in a code make it easier for users to understand and possibly easier to remember, but such a code is unfriendly for computing and numerical handling.

An analysis of the existing coding systems in different European countries (Germany, France, Denmark, Czech Republic, Norway) has been done (Brilly and Vidmar 1994) and the best one is the new German system (LAWA 1993). In English publications it is called the Pfafstetter system (Verdin and Verdin 1999).

At the first level Germany is divided into six large river basins with code ( ): (1) the Danube, (2) the Rhine, (3) the Ems, (4) the Weser, (5) the Elba, (6) the Odder, and (9) other watersheds with direct outflow to the sea.

Each large river basin is then subdivided into up to nine sub-basins. The sub-basins are numbered from the upstream to the downstream part of the main basin. Odd numbers are used for the coding of catchment areas with a direct connection to the main stream and even numbers are reserved for tributaries. A Zero (0) is used on any level when a watershed is not further subdivided because the code (a string of numbers) should have the same length on every level. On the lower level, each

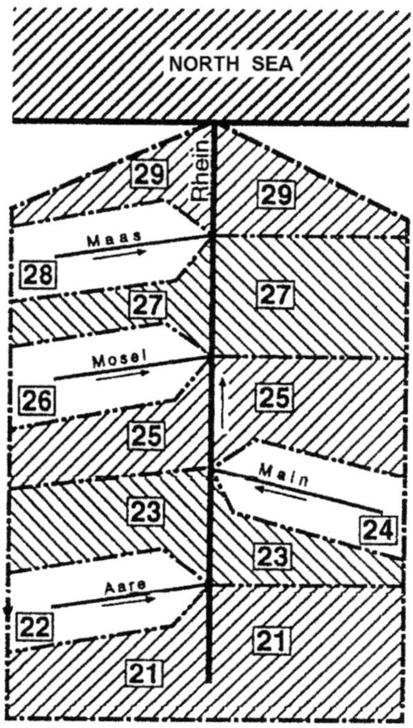

Fig. 4.1 First-level coding of the Rhine River (LAWA 1993)

sub-basin is subdivided and numbered in the same way. An example of this methodology is presented for the Rhine River in Fig. 4.1. At the state level in Bavaria, maps on a scale of 1:200,000 have been made which subdivide watersheds into four levels (Ruhs 1978).

4.3 Danube River Basin Coding

4.3.1 First Level

At the first level, the Danube basin was divided into six parts. Five parts are related to the main stream and only the Tisa River, as the largest Danube tributary, is counted as a separate part (Fig. 4.2, Table 4.1). The first upstream part (1, The Upper Danube) goes from the spring to the confluence with the river Enns. The second part (3, The Upper middle Danube) is from the Enns to the confluence with the Tisa River, the third part (4, Tisa River) is the Tisa River basin, the fourth part (5, The Lower middle Danube) is from the Tisa River to the Iron Gate, the fifth part (7, The Lower Danube) is from the Iron Gate to the Jalomita River outlet, and the sixth part is from the Jalomita outlet to the Danube inflow into the Black Sea (9, The Danube Delta).

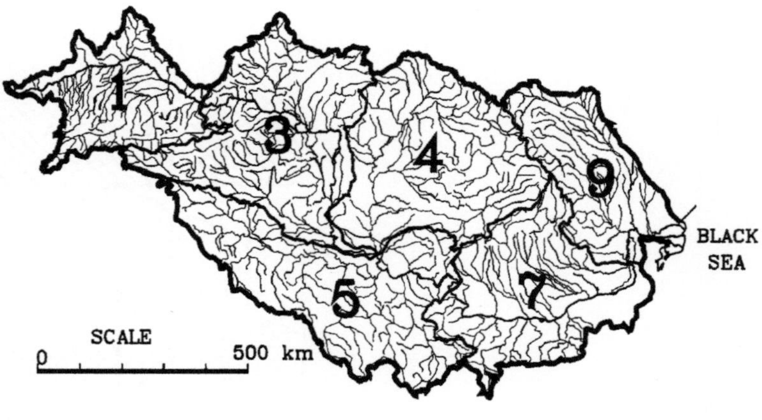

Fig. 4.2 First-level coding of the Danube river basin

4.3.2 Second Level

Coding at the second level is presented in Fig. 4.3 and Table 4.1. The upper Danube is divided into nine parts according to the German river basin numbering. The upper middle part is also divided into nine parts including the Morava (32), right Danube branch (34), Vah, Nitra and left Danube branch (36), and the Drava (38) tributaries.

Table 4.1 Dunube River Basin coding

Name	Code	Name	Code
Upper Danube	1		
The Danube down to Lech (15645)	11		
Lech (4398)	12		
The Danube from Lech to Naab	13		
Naab (5645)	14		
The Danube from Naab to Isar	15		
Isar	16		
The Danube from Isar to Inn	17		
Inn	18		
The Danube from Inn to Enns including Enns	19		
Upper middle Danube	3		
The Danube from Enns to Morava	31		
Morava (38233)	32		
		Morava to Dyje	321
		Dyje	322
		Morava from Dyje to Danube	323
Danube from The Morava to	33		
The Right Danube branch	34		
		Right Danube branch from Danube to Leitha	341
		Leitha	342
		Right Danube branch from Leitha to Raba	343
		Raba	344
		Right Danube branch from Raba to Danube	345
The Danube from The Right to the Left Branch	35		
Left Danube branch	36		
The Danube from The Left Danube Branch to The Drava	37		
		Danube from branch to Hron	371
		Hron	372
		Danube from Hron to Ipef	373
		Ipef	374
		Danube from Ipef to Sio	375
		Sio	376
		Danube from Sio to Dunavolgyi fescat	377
		Dunavolgyi fescat	378
		Danube from Sio to Dunavolgyi fescat	379
Drava	38		
		Drava to Mura	381
		Mura	382
		Drava from Mura to Danube	383
The Danube from Drava to Tisa	39		
Tisa	4		

Table 4.1 (continued)

Name	Code	Name	Code
Headwater Tisa	41		
Upper Tisa	43		
Sajo	44		
Tisa from Sajo to Keres	45		
Keres	46		
Tisa from Keres to Maros	47		
Maros	48		
Tisa from Maros to Danube	49		
Lower middle Danube	5		
Danube from Tisa to Sava	51		
Sava	52		
		Sava to Kupa	521
		Kupa	522
		Sava from Kupa to Una	523
		Una	524
		Sava from Una to Bosna	525
		Bosna	526
		Sava from Bosna to Drina	527
		Drina	528
		Sava from Drina to Danube	529
Danube from Sava to Tamis	53		
Tamis	54		
Danube from Tamis to Velika Morava	55		
Velika Morava	56		
Danube from V. Morava to Timok	57		
Timok	58		
Lower Danube	7		
Danube from Timok to Jiu	71		
Jiu	72		
Danube from Jiu to Olt	73		
Olt	74		
Danube from Olt to Arges	75		
Arges	76		
Danube from Arges to Jalomita	77		
Jalomita	78		
Danube Delta	9		
Danube from Jalomita to Siret	91		
Siret	92		
Danube from Siret to Prut	93		
Prut	94		
Danube from Prut to the Black Sea	95		

The Tisa River is divided into nine parts including Somos (42), Sajo (44), Keres (46), and Maros (48) tributaries. The lower middle part is divided into seven parts including the Sava (52), Tamis (54) and Velika Morava (56) tributaries. The lower Danube is divided into eight parts including Jiu (72), Olt (74), Arges (76), and Jalomita (78) tributaries. The lowest part is subdivided into five parts including the Siret (92) and the Prut (94).

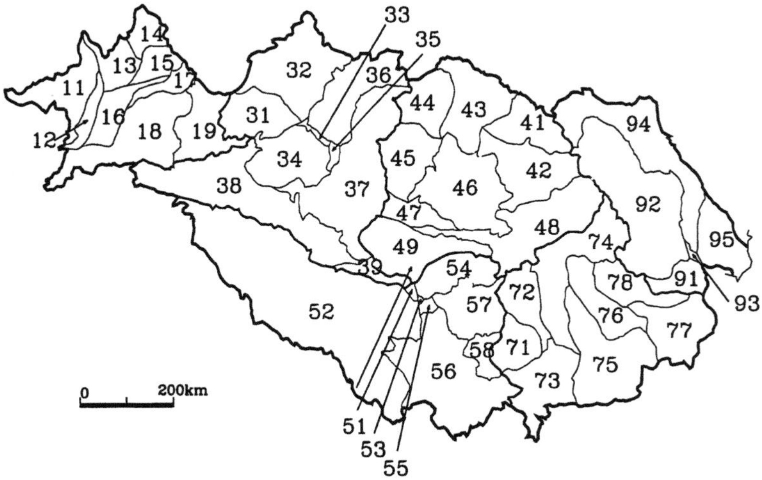

Fig. 4.3 Second-level coding of the Danube River Basin

4.3.3 Third Level

At the third level we subdivide only some of the transnational tributaries of the Danube: the Morava (32), the left Danube branch (34), the Drava (38), and the Sava (52) (Figs. 4.4, 4.5 and Table 4.1). Other parts could be subdivided at the third level according to national or bi-lateral water management coding systems. Examples of such coding for Ukraine are presented in Fig. 4.6.

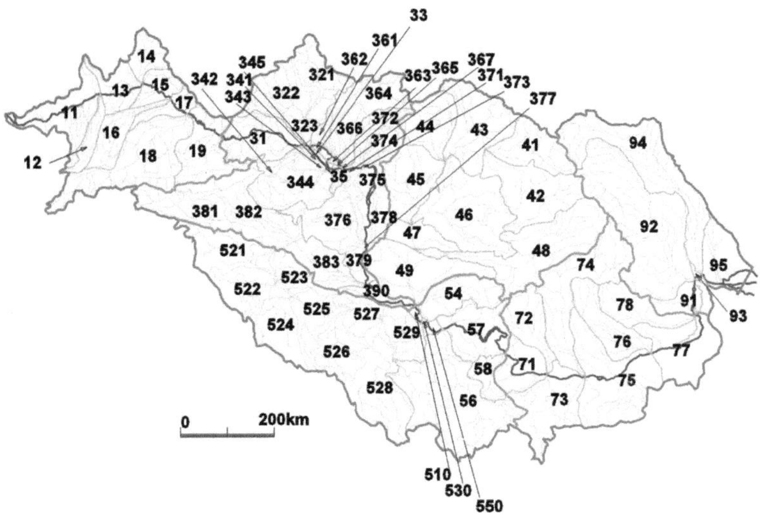

Fig. 4.4 Third-level coding of the Danube River Basin

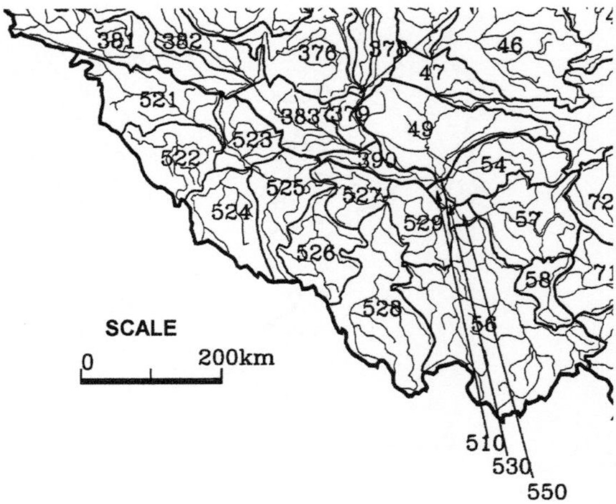

Fig. 4.5 Third level coding of the Danube River Basin for the Sava Basin

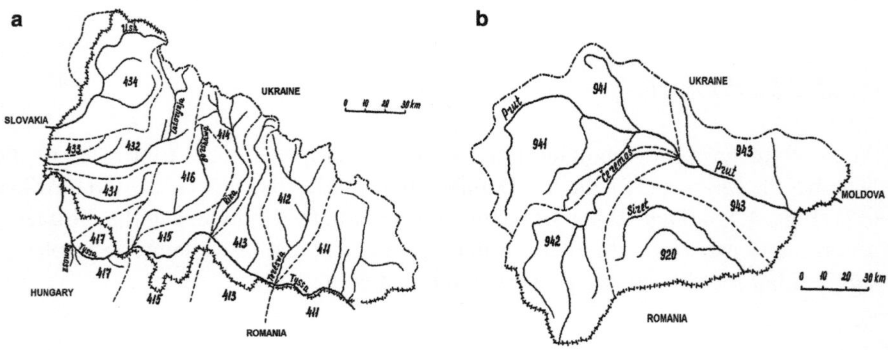

Fig. 4.6 Third-level coding of the Danube River Basin-Ukraine. (a) Tisza watershed. (b) Prut watershed

Almost all tributaries or sub-basins larger than 10,000 km^2 are extracted at the third level. The three digit codes, which are currently in use at the national level for coding the main sub-basins in various countries, give us the possibility to establish a common international database.

4.4 The Relationship Between the Proposed Coding System and Existing National Systems

4.4.1 Germany

The proposed coding system is closely related to the new German national system and the code numbers are almost identical to those found in the new German system.

4.4.2 Austria

The Austrian practice of watershed coding up to the third level is presented in Fig. 4.7. The number 2 is the code number for the first level of the Danube watershed in Austria. Between one and three digits are used for numbering at the second and third levels. The proposed system (Table 4.2) has been given the same numbers as those used in the German system for common watersheds (Table 4.1). The remaining part of the Austrian Danube watershed belongs to the upper middle part of the Danube.

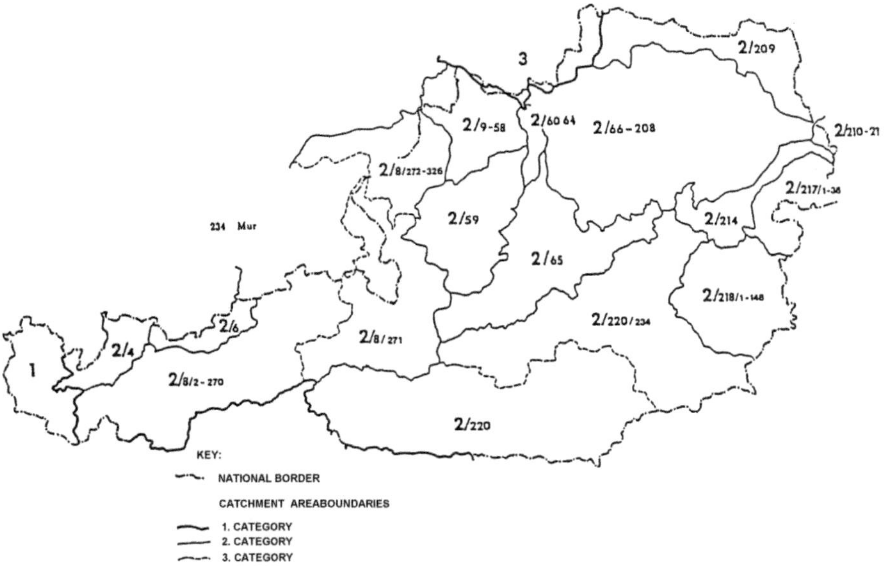

Fig. 4.7 Austrian coding system

Table 4.2 Austrian coding system

River	Present Austrian coding system	Proposed coding
Lech	2/4	12
Isar	2/6	16
Inn	2/8	18
Salzach	2/8/271	186
Traun	2/59	192
Enns	2/65	194
Morava	2/209	32
Danube Enns–Morava	2/66-208	31
Leitha	2/214	342
Raab	2/218-217	344
Drava	2/220	38
Mur	2/220/234	382

4.4.3 Czech Republic

The Czech system of watershed coding (Table 4.3) up to the third level is partially presented in Fig. 4.8. The number 4 is used to signify the first level of the Danube watershed in the Czech Republic, followed by two digits used for numbering at the second level. Most of the Danube watershed in the Czech Republic belongs to the Morava River basin (proposed number 32). Only small parts along the border with Austria belong to the Danube Enns–Morava watershed (31) and on the border with Slovakia belong to the Vah River watershed (36).

Table 4.3 Czech coding system

River	Present Czech coding system	Proposed coding
Morava	4-10, 4-11, 4-12, 4-13	323
Dyje	4-14, 15,16,17	322
Vah	4-21	26

Fig. 4.8 Czech coding system

4.4.4 Slovak Republic

The Slovak system of watershed coding up to the third level is partially presented in Fig. 4.9 and Table 4.4. The number 4 is used to signify the first level of the Danube watershed in Slovakia, followed by two digits used for numbering at the second

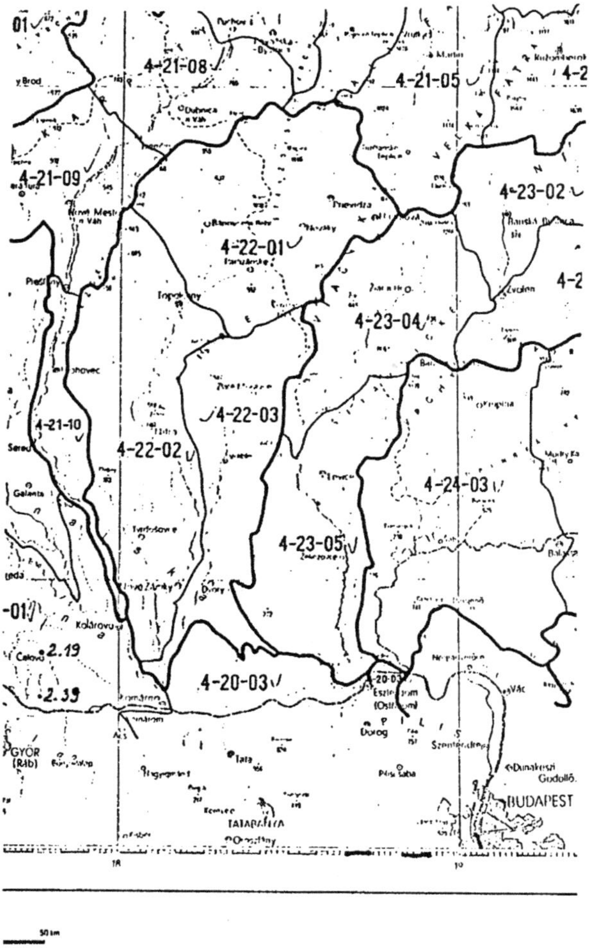

Fig. 4.9 Slovak coding system

Table 4.4 Slovak coding system

River	Slovak practice	Proposal
The Morava down to Dyje	4-13	321
The Drava to Dyje	4-17	323
The Danube from Morava to Vah	4-20-01	33, 35
Vah, Nitra and Little Danube	4-21, 4-22, 4-20-03	36
The Danube from Vah to Hron	4-20-03	371
Hron	4-23	372
The Danube from Hron to Ipel	4-20-03	373
Ipel	4-24	374

and third levels. The proposed coding (Table 4.4) for this upper middle part of the Danube is a little bit complicated. At the second level, we extract the Mali Dunaj watershed including Vah and Nitra River (36).

4.4.5 Hungary

The Hungarian system of watershed division into 26 parts is presented in Fig. 4.10 and only one or two digits are used for this coding system. The proposed system provides a connection to the Hungarian system at the third level (Table 4.5).

4.4.6 Slovenia

The Slovenian coding system was developed in 1996 according to the LAWA methodology (see Fig. 4.11 and Table 4.6). The sub-watershed borders and the coding system are closely related to the international Danube river coding system. The main difference is that the first-level numbers in the Slovenian national coding system use fewer digits for the same watershed than the coding system for the Danube River Basin.

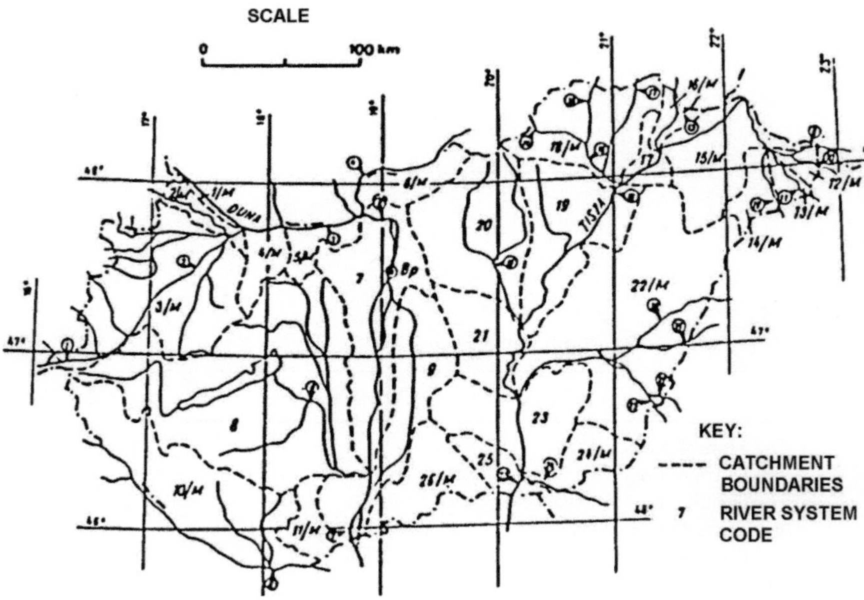

Fig. 4.10 Hungary coding system

Table 4.5 Hungarian coding system

River	Hungarian practice	Proposal
Danube	1	33
Mosoni Duna	2	341–343
Raab	3	344
Danube	4	35
Danube	5	371
Ipel	6	374
Danube	7	375
Sio	8	376
Danube	9	378, 379
Drava	10	38
Danube	11	379
Upper Tisza	12	41
Somes	13, 14	42
Tisza	15, 16, 17	43
Sajo	18	44
Tisza	19, 20, 21	45
Keres	22	46
Tisza	23	47
Maros	24	48
Tisza	25, 26	49

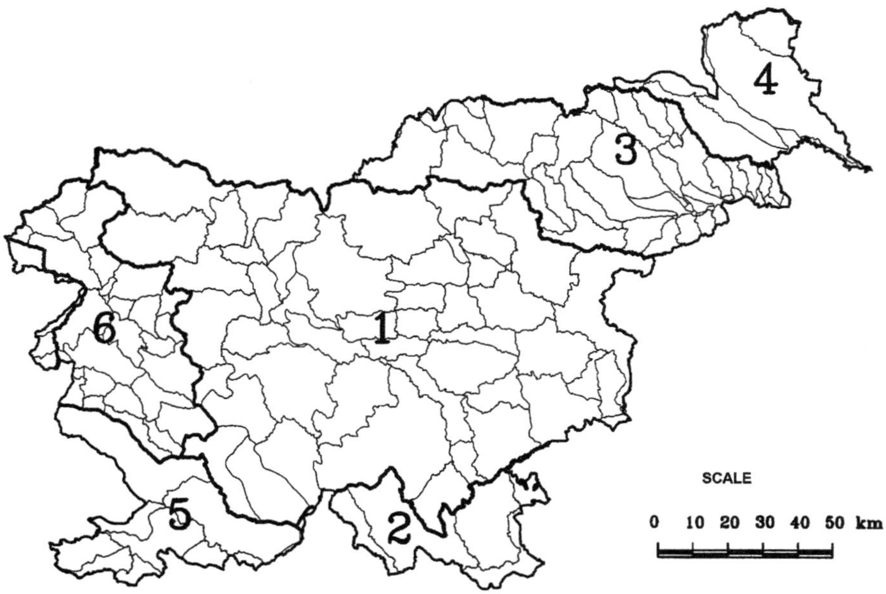

Fig. 4.11 Slovenian coding system

Table 4.6 Slovenian coding system

River	Slovenian practice	Proposal
Sava	1	521
Kolpa	2	522
Drava	3	381
Mura	4	382

4.4.7 Romania

The Romanian system of watershed division into 15 parts is presented in Fig. 4.12 and only one or two digits are used in this coding system. The proposed system will correlate with the Romanian practice at the third level (Table 4.7).

4.4.8 Croatia, Bulgaria, Bosnia and Herzegovina, Former Yugoslavia, Moldova and Ukraine

A watershed coding system does not exist in these countries.

Fig. 4.12 Romanian coding system

Table 4.7 Romanian coding system

River	Romanian practice	Proposal
Upper Tisa	I	41
Sames	II	42
Crisuri	III	46
Mures	IV	48
Timis	V	54
Danube	VI	57
Danube	XIV	71, 73, 77, 91
Jiu	VII	72
Olt	VIII	74
Danube, Vedea	XIV, IX, XIV	75
Arges	X	76
Ialomita	XI	78
Siret	XII	92
Prut	XIII	94

4.5 Coding of National Territories

The Danube River has many transnational tributaries whose watersheds are on the territories of more than one country, for example: the Drava (four countries), the Tisa (five countries) and the Sava (four countries). That is why we should also code national territories. We have chosen ISO 3166:1993 "Codes for the representation of names of countries" (see Table 4.8), which identifies countries either by two characters, three characters or a three digit number. We have used the numerical country code with a dot after the watershed code number in order to code the national territory of a particular watershed. For example, the German territory of the Inn River basin is 18.276.

Table 4.8 Codes for the representation of names of countries – ISO 3166:1993

ENTITY short name in English	Alpha-2 code	Alpha-3 code	Numerical code
AUSTRIA	AT	AUT	040
BOSNIA AND HRZEGOVINA	BA	BIH	070
BULGARIA	BG	BGR	100
CROATIA	HR	HRV	191
CZECH REPUBLIC	CZ	CZE	203
GERMANY	DE	DEU	276
HUNGARY	HU	HUN	348
MOLDOVA	MD	MDA	498
ROMANIA	RO	ROM	642
SLOVAKIA	SK	SVK	703
SLOVENIA	SI	SVN	705
UKRAINE	UA	UKR	804
YUGOSLAVIA	YU	YUG	891

4.6 Water Management and Coding System

The Danube River Basin is an international basin covering 19 countries, which is unique in the world. It is hardly possible to organise action programs at such a large and broad scale. Common bodies such as the Danube River Basin Commission on the basin-wide level with so many independent parties can produce only strategic documents that declare very broad common interests. Because of this negotiations for action programs should occur at the lower sub-basin level including fewer participants which could negotiate much more easily about common interests; however action programs are mostly derived by bi-lateral agreement at the internationally lowest possible level according to the subsidiary principle.

Analysing the map of the Danube River Basin coding system (Fig. 4.3) all the sub-basins suitable for international water management planning involving more than two countries are extracted at the second level of the coding system and then at the third level sub-basins are extracted that are suitable for providing action plans between two countries by bi-lateral agreement.

Development of a common action plan is the reason for establishing a commission for the Sava River Basin (code 52). Countries on the Tisa River sub-basin (code 4) are also planning to develop a common water management plan. Interest is also increasing for development of a common body for the Drava River sub-basin (code 38). However, no proper results can be achieved without decisions made at the operational level, involving bi-lateral commissions with a long history in negotiation between countries, and execute of operation. Action plans are derived from bi-lateral agreement driven by the interests of parties from both sides. The coding system will allow all of these broader developments at different levels to be collected in an integrated information system.

4.7 Conclusions

These various national coding systems provide a satisfactory coding of the Danube River Basin using two or three digits. There is a good correlation between the national coding systems achieved on the third level of numbering. Development of the third and lower levels, for better connection with national systems, should be done at the national level.

The coding system is suitable for integrated water management policy in the Danube River Basin.

References

Brilly M, Smith M, Vidmar, A (1993) Spatially oriented surface water hydrological modelling and GIS. In: Kovar K, Nachtnebel HP (eds) Application of geographic information systems in hydrology and water resources management. IAHS publ. No. 211, Wallingford, pp 547–557

Brilly M, Vidmar A (1994) River basin coding. In: Verwey A, Minns AW, Babovic V, Maksimovic C (eds) Hydrinformatics '94. A.A. Balkema, Rotterdam, pp 531–533

Brilly M, Vidmar A (1996) Šifrant povodij Slovenije, GIS v Sloveniji, FGG, Ljubljana, pp 54–59
Chorowicz J, Ichoku C, Riazanoff S, Kim Y, Cervell B (1992) A combined algorithm for automated drainage network extraction. Water Resour Res 28(5):1293–1302
ISO 3166, International standard, codes for the representation of names of countries, ISO 1993, Geneva
LAWA (1993) Richtlinie fur die Gebietsbezeichnung und die Verschlusselung von Fliesgewassern, Landerarbeitsgemeinschaft Wasser
Ruhs P (1978) Verzeichnis der Bach- und Flusgebiete in Bayern, Bayerisches Landsamt fur Wasserwirtschaft, Bayer, Landesvermessungsampt
Smart J (1970) Use of topologic information in processing data for channel networks. Water Resour Res 6:932–936
Strahler AN (1964) Quantative geomorphology of drainage basins and channel networks. In: Chow VT (ed) Handbook of applied hydrology. McGraw-Hill, New York, NY, Section 4, pp 39–76
Verdin KL, Verdin JP (1999) A topological system for delineation and codification of the Earth's river basins. J Hydrol 218(1–2):1–12

Chapter 5
Characterization of the Runoff Regime and Its Stability in the Danube Catchment

Péter Kovács

Abstract The main purposes of this investigation were to typify the runoff regime (i.e., the typical timely distribution of their flow discharges within the year) in the various regions of the Danube Catchment, and to determine the areal distribution within the Danube Catchment of various indices, each of which characterize the stability of one selected (or integrated) element of the interannual distribution of the flow discharges in the rivers.

The present investigation was carried out as one of the projects of the hydrological co-operation of the 13 Danube Countries in the frame of IHP (International Hydrological Programme) UNESCO.

The computations necessary for the investigations were carried out by processing the series of monthly mean discharge values of 206 gauging stations operated on the river network of the Danube Catchment. Data series covering 51 years (1950–2000) or the only slightly shorter length of 42 years could be provided for the majority (95%) of the stations.

The identification of runoff regime types and stability is based on the probability of occurrence of six particular hydrological events: the first, second and third greatest monthly mean discharges of the year, symbolized with MAX1, MAX2 and MAX3, and the first, second and third lowest monthly mean discharges of the year, min1, min2 and min3.

The numerical results of the runoff regime type identification process can be seen in Table 5.5 and in Annex 3. They are graphically displayed on the map of the Danube Catchment in Annex 2.

P. Kovács (✉)
Middle-Danube-Valley Environmental and Water Management Directorate, Budapest H-1088, Rákóczi út 41, Hungary
e-mail: kovacs.peter@kdvvizig.hu

This publication was prepared in co-operation with the following experts from respective countries: Johann Weber (Germany), Harald Kling (Austria), Dr. Eva Soukalová (Czech Republic), Jana Poórová (Slovakia), Dipl. Geogr. Kovács Péter, Miklós Domokos, Dr Nováky Béla (Hungary), Dr. Mojca Sraj (Slovenia), Bojana Horvat (Croatia), Esena Kupusovic (Bosnia and Herzegovina), Prof. Stevan Prohaska (Serbia and Montenegro), Valentina Ungureanu (Romania), Dr. Snejana Dakova (Bulgaria), Ludmila Serenko (Moldova), Mihail Sosyedko (Ukraine)

M. Brilly (ed.), *Hydrological Processes of the Danube River Basin*,
DOI 10.1007/978-90-481-3423-6_5, © Springer Science+Business Media B.V. 2010

Eight main runoff regime types and 17 subtypes were found in the Danube Catchment. Various indices characterizing the stability of the runoff regime, *N*(MAX1), *N*(MAX2), *N*(MAX3), *N*(min1), *N*(min2), *N*(min3), along with the synthesizing indices *N*MAX, *N*min and N_R, as defined, after Nováky, by the formulae Eq. (5.2) and Eq. (5.6), were computed and displayed on the maps of Annexes 4–6.

Keywords Danube River Catchment · Runoff regime types · Runoff regime stability · Flow distribution · Flow discharges

5.1 Introduction

By size, the River Danube is only the second greatest river in Europe after the River Volga, but it is one of the most international rivers in the world. The Danube and its tributaries collect their waters from the territory of 18 countries, whose life and history – in spite of the different social and political traditions – are more or less defined by this river. Because of this geographical connection, it was a natural necessity that the countries sharing the Danube Catchment – despite political disagreements – had to co-operate in the field of hydrology. The first steps of this collaboration persisting today, were in 1971, when the – at that time – eight Danube countries began to co-operate for the development of the first Hydrological Monograph of the Danube River. This work was finalized in 1986 by the publication of the German-language version of the Monograph. Since 1987, the aim of the Regional Hydrological Co-operation of the Danube Countries has been to improve and update the Monograph by jointly compiled and published follow-up volumes on selected topics of common interest. The goal of the preparation of the present chapter was to investigate the runoff regime by using a uniform methodology in the whole Danube Catchment. The main points were to typify the runoff regime and the investigation of its stability on the rivers of the Danube Basin.

The runoff regime is the fluctuation of a certain hydrological event within the year. In the characterization of the runoff regime, selected events, first of all the extreme values play an important role. Although the realization of a selected hydrological event within a particular year may differ considerably from that of other years, the typical pattern of the runoff regime over a longer period can be detected. Regime stability is the measure of deviation between the runoff regime of individual years and the typical regime pattern.

Because of spatial variability of climate and prevailing physio-geographical conditions, the regime of the watercourses is also different and has variability in space. Moreover, the regime is influenced also by impacts from anthropogenic activities (e.g. land and water uses). When factors that determine the runoff regime do change in time (climate change, change in anthropogenic activities), both type and stability of the runoff regime will change. Thus, characteristics of the runoff regime can be used as indicators of climatic changes.

This chapter – besides the introductory and closing parts – has four main sections dealing with the investigation of the runoff regime. The first presents a general description of the Danube Basin, emphasizing those natural factors, which more or less affect the runoff regime (e.g. climatic and geomorphologic conditions). The second part contains the methodological description of the investigation and presents the data collecting procedure. The next section introduces the results of the typification of the runoff regime together with a colour map and detailed tables of the regime types. The fourth part deals with runoff regime stability. The stability values of the six hydrological events to be introduced later in Annex 3 are presented in a table, while from the nine stability indices (six original events and three derived) three are displayed on colour maps in Annexes 4–6.

The base map of charts presenting the results of the investigation is USGS HYDRO1k, which was developed from a 30 arc-s digital elevation model of the world, it provides a standard suite of geo-referenced data sets (at a resolution of 1 km) that will be of value to organize, evaluate, or process hydrologic information on a continental scale (USGS 2003).

5.2 Investigations Carried Out in the Past

Various regions of the Danube Catchment have been investigated several times in regards to runoff regime characterization and stability calculation, but the catchment was never analyzed as a whole system. The main projects involved in the regional analyses are noted below.

Stanescu and Ungureanu (1997) within the frame of the FRIEND-AMHY co-operation carried out a runoff regime stability investigation of the Tisza River Basin. For the analysis the data were collected from the FRIEND-AMHY database, from yearbooks and from the national co-ordinators of III-AMHY. Relying upon the available data from stations in Romania, Spain, Yugoslavia, Greece and Switzerland, the discriminating periods which define a particular river flow regime were determined. The existence of different zones expressed by their mean altitudes that from the stand point of physiographical properties are quasi-homogeneous, allows hydrological regionalization to be carried out. Stability investigations allowed the determination of very stable or stable mountainous zones, relatively stable zones in medium altitudes and unstable or relatively unstable zones in low territories (Stanescu and Ungureanu 1997). The investigation was continued later by the development of the methodological part (Stanescu and Corbus 2004).

In the framework of a bi-lateral hydrological co-operation between the Water Resources Research Centre (VITUKI, Budapest) and Technical University of Graz, 25 discharge-measuring stations of the catchment of the Upper Rába River (belonging to the section of Sárvár) were processed for typifying the runoff regime and to determine its stability (Bergmann et al. 2001). Runoff regime could be clustered, depending on climatic and hypsographic conditions, into three types. Stability investigations were carried out both on the basis of the index of Corbus and that of the Nováky index *N*. Mapping was carried out on the basis of the index of Corbus: one map was created for the whole runoff regime (six events), one for MAX1 event and one for min1 too. There was a definite clustering in the case of three events. For the Rába River, also the longitudinal profile of the stability index was plotted (Nováky et al. 2001).

In the framework of an investigation of 30 watercourses in Hungary, also a stability investigation was carried out (Nováky et al. 2001). It was determined that the most stable was the runoff regime of watercourses originating from karst regions. The least stable was the runoff regime of watercourses heavily affected by anthropogenic impacts. The runoff conditions of the western part of the country (Transdanubia) are more stable, due to climatic effects, than those of the eastern part (Tisza Catchment). With some exceptions (Lajta River/Hegyeshalom and Mura River/Letenye), the stability of the flood regime is higher than that of the low flow regime. This might be due in part to higher climatic stability, and partly because of special riverbed conditions of the two rivers mentioned. The Raba River originating in the Alps is the only exception with the lowest runoff regime stability among all the investigated watercourses.

As a first step of the evaluation of runoff regime in the whole Danube Basin, the hydrological regime in one of the most important subcatchments of the Danube River, the Tisza River Basin was investigated (Kovács and Nováky 2004). After the examination of 40 data series, it can be declared, that the runoff regime types of the area relate well to territorial climatic changes connected with elevation, while the runoff regime stability of the low flow events is higher than that of the flood events. The most stable part of the basin is the mountainous area around the springs of the Tisza and the most unstable territory is in the western part, the Zagyva River Basin. The results of this work were presented at the XXIInd Danube Conference, held in Brno, Czech Republic, 2004.

5.3 General Description of the Danube River Basin

The River Danube, crossing Middle and Southeastern Europe along its 2,857-km long course, with its multi-annual mean discharge of 6,855 $m^3 s^{-1}$ is the 21st greatest river of the Earth and – after the River Volga – the second greatest of Europe. Its catchment, covering an area of 817,000 km^2, is situated south of the European main continental divide, running from Gibraltar to the Ural Mountains, the central band of this southern part, between the springs of the Rivers Rhine and Dnepr (RZD 1986).

The Danube Catchment not only crosses the European continent geographically, but it also provides historical and political contact between its western and eastern regions too. At the beginning of the hydrological co-operation of the Danube Countries – in 1971 – 12 European states shared the Danube Catchment; moreover these states were separated by the Iron Curtain. Today, after political changes in East-Central Europe, 18 countries share in the Danube Basin. The most important data related to these countries are presented in Table 5.1.

A detailed general description of the Danube River Basin has been published in previous issues within the frame of the co-operation of the Danube Countries (RZD 1986, RCDC 1999), so in this publication only a short introduction to the most important geographical factors is presented. For every factor, the authors have tried to use the most recent maps, compiled on the basis of the latest databases at that time.

Table 5.1 Countries sharing the Danube catchment in 2005

No.	Country's Symbol	Country's Name	Area (1,000 km^2) Total	Area (1,000 km^2) In the catchment	Share of Country in the catchment %	Share of Catchment in the country %	Inhabitants (2003) Total (10^6)	Inhabitants (2003) In the catchment (10^6)	Inhabitants (2003) In the catchment %
"Danube countries" (with major areal shares), approx. in hydrographic order									
1.	D	Germany	357.0	59.6	7.30	16.8	82.1	9.1	11.1
2.	A	Austria	83.9	80.7	9.88	96.4	8.1	7.7	9.4
3.	CZ	Czech Republic	78.9	24.5	3.00	31.1	10.3	2.8	2.5
4.	SK	Slovakia	49.0	48.5	5.94	99.0	5.4	5.2	6.4
5.	H	Hungary	93.0	93.0	11.39	100.0	10.0	10.0	12.2
6.	SLO	Slovenia	20.3	18.0	2.19	88.8	2.0	1.7	2.1
7.	HR	Croatia	56.6	35.4	4.33	62.5	4.8	3.2	3.8
8.	BIH	Bosnia and Herzegovina	51.1	38.3	4.66	74.9	3.8	2.9	3.5
9.	SCG	Serbia and Montenegro	102.2	91.4	11.19	90.0	10.4	9.0	11
10.	RO	Romania	237.5	232.2	28.43	97.6	22.6	22.6	27.6
11.	BG	Bulgaria	110.9	48.2	5.90	43.6	8.3	3.9	4.8
12.	MD	Moldova	33.7	12.0	1.46	35.6	4.3	1.1	1.2
13.	UA	Ukraine	603.7	32.5	3.96	5.4	50.9	3.1	3.7
"Peripheral countries" (with minor areal shares)									
14.	CH	Switzerland	41.3	1.8	0.22	4.4	6.7	0.3	0.4
15.	I	Italy	301.3	0.5	0.06	0.2	57.5	0.1	0.1
16.	PL	Poland	312.7	0.3	0.03	0.1	37.8	0.04	
17.	AL	Albania	28.7	0.1	0.01	0.01	3.2		
18.	MK	Macedonia	25.7	0.4	0.05	0.2	2.1		
1–18.	Danube Catchment total			817.0	100.00			82.74	100.0

5.3.1 Topography and the River Network

The springs of the Danube are in the Black Forest, in Western Europe, while its mouth at the Black Sea is in Southeastern Europe. The northernmost point of its divide is, near the springs of its tributary Morava/March, while the southernmost one is in the Rila Mountains, at the origin of its tributary Isker. The length of the longitudinal axis of the catchment is 1,630 km, that of its watershed 6,320 km. The orography of the latter however is uneven: while it reaches on the southern stretch a height of 4,052 m above sea level (Piz Bernina), the highest point on its northern stretch is the peak Krivan, in the High Tatra Mountains, at only 2,496 m a.s.l. The average elevation of the catchment is 475 m above sea level (RCDC 1999).

The mountain ridges articulating the surface of the Danube Catchment – the Alps, the Carpathians, the Dinarides and the Balkan Mountains – subdivide it into three characteristic units. Its upper part between the riverhead and the Devín Gate is the Upper Danube Region. Within the latter, the Uppermost Danube Region, between the springs and the town of Ulm, is crossed by the non-navigable stretch of the Danube.

The most outstanding tributary in the Upper Danube Region is the River Inn, due not only to its catchment area, but also particularly to the fact that it collects the waters of the highest part of the Danube Catchment.

The Central Danube Region is the subcatchment situated between the Devín Gate and the Iron Gate. This area is sometimes referred to as the Carpathian Basin, although the inclusion of the area south of the River Sava into that Basin is – partly for political considerations – not generally accepted.

The two most important tributaries reach the Danube in this region, not so far from each other, the Tysa/Tisza and the Sava. The River Tysa/Tisza collects the most runoff from the eastern part of the Carpathian Basin, including Transylvania, while most of the discharges of the River Sava are supplied by its southern tributaries originating in the Dinarides.

The Lower Danube Region is between the Iron Gate and the mouth into the Black Sea. This territory can be divided into two parts. Above Brăila, the Danube and its tributaries cross the basin in wide, terraced valleys; this is the area called the Romanian Lowland. The Danube Delta follows downstream Brăila, and this is sometimes also called the Maritime Danube Region (RCDC 2004).

The left side tributaries of the Danube, in this region, collect their waters from the southern slopes of the Southern Carpathians. Numerous watercourses originating in the Eastern Carpathians are collected by the two greater north-eastern tributary streams, the rivers Siret and Prut. The southern tributaries of the Danube originate from the northern slopes of the Balkan Mountains. An overview of the most important parameters of the main tributaries of the River Danube is presented in Table 5.2.

5.3.2 Climatic Conditions, Hydrometeorology

Because of the elongated shape of the Danube Basin in the west–east direction and diverse relief features the climatic conditions are variable. The Danube Catchment

Table 5.2 Overview of the main tributaries of the River Danube (RZD 1986)

River	Catchment area (km^2)	Partial discharge from subcatchment ($m^3 s^{-1}$)
Danube (from the spring to the mouth of Lech)	15,654	184
Lech	4,398	107
Naab	5,645	52
Isar	8,369	163
Inn	26,976	742
Enns	5,940	191
Morava/March	27,633	124
Raab/Rába	14,702	61
Váh	9,714	128
Nitra	5,415	27
Hron	5,251	48
Ipel/Ipoly	5,494	20
Sió	15,129	39
Drau/Drava/Dráva	41,810	577
Tysa/Tisza	158,182	888
Sava	94,778	1545
Velika Morava	38,233	262
Jiu	10,731	95
Iskar	7,811	57
Olt	24,810	184
Lom	3,380	7
Arges	11,814	71
Ialomita	10,305	48
Siret	45,420	226
Prut	28,954	88
Danube (at the mouth to the Black Sea)	817,000	6,857

extends from the western regions of the Upper Danube with high Atlantic influence, to the eastern territories affected by a continental climate. In the Upper and Central Danube Region, especially in the Drau/Drava/Dráva and Sava basins, the climate is influenced by the Mediterranean. This basic character of the climate is varied and modified into natural regions by the great mountain systems, influenced by the height above sea level and the relief features (exposure, leeward and windward position) too (Stančík and Jovanović 1988).

The summarized effects of the climatic components show great differences over the sub-basins of the Danube Catchment. In the case of this investigation, the exact climate types of the sub-basins are not really important, only the tone of the colours in every patch is interesting on the map. This is the reason for presenting the map without a legend, while the original printed version – with its legend – can be found near the publication of the climate types (RZD 1986). On the climate map, the colours change from dark (blue) to bright (yellow), as the measure of the precipitation in that climate type decreases.

5.3.2.1 Air Temperature

The pronounced air temperature differences are also determined by the extensive area and elongated character of the Danube Basin from west to east. Average annual air temperature within the basin ranges from –6.2°C to +12°C. The lowest value originates from Sonnblick, the highest mean annual temperature was observed in the northern part of the Hungarian Lowland and at the Black Sea coast. In the entire Danube Basin July is the warmest month, January being the coldest one (Stančík and Jovanović 1988).

In the upper reach of the Danube Basin the Winter period usually lasts from December to February. The average January temperature in the plains being –0.8 to –3°C, and in the mountains –6 to –13°C. Summers are warm and last here from June to August. Mean July temperatures are 17–20°C, the mean July isotherm in the high mountains being 0°C at heights of about 3,500 m.

In the Central Danube Region Winter only lasts for 1.5–2 months, the mean January temperatures in the lowlands being –0.3 to –2°C and on the highest points about –10°C, but in some places even lower. In July in the middle reach of the Danube basin the average air temperatures in the valleys rise to 20–23°C, but in the higher mountain regions it is only 4–5°C.

The Winter period of the Lower Danube Region usually begins 2 weeks later than the westernmost parts of the Danube Catchment and lasts from the second half of December to the end of February. Mean January temperatures fluctuate within the range –1.2 to –3°C and in the mountains –8 to –9°C. Summer starts in late May and ends in September with maximum monthly temperatures of 22–24°C in July (Stančík and Jovanović 1988).

5.3.2.2 Precipitation

Average annual precipitation fluctuates within the range of 2,300 mm in the high mountains to 400 mm in the delta region. The regional variation of the annual precipitation in the Danube Catchment is shown in Annex 1, where the mean annual values are interpreted for the time period 1961–1990 (Holko et al. 2005).

The Upper Danube Basin shows an astonishingly variable character for precipitation. In the high Alpine regions values of 2,000 mm are sometimes exceeded, the mountain marginal belts being extraordinarily rich in precipitation. The increment of mean annual precipitation values amounts to about 50 mm per 100 m height in the northern Alpine promontories and in the Alps. A further contribution is the distant influence of the mountains, affecting precipitation on windward slopes and increasing precipitation on approach to the mountains. Thus, the lines of uniform precipitation follow the contours of the mountains. In the northern Alpine Foothills the amount of precipitation decreases in this way from about 1,500 mm per year on the periphery of the mountains to 700 mm per year in the Danube valley. About 1,500 mm precipitation per year also falls in the Danube source area, in the Schwarzwald Massif and in the higher regions of the Bavarian and Bohemian

Forests. Other territories show average precipitation values 600–1,000 mm, the valleys and basins being relatively dry with about 700 mm. The intermountain valleys are also relatively dry (Stančík and Jovanović 1988).

In the Central Danube Region the highest values of mean annual precipitation occur on the outskirts of mountains surrounding the lowlands. The highest precipitation values above 2,000 mm occur on the southern-oriented mountain chains of the Julian Alps and the Dinaric system, which are exposed to the effect of humid-warm air masses coming from the Mediterranean. In the Carpathians the mean precipitation values vary between 1,000 and 1,500 mm. In the shelter of those mountains in the east Bohemian-Moravian Uplands, as well as in the Carpathian Foothills, the average precipitation amounts to 600 to 1,000 mm. In the southern part of the Central Danube Lowland the annual precipitation falls to 600–800 mm, in Alföld to 550 mm and in the region of the middle Tysa/Tisza to 500 mm.

This dry climate regime increases in the middle Danube Lowland in the northerly and easterly direction. In the plains of the Lower Danube Basin, the precipitation is only 500–600 mm, though the lowest precipitation values of less than 400 mm are recorded within the Danube estuary. In some years there is no precipitation over the Summer period. Because of the low precipitation and high Summer temperatures, the region of the Danube estuary may be considered as a region with a steppe climate.

When the annual distribution of the precipitation is studied, it can be seen that the maximum occurs regularly in the Summer months. This is especially true in low-lying parts of the Danube Basin, where convective precipitation constitutes a considerable contribution to the total precipitation (Stančík and Jovanović 1988).

In those regions the maximum is shifted with increasing continentality from July to June or May, since in mid-Summer the low air humidity is not sufficient for the development of the showers. The minimum precipitation occurs there in February and sometimes in January in mid-Winter when the Asiatic region of high pressure blocks the transfer of Atlantic air masses to the east.

The conditions in mid-mountain and high mountain ranges under the influence of a maritime climate are different. They usually have their maximum in Summer but due to their orography enabling gliding precipitation in the Winter months of December and January, they frequently show a secondary maximum, in some places even an absolute maximum of the precipitation activity, as for instance in the Bavarian and Bohemian Forest.

A deviation from this basic pattern is shown by regions influenced by the Mediterranean, where the months of October–December show maximum precipitation and the Summer is relatively dry.

The number of days with snow cover, duration and thickness increase with altitude. The shortest duration of snow cover (9–12 days) is on the Black Sea coast. The snow cover lasts for only 20–30 days in the Central Danube Region, 40–60 days in the Upper Danube Basin and the mean proportion of the total annual precipitation is about 10–15% (Stančík and Jovanović 1988).

5.4 Data Collection for the Investigation

For characterization and regionalization of the runoff regime and its stability, we used the data series of 206 discharge gauging stations from the countries of the Danube catchment. Investigating the number of data series sent from the Danube Countries (see Table 5.3), three casts of countries can be separated by using the index of the station density in this investigation. The average over the whole basin is 0.25 stations per 1,000 km^2; around this value are five countries out of the 13, Germany, the Czech Republic, Slovakia, Serbia and Montenegro and Moldova. There are another five countries with values between 0.13 and 0.17 which includes Slovenia, Croatia, Bosnia and Herzegovina, Romania and Bulgaria. The highest values for station density are above 0.40 in three countries of the Danube Basin: Austria, Hungary and Ukraine.

We planned to use rather extended catchments (>500 km^2), which limits the resolution of the investigation. We used 206 stations for the project, and there were five data series with a watershed size under this limit. We used them for the investigation, because they are located in mountainous regions; all of them are above 450 m a.s.l. They have a determinative function for the set of the runoff regime of their receiver.

The limitation at 500 km^2 means that some features of smaller catchments are merged into the value of the bigger area. But there are some great basins, which contain some smaller areas: for example the basin of Sajó at Felsőzsolca includes

Table 5.3 The number of stations for the Danube Countries

Country	Number of stations	Territory in the Danube Catchment (1,000 km^2)	Station density of the investigation (station/1,000 km^2)
Germany	16	59.6	0.27
Austria	34	80.7	0.42
Czech Republic	6	24.5	0.24
Slovakia	12	48.5	0.25
Hungary	38	93.0	0.41
Slovenia	3	18.0	0.17
Croatia	5	35.4	0.14
Bosnia and Herzegovina	5	38.3	0.13
Serbia and Montenegro	26	91.4	0.28
Romania	37	232.2	0.16
Bulgaria	8	48.2	0.17
Moldova	3	12.0	0.25
Ukraine	13	32.5	0.40
Total	206	814.3[a]	0.25

[a]The total area of the Danube Catchment is 817,000 km^2, this number is the territory of the countries with a major areal share in the Danube Catchment. The other 2,700 km^2 is divided between five countries, Switzerland, Poland, Italy, Albania and Macedonia, which each have a minor share in the Danube Basin (under 2,000 km^2/country).

the area of Slana/Lenartovce and Bodva/Turna. Of course the downstream stations contain all the investigated areas above them.

The time period of the investigation is 1950–2000. The length of most of the data series matches this – 131 stations from the total, which is almost 64% (51 years). The other 75 are shorter. As the length of the data series influences the results of the computation we had to make a compromise and we decided to apply an allowance of an 8-year-long time section to the data to increase the number of usable data series. This means, that if the data series is at least 42 years long, no limitation is applied in its utilization. After this, 196 stations were available, which is more than 95% of the total number.

The lengths of the remaining ten data series are between 11 and 41 years. These significantly shorter data series had to be included in the investigations as their exclusion would have resulted in extended empty spaces in the map of gauging stations within the Danube Catchment, thus enabling the investigation of only a part of that Catchment. In these cases we did not use any extending statistical operation, so the collation of the results of individual rivers is restricted, while the separation of the discriminant period – the basis of the calculation – is hardly ever influenced by them. In terms of geography for the stations with much shorter data series, seven out of the ten are on the territory of the former Yugoslavia. Reasons for data shortcomings in these cases should be looked for in the historical and political circumstances of the country involved. Mostly the missing data are at the end of the time period of the investigation, usually starting around 1990.

Of the other three stations, two are in the Republic of Moldova, in which there were also remarkable political changes at the beginning of the last decade of the twentieth century. The gauging station of the last short data series is located on the River Hornád/Hernád in Slovakia close to the Hungarian border. It is possible that the beginning of the data series is the starting point of the data collection in this location. It was possible here to check the geographical correctness of this investigation, because there is a gauging station on the Hungarian side – Hidasnémeti, not far down from the Slovak station, with a reliably long data series. After comparing the outcomes of the computations for these two stations, it became apparent there were no differences between the results for runoff regime types or stability.

As the number of stations with shorter then the 42-year-long data series was low and because their runoff patterns did not differ sharply from that of nearby stations, artificial extension of the short data series was not necessary.

5.5 Methodology for Determination of Runoff Regime Types, Stability Index and Stability Categories

The method of identifying the runoff regime type is based on the investigation of so-called discriminant periods within the years, for six selected hydrological events which represent the first, second and third highest as well as the first, second and third lowest monthly (mean) discharges (symbolized: MAX1, MAX2, MAX3 and min1, min2, min3). The discriminant period is defined as that partial period of pre-fixed length of the year, in which at a given gauging station, one particular (hydrological) event – for example the first highest monthly mean discharge or the second lowest monthly mean discharge – has occurred in most years (i.e., with the highest frequency) during the observation period. In this investigation the discriminant period was chosen as 3 months.

Using the time series of monthly flow the discriminant periods were evaluated for all hydrological stations investigated. Two hydrological stations have the same runoff regime type if the discriminant periods related to all hydrological events are the same or differ with only some limitations. The runoff regime of two stations may be accepted as identical if the discriminant periods of one or two hydrological events, especially of MAX3 or min3, rarely MAX2 or min2, are different by not more than 1 month (Nováky et al. 2001).

After having determined the type of runoff regime for all hydrological stations, the runoff regime stabilities of the individual stations (or else of regional station groups) can be investigated.

Stability can be characterized by adopting the following index: the index H, measuring the entropy as defined by Shannon, is the sum $H = \Sigma H\ (E_j)$ of the index $H(E_j)$, characterizing the individual stabilities of the six hydrological events listed above, as defined by the following equation (Nováky et al. 2001):

$$H\left(E_j\right) = p_i \times \ln p_i + (1 - p_i) \times \ln\left(1 - p_i\right) \tag{5.1}$$

where p_i is the probability of occurrence of the given hydrological event within the selected discriminant period of the year and $(1 - p_i)$ is the probability of occurrence of the same event within the complementary period. The value of entropy $H(E_j)$ depends on the length of the observation period, namely the number of the investigated years, and the position of this period in absolute time. The entropy decreases with the growth of the investigated period if its length is at least 30 years or more.

Function (5.1) is symmetrical, that is $H\,(p_i) = H\,(1 - p_i)$. Thus, the stability index H can be used only in the case where $p_i > 0.5$. This limitation can be lifted by introducing a modified version as was proposed by Nováky and Szalay (2001):

$$N = -\sum p_i \times \ln p_i \tag{5.2}$$

where p_i is the probability of occurrence of a given hydrological event within the ith period, with i ranging from 1 to 4. This means that the whole year is divided into four equal periods, consisting of 3 months. One of the periods will become the discriminant period. Obviously, the equality

$$p_1 + p_2 + p_3 + p_4 = 1 \tag{5.3}$$

is valid.

On the basis of the index of the runoff regime stability, the stability can be qualified or categorized. The selection of the category limits needs to be made and may

Table 5.4 Empirical classes of runoff regime stability on the basis of the Nováky index

N(MAX1), ..., N(min3)	N_{MAX}; N_{min}	N_{R}	Stability grade
<0.28	<0.84	<1.68	Very stable
0.28–0.92	0.84–2.76	1.68–5.52	Stable
0.92–1.24	2.76–3.72	5.52–7.44	Relatively stable
1.24–1.39	3.72–4.17	7.44–8.34	Relatively unstable
>1.39	>4.17	>8.34	Unstable

even be modified in the course of the investigation. Not only the yearly runoff regime (runoff regime as a whole) can be qualified, but also individual hydrological events or selected groups (e.g.: the group of maximum monthly flow) as well (Nováky et al. 2001). The stability of only the high flow regime can be characterized by the index

$$N_{\mathrm{MAX}} = N(\mathrm{MAX1}) + N(\mathrm{MAX2}) + N(\mathrm{MAX3}), \tag{5.4}$$

the stability of only the low flow regime by the index

$$N_{\mathrm{min}} = N(\mathrm{min}\,1) + N(\mathrm{min}\,2) + N(\mathrm{min}\,3) \tag{5.5}$$

and the stability of the yearly flow regime by the index

$$N_{\mathrm{R}} = N_{\mathrm{MAX}} + N_{\mathrm{min}}. \tag{5.6}$$

On the basis of the N index the stability of a given hydrological event or the flow regime can be classified as by Nováky and Szalay (2001) as is shown in Table 5.4.

The stability index can be displayed on a map, on which (1) the isolines of the identical stability indices can be plotted and (2) the regions belonging to the same stability categories can be identified. It is proposed to compile isoline maps (1) for all the six events considered (N_{R}), (2) for the three flood events (N_{MAX}), (3) for the three low flow events (N_{min}), (4) for MAX1 and (5) for min1. Both for the identification of discriminant periods and for the computation of stability indices, software have been developed.

5.6 Runoff Regime Typization

The characterization of the runoff regime was executed by using the discriminant periods, selected as a basis for computing the Nováky index. The discriminant periods are in accordance with the statement, according to which the highest monthly runoff usually occurs during the period between the end of the Winter to the dawn of the Summer, and the lowest monthly runoff falls between the end of the Summer to the end of the Autumn.

The investigation was executed by using the so-called "closest neighbour" principle. This means that the catchments are graded by the discriminant periods to obtain an order, where the basins with identical discriminant period come close together by using a chosen power sequence of the six hydrological events. This settlement insures that the discriminant periods of the directly neighbouring catchments would be the closest together. Our chosen power sequence of the discriminant periods – as was used by Nováky et al. (2001) – is:

$$\text{MAX1} \rightarrow \text{min}\,1 \rightarrow \text{MAX2} \rightarrow \text{min}\,2 \rightarrow \text{MAX3} \rightarrow \text{min}\,3$$

In this case we try to grade the rivers by setting together the stations with similar discriminant periods of the MAX1 event. Within the group of similar discriminant periods of the MAX1 event the order is then defined by the identity of the discriminant periods of the min1 event. The principle of alignment hereafter is based on the similarity of the discriminant periods of events in the order: MAX2, min2, MAX3, min3.

The grouping of the catchments is executed by the changing of the discriminant period of a given group and event. The increase of the period is as big as the length of the time space between the earliest and latest discriminant periods of that event in a given group. So the new discriminant period covers all the individual periods of the catchments belonging to the group.

The number of stations with less accommodation to their groups is 29, which is 14% of the total number of data series. For the runoff regime types of these stations the length of the discriminant period of the min1 event is extended to 4 months, because of the 1 month variance of the given discriminant periods. There are some catchments in the contracted runoff regime types, which occur as outliers to their group. In these cases the obligate increase of the discriminant period at the designation to a given type is at least 2 months. Possible reasons for these discrepancies can be found in the smaller length of the given data series, and in some cases the anthropogenic modifying effects or the limits of the characterization methodology.

Considering the above-mentioned observations – more or less subjectively – the 206 stations of the Danube River Basin are classified by their runoff regimes. As seen in Table 5.5, eight main types were defined, and together with the subtypes in total 17 runoff regime types were identified within the whole Danube Catchment. After the investigation of the runoff regime types, it can be seen, that the high-water period is drifting from the beginning of the year to the Summer time section. The high values of discharge occurrences follow well the 3-month time sections of the discriminant periods of the MAX1 event in each runoff regime type. Most of the

Table 5.5 Runoff regime types in the Danube Catchment

Runoff regime type	MAX1	MAX2	MAX3	min3	min2	min1	Number of catchments
1	X–XII	X–VI	X–VI	VI–IX	VI–IX	VIII–X	4
2	XII–II	XII–IV	II–V	VI–XI	VII–XI	VII–XI	4
3/a	I–III	I–V	XII–V	VII–XI	VII–X	VII–X	4
3/b	I–III	II–IV	II–IV	VI–X	VIII–XI	IX–XI	4
4/a	II–IV	XII–VI	XI–VI	VI–XII	VII–XI	VII–X	48
4/b	II–IV	II–IV	I–VI	VIII–I	VIII–XI	IX–XI	5
4/c	II–IV	II–IV	III–V	VIII–XI	IX–I	XI–II	2
5/a	III–V	XI–VI	I–VII	VII–I	VII–XI	VII–X	27
5/b	III–V	II–VII	I–VII	VII–I	VIII–XII	IX–XII	15
5/c	III–V	III–VI	II–VII	VII–III	VIII–I	XII–II	5
6/a	IV–VI	II–VI	XI–VI	VI–XII	VII–XI	VIII–X	8
6/b	IV–VI	II–VIII	II–VII	VII–II	VIII–XII	IX–XI	13
6/c	IV–VI	III–VII	III–VIII	IX–III	VII–III	XII–II	28
7/a	V–VII	IV–VII	IV–VII	X–III	X–I	X–I	6
7/b	V–VII	IV–VII	IV–IX	X–IV	IX–III	XII–III	21
8/a	VI–VIII	III–VI	IV–VI	IX–I	IX–I	IX–XII	5
8/b	VI–VIII	III–VIII	IV–VIII	IX–III	IX–III	XII–III	7
Total number of catchments							206

runoff regime types have one peak in their runoff regime, with the exception of Type 1, where complex climatic effects cause a double peak.

It can be determined that the dates of the hydrological events change widely over the year. The occurrence of the two most important events, the first highest and first lowest monthly mean discharge, follows well the specifications of the defined global climate type of the whole Danube Basin: the high water period usually occurs in the first half of the year and the low water period is usually the second half of the year. This general idea is modified by many local effects, when the highest discharges occur from October to August, while the lower water period is from July to March in the catchments of the Danube and its tributaries.

The territorial delimitation of the runoff regime types is displayed in Annex 2. On the map, in the headwater regions of the rivers, the types represent the catchments, but downstream of the main rivers (Danube, Sava, Tysa/Tisza, Mures/Maros and Crisul/Körös) the runoff regime classes represent the river sections, because of the complexity of influencing factors. In these cases the letters and the colour of the line on the river section shows the regime type.

There are some white patches on the map of Annex 2. Basically this means that we did not have any runoff regime information concerning those sections of the rivers (e.g. in the case of the downstream sections of the tributaries). On the other hand, at the gauging stations of the downstream sections of the main rivers, where the regime types are assigned by letters, the catcments could not be pictured on the map, so these territories must be uncoloured too.

5.6.1 Runoff Regime Types Defined by River Catchments

In runoff regime Type 1, there are the four headwater catchments of the River Sava, in the territory of Slovenia and the western part of Croatia. Here the first maximum of the monthly mean discharge is the earliest for the whole Danube Basin; it usually arrives between October and December. The other two maxima show significant discrepancies in these four subcatchments: there is a 6-month difference in the occurrence of the second or third maximum between the different territories. The MAX2 and MAX3 events usually occur from March or April to June, but in some cases they occur in the Autumn, as for the MAX1 event. In this runoff regime type the first lowest monthly mean discharge occurs between August and October, and the two other minima are also in the Summer period (from June to September). The runoff regime of this type is two-faced, as it lies at the border of two climatic areas. The Mediterranean effects mostly define the high flow characteristic – the water surplus in Winter – but sometimes the springtime maximum of the continental climate prevails. The continental climate determines the low flow conditions – the water deficit in late Summer or in Autumn – but in some places low flow occurs in mid-Summer, just as in the real Mediterranean territories.

Runoff regime Type 2 is a specific group, because four individual catchments belong to it. Three of them are in the marginal area of the Carpathian Basin and the fourth is the spring area of the River Danube, in the Black Forest region. They are not located that close to each other. Here the first maximum usually occurs in Winter, between December and February, and the other two maxima occur in early springtime or in Winter too. All the low flow events are mostly in Autumn, sometimes in late Summer. This phenomenon is possibly caused by local effects.

Runoff regime Type 3 has two subtypes. It can be seen – just like the catchments of runoff regime Type 2 – that the basins belonging to this type are scattered throughout the Danube Basin. In this group the first highest monthly mean discharge is usually detected between January and March.

In runoff regime subtype 3/a, the date of the second and third maxima is highly variable; these periods can be 6 months long. The three minima are usually in the same time period, between June–July and October. Four catchments belong to this subtype: three from the marginal area of the Bavarian Basin and one from the Carpathian Basin.

Runoff regime subtype 3/b contains four catchments again, from three "edges" of the Danube Basin: the Naab catchment from Bavaria, the two subcatchments of Crisul Negru/Fekete-Körös from the Carpathian Basin and the Kolubara from the Dinarides. In this group, the occurrence of the hydrological events is more uniform than it was in the previous types. The MAX1 event is between January and March; the other two maxima occur in the same period, between February and April. The lowest monthly mean discharge is usually detected from September to November, and the other two minima occur in late Summer or in Autumn.

The geographical distance of the subcatchments belonging to runoff regime Type 3 is possibly caused by local effects, because there are no basin-wide reasons for it on the thematic maps of the Danube Catchment.

Runoff regime Type 4 is the most populous group in this investigation; it contains 55 subcatchments in its three subtypes. A general feature of this group is that the first maximum flow occurs between February and April and most of the subcatchments are creating bigger continuous territories in the Danube Basin.

Runoff regime subtype 4/a contains the most elements among the runoff regime types, it has 48 subcatchments. This subtype has the earliest occurrence of the hydrological events in this main group, where MAX1 occurs between February and April. The time period of the other two maxima has more than a 6-month long section from November to June. The low flow events occur naturally in the other part of the year, from June to December, while min1 is usually between July and October in this group. Geographically the catchments belonging to this subtype – besides some individual basins – creating some cohesive zones in the Danube catchment: all the Moravian basins, most of the catchments of the Little Hungarian Plain, some rivers of the Northern Carpathians (Nitra, Ipel/Ipoly, Slaná/Sajó and Bodrog), the western part of Transylvania and the eastern region of the Dinarides, the Serbian catchments.

In runoff regime subtype 4/b, there are five basins: Regen, Somes/Szamos, Crasna/Kraszna, Ondava and Kysuca. Here the MAX2 event occurs in the same time period as MAX1, between February and April, while MAX3 – similar to the previous subtype – has a 6-month long section. The first lowest monthly mean discharges are usually detected between September and November, 1 or 2 months later than for subtype 4/a. The other two minima usually occur after August, and their period is 4–6 months long.

The smallest group in the investigation is runoff regime subtype 4/c, which contains only two elements: Rika and Jijia catchments. For these two basins all the high flow events arrive at the same time, the first two maxima occur between February and April, MAX3 occurs 1 month later. The three minima are in Autumn or in Winter, the earliest is min3 from August, the latest is min1 from November to February. Possibly local effects cause this significant variance from surrounding territories.

Runoff regime Type 5 is a more or less mountainous group. The highest ridges of the Northern Carpathians, the eastern part of Transylvania, most of the Dinarides and the catchments of the Balkan Mountains are represented under the three subtypes. The common features of these areas are the time period of the first maximum discharge, from March to May, and the occurrence of the second and third events, which last over a long period.

Twenty-seven subcatchments create the runoff regime subtype 5/a. The first maximum of this group is 3 months long (from March to May); the first minimum lasts from July to October. The time period of the other hydrological events (the second and third maxima and minima) lasts for a longer time, usually 5–8 months. In this subtype there are the southern mountainous territories, the western ridges of the Dinarides and the Balkan Mountains, while there are some marginal parts of plains, for example at the Austro-Hungarian border (Leitha/Lajta or Raab/Rába) and Tysa/Tisza upstream in the northern part of the Great Hungarian Plain.

In runoff regime subtype 5/b, there are 15 catchments. Here the first maximum is between March and May, of course, while the first minimum is detected from September to December. Here the second and third events have a longer period again, 5–7 months. The catchments of the High Tatras, the upstream area of the River Mures/Maros and some smaller direct tributaries of the Danube in its Austrian section belong to this subtype.

Runoff regime subtype 5/c is a group with individual catchments again. There are five basins in it, the upstream area of the River Olt, and four upstream river sections from the Alps. In the latter cases climatic reasons may include that the intermountain valleys of these rivers are relatively dry (see Annex 1). Here the second and third events have a long period again – 4–9 months, while the length of the time section of the first events is only 3 months. MAX1 is from March to May, min1 is observed between December and February.

The subcatchments of runoff regime Type 6 are mainly in the eastern part of the Danube Basin, especially around the Eastern and Southern Carpathian Mountains. The first monthly mean discharge is between April and June in the three subtypes of this group.

In runoff regime subtype 6/a, the first lowest monthly mean discharges were detected in the three-month period of August–October. The second events have a 2-month longer time section (MAX2 from February to June, min2 from July to November), while the third events have a 6–8-month long period. Three catchments of this subtype are in the Dinarides, another three are on the southern slopes of the Southern Carpathians, and one is in the Balkan Mountains.

In runoff regime subtype 6/b, the min1 events usually occur between September and November. For this group the other four events (MAX2, MAX, min2, min3) have a much longer, 5–8-month long time period. The subcatchments belonging to this group are on the inner side, on the slopes looking towards the Great Hungarian Plain, of the Eastern Carpathian Mountains. These include the Timis watershed, the upstream area of River Tysa/Tisza and the whole catchment of the Crisul Repede/Sebes-Körös, which collects the waters of the northern slopes of the Apuseni Mountains. There is a specific watershed belonging to this group in the Northern Carpathians, the Hnilec basin. It is completely east-facing and more or less closed in by mountain ridges. Possibly this is the reason for the slight difference from the regime type of its surrounding territories.

The most populous subgroup of Type 6 is runoff regime subtype 6/c. It has the second highest number of elements of the 17 regime types. The low flow is usually detected in wintertime, between December and February in its subbasins, while the other events have 5–9-month long intervals. This group is the easternmost type of the Danube Basin; most of its catchments are located on the eastern or south-eastern slopes of the Carpathian Mountains. The upstream sections of the Rivers Siret and Prut and its tributaries belong to this subtype. Moreover, there are four northeast-facing smaller basins in the Alpine region pertaining to this group too.

Runoff regime Type 7 is the group of the Alpine region. Here the high flow events always occur after April and are usually finalized by mid-Summer. The low flow events generally begin in October and the rivers reach their lowest discharges in the Winter months.

The first group in this type is runoff regime subtype 7/a with three basins of the Iller and Lech rivers, where the flood events are always between April and July and the low flow events usually cover an approximately 4–6-month long time period.

The other subtype is runoff regime subtype 7/b, which forms the main body of the Alpine region with its more than 20 elements. The catchments in this group are under the climatic and geomorphological influence of the Alps. Here the high flow events are usually finalized by July – as in the previous subtype, but they can sometimes extend to September. The time section of the low flow events fluctuates between September and March.

The watersheds of runoff regime Type 8 are in the Alps again. It has two subtypes, in which the first maximum monthly mean discharge usually occurs in Summer. It is highly controlled by snow-melt, as a climatic effect of the highest regions of the Alpine Mountains.

Runoff regime subtype 8/a contains only one catchment, the Isar Basin. Here the MAX1 event is possibly defined by the melting of snow of the spring area (between June and August), while the other two flood events are modified by the effects of the lower plains of the downstream section (beginning in March or April). The low flow events generally arrive in September and usually finish before January.

The real high mountainous regime type is runoff regime subtype 8/b in the Danube Basin. In those catchments belonging to this subtype, the first maximum is detected between June and August and the first minimum is observed from December to March. Flood events can occur from March, but not later than August, while the earliest time for low flow events is September and the high water always arrives by the beginning of April. This is a specific snow-melt regulated runoff regime, which can be seen on rivers like the upstream section of the Rivers Inn, Salzach or some smaller watercourses. Moreover it can be seen in Annex 1, that the upper Inn valley, extending in a west-east direction is a characteristic interalpine dry valley with a precipitation of only 600–700 mm.

5.6.2 The Runoff Regime Types of the Main River Sections

At 18 measuring stations and data series, the runoff regime types do not belong to catchments, but only to the river sections, because of the complex influence of the tributaries. These rivers are the Danube – except for some subcatchments around the spring area, the downstream part of the River Sava, and River Tysa/Tisza together with its main tributaries, downstream of Crisul/Körös and Mures/Maros. On the one hand, here the runoff regime types are noted by letters, located to that section of the river, where the investigated data series were observed. Secondly, the type of runoff regime is also assigned a colour. The lines of the river sections are toned according to the colour assigned to a particular type. The lengths of river sections, belonging to one type, are defined by the availability of information: (1) one type section lasts until the next upstream gauging station – from which we have the data series – above the investigated one; (2) the type section lasts until the mouth of the closest considerable tributary, which is delineated on the map of Annex

2, where the different runoff regime presumably changes the regime type of the main stream.

Runoff regime types continuously change down the rivers. If we investigate longitudinally the main stream, the Danube River, a linear change can be observed above the mouth area of the two greatest tributaries, the Sava and Tysa/Tisza. While in the spring area the regime type is Type 2 or 3, down the river – from the station Hofkirchen – the regime types change permanently from Type 6/a to 7/a and 8/a. After receiving the two main tributaries, the regime type of the Danube is altered to Type 5/a. Under the river section of the Iron Gate, on the territory of the Romanian Lowland, the regime type is 6/b, and at the bottommost station (from which we have data series), at Giurgiu, Type 5/b is observed.

The regime type of the downstream part of the River Sava is 5/a, which is possibly defined by the runoff regime of its tributaries coming from the Dinarides. At the River Tysa/Tisza, the matter is a little bit more complicated. The system is confused by the effects of its significant number of great tributaries. While the type of the main stream individually is 5/a, the Type 6/b runoff regime of the downstream sections of Rivers Crisul/Körös and Mures/Maros change the regime type of Tysa/Tisza at Szeged to 6/b. The category of the bottommost station of the Tysa/Tisza (of our investigation) becomes regime Type 5 again.

Where the regime type belongs to the river sections, it is interesting to observe, that different subtype categories are able to connect to longer watercourses. Above the Iron Gate, the main stream, the Danube, has all the subtypes assigned to a, which means, that the hydrological events usually occur at the first section of the time period of their runoff regime main type. On the Romanian Lowland, under the Iron Gate, there are only subtypes assigned to b, where the occurrence of the investigated events is in the latter part of the time period belonging to the main regime type. Moreover the most important tributary, the River Tysa/Tisza has only subtypes assigned to b. Probably this phenomenon is justifiable because of the subjectivity of the categorization, but there can be some other – later researchable – reasons too.

5.7 Characterization of the Runoff Regime Stability

The stability of the runoff regime was computed for 206 catchments of the Danube River Basin by using the Nováky stability index of Eq. (5.2). The runoff regime stability was calculated finally for nine hydrological events:

- the first (1), second (2) and third (3) highest (MAX1, MAX2, MAX3)
- and the first (4), second (5) and third (6) lowest monthly mean discharges (min1, min2, min3), which are the six basic events of the investigation;
- and for three cumulative events [as is defined by Eqs. (5.4), (5.5) and (5.6)];
- the stability of the high flow regime (N_{MAX}), which is the sum of the stabilities of the three flood events (7),
- the stability of the low flow regime (N_{min}), which is the sum of the stabilities of the three low flow events (8)

- and the stability of the annual flow regime (N_R), which is the sum of the stabilities of all the six basic hydrological events (9).

The results are presented in the table of Annex 3, the graphical presentation of the three most important indices [N_R, N(MAX1), N(min1)] are in Annexes 4–6.

By the annual runoff regime stability (N_R), most of the rivers belong to the relatively stable category. From the 206 stations, about 130 are in this rate, which is more than 60% of the investigated places. A little bit more than one third of the Danube Catchment is rated to the stable section and eight measuring stages were placed in the relatively unstable group.

The geographical spread of these categories in relation to the stability of the annual flow regime can be seen in Annex 4. As is shown on the figure, the stable parts of the Danube Catchment are all in the mountainous regions, the Alps, the Eastern Carpathian Mountains and the Dinarides, because of their less variable climate. From the three territories, the most stable is the westernmost Alpine area, possibly because the regular oceanic climatic effect determines the runoff regime of its rivers. The stability rate of the other two mountainous regions is almost the same.

The most unstable part of the Danube Catchment – from the point of view of N_R – is the western side of the Carpathian Basin, consisting of subcatchments of the river systems Raab/Rába and Mur/Mura. A possible reason for this phenomenon is climatic again, because the normally continental climate of this area is often influenced by Mediterranean or sometimes oceanic climatic impacts. These effects are reflected in the runoff regime of the rivers (e.g. the two peaks of the runoff regime curve).

At the maximum events (N_{MAX}), only one station – Inn-Magerbach – is in the very stable category. About a quarter of the stages are stable, but generally the observed measuring places – more than 140 stations – are in the relatively stable group. There are 11 relatively unstable stations in the Danube Catchment based on the flood events.

Geographically the more stable regions are located in the mountains again. In the Alps it is the same territory as for the stability of the annual flow regime. In the Carpathians the size of the stable area is increased, but in the Dinarides it is decreased according to the previous index. The relatively unstable sector is the same as for the annual flow regime, complemented with some smaller territories from the Alpine region.

Investigating the stability of the MAX1 event, we can establish that almost half of the Danube Basin is in the stable category (Annex 5). For the western side of the Danube Catchment, the Alps and its wider countryside frame is a stable area, while in the eastern south-eastern part, the Carpathians together with the Dinarides create a far-flung continuous stable territory. About 5% of the stations have a very stable runoff regime, and these are located mainly in the Upper Inn catchment, in addition to some stages from the north-eastern Carpathians, around the spring area of the Rivers Tysa/Tisza and Prut. The inner parts of the Carpathian Basin and the Western Dinarides form the relatively stable category for the stability of the first highest monthly mean discharge. In the relatively unstable group, there are almost the same rivers as for the previous indices. The only new element in this category is station Hegyeshalom on the River Leitha/Lajta, where the runoff regime is substantially influenced by anthropogenic activities.

At the minimum events (N_{min}), only the Ötztaler Ache-Tumpen station is in the very stable grade. Forty percent of the Danube Basin is stable, seven stages are relatively unstable, while relatively stable results were calculated in almost 120 cases. The mountainous regions are the most stable parts again: the stable area sensibly increased in the Dinarides and in the Alps, while it is decreased in the Carpathians according to its range in terms of stability of the flood events. It can be seen that there are some new stable locations in the middle of the Carpathian Basin and between the ridges of the Balkan Mountains too. It is remarkable that the two longer sections of the Danube River – between Achleiten and Bratislava and from Giurgiu to the mouth – belong to the stable category. The territory of the relatively unstable group – which is in the same area as it was before – decreased slightly, creating a homogene patch in the middle section of the Raab/Rába catchment.

At the min1 event (Annex 6), more than 65% of the Danube Catchment is at least stable, together with the very stable values of some stations in the Alps. In the western part of the Danube Basin, there are only three greater districts with a relatively stable rate: the northern part of the Bavarian Basin, Moravia and in a lane that extends from the Western Dinarides to the North. In the eastern part there are more smaller patches with relatively stable values: for example around the ridges of the High Tatras, in the northeastern Carpathians in Ukraine, around the mouth section of Crisul/Körös and Mures/Maros, or the area between the Rivers Siret and Prut. There are three stages with relatively unstable runoff regime, again from the watershed of the River Raab/Rába, with a significantly reduced, small and fettered area.

5.8 Conclusions

It can be declared that the runoff regime types – basically defined by using the first maximum and minimum values of the monthly mean discharges (MAX1 and min1) – follow well territorial climatic changes connected with elevation, and this system in some places may be modified by significant aerial impacts, mainly the delayed effects of underground storage basins or anthropogenic interferences.

As is seen in Annex 2, eight main runoff regime types were defined, and together with the subtypes in total 17 runoff regime types were identified within the whole Danube Catchment. In the headwater regions of the rivers, the regime types are associated with the catchments, but downstream of the main rivers the runoff regime classes are associated with the river sections. On some rivers longitudinal changes in runoff regime type can also be investigated. The regime type of the main stream may be modified by its incoming tributaries, like at the Tysa/Tisza or at the Danube River.

It should be noted, that the runoff regime stability of the minimum events is larger than that of the maximum events. This is especially true at the first maximum (MAX1) and minimum (min1) events, where the difference is the most spectacular (see Annexes 5 and 6). The same phenomenon can be identified in the case of the cumulative values of the flood and low flow events too. The maps show that the regime stability of the flood events is more influenced by geographical conditions than at the minima. At the flood events, the stability is slightly larger in the mountainous areas, while at the low flow events this effect does not appear. The regional variation of the stability of the annual flow regime is mostly affected by the same geomorphologic criteria as the flood events, so on the map of Annex 4, the three mountainous regions of the Danube Basin (the Alps, the Dinarides and the Carpathians) belong to the stable category.

The most stable part of the Danube Catchment is the territory of the highest ridges of the Eastern Alps – in the area of runoff regime subtypes 7/b and 8/b (see Annexes 2 and 4). Here are almost all of the very stable parts of the whole basin for all the flood and low flow events, and this area is within the stable category for the annual values too. The possible reason – together with the high elevation – is climatic; this is one of the most stable wet areas of the catchment (defined by precipitation and snow melt). The most unstable territory of the Danube Catchment – which belonging to the relatively unstable category, is in the Raab/Rába area and – in some cases – in the Mur/Mura Basin. The most likely reason is climatic, because this area is the crossing point of three determining climate types, i.e. Atlantic, continental and Mediterranean effects.

The results of this investigation are fitting in well with the outcomes of the preliminary research (Stanescu and Ungureanu 1997, Nováky et al. 2001). The purpose of the present work was to investigate the whole Danube Catchment and to increase our knowledge of this international river.

Annex 1

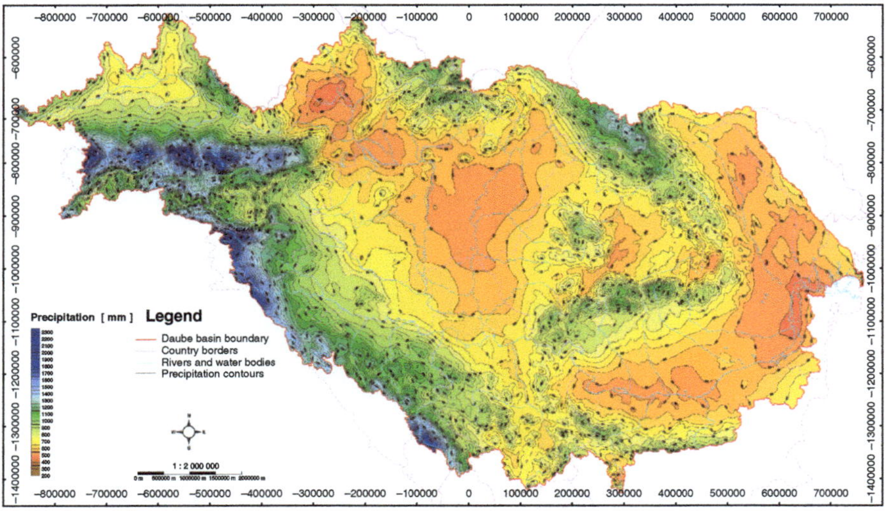

Mean annual precipitation in the Danube Basin (1961–1990)

Annex 2

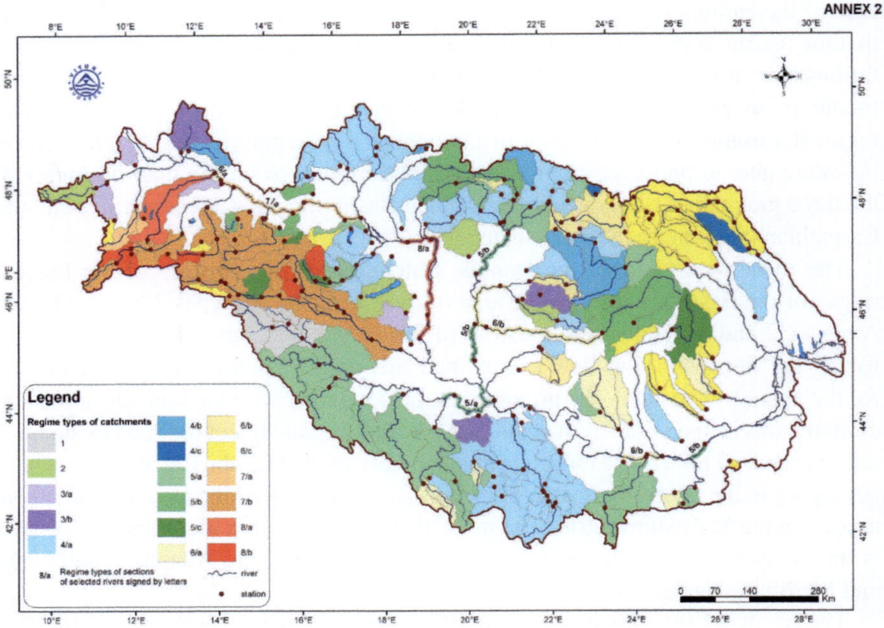

Runoff regime types in the Danube catchment

Annex 3

Country code	No.	River	Station	N(MAX1)	N(MAX2)	N(MAX3)	N(min3)	N(min2)	N(min1)	N_{MAX}	N_{min}	N_R
				computed by								
				Eq. (2)						Eq. (4)	Eq. (5)	Eq. (6)
D	1.	Donau	Berg	0,998	**0,872**	1,205	1,197	**0,900**	**0,613**	3,075	**2,710**	5,785
	2.	Iller mit Kanal	Wiblingen	1,002	0,954	**0,891**	1,077	1,069	**0,649**	2,847	2,795	5,642
	3.	Iller	Kempten	**0,848**	**0,882**	**0,916**	1,030	1,018	**0,783**	2,646	2,831	5,477
	4.	Donau	Neu Ulm-Bad Held	1,188	1,044	1,234	1,226	**0,919**	**0,836**	3,466	2,981	6,447
	5.	Lech	Landsberg	**0,513**	**0,673**	**0,693**	1,094	**0,780**	**0,679**	**1,879**	**2,553**	**4,432**
	6.	Naab	Heitzenhofen	**0,817**	0,935	**0,900**	1,152	**0,905**	0,967	**2,652**	3,024	5,676
	7.	Wörmitz	Harburg	**0,794**	1,040	1,183	1,109	**0,888**	**0,595**	3,017	**2,592**	5,609
	8.	Regen	Regenstauf	1,091	1,082	1,202	1,038	1,032	1,014	3,375	3,084	6,459
	9.	Ammer	Weilheim	1,074	1,007	*1,247*	1,109	**0,913**	0,987	3,328	3,009	6,337
	10.	Isar	Mittenwald	**0,287**	**0,544**	**0,608**	**0,671**	**0,287**	**0,334**	**1,439**	**1,292**	**2,731**
	11.	Isar	Plattling	1,050	1,005	1,204	0,998	1,153	0,998	3,259	3,149	6,408
	12.	Vils	Grafenmühle	1,062	*1,275*	*1,283*	*1,242*	**0,857**	0,952	3,620	3,051	6,671
	13.	Inn	Oberaudorf	**0,196**	**0,227**	**0,611**	**0,761**	**0,636**	**0,377**	**1,034**	**1,774**	**2,808**
	14.	Salzach	Burghausen	**0,475**	**0,704**	**0,732**	**0,679**	**0,692**	**0,518**	**1,911**	**1,889**	**3,800**
	15.	Donau	Hofkirchen	1,173	1,189	1,165	1,168	1,105	0,977	3,527	3,250	6,777
	16.	Donau	Achleiten	**0,639**	1,027	1,023	1,034	**0,840**	**0,731**	**2,689**	**2,605**	**5,294**

Stability indices according to equations of the methodology chapter, characterizing the runoff regime at selected gauging stations of the Danube River Basin

Country code	No.	River	Station	N(MAX1)	N(MAX2)	N(MAX3)	N(min3)	N(min2)	N(min1)	N_{MAX}	N_{min}	N_R
				computed by								
				Eq. (2)						Eq. (4)	Eq. (5)	Eq. (6)
	17.	Ötztaler Ache	Tumpen	**0,372**	**0,646**	**0,683**	**0,533**	**0,100**	**0,100**	**1,701**	**0,733**	**2,434**
	18.	Sill	Innsbruck-Reichenau	**0,171**	**0,100**	**0,632**	**0,650**	**0,230**	**0,171**	**0,903**	**1,051**	**1,954**
	19.	Ziller	Zell am Ziller-Zellbergeben	**0,171**	**0,427**	**0,450**	**0,789**	**0,771**	**0,636**	**1,048**	**2,196**	**3,244**
	20.	Salzach	Bruck-Salzach	**0,100**	**0,372**	**0,507**	**0,687**	**0,664**	**0,658**	**0,979**	**2,009**	**2,988**
	21.	Saalach	Weissbach	**0,603**	**0,643**	**0,827**	**0,846**	**0,790**	**0,517**	**2,073**	**2,153**	**4,226**
	22.	Traun	Steeg	**0,100**	**0,398**	**0,901**	**0,771**	**0,786**	**0,518**	**1,399**	**2,075**	**3,474**
	23.	Ager	Schachham	1,159	*1,302*	*1,273*	1,176	1,140	0,933	*3,734*	3,249	6,983
	24.	Steyr	Pergern	**0,774**	**0,713**	**0,894**	0,980	**0,783**	**0,791**	**2,381**	**2,554**	**4,935**
	25.	Ybbs	Opponitz-Mirenau	1,008	0,941	1,061	1,215	1,005	**0,920**	3,010	3,140	6,150
	26.	Erlauf	Niederndorf	0,987	0,982	1,059	1,154	1,021	1,023	3,028	3,198	6,226
	27.	Kamp	Stiefern	1,064	1,074	1,099	1,063	1,018	1,036	3,237	3,117	6,354
A	28.	Fischa	Fischamend-Rohrbrücke	0,949	1,050	1,162	1,173	1,194	1,073	3,161	3,440	6,601
	29.	Thaya	Raabs an der Thaya	0,982	1,075	1,164	1,116	1,031	1,022	3,221	3,169	6,390
	30.	Lafnitz	Dobersdorf	*1,340*	1,195	*1,347*	*1,280*	*1,301*	*1,259*	*3,882*	*3,840*	*7,722*
	31.	Enns	Schladming	**0,450**	**0,437**	**0,658**	**0,736**	**0,598**	**0,280**	**1,545**	**1,614**	**3,159**
	32.	Salza	Wildalpen	**0,533**	**0,786**	0,956	**0,823**	**0,881**	**0,644**	**2,275**	**2,348**	**4,623**
	33.	Raab	Feldbach	*1,319*	*1,323*	*1,291*	*1,343*	*1,351*	*1,269*	*3,933*	*3,963*	*7,896*
	34.	Kainach	Leiboch	1,223	*1,303*	1,175	*1,356*	*1,351*	1,202	3,701	*3,909*	*7,610*
	35.	Sulm	Leibnitz	*1,305*	*1,241*	*1,326*	*1,323*	*1,286*	1,233	*3,872*	*3,842*	*7,714*
	36.	Isel	Lienz	**1,100**	**0,372**	**0,450**	**0,445**	**0,283**	**0,230**	**1,922**	**0,958**	**2,880**
	37.	Möll	Möllbrücke	**0,664**	**0,840**	1,210	1,228	1,154	1,183	**2,714**	3,565	6,279
	38.	Lieser	Spittal-Fasan	**0,507**	**0,507**	**0,891**	**0,703**	**0,572**	**0,280**	**1,905**	**1,555**	**3,460**
	39.	Gail	Nötsch	**0,715**	0,932	1,128	1,038	**0,782**	**0,576**	2,775	**2,396**	**5,171**
	40.	Gurk	Gumisch	*1,246*	*1,259*	*1,305*	*1,242*	1,102	**0,819**	*3,810*	3,163	6,973
	41.	Lavant	Krottendorf	**0,876**	0,951	1,228	1,093	**0,853**	**0,829**	3,055	2,775	5,830
	42.	Drau	Villach	**0,287**	**0,386**	**0,685**	**0,634**	**0,513**	**0,334**	**1,358**	**1,481**	**2,839**
	43.	Inn	Kajetansbrücke	**0,101**	**0,334**	**0,662**	**0,725**	**0,703**	**0,609**	**1,097**	**2,037**	**3,134**
	44.	Inn	Magerbach	**0,101**	**0,101**	**0,548**	**0,582**	**0,512**	**0,599**	**0,750**	**1,693**	**2,443**
A	45.	Leitha	Deutsch Brodersdorf	1,074	1,227	1,119	*1,245*	*1,269*	1,144	3,420	3,658	7,078
	46.	Mur	Leoben	**0,433**	**0,513**	0,983	**0,804**	**0,548**	**0,135**	**1,929**	**1,487**	**3,416**
	47.	Salzach	Golling	**0,287**	**0,386**	**0,634**	**0,685**	**0,662**	**0,512**	**1,307**	**1,859**	**3,166**
	48.	Enns	Steyr	**0,602**	**0,712**	**0,608**	1,011	**0,780**	**0,772**	**1,922**	**2,563**	**4,485**
	49.	Traun	Wels	**0,858**	**0,849**	1,062	1,024	**0,869**	**0,795**	2,769	**2,688**	**5,457**
	50.	Donau	Kienstock	**0,764**	1,032	1,007	1,007	**0,775**	**0,767**	2,803	**2,549**	**5,352**
	51.	Morava	Komeríz	0,993	0,977	1,185	1,197	**0,876**	0,948	3,155	3,021	6,176
	52.	Morava	Straznice	0,954	1,074	1,176	*1,244*	**0,846**	0,922	3,204	3,012	6,216
CZ	53.	Dyje	Podhradí	**0,797**	1,105	1,165	1,140	1,049	0,945	3,067	3,134	6,201
	54.	Svratka	Zidlochovice	**0,874**	0,990	1,093	1,103	1,022	1,017	2,957	3,142	6,099
	55.	Jihlava	Ivancice	**0,810**	**0,769**	*1,273*	1,156	1,094	1,018	2,852	3,268	6,120
	56.	Becva	Dluhonice	1,042	1,171	1,104	1,116	1,066	1,081	3,317	3,263	6,580
	57.	Dunaj	Bratislava	**0,883**	0,955	1,059	1,039	**0,830**	**0,767**	2,897	**2,636**	5,533
	58.	Vah	Zilina	1,160	1,092	1,178	1,133	1,010	**0,873**	3,430	3,016	6,446
	59.	Kysuca	Cadca	0,936	1,206	1,230	*1,259*	1,203	1,116	3,372	3,578	6,950
	60.	Nitra	Nové Zámky	**0,827**	0,960	0,965	0,939	**0,908**	**0,777**	**2,752**	**2,624**	**5,376**
	61.	Hron	Brezno	1,049	**0,864**	1,048	1,034	1,009	0,994	2,961	3,037	5,998
SK	62.	Hron	Brehy	1,081	**0,814**	0,994	1,073	0,998	**0,849**	2,889	2,920	5,809
	63.	Ipel	Ipelsky Sokolec	0,924	0,970	0,936	1,100	**0,729**	**0,714**	2,830	**2,543**	**5,373**
	64.	Slana	Lenartovce	1,166	1,132	0,985	1,153	1,046	0,955	3,283	3,154	6,437
	65.	Bodva	Turna nad Bodvou	1,004	1,115	1,098	1,220	1,028	1,019	3,217	3,267	6,484
	66.	Hnilec	Jaklovce	**0,916**	1,129	1,151	1,206	0,976	0,959	3,196	3,141	6,337
	67.	Hornad	Zdana	1,055	**0,810**	1,138	1,073	0,979	0,930	3,003	2,982	5,985
	68.	Ondava	Horovce	**0,814**	1,192	1,168	1,226	1,069	**0,775**	3,174	3,070	6,244

Annex 3 (continued)

Country code	No.	River	Station	N(MAX1)	N(MAX2)	N(MAX3)	N(min3)	N(min2)	N(min1)	N_{MAX}	N_{min}	N_R
				computed by Eq. (2)						Eq. (4)	Eq. (5)	Eq. (6)
H	69.	Lajta	Hegyeshalom	*1,247*	1,157	1,129	1,239	1,170	1,196	3,533	3,605	7,138
	70.	Rábca	Lébény	1,226	1,151	1,046	1,072	1,069	0,990	3,423	3,131	6,554
	71.	Rába	Körmend	*1,314*	*1,320*	*1,345*	*1,244*	*1,264*	1,237	*3,979*	*3,745*	*7,724*
	72.	Rába	Sárvár	*1,318*	*1,249*	*1,360*	*1,327*	1,227	*1,247*	*3,927*	*3,801*	*7,728*
	73.	Rába	Árpás	*1,280*	*1,312*	*1,273*	*1,299*	1,224	1,038	*3,865*	3,561	7,426
	74.	Rába	Szentgotthárd	*1,342*	*1,323*	*1,307*	*1,301*	*1,269*	1,141	*3,972*	3,711	*7,683*
	75.	Pinka	Felsőcsatár	*1,315*	1,225	*1,276*	*1,280*	*1,302*	1,148	*3,816*	*3,730*	*7,546*
	76.	Marcal	Mórichida	1,008	1,046	1,200	1,011	1,016	**0,830**	3,254	2,857	6,111
	77.	Zala	Zalaapáti	1,157	1,166	1,122	0,984	0,935	**0,503**	3,445	**2,422**	5,867
	78.	Kapos	Kurd	1,231	1,196	1,119	1,154	**0,826**	**0,584**	3,546	**2,564**	6,110
	79.	Sió	Simontornya	1,040	1,041	*1,291*	1,213	1,045	**0,857**	3,372	3,115	6,487
	80.	Mura	Letenye	0,942	**0,799**	1,129	0,992	0,990	**0,835**	2,870	2,817	5,687
	81.	Dráva	Barcs	**0,685**	1,003	1,052	1,056	0,998	**0,695**	**2,740**	**2,749**	**5,489**
	82.	Duna	Dunaalmás	1,008	**0,902**	1,046	1,089	1,011	**0,778**	2,956	2,878	5,834
	83.	Duna	Nagymaros	0,961	**0,889**	1,077	0,980	1,038	**0,771**	2,927	2,789	5,716
	84.	Duna	Dombori	**0,900**	**0,840**	1,157	1,006	1,019	**0,752**	2,897	2,777	5,674
H	85.	Duna	Mohács	**0,893**	**0,901**	1,081	0,949	1,091	**0,873**	2,875	2,913	5,788
	86.	Karasica	Villány	1,198	1,176	1,128	1,100	0,936	**0,501**	3,502	**2,537**	6,039
	87.	Cuhai Bakony ér	Bakonybánk	1,116	*1,276*	1,065	1,222	**0,823**	**0,648**	3,457	**2,693**	6,150
	88.	Ipoly	Nógrádszakál	0,935	1,116	1,088	1,134	**0,814**	**0,686**	3,139	**2,634**	5,773
	89.	Tisza	Tivadar	1,046	1,160	1,181	1,172	1,092	**0,916**	3,387	3,180	6,567
	90.	Tisza	Záhony	0,981	1,075	1,053	1,135	1,021	**0,796**	3,109	2,952	6,061
	91.	Tisza	Tiszapalkonya	0,978	1,084	**0,903**	1,105	1,059	**0,829**	2,965	2,993	5,958
	92.	Tisza	Szolnok	0,972	0,977	1,091	1,124	1,029	1,001	3,040	3,154	6,194
	93.	Tisza	Szeged	0,989	**0,877**	1,035	1,066	1,071	**0,883**	2,901	3,020	5,921
	94.	Szamos	Csenger	0,960	1,070	0,959	1,083	1,013	**0,875**	2,989	2,971	5,960
	95.	Kraszna	Ágerdőmajor	**0,894**	1,202	1,196	0,995	1,042	0,993	3,292	3,030	6,322
	96.	Bodrog	Felsőberecki	**0,916**	1,020	0,921	1,102	1,105	0,960	2,857	3,167	6,024
	97.	Hernád	Hidasnémeti	1,109	**0,759**	1,164	1,060	0,987	0,931	3,032	2,978	6,010
	98.	Hernád	Gesztely	1,141	1,001	1,188	1,110	**0,908**	0,966	3,330	2,984	6,314
	99.	Sajó	Felsőzsolca	1,192	1,120	1,133	1,114	1,028	1,074	3,445	3,216	6,661
	100.	Zagyva	Jásztelek	1,109	1,184	1,180	1,038	**0,801**	**0,375**	3,473	**2,214**	5,687
	101.	Berettyó	Berettyóújfalu	1,025	1,065	1,128	1,043	**0,849**	**0,718**	3,218	**2,610**	5,828
	102.	Sebes-Körös	Körösszakál	1,157	1,128	1,098	1,143	1,062	**0,796**	3,383	3,001	6,384
	103.	Fekete-Körös	Sarkad	1,123	1,117	1,142	1,142	1,027	**0,830**	3,382	2,999	6,381
	104.	Fehér-Körös	Gyula	1,082	1,085	1,218	1,069	1,003	**0,779**	3,385	2,851	6,236
	105.	Hármas-Körös	Gyoma	1,045	1,102	1,189	1,094	1,226	1,078	3,336	3,398	6,734
	106.	Maros	Makó	**0,694**	**0,883**	1,126	1,048	0,998	0,938	**2,703**	2,984	5,687
SLO	107.	Sava	Litija	1,130	1,122	1,178	1,130	1,085	1,171	3,430	3,386	6,816
	108.	Sava	Catez	1,141	*1,264*	1,177	1,204	1,018	1,101	3,582	3,323	6,905
	109.	Savinja	Lasko	*1,256*	*1,258*	1,219	1,134	*1,273*	1,159	*3,733*	3,566	7,299
HR	110.	Drava	Donji Miholjac	**0,816**	**0,877**	1,045	1,077	1,045	**0,885**	**2,738**	3,007	5,745
	111.	Mura	Mursko Sredisce	0,938	0,971	0,948	1,169	0,943	**0,795**	2,857	2,907	5,764
	112.	Sava	Zagreb	1,190	1,274	1,071	0,973	1,094	1,129	3,535	3,196	6,731
	113.	Sava	Jasenovac	**0,829**	1,083	1,153	1,041	**0,653**	0,943	3,065	**2,637**	5,702
	114.	Kupa	Sisinec	0,996	0,967	1,149	1,219	**0,848**	**0,819**	3,112	2,886	5,998

Annex 3 (continued)

Country code	No.	River	Station	N(MAX1)	N(MAX2)	N(MAX3)	N(min3)	N(min2)	N(min1)	N_{MAX}	N_{min}	N_R
				computed by								
				Eq. (2)						Eq. (4)	Eq. (5)	Eq. (6)
	115.	Una	Kostajnica	1,064	**0,847**	1,064	**0,604**	**0,652**	**0,815**	2,975	**2,071**	**5,046**
	116.	Una	Bosanski Novi / Novi Grad	**0,880**	1,018	1,122	0,977	**0,697**	**0,719**	3,020	**2,393**	**5,413**
BIH	117.	Vrbas	Delibasino selo	**0,832**	1,103	1,126	1,018	**0,652**	**0,576**	3,061	**2,246**	**5,307**
	118.	Bosna	Modrica	1,022	**0,813**	0,949	**0,869**	**0,908**	**0,522**	2,784	**2,299**	**5,083**
	119.	Drina	Zvornik	**0,617**	1,010	**0,831**	**0,831**	**0,540**	**0,536**	2,458	**1,907**	**4,365**
	120.	Juzna Morava	Grdelica	0,939	**0,876**	1,040	**0,736**	**0,807**	**0,661**	2,855	**2,204**	**5,059**
	121.	Vlasina	Vlasotince	**0,844**	0,997	1,055	1,023	**0,706**	**0,829**	2,896	**2,558**	**5,454**
	122.	Nisava	Nis	**0,830**	1,027	**0,792**	**0,873**	**0,816**	**0,324**	**2,649**	**2,013**	**4,662**
	123.	Veternica	Leskovac	**0,905**	0,950	1,088	0,972	**0,600**	**0,567**	2,943	**2,139**	**5,082**
SCG	124.	Jablanica	Pecenjevce	**0,902**	**0,896**	1,079	**0,820**	**0,631**	**0,405**	2,877	**1,856**	**4,733**
	125.	Pusta	Pukovac	**0,849**	0,990	0,980	**0,861**	**0,712**	**0,628**	2,819	**2,201**	**5,020**
	126.	Toplica	Doljevac	**0,861**	0,926	0,997	1,021	**0,836**	**0,462**	2,784	**2,319**	**5,103**
	127.	Rasina	Bivolje	**0,876**	0,967	1,137	**0,902**	**0,891**	**0,656**	2,980	**2,449**	**5,429**
	128.	Ibar	Leposavic	**0,914**	1,059	1,024	1,105	0,966	**0,742**	2,997	2,813	5,810
	129.	Raska	Raska	**0,871**	0,927	1,046	1,050	1,045	**0,743**	2,844	2,838	5,682
	130.	Studenica	Usce	**0,733**	0,993	1,169	1,105	**0,905**	0,927	2,895	2,937	5,832
	131.	Moravica	Arilje	**0,861**	0,955	1,032	1,015	**0,810**	**0,902**	2,848	**2,727**	5,575
	132.	Zapadna Morava	Kraljevo	0,933	**0,784**	1,112	**0,843**	0,929	**0,749**	2,829	**2,521**	**5,350**
	133.	Beli Timok	Zajecar	**0,660**	1,062	**0,878**	0,978	**0,772**	**0,265**	**2,600**	**2,015**	**4,615**
	134.	Crni Timok	Zajecar	**0,564**	**0,756**	0,980	**0,500**	**0,764**	**0,334**	**2,300**	**1,598**	**3,898**
	135.	Mlava	Rasanac	0,959	0,999	1,118	0,968	**0,753**	**0,600**	3,076	**2,321**	**5,397**
	136.	Pek	Kusici	**0,768**	**0,851**	1,024	0,949	**0,666**	**0,623**	**2,643**	**2,238**	**4,881**
SCG	137.	Kolubara	Drazevac	0,991	0,990	1,156	1,039	1,035	0,977	3,137	3,051	6,188
	138.	Piva	Scepan Polje	**0,918**	0,958	1,074	1,043	**0,739**	**0,421**	2,950	**2,203**	**5,153**
	139.	Tara	Scepan Polje	**0,769**	**0,709**	1,079	1,081	**0,623**	**0,620**	**2,557**	**2,324**	**4,881**
	140.	Lim	Prijepolje	**0,766**	**0,718**	1,121	0,998	**0,613**	**0,516**	**2,605**	**2,127**	**4,732**
	141.	Cehotina	Vikoc	**0,846**	1,112	0,922	1,005	**0,843**	**0,872**	2,880	**2,720**	5,600
	142.	Dunav	Pancevo	**0,829**	1,124	1,192	1,139	1,042	**0,842**	3,145	3,023	6,168
	143.	Tisa	Novi Becej	1,011	**0,898**	1,077	1,139	1,102	**0,911**	2,986	3,152	6,138
	144.	Sava	Sremska Mitrovica	**0,877**	0,983	1,150	1,030	**0,714**	**0,546**	3,010	**2,290**	**5,300**
	145.	Velika Morava	Ljubicevski Most	**0,914**	**0,904**	1,010	1,029	**0,738**	**0,533**	2,828	**2,300**	**5,128**
	146.	Viseu	Bistra	**0,546**	**0,889**	1,217	1,203	**0,836**	0,968	**2,652**	3,007	5,659
	147.	Tur	Turulung	1,022	1,118	1,095	1,079	1,021	**0,888**	3,235	2,988	6,223
	148.	Somesul Mare	Beclean	0,973	1,062	1,009	1,018	0,940	**0,870**	3,044	2,828	5,872
	149.	Lapus	Lapusel	0,984	1,075	1,000	1,021	1,062	**0,864**	3,059	2,947	6,006
	150.	Somes	Satu Mare	0,952	1,058	**0,911**	1,006	0,999	**0,911**	2,921	2,916	5,837
	151.	Crisul Repede	Vadu Crisului	0,970	1,133	1,057	1,213	1,140	**0,851**	3,160	3,204	6,364
	152.	Crisul Negru	Tinca	1,083	1,142	1,109	1,144	**0,796**	**0,905**	3,334	2,845	6,179
	153.	Crisul Alb	Gurahont	0,948	1,112	0,945	1,142	0,948	**0,732**	3,005	2,822	5,827
	154.	Mures	Alba Iulia	**0,716**	1,016	1,157	1,054	**0,732**	**0,883**	2,889	**2,669**	5,558
	155.	Aries	Turda	**0,759**	0,992	1,031	0,999	0,951	0,952	2,782	2,902	5,684
	156.	Tarnava Mare	Medias	**0,752**	**0,876**	1,225	0,991	**0,907**	**0,868**	2,853	2,766	5,619
	157.	Bega	Balint	1,085	1,008	1,207	1,074	0,994	**0,803**	3,300	2,871	6,171
	158.	Timis	Lugoj	**0,749**	0,950	1,129	1,144	0,971	0,936	2,828	3,051	5,879
RO	159.	Barzava	Gataia	1,072	1,019	1,136	1,082	1,018	0,967	3,227	3,067	6,294
	160.	Jiu	Podari	0,929	0,922	1,196	1,083	**0,891**	**0,803**	3,047	2,777	5,824
	161.	Motru	Fata Motrului	0,926	1,077	0,962	1,051	1,084	**0,759**	2,965	2,894	5,859
	162.	Olt	Hoghiz	**0,594**	1,000	1,053	**0,854**	1,033	**0,815**	**2,647**	**2,702**	**5,349**
	163.	Cibin	Talmaciu	**0,690**	**0,914**	1,187	1,108	0,968	0,965	2,791	3,041	5,832
	164.	Oltet	Bals	1,105	1,088	1,031	1,078	**0,893**	**0,778**	3,224	**2,749**	5,973
	165.	Vedea	Alexandria	1,088	0,980	1,231	0,948	0,955	**0,794**	3,299	**2,697**	5,996
	166.	Arges	Malu Spart	**0,362**	**0,650**	0,998	1,095	0,961	**0,793**	**2,010**	2,849	**4,859**
	167.	Raul Targului	Piscani	**0,546**	**0,890**	1,106	0,927	1,013	**0,904**	**2,542**	2,844	**5,386**
	168.	Dambovita	Contesti-Lunguletu	**0,546**	**0,781**	1,056	1,039	0,958	0,975	**2,383**	2,972	**5,355**
	169.	Ialomita	Cosereni	0,977	0,979	1,109	0,927	1,025	1,008	3,065	2,960	6,025
	170.	Prahova	Adancata	0,936	0,955	1,184	0,961	0,984	1,022	3,075	2,967	6,042
	171.	Siret	Lespezi	**0,749**	0,973	1,130	1,157	1,073	**0,786**	2,852	3,016	5,868
	172.	Suceava	Itcani	**0,565**	0,918	**0,878**	0,977	1,020	**0,628**	**2,361**	**2,625**	**4,986**

Annex 3 (continued)

Country code	No.	River	Station	N(MAX1)	N(MAX2)	N(MAX3)	N(min3)	N(min2)	N(min1)	N_{MAX}	N_{min}	N_R
				computed by								
				Eq. (2)						Eq. (4)	Eq. (5)	Eq. (6)
	173.	Moldova	Tupilat	**0,551**	**0,854**	**0,889**	1,020	**0,915**	**0,570**	**2,294**	**2,505**	**4,799**
	174.	Bistrita	Frumosu	**0,516**	**0,620**	0,955	**0,909**	**0,636**	**0,616**	**2,091**	**2,161**	**4,252**
	175.	Trofus	Targu Ocna	**0,500**	**0,761**	0,955	**0,885**	**0,867**	**0,572**	**2,216**	**2,324**	**4,540**
	176.	Barlad	Barlad	1,045	1,028	*1,277*	*1,249*	0,998	1,042	3,350	3,289	6,639
	177.	Buzau	Nehoiu	**0,628**	**0,706**	*1,263*	1,094	0,925	**0,891**	**2,597**	2,910	**5,507**
RO	178.	Prut	Radauti	**0,565**	**0,872**	1,097	0,978	1,033	**0,753**	**2,534**	2,764	**5,298**
	179.	Jijia	Victoria	1,008	0,942	1,223	1,212	*1,271*	1,154	3,173	3,637	6,810
	180.	Dunarea	Corabia	**0,889**	1,050	1,016	1,081	**0,901**	**0,905**	2,955	2,887	5,842
	181.	Dunarea	Turnu Magurele	0,930	0,971	1,079	1,107	**0,814**	**0,883**	2,980	2,804	5,784
	182.	Dunarea	Giurgiu	0,973	0,955	0,932	0,988	**0,852**	**0,851**	2,860	**2,691**	5,551
	183.	Ogosta	Mizija	0,905	1,022	1,132	1,116	1,056	**0,879**	3,059	3,051	6,110
	194.	Iskar	Orjahovo	**0,750**	0,983	1,043	**0,863**	**0,811**	**0,651**	2,776	**2,325**	**5,101**
	185.	Iskar	Novi Iskar	1,042	0,997	1,033	1,142	**0,836**	**0,759**	3,072	**2,737**	5,809
BG	186.	Jantra	Tarnovo	**0,917**	1,107	1,141	1,077	**0,813**	**0,765**	3,165	**2,655**	5,820
	187.	Jantra	Gabrovo	0,987	1,037	1,152	1,061	**0,829**	**0,839**	3,176	**2,729**	5,905
	188.	Rositza	Sevlievo	0,957	**0,905**	1,141	1,038	**0,875**	**0,780**	3,003	**2,693**	5,696
	189.	Vit	Tarnjane	**0,893**	1,026	**0,903**	1,184	1,174	1,124	2,822	3,482	6,304
	190.	Rusenski Lom	Razgrad	0,972	1,161	1,186	1,197	1,155	1,017	3,319	3,369	6,688
	191.	Prut	Shireutsi	**0,325**	**0,500**	1,089	**0,898**	**0,500**	**0,572**	**1,914**	**1,970**	**3,884**
MD	192.	Prut	Costeshti	**0,655**	0,958	**0,855**	**0,668**	**0,888**	**0,854**	**2,468**	**2,410**	**4,878**
	193.	Prut	Ungheni	**0,783**	1,011	1,061	1,183	1,058	**0,773**	2,855	3,014	5,869
	194.	Tysa	Rahiv	**0,227**	0,973	1,100	1,137	**0,779**	**0,856**	**2,300**	2,772	**5,072**
	195.	Teresva	Ust Chorna	**0,450**	**0,843**	*1,290*	1,208	1,073	**0,901**	**2,583**	3,182	5,765
	196.	Rika	Mizhhirya	1,161	1,183	*1,281*	1,121	1,218	1,053	3,625	3,392	7,017
	197.	Latorycya	Mucachove	1,022	0,982	1,160	1,124	1,197	0,989	3,164	3,310	6,474
	198.	Latorycya	Chop	1,010	0,987	0,947	1,129	1,194	0,956	2,944	3,279	6,223
	199.	Uzh	Uzhhorod	0,994	**0,760**	*1,267*	1,119	1,146	1,052	3,021	3,317	6,338
UA	200.	Uzh	Zarichevo	**0,864**	**0,849**	1,203	1,074	1,200	0,925	2,916	3,199	6,115
	201.	Tysa	Vylok	**0,908**	0,971	1,207	1,055	1,028	1,024	3,086	3,107	6,193
	202.	Siret	Storozhinec	**0,856**	**0,919**	1,178	1,160	1,001	**0,758**	2,953	2,919	5,872
	203.	Prut	Jaremcha	**0,462**	**0,889**	0,999	**0,901**	0,948	**0,594**	**2,350**	**2,443**	**4,793**
	204.	Cheremosh	Usteriky	**0,253**	**0,569**	**0,858**	1,070	**0,660**	**0,649**	**1,680**	**2,379**	**4,059**
	205.	Chorny Cheremosh	Verkhovyna	**0,257**	**0,575**	**0,703**	0,939	**0,754**	**0,598**	**1,535**	**2,291**	**3,826**
	206.	Bily Cheremosh	Yablunyca	**0,519**	**0,732**	**0,765**	**0,909**	**0,777**	**0,583**	**2,016**	**2,269**	**4,285**

Annex 3 (continued)

Annex 4

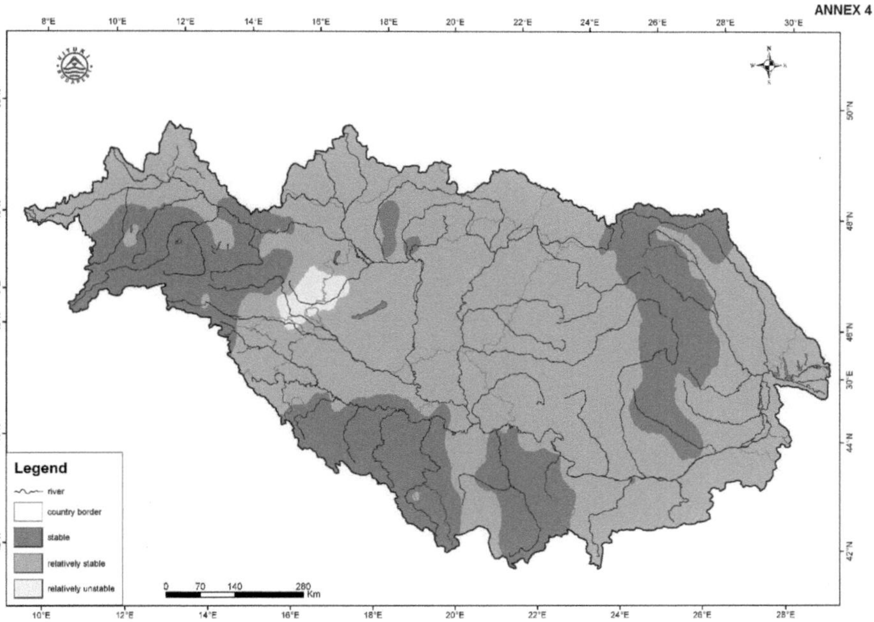

The regional variation of the stability of the annual flow regime N_R

Annex 5

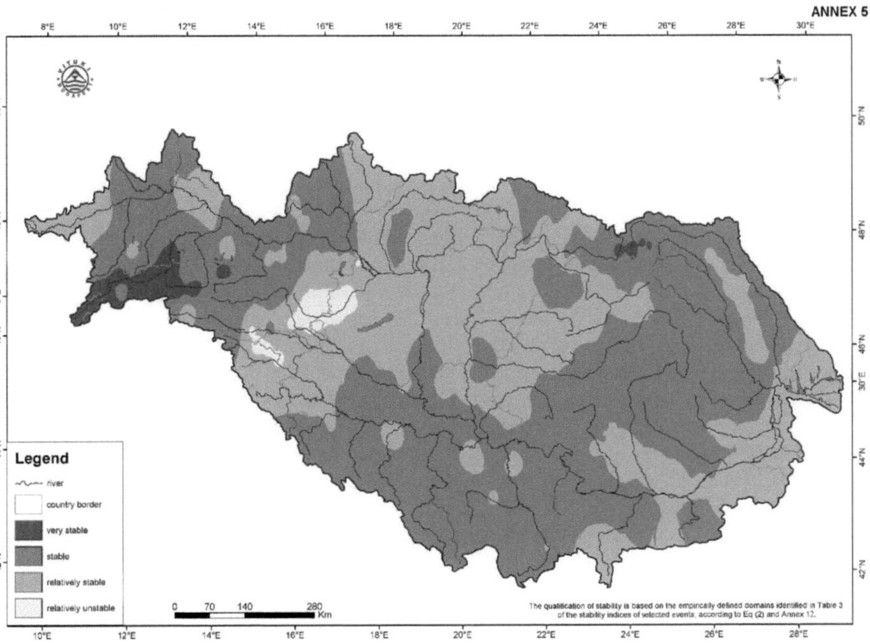

The regional variation of the stability of the first highest monthly mean discharge N(MAX1)

Annex 6

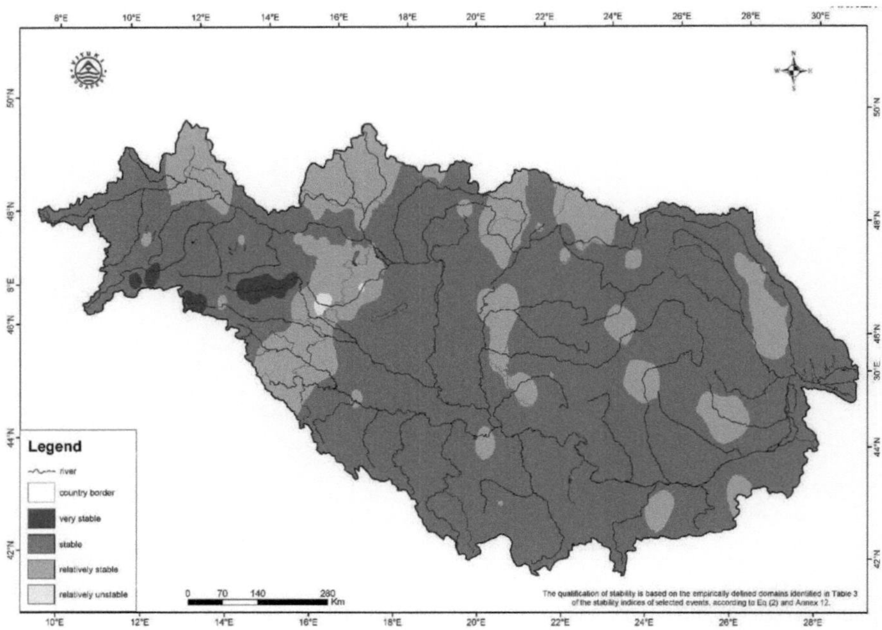

The regional variation of the stability of the first lowest monthly mean discharge N(min 1)

References

Bergmann H, Domokos M, Krainer R, Krall E, Goda L, Hamza I, Neppel F, Nováky B (2001) Untersuchungen im Einzugsgebiet der oberen Raab über hydrologische Folgen einer möglichen Klimaänderung. Schriftenreihe zur Wasserwirtschaft No. 36, Technische Universität Graz, Graz – Budapest

Holko L, Kostka Z, Miklánek P (2005) Basin-Wide Water Balance of the Danube River Basin – Maps of mean annual precipitation, actual evapotranspiration and runoff 1961–1990, Final report. Institute of Hydrology SAS, Liptovský Mikuláš, Slovakia

Kovács P, Nováky B (2004) Characterization of the runoff regime and its stability in the Tisza Catchment. Proceedings of the 22nd conference of Danubian countries on the hydrological forecasting and hydrological bases of water management. Brno, Czech Republic

Nováky B, Goda L, Domokos M, Bergmann H (2001) Looking for the impacts of climate change on the hydrometeorological data series of the Upper Raab Catchment. Vízügyi Közlemények, LXXXIII, vol 3. Hungarian National Water Authority, Budapest, p 393

Nováky B, Szalay M (2001) The runoff regime stability of the Hungarian rivers. Szent István University, Gödöllő

RCDC (Regional Cooperation of the Danube Countries) (1999) Palaeogeography of the Danube and its catchment [Follow-up volume No.V/2 to the Danube Monograph]. VITUKI, Budapest

RCDC (Regional Cooperation of the Danube Countries) (2004) Inventory of the main hydraulic structures in the Danube Basin [Follow-up volume No.VIII/1 to the Danube Monograph]. NIMH, Bucharest

RZD (Regionale Zusammenarbeit der Donauländer) (1986) Die Donau und ihr Einzugsgebiet. Eine hydrologische Monographie. Teil 1: Texte, Teil 2: Tabellen, Teil 3: Karten [Danube Monograph, German version]. Bayerisches Landesamt für Wasserwirtschaft, München

Stančík A, Jovanović S et al (1988) Hydrology of the River Danube [Quadrilingual, abridged version of the Danube Monograph]. Príroda, Bratislava

Stanescu VA, Corbus C (2004) Discriminant descriptors and stability of the river flow regime; A methodological attempt. Proceedings of the 22nd Conference of Danubian countries on the hydrological forecasting and hydrological bases of water management. Brno, Czech Republic

Stanescu VA, Ungureanu V (1997) Hydrological regimes in the FRIEND-AMHY area: space variability and stability. FRIEND '97 – Regional hydrology: Concepts and Models for Sustainable Water Resource Management. IAHS Publ. No. 246. p 67

USGS (United States Geological Survey – EROS Data Center) (2003) HYDRO1k Documentation. Sioux Falls, SD. http://edcdaac.usgs.gov/gtopo30/hydro/readme.asp

Chapter 6
Coincidence of Flood Flow of the Danube River and Its Tributaries

Stevan Prohaska and Aleksandra Ilic

Abstract The flood coincidence methodology within this contribution gives a statistically sound analysis concerning an important feature of flood genesis. The whole Danube reach is covered by such research, which reveals flood behaviour along the Danube impressively. The practical results could serve as a tool for experts and decision-makers, and aid in determining how to improve flood control and flood plain management and how to arrange this task for the greatest benefit. The fact that, more or less, no significant trends in flood behaviour were detected along the Danube River is worth mentioning. It should be considered that human action and climate change as well could change this substantially. Besides other flood characteristics, this could well change the coincidence behaviour of the Danube and its tributaries. To judge this, the evaluation of historical data is of great importance.

Keywords Flood · Discharge · Flood wave volume · Time lag · Coincidence · Two-dimensional events · Time series · River sectors · Observed flood hydrograph · Simulated flood hydrograph

6.1 Introduction

6.1.1 Preliminary Remarks

The term "high flow" is used to describe hydrological conditions where a water level and/or discharge exceed the capacity of a river's main channel, causing the riverside to become flooded. The water level and discharge rise are usually sharp, whereas, after reaching the maximum, their decrease becomes more gradual. This event is

S. Prohaska (✉)
Jaroslav Cerni Institute for the Development of Water Resources, Pinosava 11226, Belgrade, Serbia
e-mail: stevan.prohaska@jcerni.co.rs

M. Brilly (ed.), *Hydrological Processes of the Danube River Basin*,
DOI 10.1007/978-90-481-3423-6_6,

frequently referred to as a flood wave. Its graphical presentation is known as a flood hydrograph. Both terms are synonymously used to denote a flood.

Flood damage has increased all over the world. Flood protection and mitigation programs, including both preventative and remedial measures, can consume a considerable portion of a country's income. It is widely accepted that the development of an efficient flood protection system requires the optimum allocation of relevant resources. As such, there is a need for a comprehensive evaluation of all the causes and consequences of floods.

The causes of increased flood damage are primarily related to human activities, and can be arbitrarily divided into two groups. In the first group there are flood characteristics which can be significantly affected by urban development, inadequate land use, deforestation, channel improvement, etc. Namely, such activities shorten the flood concentration time, which consequently forces flood peaks to become larger. In the second group there are factors such as the increasing value of goods within the areas exposed to flooding (development of housing projects, industrial and transportation facilities). At the same time, the number of people living within flood-prone areas is continuing to grow rapidly.

In addition, there are also floods caused by dam failure, or initiated by land slides. In these cases a large amount of impounded water is suddenly released. The downstream river channel and/or flood control structures are unable to accommodate the additional discharge.

The cost of flood damage can be very high in comparison to the national income. It is understandable that developing countries are in a peculiar position. Namely, in undeveloped countries the value of goods within a flood-exposed area is relatively low. On the other hand, the developed countries have managed, to a large extent, to develop flood protection systems and to implement other efficient measures required to cope with floods and/or to live with them.

An efficient flood protection system is composed of a very complex set of complementary structural and non-structural measures (development and maintenance of engineering structures, optimal system operation, zoning, insurance, flood forecasting, issuance of warnings, etc.). The implementation of associated tasks requires large expenditures. Average annual investment in flood protection should, obviously, not be lower than the average flood damage incurred per annum. For human, social and other development reasons, significantly higher investments can be easily justified.

It is obvious that a complete flood protection system is impossible to set up. Even assuming that anthropogenic factors can be eliminated, there is always the risk that the capacity of the developed flood protection systems, which have been designed to provide a given degree of safety, will be exceeded. Additional levels of protection may demand significantly higher levels of investment. For that reason it is necessary to select an optimum level of protection. In other words, the cost of all measures to be implemented should be reasonable, and the relationship between such measures and potential damage optimal. Human, social and other conditions must be fully accounted for, to prevent the loss of lives.

Routine engineering practice selects the level of protection based on empirical criteria that address the value of the goods which can be protected and the number of people saved. Namely, a system is designed and appropriate measures implemented to provide protection from a flood magnitude of a pre-selected return period. Naturally, where there is a higher concentration of population or additional property value to be protected, a longer return period should be selected.

In view of the high cost of flood protection system development and operation, a comprehensive and reliable assessment of flood characteristics deserves adequate attention. This issue is being addressed by a large number of participants in flood protection: scientists, engineers, government officials, planners, and others, all the way down to system operators. Basically, the overall effectiveness of the measures implemented at all stages (planning, development and implementation) is dependent on a large set of hydrological and other relevant data, which must be collected over time. A proper methodical approach to data analysis and an appropriate economic evaluation are also required for the same purpose.

6.1.2 The Study of Objectives

The assessment of high flow characteristics related to the design of hydraulic structures plays the most important role in the selection of the optimum size for a flood protection system, which needs to satisfy both economic and safety criteria. It is clear that structures designed to convey underestimated flows are at high risk of structural damage and collapse, which may in turn cause other adverse effects. Overestimated flood characteristics diminish the risk, but result in the development of costly systems and inefficient use of limited financial resources.

The capacity of hydraulic structures within flood control systems should be equal to the maximum discharge of the design flood. When dealing with the assessment of the effects of impounding reservoirs and/or the effects of a river's channel on flood wave attenuation, as well as with the maximum discharge, the design volume and shape of the flood wave must be known. Since the system must provide the required level of safety, respective values are selected to correspond to the relevant hydrograph characteristics associated with the chosen return period, which is actually a function of the reciprocal value of the acceptable risk.

A conventional approach to risk assessment is to determine the probability that a pre-selected value of the flood characteristic will be exceeded by the flood event. This is actually equivalent to the determination of the return period of that event. The usual procedure applied consists of statistical analyses of hydrological data from nearby gauging stations. This approach, from an engineering point of view, gives satisfactory results for a large number of tasks. In particular, it appears to be the most frequently applied approach when dealing with reasonably simple river systems (i.e., when the river does not receive large tributaries within the studied reach).

However, when the system encompasses the confluences of large rivers, the described approach cannot provide a reliable flood characteristic estimate. Namely, for various reasons, flood wave geneses in two watersheds are different, so that flood peaks do not occur simultaneously at both the main river and its tributary. Yet, the flood wave of one river may greatly affect the water level of the other stream. In addition, hydrological data are usually collected at gauging stations located outside of the reach where the influence can be realized, so that the interaction between one river and another is not reflected in the gathered data. Because of this, reliable estimation of flood event coincidence on both the main river and its tributary is a vital issue.

The hydrological field is abundant with examples of the application of regression analyses, which basically describe the interdependence between two events. However, when dealing with flood events, this approach is insufficient, for it gives the "expected conditional value" [i.e., the expected value of the dependent random variable, provided that the value of the independent variable is given (either known or assumed)]. Hence, a more comprehensive approach is needed to evaluate the probability of the simultaneous occurrence of the events in question.

Statistical analysis using a two-dimensional variable appears to be a suitable tool for overcoming the difficulties encountered. Using such analysis, as will be discussed later, makes it feasible to arrive at more reliable results regarding the flood stages which should be used in designing a flood protection system.

The objective of this study was to develop a bi-variate statistical model for the assessment of the most important flood features in situations where there is a strong interaction between two rivers. To that end, the following goals applied to this study:

1. to outline a methodology for assessing the probability of a bi-variate process;
2. to conduct two-dimensional probability law analyses of various characteristics of a flood event;
3. to draw a conclusion regarding the suitability of the proposed approach; and,
4. to make recommendations for future research needs with respect to this issue.

6.2 Methodology

6.2.1 Theoretical Background

Hydrological events are frequently viewed as randomly occurring events. Consequently, the theory of probability is used in their analysis. This section provides a brief reminder of the theoretical background for using the probabilistic approach in the analysis of flood hydrograph characteristics. It is assumed that the reader is well acquainted with these techniques. A more detailed description can be found in various works (e.g., Yevjevich 1972, Jovanovic 1981).

One-Dimensional Random Variable. A probabilistic analysis of a one-dimensional random variable is the technique most often applied in hydrologic investigations. The probability density function of a continuous random variable

$f(x,\theta)$ is defined in finite or infinite space (where θ stands for the parameter vector of the considered function type).

In hydrological practice, several well-known types of the density distribution function, $f(x)$, are used, such as: Normal, Log-Normal, Gumbel, Pearson III, Log-Pearson III, and so on. The parameter set in each calculation is constant, and for reasons of simplicity will be omitted in further considerations, except where it is of particular interest.

In highly general terms, the cumulative distribution function $F(x)$ is defined by the following expression:

$$F(x) = P[X \leq x] = \int_{-\infty}^{x} f(t)\mathrm{d}t \tag{6.1}$$

The probability of exceedance, $\Phi(x)$, can be obtained from:

$$\Phi(x) = 1 - F(x) = P[X > x] = \int_{t=x}^{t=\infty} f(t)\mathrm{d}t \tag{6.2}$$

Basically, the use of any distribution function implies the fitting of the theoretical function to the observed/measured hydrologic sample. This is accomplished by using the estimated set of parameters, θ, – usually 2–3 parameters. The reliability of the design flood estimate depends on both the parameter set, θ, and the type of the applied distribution function, $f(x)$. As such, various procedures can be used to evaluate the reliability of the calculated results, and to that end, testing of statistical hypotheses and confidence assessments are performed.

Several statistical tests, usually known as parametric tests, are available for testing the parameter θ. They can be "one-sided" or "two-sided" tests. Most often the normalized test (Z-test), Fisher's test (F-test) and Student's test (T-test) are employed.

Non-parametric tests (the Kolmogorov–Smirnof test and the chi-square test) are used to evaluate the goodness of fit of the distribution function.

In both cases, the essence of statistical testing is to find out the level of confidence which applies to the obtained results. In other words, the testing procedure determines the probability that future events will fall within a defined interval, whose width depends on sample properties, primarily length and variance.

On the basis of the results of statistical tests (or for other reasons), hydrologists sometimes arrive at the conclusion that not much confidence can be associated with the estimated values. In such cases, for practical reasons stemming from safety requirements of a flood protected area, the magnitude which corresponds to the upper limit of the confidence interval of a given probability is adopted as the design value.

In the same way, the confidence interval appears as a range within which future events will fall with a desired (adapted) probability range.

More details regarding tests of statistical hypotheses using confidence intervals can be found in the previously quoted literature and in many other books.

Two-Dimensional Random Variable. Many hydrological processes are the product of two distinct, inter-connected systems (i.e. they can be the compound result of two or more random processes). The degree of interaction between the systems can be different. Their dependence is most often described by the correlation coefficient. This approach, however, does not provide satisfactory results when flood wave characteristics at a confluence are examined, since the correlation coefficient only relates events at two specific points, without the possibility of assigning a probability level for their simultaneous occurrence. However, under such circumstances, the water level at gauging stations on one stream can greatly influence the water level of the adjoining river. Hence, the complete reach, for which flood control measures are being analyzed, is affected by the events occurring simultaneously at all observation points. For this reason, a more comprehensive approach is needed, which can account for the complexity of hydrologic and hydraulic conditions and which can enable the evaluation of an overall risk level, based on the criterion being used for the design of flood control measures. A multi-dimensional statistical analysis appears to offer an opportunity to overcome encountered drawbacks. More precisely, it provides a method by which to determine the probability distribution for the simultaneous occurrence of flood events at the rivers under consideration. Therefore, the flood protection system satisfying the design criteria should be developed within the zone of confluence (i.e., to withstand floods characterized by the return period calculated using multi-variate analyses).

To that end, the methodology to be presented below pertains to the estimation of the coincidence of flood flows at the main river and at its tributary. The term "coincidence" is used to denote the probability of a simultaneous occurrence of two random variables, X and Y, which stand for random events at the main stream and at its tributary.

On the basis of statistical theory, a two-dimensional probability distribution function for a normally distributed bi-variate random process, X and Y, is defined as follows:

$$f(x,y) = \frac{1}{2\pi \cdot \sigma_x \cdot \sigma_y \cdot \sqrt{1-\rho^2}} \cdot \exp\left\{ -\frac{1}{2(1-\rho^2)} \left[\frac{(x-\mu_x)^2}{\sigma_x^2} - \frac{2\rho(x-\mu_x)(y-\mu_y)}{\sigma_x \sigma_y} + \frac{(y-\mu_y)}{\sigma_y^2} \right] \right\} \quad (6.3)$$

The symbols in expression (6.3) stand for the following:

X and Y random variables (flood flow characteristics at the main river and its tributary or at two adjacent profiles of the main river);
x and y simultaneous realization of the random variables X and Y, respectively;
μ_x and μ_y expected values of the X and Y variables;

σ_y and σ_x standard deviations of the X and Y variables;
ρ coefficient of correlation between the X and Y variables.

For a joint-probability density function – j.p.d.f., $f(x, y)$, the marginal densities, $f(x,\bullet)$ and $f(\bullet,y)$, are defined by:

$$f(x,\bullet) = \int_{y=-\infty}^{y=\infty} f(x,y)\mathrm{d}y \tag{6.4}$$

$$f(\bullet,y) = \int_{x=-\infty}^{x=\infty} f(x,y)\mathrm{d}x \tag{6.5}$$

The marginal cumulative probability functions are determined using:

$$F(x,\bullet) = \int_{t=-\infty}^{t=x} f(t,\bullet)\mathrm{d}t \tag{6.6}$$

and

$$F(x,\bullet) = \int_{t=-\infty}^{t=x} f(t,\bullet)\mathrm{d}t \tag{6.7}$$

The cumulative probability function – c.d.f., $F(x,y)$, is evaluated using the equation:

$$F(x,y) = P[X \leq x \cap Y \leq y] = \int_{t=-\infty}^{t=x} \int_{z=-\infty}^{z=y} f(t,z)\mathrm{d}t\mathrm{d}z \tag{6.8}$$

The cumulative exceedance probability, $\Phi(x,y)$, can be obtained from the following relation:

$$\Phi(x,y) = \int_{t=x}^{t=+\infty} \int_{z=y}^{z=+\infty} f(t,z)\mathrm{d}t\mathrm{d}z = P\left[X > x \cap Y > y\right] = 1 - P\left[X < x \cup Y < y\right] = \\ = 1 - F(x,\bullet) - F(\bullet,y) + F(x,y) \tag{6.9}$$

6.2.2 Flood Coincidence Computation Method

In bi-variate statistical analyses of flood characteristics, hydrologists encounter two basic obstacles which must be overcome in a practical implementation of the proposed model. The first obstacle stems from the fact that most flood characteristics are not normally distributed. It is, however, customarily assumed that the variables follow the log-normal distribution. Therefore, their logarithmic transformations:

$$\begin{aligned} U &= \log X \\ W &= \log Y \end{aligned} \tag{6.10}$$

are said to be normally distributed.

The previously described model for the evaluation of cumulative distribution functions appears to be a direct method. Nevertheless, it can be lengthy and cumbersome, since it involves extensive calculation in a three-dimensional space, X, Y and ρ. For most current computers, it does not present a severe obstacle. However, at the time when this project was first conceived, the method of direct computation was virtually impracticable. For that reason, a more convenient, grapho-analytical scheme was implemented. This scheme was described by Abramowitz and Stegun (1972), and is briefly discussed in the ensuing text.

The scheme uses standard normal variables. A transformation of non-standard variables into standardized variables can be achieved using a well-known procedure, namely:

$$\psi = (u - \overline{U})/\sigma_u$$

$$\xi = (w - \overline{W})/\sigma_w \tag{6.11}$$

The variables ψ and ξ are, based on the above assumption, normally distributed with the expected values $\mu\psi = \mu\xi = 0$, and the standard deviations $\sigma_\psi = \sigma_\xi = 1$.

With the above transformations, the joint probability density function can be written as:

$$f(\psi,\xi) = \frac{1}{2\pi\sqrt{(1-\rho^2)}} \exp\left\{\frac{1}{2(1-\rho^2)}\left[\psi^2 - 2\rho\psi\xi + \xi^2\right]\right\} \tag{6.12}$$

The values of the correlation coefficient, ρ, should be replaced by R, which can be calculated from the observed data using the standardized series ψ and ξ. With this parameter, and after simplifying the notation in Eq. (6.12), the following relation is obtained:

$$\iint_A f(\psi,\ \xi)\mathrm{d}\psi\mathrm{d}\xi = 1 - \exp\left\{-\frac{\lambda^2}{2(1-\rho^2)}\right\} \tag{6.13}$$

Namely, the integral provided by Eq. (6.13) over an area A, i.e. the integral over the space $\psi, \xi \in A$, represents the probability that the realization of events $\psi = h$ and $\xi = k$ will fall into the area A, which is contoured by an ellipse indicated by the following equation

$$\psi^2 - 2\rho\psi\xi + \xi^2 = \lambda^2 \tag{6.14}$$

The newly introduced symbol λ, is obviously related to the constant value of the integral (6.13). Consequently, it is related to the variables ψ and ξ, as well as to the correlation coefficient.

Hence, for each value of $\lambda = \text{const}$, the probability contained within the ellipse of Eq. (6.14) can be calculated.

Equating the variable part of the exponent in Eq. (6.12) to the exponent in Eq. (6.13), the following relation is obtained:

$$\psi^2 - 2\rho\psi\xi + \xi^2 = \lambda^2 \tag{6.15}$$

$$\xi^2 - 2\rho\psi\xi + (\psi^2 - \lambda^2) = 0 \tag{6.16}$$

As previously stated, any value $\lambda = \text{const.}$ corresponds to an ellipse. Furthermore, any given value, $\psi = h$, intersects the ellipse at two different values of ξ, say, $\xi = k_1$ and $\xi = k_2$.

Hence, solving the quadratic equation (6.16) for any specific value of $\lambda = \text{const}$ corresponding to the required level of probability given by Eq. (6.13), one obtains two coordinates (values of ξ, say $\xi = k_1$ and $\xi = k_2$) representing the intersection of the ellipse and the straight line $\psi = h_0$. Repeating the calculation for several selected values of λ while changing the values of $\psi = h_0$, a series of ellipses can be constructed. It should be noted that after each calculation, an appropriate transformation should be performed to correspond with Eqs. (6.10) and (6.11), in order to obtain non-standardized, natural values of the flood characteristics instead of standardized logarithmic values.

The outlined computational scheme is rather direct. However, the results are not of much use, except to provide the analyst with general insight into the relation of the considered flood characteristics.

When it comes to the evaluation of the cumulative distribution function, the direct method, as previously outlined, is not convenient. To avoid computational difficulties, the Abramowitz & Stegun procedure was implemented in this study. This computational scheme uses a grapho-analytical procedure that defines the cumulative probability, $\Phi(h,k,\rho)$, in terms of the probability $\Phi(h,0,r)$ and $\Phi(k,0,r)$, where instead of the correlation coefficient, ρ, the value $r = r(h,k,\rho)$ is used. The value r is related to h and k, as well as to ρ itself. More specifically, the probability $\Phi(h,k,\rho)$ can be determined from:

$$\Phi(h,k,\rho) = \Phi\left(h, 0, \frac{(\rho h - k)\cdot \operatorname{sgn} h}{\sqrt{h^2 - 2\rho hk + k^2}}\right) + \Phi\left(k, 0, \frac{(\rho k - h)\cdot \operatorname{sgn} k}{\sqrt{h^2 - 2\rho hk + k^2}}\right) - \begin{cases} 0 \text{ if } hk \geq 0 \text{ and } h+k \geq 0 \\ \frac{1}{2} \text{ for all other cases} \end{cases} \tag{6.17}$$

Where (sgn h) and (sgn k) are equal to 1 when h or k, respectively, are greater than or equal to zero, while they become –1 whenever h and k are smaller than zero.

The procedure, through which the required probabilities are obtained, consists of the following steps:

1. Take the logarithm of the sample data using Eq. (6.10);
2. Calculate the set of parameters (average values, $\overline{U}$and $\overline{W}$, variances, σ_u^2 and σ_w^2, and correlation coefficient R) which define the probability density function, using the well known procedure not discussed here;
3. Standardize the U and W variables using Eq. (6.11) to obtain the standardized normal variables, ψ and ξ;
4. Construct a grid in the coordinate system to encompass all interesting values of X and Y, for which probabilities should be calculated;
5. Choose specific values $\psi = h_0 = g(x)$ and $\xi = k_0 = g(y)$ for which the exceedance probability will be calculated;
6. Using the appropriate charts from Abramowitz and Stegun (1972), calculate the probability $\Phi(h_0, k_0)$ using Eq. (6.17);
7. Plot the calculated probability into the x–y coordinate system;
8. Repeat steps 5–7 until a sufficient number of points are obtained and plotted;
9. Make isolines of equal exceedance probability (i.e., the lines showing the probability that the indicated X and Y values will simultaneously be exceeded).

On the basis of the above calculations, several lines of equal (rounded-values) exceedance probability can be drawn in the X–Y coordinate system.

It should be emphasized once again that this procedure requires that the variables X and Y be transformed logarithmically into standard normal variables. Therefore, the specific values of h_0 and k_0, for which exceedance probability is calculated, must consequently be converted into natural values; namely:

$$x = 10^u = 10^{(\sigma_u \psi + \overline{U})} = 10^{(\sigma_u h_0 + \overline{U})}$$

$$y = 10^w = 10^{(\sigma_w \xi + \overline{W})} = 10^{(\sigma_w k_0 + \overline{W})} \tag{6.18}$$

Any value obtained based on the above model represents the probability that a flood event, which corresponds to the specific magnitudes x_0 and y_0, will exceed a chosen combination of X and Y.

A computer program, based on the described procedure and utilizing the charts presented in Abramowitz and Stegun (1972), has been developed to perform the above calculations related to the two-dimensional distribution function.

As shown by the above equations, the model uses the correlation coefficient as a measure of the dependence of the flood events in question. In order to evaluate the strength of that relationship, the error of the computed correlation coefficients should be estimated. The following relation was used for this purpose:

$$\sigma_R = \frac{1 - R^2}{\sqrt{N}} \tag{6.19}$$

Where the symbols denote:

σ_R – error of the correlation coefficient R;
N – total number of data points.

In this study the following criterion was adopted: the correlation coefficient, R, is significantly different from zero if its absolute value is greater than the triple value of the error, σ_R; that is:

$$|R| \geq 3 \cdot \sigma_R \tag{6.20}$$

6.2.3 Selection of Variable Combinations Required for Coincidence Assessment

The analysis of the flood wave coincidence on the main river and its tributary is based on a bi-variate statistical distribution of the following variable combinations:

(a) Maximum annual flood on the main river, X, and maximum annual flood on the tributary, Y;
(b) Maximum annual flood on the main river, X, and the corresponding (simultaneous) flood on the tributary, Ycor;
(c) Maximum annual flood on the tributary, Y, and the corresponding (simultaneous) flood on the main river, Xcor.

It should be noted that in the three preceding paragraphs the term flood is used to denote various flood wave characteristics, such as: flood peak discharge, flood wave volume, flood wave duration, and time lag between the occurrences of two flood waves.

In a similar manner, instead of considering the main river and its tributary, the flood characteristics at two profiles of the main river (upstream and downstream from the tributary) can be investigated.

The lines of equal probabilities (bi-variate joint probability density function), as well as the lines which define the exceedance probability (cumulative distribution function) of the mentioned variables/characteristics of floods are the result of a coincidence calculation of the variable combinations in question. The computation for the cumulative exceedance probabilities is described by the following equations:

$$P\left[X > X_1; Y > Y_1\right] = \int_{X_1}^{\infty}\int_{Y_1}^{\infty} g(X, Y, R)\mathrm{d}x\mathrm{d}y$$

$$P\left[X > X_1; Y_{\text{cor}} > Y_1\right] = \int_{X_1}^{\infty}\int_{Y_1}^{\infty} g(X, Y_{\text{cor}}, R)\mathrm{d}x\mathrm{d}y_{\text{cor}}$$
$$P\left[X_{\text{cor}} > X_1; Y > Y_1\right] = \int_{X_1}^{\infty}\int_{Y_1}^{\infty} g(X_{\text{cor}}, Y, R)\mathrm{d}x_{\text{cor}}\mathrm{d}y \tag{6.21}$$

Flood events are considered to be flood waves in which the peak flow exceeds a pre-defined magnitude. That magnitude can be taken from the flow duration curve or selected in some other way. As mentioned, the following characteristics of floods can be analyzed:

1. Extreme value (flood peak) – Qmax;
2. Flood hydrograph volume above a pre-defined flow – W;
3. Duration of flood wave above a pre-defined flow – T;
4. Time lag between the occurrences of flood events – τmax.

Depending on the overall hydrological regime, a pre-defined value can be exceeded more than once during a single calendar year. This implies that the number of previously defined events changes from year to year. Hence, the annual frequency of floods appears as an important flood characteristic to be analyzed in due course. In addition, the time of year when the flood occurs is also of interest.

When analyzing the flood coincidence of the main stream and its tributary, or of the main river upstream and downstream from its tributary, attention should be paid to flood wave genesis. The following characteristics are important in this regard: snowmelt, heavy rainfall, concentration time, etc.

In flood coincidence analyses, it is necessary to consider gauging stations immediately downstream and upstream from the tributary. The most interesting characteristics generally included in coincidence analyses are described in Figs. 6.1, 6.2, 6.3 and 6.4. The ensuing text offers a rather extensive list of flood variables to be discussed.

I. Discharge coincidence (Fig. 6.1)

- QINmax maximum annual discharge on the main river upstream from the tributary,
- QOUTmax maximum annual discharge on the main river downstream from the tributary,
- qTRmax maximum annual discharge on the tributary,
- QINcor1 corresponding discharge on the main river, upstream from the tributary, at the time of occurrence of the maximum annual discharge on the main river downstream from the tributary,
- QOUTcor1 corresponding discharge on the main river, downstream from the tributary, at the time of occurrence of the maximum annual discharge on the main river, upstream from the tributary,

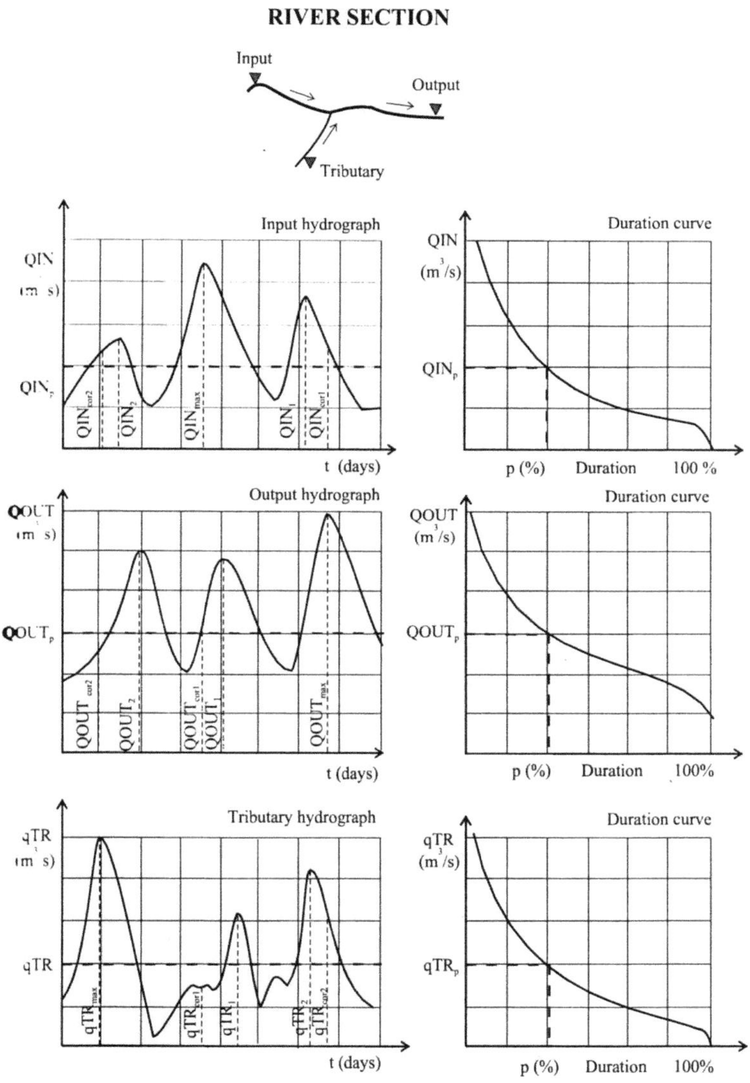

Fig. 6.1 Characteristic parameters of flood wave hydrographs – discharges

*q*TRcor1 corresponding discharge on the tributary at the time of occurrence of the maximum annual discharge on the main river downstream from the tributary,

*Q*INcor2 corresponding discharge on the main river, upstream from the tributary, at the time of occurrence of the maximum annual discharge on the tributary,

*Q*OUTcor2 corresponding discharge on the main river, downstream from the tributary, at the time of occurrence of the maximum annual discharge on the tributary,

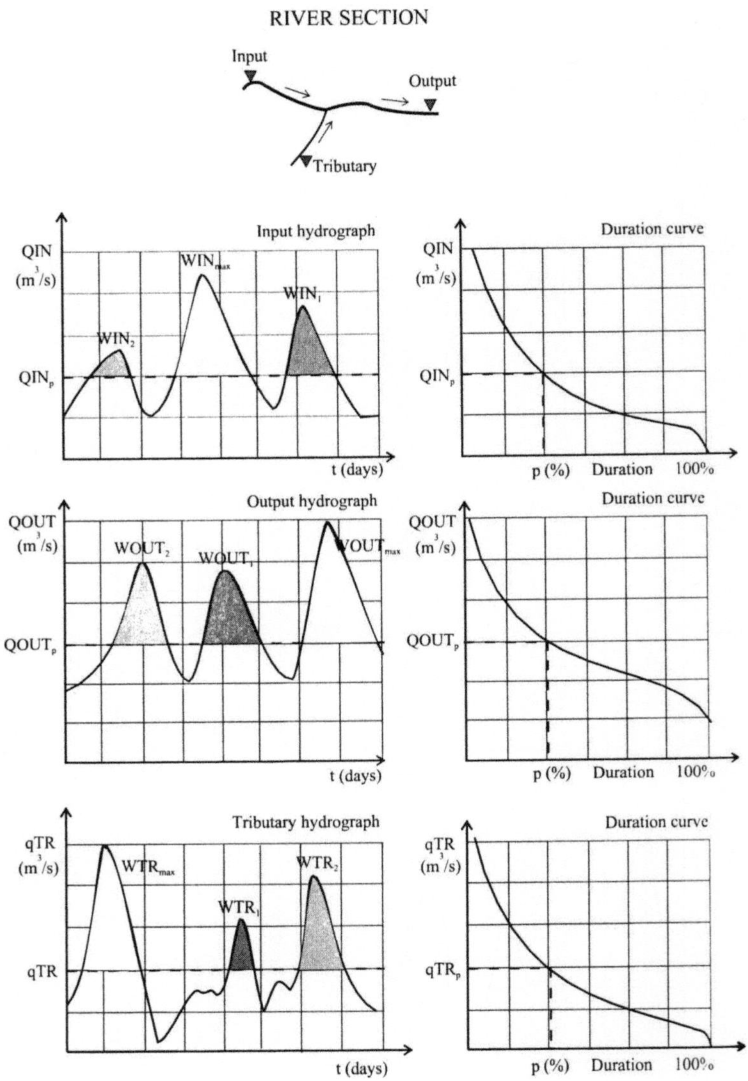

Fig. 6.2 Characteristic parameters of flood wave hydrographs – volume

*q*TRcor2 corresponding discharge on the tributary at the time of occurrence of the maximum annual discharge on the main river upstream from the tributary,

*Q*IN1 maximum discharge on the main river, upstream from the tributary, at the time of occurrence of the maximum annual discharge on the main river, downstream from the tributary,

*Q*OUT1 maximum discharge on the main river, downstream from the tributary, at the time of occurrence of the maximum annual discharge on the main river, upstream from the tributary,

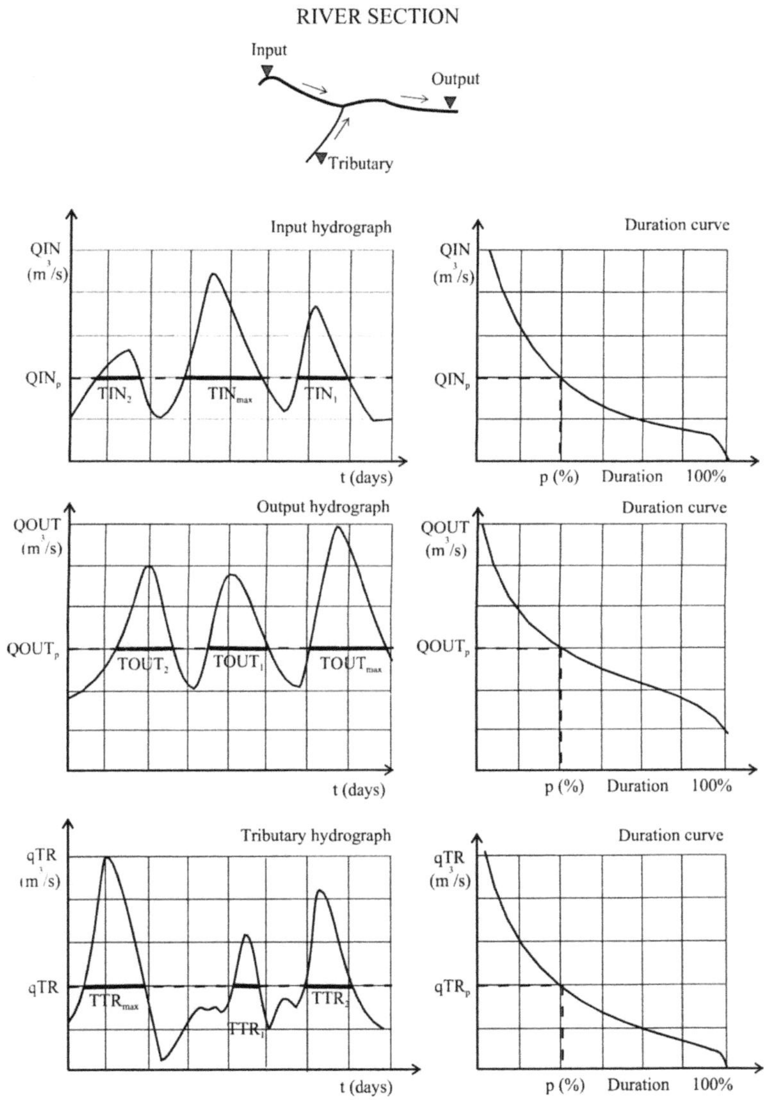

Fig. 6.3 Characteristic parameters of flood wave hydrographs – duration

- *q*TR1 maximum discharge on the tributary at the time of occurrence of the maximum annual discharge on the main river upstream from the tributary,
- *Q*IN2 maximum discharge on the main river, upstream from the tributary, at the time of occurrence of the maximum annual discharge on the tributary,
- *Q*OUT2 maximum discharge on the main river, downstream from the tributary, at the time of occurrence of the maximum annual discharge on the tributary,

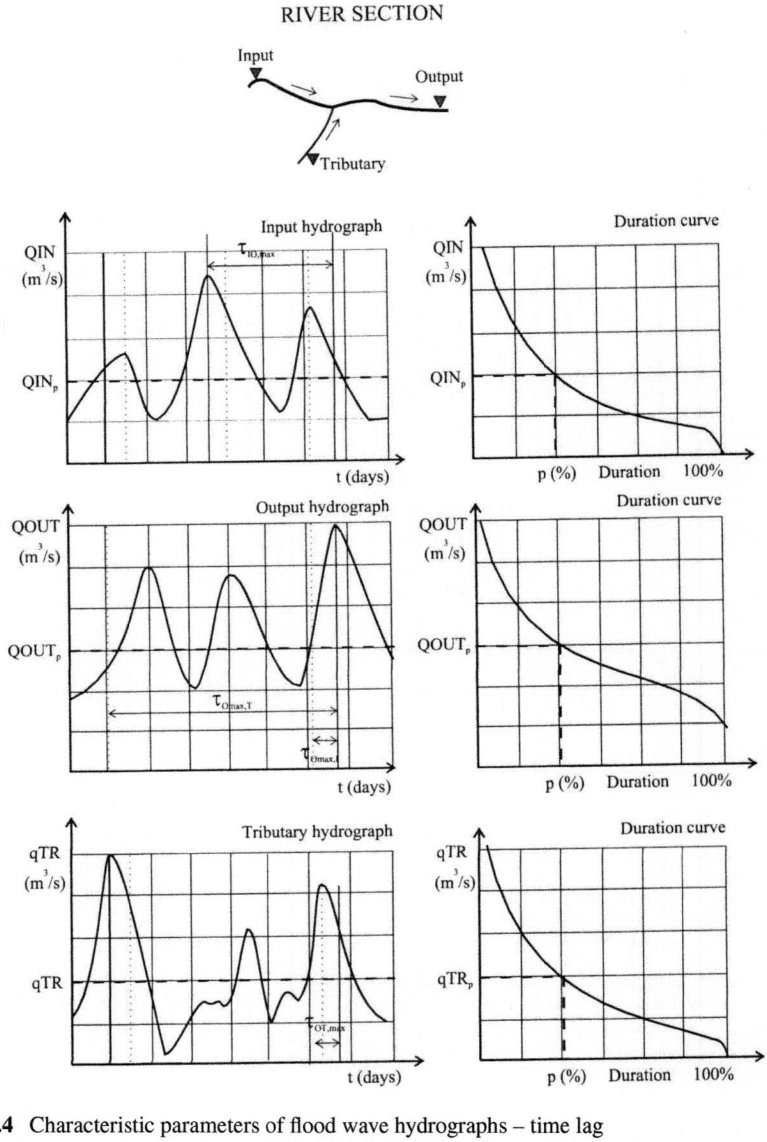

Fig. 6.4 Characteristic parameters of flood wave hydrographs – time lag

*q*TR2 maximum discharge on the tributary at the time of occurrence of the maximum annual discharge on the main river downstream from the tributary.

II. Volume coincidence (Fig. 6.2)

*W*INmax maximum annual volume of the main river upstream from the tributary, above the pre-defined flow *Q*IN

*W*OUTmax maximum annual volume of the main river downstream from the tributary, above the pre-defined flow *Q*OUT
*W*TRmax maximum annual volume of the tributary, above the pre-defined flow *q*TR
*W*IN1 volume of the main river above the pre-defined flow *Q*IN upstream from the tributary, at the time of occurrence of the maximum annual discharge on the main river downstream from the tributary,
*W*OUT1 volume of the main river above the pre-defined flow *Q*OUT downstream from the tributary, at the time of occurrence of the maximum annual discharge on the main river upstream from the tributary,
*W*TR1 volume of the tributary above the pre-defined flow *q*TR, at the time of occurrence of the maximum annual discharge on the main river upstream from the tributary,
*W*IN2 volume of the main river above the pre-defined flow *Q*IN upstream from the tributary, at the time of occurrence of the maximum annual discharge on the tributary,
*W*OUT2 volume of the main river above the pre-defined flow *Q*OUT downstream from the tributary, at the time of occurrence of the maximum annual discharge on the tributary,
*W*TR2 volume of the tributary above the pre-defined flow *q*TR, at the time of occurrence of the maximum annual discharge on the main river downstream from the tributary.

III. Duration coincidence (Fig. 6.3)

*T*INmax maximum annual duration of the flood wave above the pre-defined flow *Q*IN on the main river upstream from the tributary,
*T*OUTmax maximum annual duration of the flood wave above the pre-defined flow *Q*OUT on the main river downstream from the tributary,
*T*TRmax maximum annual duration of the flood wave above the pre-defined flow *q*TR on the tributary,
*T*IN1 duration of flood wave on the main river upstream from the tributary, above the pre-defined flow *Q*IN, at the time of occurrence of the maximum annual discharge on the main river downstream,
*T*OUT1 duration of the flood wave above the pre-defined flow *Q*OUT on the main river downstream from the tributary, at the time of occurrence of the maximum annual discharge on the main river upstream from the tributary,
*T*TR1 duration of flood wave above the pre-defined flow *q*TR on the tributary, at the time of occurrence of the maximum annual discharge on the main river upstream from the tributary,

*T*IN2 duration of flood wave above the pre-defined flow *Q*IN on the main river upstream from the tributary, at the time of occurrence of the maximum annual discharge on the tributary,

*T*OUT2 duration of flood wave above the pre-defined flow *Q*OUT on the main river downstream the tributary, at the time of occurrence of the maximum annual discharge on the tributary,

*T*TR2 duration of flood wave above the pre-defined flow *q*TR on the tributary, at the time of occurrence of the maximum annual discharge on the main river downstream from the tributary.

IV. Time lag coincidence (Fig. 6.4)

τIO,max time lag between the occurrence of the maximum annual discharge on the main river upstream and downstream from the tributary,

τO*T*,max time lag between the occurrence of the maximum annual discharge on the main river downstream from the tributary and the maximum discharge on the tributary,

τOmax,*I* time lag between the occurrence of the maximum annual discharge on the main river downstream from the tributary and the maximum discharge of the same wave on the main river upstream from the tributary,

τOmax,*T* time lag between the occurrence of the maximum annual discharge on the main river downstream from the tributary and the maximum discharge of the same wave on the tributary.

6.3 Application of the Proposed Method

6.3.1 Combination of Events

In analyzing the coincidence of the flood flow on the main river and its tributary, the most interesting combination of random variables is:

I. Discharge coincidence

(a) Simultaneous occurrence

(I.a.1) Maximum annual discharge on the main river upstream from the tributary – corresponding discharge on the main river downstream from the tributary.

– (*Q*INmax; *Q*OUTcor1)

(I.a.2) Maximum annual discharge on the main river upstream from the tributary – corresponding discharge on the tributary.

– (*Q*INmax; *q*TRcor1)

(I.a.3) Maximum annual discharge on the main river downstream from the tributary – corresponding discharge on the main river upstream from the tributary.

– (*Q*OUTmax; *Q*INcor1)

(I.a.4) Maximum annual discharge on the main river downstream from the tributary – corresponding discharge on the tributary.

– (*Q*OUTmax; *q*TRcor2)

(I.a.5) Maximum annual discharge on the tributary – corresponding discharge on the main river downstream from the tributary.

– (*q*TRmax; *Q*OUTcor2)

(I.a.6) Maximum annual discharge on the tributary – corresponding discharge on the main river upstream from the tributary.

– (*q*TRmax; *Q*INcor2)

(b) Flood waves of the same genesis

(I.b.1) Maximum annual discharge on the main river upstream from the tributary – maximum discharge of the flood wave of the same genesis on the main river downstream from the tributary.

– (*Q*INmax; *Q*OUT1)

(I.b.2) Maximum annual discharge on the main river upstream from the tributary – maximum discharge of the flood wave of the same genesis on the tributary.

– (*Q*INmax; *q*TR1)

(I.b.3) Maximum annual discharge on the main river downstream from the tributary – maximum discharge of the flood wave of the same genesis on the main river upstream from the tributary.

– (*Q*OUTmax; *Q*IN1)

(I.b.4) Maximum annual discharge on the main river downstream from the tributary – maximum discharge of the flood wave of the same genesis on the tributary.

– (*Q*OUTmax; *q*TR2)

(I.b.5) Maximum annual discharge on the tributary – maximum discharge of the flood wave of the same genesis on the main river upstream from the tributary.

– (*q*TRmax; *Q*IN1)

(I.b.6) Maximum annual discharge on the tributary – maximum discharge of the flood wave of the same genesis on the main river downstream from the tributary.

– (*q*TRmax; *Q*OUT1)

(c) Extreme annual values

(I.c.1) Maximum annual discharge on the main river upstream from the tributary – maximum annual discharge on the main river downstream from the tributary.

– (*Q*INmax; *Q*OUTmax)

(I.c.2) Maximum annual discharge on the main river upstream from the tributary – maximum annual discharge on the tributary.

– (*Q*INmax; *q*TRmax)

(I.c.3) Maximum annual discharge on the main river downstream from the tributary – maximum annual discharge on the tributary.

– (*Q*OUTmax; *q*TRmax)

(II) Volume coincidence

(a) Simultaneous occurrence

(II.a.1) Maximum annual volume on the main river upstream from the tributary – corresponding volume on the main river downstream from the tributary.

– (*W*INmax; *W*OUT)

(II.a.2) Maximum annual volume on the main river upstream from the tributary – corresponding volume on the tributary.

– (*W*INmax; *W*TR)

(II.a.3) Maximum annual volume on the main river downstream from the tributary – corresponding volume on the main river upstream from the tributary.

– (*W*OUTmax; *W*IN)

(II.a.4) Maximum annual volume on the main river downstream from the tributary – corresponding volume on the tributary.

– (*W*OUTmax; *W*TR)

(II.a.5) Maximum annual volume on the tributary – corresponding volume on the main river upstream from the tributary.

– (*W*TRmax; *W*OUT)

(II.a.6) Maximum annual volume on the tributary – corresponding volume on the main river downstream from the tributary.

– (*W*TRmax; *W*IN)

(b) Extreme annual values

(II.b.1) Maximum annual volume on the main river upstream from the tributary – maximum annual volume on the main river downstream from the tributary.

– (*W*INmax; *W*OUTmax)

(II.b.2) Maximum annual volume on the main river upstream from the tributary – maximum annual volume on the tributary.

– (*W*INmax; *W*TRmax)

(II.b.3) Maximum annual volume on the main river downstream from the tributary – maximum annual volume on the tributary.

– (*W*OUTmax; *W*TRmax)

III. Duration of flood waves

(a) Simultaneous occurrence

(III.a.1) Maximum annual flood duration on the main river upstream from the tributary – corresponding flood duration on the main river downstream from the tributary.

– (*T*INmax; *T*OUT)

(III.a.2) Maximum annual flood duration on the main river upstream from the tributary – corresponding flood duration on the tributary.

– (*T*INmax; *T*TR)

(III.a.3) Maximum annual flood duration on the main river downstream from the tributary – corresponding flood duration on the main river upstream from the tributary.

– (*T*OUTmax: *T*IN)

(III.a.4) Maximum annual flood duration on the main river downstream from the tributary – corresponding flood duration on the tributary.

– (*T*OUTmax; *T*TR)

(III.a.5) Maximum annual flood duration on the tributary – corresponding flood duration on the main river upstream from the tributary.

– (*T*TRmax; *T*IN)

(III.a.6) Maximum annual flood duration on the tributary – corresponding flood duration on the main river downstream from the tributary.

– (*T*TRmax; *T*OUT)

(b) Extreme annual values

(III.b.1) Maximum annual flood duration on the main river upstream from the tributary – maximum annual flood duration on the main river downstream from the tributary.

– (*T*INmax; *T*OUTmax)

(III.b.2) Maximum annual flood duration on the main river upstream from the tributary – maximum annual flood duration on the tributary.

– (*T*INmax; *T*TRmax)

(III.b.3) Maximum annual flood duration on the main river downstream from the tributary – maximum annual flood duration on the tributary.

– (*T*OUTmax; *T*TRmax)

IV. Time lag coincidence

(IV.1) Maximum annual discharge on the main river downstream from the tributary – time between the occurrence of maximum annual discharge on the main river upstream and downstream from the tributary.

– (*Q*OUTmax; τIO,max)

(IV.2) Maximum annual discharge on the main river downstream from the tributary – time between the occurrence of maximum annual discharge on the main river downstream of the tributary and the maximum annual discharge on the tributary.

– (*Q*OUTmax; τOT,max)

(IV.3) Maximum annual discharge on the main river downstream from the tributary – time between the occurrence of maximum annual discharge on the main river downstream and the maximum discharge of the same wave on the main river upstream from the tributary.

– (*Q*OUTmax; τOmax,*T*)

(IV.4) Maximum annual discharge on the main river downstream from the tributary – time between the occurrence of maximum annual discharge on the main river downstream and the maximum discharge of the same wave on the tributary.

– (*Q*OUTmax; τOmax,*T*)

6.3.2 Scope of Application

The most interesting points at which flood coincidence should be assessed along the Danube River are those where several significant tributaries merge with the main river. The following nine tributaries were chosen: the Inn, Enns, Morava, Drava, Tisa, Sava, Velika Morava, Siret and Prut.

The analyses were applied to those sections of the Danube River that are bound by adjacent gauging stations where long-term monitoring data are available. In other words, the sections consist of the river reach from the nearest upstream gauge to the first downstream gauging station, as well as to the first gauging station at each tributary. On the basis of data availability, the following sections of the Danube River were studied:

1. The Danube River (Hofkirchen–Achleiten) – the Inn River (Passau-Ingling)
2. The Danube River (Linz–Krems) – the Enns River (Steyr)
3. The Danube River (Vienna–Bratislava) – the Morava River (Moravsky Jan)
4. The Danube River (Bezdan–Bogojevo) – the Drava River (Donji Miholjac)
5. The Danube River (Bogojevo–Slankamen) – the Tisa River (Senta)
6. The Danube River (Slankamen–Pancevo) – the Sava River (Sremska Mitrovica)
7. The Danube River (Pancevo–Veliko Gradiste – the Velika) Morava River (Ljubicevski Most)
8. The Danube River (Vadu Oii–Ceatal Izmail) – the Siret River (Lungoci)
9. The Danube River (Vadu Oii–Ceatal Izmail) – the Prut River (Tchernovtsy)

6.3.3 Available Data

With regard to selecting representative river reaches, as stated in Section 6.3.2, for analyses of flood flow coincidence on the Danube River and its tributary, hydrological observations are needed from the following gauging stations:

– The Danube River:
at Hofkirchen
at Achleiten
at Linz
at Stein-Krems
at Vienna
at Bratislava
at Bezdan
at Bogojevo
at Slankamen
at Pancevo
at Veliko Gradište
at Vadu Oii
at Ceatal Izmail

– The Inn River:	at Passau-Ingling
– The Enns River:	at Steyr
– The Morava River:	at Moravsky Jan
– The Drava River:	at Donji Miholjac
– The Tisa River:	at Senta
– The Sava River:	at Sremska Mitrovica
– The V. Morava River:	at Ljubicevski Most
– The Sirit River:	at Lungoci
– The Prut River:	at Tchernovtsy.

Records of average daily flow, as well as daily extreme values, were obtained from all 13 locations on the Danube River and nine gauges on the tributaries. The data were supplied by the national committees of the following countries: Germany, Austria, Slovakia, Yugoslavia, Romania and Ukraine. The lengths of available time series were different, and mostly covered the 1901–1993 period.

The lengths of available flow data can be seen in the presented illustrations. In addition, a general pattern of long-term flow changes can be observed. On the basis of a visual assessment, it can be concluded that there have been no significant changes. All changes are reasonable and can be attributed to the stochastic nature of the examined process.

6.3.4 Preliminary Data Analyses

Basic Characteristics of the Time Series. The procedure followed in selecting and defining random variables for analyses of coincidence of flood flow events on the main stream and its tributaries was explained in Sections 6.2.3 and 6.3.1. The described scheme has been consequently implemented on the Danube River. The criteria used for defining the hydrograph base, or the threshold for both the main river and its tributary, should correspond to a 50% duration discharge ($Q_{50\%}$), from the average annual duration curve.

The relevant data (drainage area, observed discharge – average, minimum and maximum, as well as 50% duration discharge) at all gauging stations in question are presented in Table 6.1.

To assemble the relevant series needed to study various input–output combinations of variables, the available data collected at gauging stations during the selected time period were used. Actually, only the same periods were used, whenever data for all locations in question were available.

The series (consisting of maximum discharge, flood wave volume, flood wave duration and lag-time of maximum discharge occurrence), used to analyze the identified river reaches of the Danube River are depicted in Chapter 5.

Time Series Homogeneity and Trends. The previously described theoretical approach to analyses of flood flow coincidence can be used only when the series are homogenous and when there are no significant trends. For this reason, the properties of these time series had to be investigated.

Table 6.1 Basic hydrological characteristics at selected gauging stations in the Danube River Basin

No	Gauging station	River	Drainage area A (km^2)	Realized discharge (m^3 s^{-1})			
				Q_{av}	Q_{min}	Q_{max}	$Q_{50\%}$
1	Hofkirchen	Danube	47,496	645	165	3,830	568
2	Achleiten	Danube	76,597	1,430	349	9,100	1,302
3	Linz	Danube	79,490	1,509	412	8,800	1,320
4	Stein-Krems	Danube	96,045	1,864	508	10,200	1,711
5	Wien	Danube	101,700	1,943	497	9,600	1,715
6	Bratislava	Danube	131,338	2,020	582	9,998	1,843
7	Bezdan	Danube	210,250	2,479	505	7,689	2,162
8	Bogojevo	Danube	251,593	3,060	680	8,601	2,714
9	Slankamen	Danube		3,694	883	10,076	3,455
10	Pancevo	Danube	525,009	5,490	1,273	13,378	4,954
11	V. Gradiste	Danube	570,375	5,746	1,303	14,760	5,169
12	Vadu Oii	Danube	709,100	6,216	1,540	14,950	5,883
13	Ceatal Izmail	Danube	807,000	6,550	1,790	15,540	6,214
14	Passau-Ingling	Inn		741	195	6,700	636
15	Steyr	Enns		205	22.2	2,560	166
16	Moravsky Jan	Morava	24,129	110	7.7	1,573	81
17	Donji Miholjac	Drava	37,142	554	123	2,281	487
18	Senta	Tisa	141,715	766	79	3,730	577
19	S. Mitrovica	Sava	87,996	1,613	194	6,638	1,361
20	Ljubicevski Most	V. Morava	37,320	238	23.6	2,355	157
21	Lungoci	Siret	36,036	172	15.6	2,825	152
22	Tchernovtsy	Prut	6,890	71.5	1.48	3,880	85

Series homogeneity and trend analyses were performed only for maximum annual discharge series at all gauging stations along the Danube River. As such, the basic series with the longest period of observation were used.

An analysis of the homogeneity of a mean value and dispersion was performed in such a way that the available series was split into two subseries of approximately equal length. Student's test (T-test) and the normalized test (Z-test) were used to assess homogeneity, while Fisher's test (F-test) was applied to examine the dispersion. The results of homogeneity testing, for confidence levels $1 - \alpha = 0.95$, are shown in Table 6.2. On the basis of the results, it can be inferred that the analyzed series of maximum annual discharge levels are homogeneous, since no statistically significant changes in average values and dispersion were detected.

The presence of trends in mean values was tested using Mann's test and the linear regression model test. The required significance level was $1 - \alpha = 0.95$. The results of the computation are shown in Table 6.3 for the linear regression model test. It is obvious that Mann's statistics, throughout the investigated region, fall between the critical values, which implies that no significant changes in the average values are present. Similar results were obtained using the linear regression test. An exception was found at the Bogojevo gauging station, where the alternative hypothesis, Ha,

Table 6.2 Tests of homogeneity of average values and dispersion of maximum annual discharge series

			Statistics						Test of mean value								Test of dispersion		
			ISeries			*IISeries*			Student's *T*-test				Normalized *Z*-test						
No	Station	River	*Q*sr1	*S*1	*n*1	*Q*sr2	*S*2	*n*2	σt	*t*	$T\alpha$		σz	*z*	z1−α		*F*	$F\alpha$	
1	Hofkirchen	Danube	1,860.29	508.83	45	1,882.44	590.24	45	255.9	−0.087	1.991	H_0	116.2	−0.191	1.96	H_0	1.35	1.69	H_0
2	Achleiten	Danube	4,092.67	1,011.19	45	4,200.67	1,194.35	45	513.8	−0.210	1.991	H_0	233.3	−0.463	1.96	H_0	1.40	1.69	H_0
3	Linz	Danube	4,521.52	1,408.97	21	4,044.18	1,041.03	22	698.3	0.684	2.020	H_0	379.2	1.259	1.96	H_0	1.83	2.10	H_0
4	Steir-Krems	Danube	5,912.00	1,539.79	20	5,768.60	1,492.08	20	874.7	0.164	2.023	H_0	479.4	0.299	1.96	H_0	1.06	2.16	H_0
5	Wien	Danube	5,711.62	1,474.11	21	5,636.00	1,304.50	22	786.2	0.096	2.020	H_0	425.2	0.178	1.96	H_0	1.28	2.10	H_0
6	Bratislava	Danube	5,602.23	1,389.70	30	5,539.00	1,490.15	30	744.6	0.085	2.002	H_0	372.0	0.170	1.96	H_0	1.15	1.84	H_0
7	Bezdan	Danube	4,957.87	811.04	30	4,599.39	1,040.09	31	480.9	0.745	2.001	H_0	238.4	1.504	1.96	H_0	1.64	1.84	H_0
8	Bogojevo	Danube	5,776.55	1,036.01	31	5,266.56	1,191.58	32	570.2	0.894	2.000	H_0	281.1	1.815	1.96	H_0	1.32	1.84	H_0
9	Slankamen	Danube	6,964.87	1,231.33	31	6,762.84	1,371.21	31	667.6	0.303	2.000	H_0	331.0	0.610	1.96	H_0	1.24	1.84	H_0
10	Pancevo	Danube	9,897.61	1,707.45	31	9,845.69	1,800.40	32	895.5	0.058	2.000	H_0	442.0	0.117	1.96	H_0	1.11	1.84	H_0
11	Veliko Gradste	Danube	1,0468.71	1,965.62	31	1,0512.47	1,856.70	32	975.0	−0.045	2.000	H_0	482.0	−0.091	1.96	H_0	1.12	1.84	H_0
12	Vadu Oii	Danube	1,0679.66	2,143.06	29	11,094.67	1,900.08	30	1,050.4	−0.395	2.003	H_0	527.9	−0.786	1.96	H_0	1.27	1.84	H_0
13	Ceatal Izmail	Danube	10,505.86	2,004.08	29	11,357.67	1,885.78	30	1,009.7	−0.844	2.003	H_0	507.0	−1.680	1.96	H_0	1.13	1.84	H_0
14	Passau-Ingling	Inn	2,932.86	953.03	35	2,940.29	846.73	35	447.2	0.017	1.997	H_0	215.5	−0.034	1.96	H_0	1.27	1.78	H_0
15	Steyr	Enns	1,376.41	496.24	17	1,302.06	514.88	18	303.0	0.245	2.036	H_0	170.9	0.435	1.96	H_0	1.08	2.31	H_0
16	Moravsky Jan	Morava	618.67	331.29	30	522.13	220.03	30	145.3	0.664	2.002	H_0	72.6	1.329	1.96	H_0	2.27	1.84	*Ha*
17	Donji Miholjac	Drava	1,333.66	270.53	29	1,360.57	377.89	30	171.1	−0.157	2.003	H_0	85.3	−0.315	1.96	H_0	1.95	1.84	*Ha*
18	Senta	Tisa	2,041.61	640.11	31	2,141.19	470.14	32	285.8	−0.348	2.000	H_0	141.9	−0.702	1.96	H_0	1.85	1.84	*Ha*
19	S. Mitrovica	Sava	4,171.32	680.54	34	4,050.94	835.36	34	380.9	0.316	1.998	H_0	184.8	0.651	1.96	H_0	1.51	1.80	H_0
20	Lj Most	V. Morava	1,281.03	443.37	30	1,187.73	462.69	31	233.3	0.400	2.001	H_0	116.0	0.804	1.96	H_0	1.09	1.84	H_0
21	Lungoci	Siret	1,063.45	641.49	20	1,374.95	676.24	20	380.3	−0.819	2.023	H_0	208.4	−1.495	1.96	H_0	1.11	2.16	H_0
22	Tchernovtsy	Prut	951.50	790.16	20	1,386.60	868.02	20	478.9	−0.909	2.023	H_0	262.5	−1.658	1.96	H_0	1.21	2.16	H_0

Table 6.3 Testing of trends in average values of maximum annual discharge along the Danube River

No	Station	River	n	Parameters b	σb	$t = b/\sigma b$	$T\alpha$	Accepted hypothesis
1	Hofkirchen	Danube	90	0.7360	2.235	0.3293	1.991	H_0
2	Achleiten	Danube	90	0.8930	4.494	0.1987	1.991	H_0
3	Linz	Danube	43	−13.8440	15.310	−0.9042	2.020	H_0
4	Steir-Krems	Danube	40	−12.9480	20.443	−0.6334	2.023	H_0
5	Wien	Danube	43	−7.7600	17.043	−0.4553	2.020	H_0
6	Bratislava	Danube	60	5.9700	10.715	0.5572	2.002	H_0
7	Bezdan	Danube	61	−9.8050	6.782	−1.4457	2.001	H_0
8	Bogojevo	Danube	63	−16.3170	7.670	−2.1274	2.000	Ha
9	Slankamen	Danube	62	−10.9700	9.168	−1.1966	2.000	H_0
10	Pancevo	Danube	63	−16.4750	11.979	−1.3753	2.000	H_0
11	Veliko Gradiste	Danube	63	−14.2860	13.114	−1.0894	2.000	H_0
12	Vadu Oii	Danube	59	1.0500	15.210	0.0690	2.003	H_0
13	Ceatal Izmail	Danube	59	15.5230	14.757	1.0519	2.003	H_0
14	Passau-Ingling	Inn	70	0.2120	5.332	0.0398	1.997	H_0
15	Steyr	Enns	35	−1.5870	8.136	−0.1951	2.036	H_0
16	Moravsky Jan	Morava	60	−3.1740	2.087	−1.5208	2.002	H_0
17	Donji Miholjac	Drava	59	−0.0060	2.416	−0.0025	2.003	H_0
18	Senta	Tisa	63	−2.1330	3.888	−0.5486	2.000	H_0
19	S. Mitrovica	Sava	68	−3.0950	4.707	−0.6575	1.998	H_0
20	Lj. Most	V. Morava	61	−3.1670	3.180	−0.9959	2.001	H_0
21	Lungoci	Siret	40	13.9170	9.011	1.5444	2.023	H_0
22	Tchernovtsy	Prut	40	20.5280	10.862	1.8899	2.023	H_0

should be accepted; however, this is negligible considering the number of examined locations/observations.

On the basis of testing of homogeneity and trends in high flows series, it can be concluded that the annual maximum discharge at all examined locations satisfies the conditions required for the implementation of a complex procedure pertaining to an assessment of flood flow coincidence along the Danube River and its tributaries.

Uni-variate Probability Estimation. In order to gain insight into the properties of the analyzed time series (maximum annual discharge, volume and duration of the flood wave, time-lag of flood occurrence), probabilistic analyses of all annual maximum values series were carried out for each considered location (gauging station). The Pearson III and Log-Pearson III probability distribution functions were fitted to the observed data. The goodness of fit was investigated by χ^2 and Kolmogorov tests.

Calculations were performed for the most frequently used probability level of 1% (return period 100 years). Table 6.4 shows results related to the maximum annual

Table 6.4 Characteristic values of maximum flood waves

No	Station	River	Qmax,p (m^3 s–^{1}) Qmax,1% (m^3 s^{-1})	Wmax,1% (10^6 m^3)	Tmax,1% (days)	τmax,1% (days) τ(+)max	τ(–)max
1	Hofkirchen	Danube	3,522	12,357	235	284	–127
2	Achleiten	Danube	7,561	22,112	243	72	–198
3	Linz	Danube	8,347	35,206	238	57	–173
4	Stein-Krems	Danube	10,525	57,662	234	170	–338
5	Wien	Danube	9,402	46,123	232	38	–31
6	Bratislava	Danube	10,324	54,658	228	273	–127
7	Bezdan	Danube	7,487	74,383	296	45	–88
8	Bogojevo	Danube	8,596	73,516	272	142	–227
9	Slankamen	Danube	10,574	86,856	270	83	–205
10	Pancevo	Danube	14,231	87,004	258	32	–150
11	Veliko Gradiste	Danube	15,559	93,150	257	75	–182
12	Vadu Oii	Danube	16,124	137,495	270	126	–387
13	Ceatal Izmail	Danube	15,887	167,651	266	11	–15
14	Passau-Ingling	Inn	5,964	13,283	202	189	–210
15	Steyr	Enns	3,112	8,484	209	153	–115
16	Moravsky Jan	Morava	1,518	4,945	247	258	–188
17	Donji Miholjac	Drava	2,282	14,995	229	153	–115
18	Senta	Tisa	3,787	27,036	259	205	–241
19	S. Mitrovica	Sava	6,597	57,692	171	284	–127
20	Lj. Most	V. Morava	2,564	8,972	229	72	–198
21	Lungoci	Siret	3,462	6,965	253	57	–173
22	Tchernovtsy	Prut	4,372	1,684	108	170	–338

discharge series, the flood wave volume, and the corresponding flood duration (over $Q_{50\%}$ discharge value). Table 6.4 also depicts the results of the time-lag analyses. It should be noted that lag time is measured from the moment of occurrence of a maximum value at the downstream station. Positive values of lag-time indicate that maximum values at the upstream location occur before maximums at the downstream profile. The negative lag-time values indicate the opposite: maximum values at the downstream profile occur before corresponding ones at the upstream station.

6.4 Results

Quantitative indicators of strength of established relations and their graphic dependences were defined using the described methodology for the calculation of flood coincidence of significant (basic) hydrological features of the flood hydrograph on the main river and its tributaries. Because of the large volume of resulting data, only numerical indicators of strength of established relations on all considered sectors will be shown in the following text, related to the coincidence of flood flow of the basic hydrograph characteristics – maximum annual discharges and maximum flood volume. Graphic interpretations of the results of performed calculations will be given only for the coincidence of maximum flows studies for inflowing profiles of the following significant sectors:

1. The Danube River (Hofkirchen) – the Inn River (Passau-Ingling),
2. The Danube River (Vienna) – the Morava River (Moravsky Jan),
3. The Danube River (Slankamen) – the Sava River (Sremska Mitrovica).

Sector 1 is unique because it contains an equal amount of water from the main river (the Danube) and the tributary (the Inn). Sector 2 includes the smallest tributary, while Sector 3 includes the largest tributary in the Danube River Basin.

6.4.1 The Danube River (Hofkirchen–Achleiten) – Inn (Passau-Ingling) Sector

The results of calculations of the parameters related to coincidence of flood flows of the Danube and Inn rivers are presented in Table 6.5 and illustrated in Fig. 6.5.

These results show that all combinations of maximum discharges of the Danube and Inn rivers are statistically significant, with the exception of the combination accounting for maximum discharge of the Danube River at Hofkirchen and the corresponding discharge of the Inn River at Passau-Ingling, where high flow correlation is not statistically significant.

Flood volume results at the examined profiles are shown in Table 6.6. The calculated parameters indicate statistically significant coincidence of flood volumes of the Danube River (realized at Achleiten) and the Inn River (at Passau-Ingling).

Table 6.5 Correlation of the maximum discharge of the Danube and Inn rivers

Stations	Combination of variables	R	N	σ	3σ	Significance
Achleiten–Hofkirchen	max–max	0.779	90	0.041	0.124	+
	max–cor	0.575	90	0.071	0.212	+
	cor–max	0.802	90	0.038	0.113	+
Achleiten–Passau-Ingling	max–max	0.830	70	0.037	0.112	+
	max–cor	0.678	70	0.065	0.194	+
	cor–max	0.756	70	0.051	0.154	+
Hofkirchen–Passau-Ingling	max–max	0.512	70	0.088	0.265	+
	max–cor	0.201	70	0.115	0.344	–
	cor–max	0.475	70	0.093	0.278	+

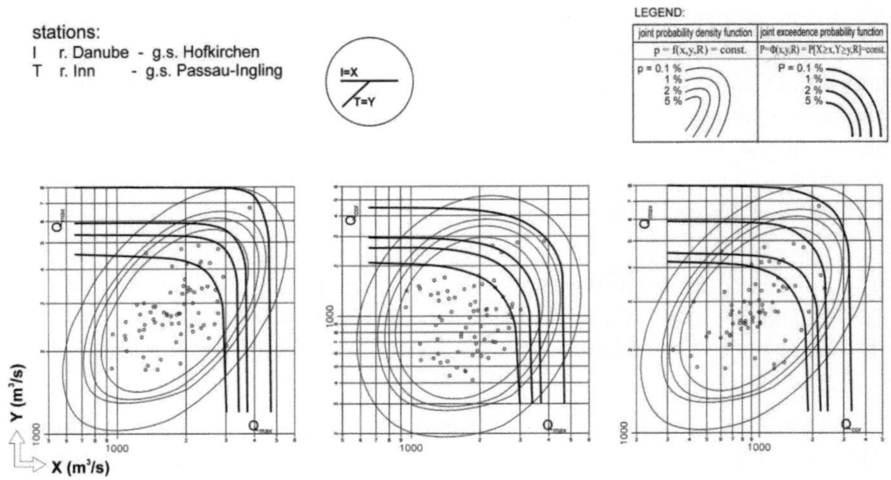

Fig. 6.5 Discharge coincidence of maximum annual flow

Table 6.6 Correlation of flood volume of the Danube and Inn rivers

Stations	Combination of variables	R	N	σ	3σ	Significance
Achleiten–Hofkirchen	max–max	0.554	74	0.081	0.242	+
	max–cor	0.643	72	0.069	0.207	+
	cor–max	0.581	74	0.077	0.231	+
Achleiten–Passau-Ingling	max–max	0.581	69	0.080	0.239	+
	max–cor	0.585	69	0.079	0.238	+
	cor–max	0.601	68	0.078	0.233	+
Hofkirchen–Passau-Ingling	max–max	0.341	60	0.114	0.342	–
	max–cor	0.049	48	0.144	0.432	–
	cor–max	0.305	54	0.123	0.370	–

However, flood volume coincidence is not statistically significant for the Danube River (at Hofkirchen) and the Inn River (at Passau-Ingling).

6.4.2 The Danube River (Linz–Stein-Krems) – Enns River (Steyr) Sector

The evaluation of discharge coincidence of the Danube and Enns rivers is presented in Table 6.7.

Correlation coefficients indicate that the high flow coincidences of the Danube River at Linz and Krems are statistically significant. However, the high flows of the Danube and Enns rivers do not coincide in most cases, except for the maximum

Table 6.7 Correlation of maximum discharge of the Danube and Enns rivers

Stations	Combination of variables	R	N	σ	3σ	Significance
Stein-Krems–Linz	max–max	0.873	40	0.038	0.113	+
	max–cor	0.706	40	0.079	0.238	+
	cor–max	0.840	40	0.047	0.140	+
Stein-Krems–Steyr	max–max	0.549	32	0.124	0.371	+
	max–cor	0.337	32	0.157	0.470	–
	cor–max	0.241	32	0.166	0.499	–
Linz–Steyr	max–max	0.417	35	0.140	0.419	–
	max–cor	0.504	35	0.126	0.378	+
	cor–max	0.175	35	0.164	0.492	–

Table 6.8 Correlation of flood volumes of the Danube and Enns rivers

Stations	Combination of variables	R	N	σ	3σ	Significance
Stein-Krems–Linz	max–max	0.833	36	0.051	0.153	+
	max–cor	0.922	36	0.025	0.075	+
	cor–max	0.936	36	0.021	0.062	+
Stein-Krems–Steyr	max–max	0.672	30	0.100	0.301	+
	max–cor	0.828	30	0.057	0.172	+
	cor–max	0.817	29	0.062	0.186	+
Linz–Steyr	max–max	0.491	32	0.134	0.402	+
	max–cor	0.742	32	0.079	0.238	+
	cor–max	0.747	30	0.081	0.242	+

discharge values at Stein-Krems and Steyr, and the maximum discharge at Linz and the corresponding discharge at Steyr.

In contrast to the above, flood volumes of the considered rivers show statistically significant coincidence, as described in Table 6.8.

6.4.3 The Danube River (Wien–Bratislava) – Morava River (Moravsky Jan) Sector

Discharge coincidence analyses of the Danube and Morava rivers are presented in Table 6.9 and partially illustrated in Fig. 6.6.

The results indicate that high flows of the Danube River mainly coincide with flood flows of the Morava River. The variable combination consisting of the maximum annual discharge of the Danube River at Vienna and the maximum annual flow of the Morava River at Moravsky Jan, does not demonstrate a statistically significant coincidence.

Similar results were obtained for flood volume coincidence of the examined rivers (see Table 6.10).

Table 6.9 Correlation of maximum discharges of the Danube and Morava rivers

Stations	Combination of variables	*R*	*N*	*σ*	3σ	Significance
Bratislava–Wien	max–max	0.928	43	0.021	0.064	+
	max–cor	0.882	43	0.034	0.102	+
	cor–max	0.665	43	0.085	0.255	+
Bratislava–Moravsky Jan	max–max	0.394	60	0.109	0.327	+
	max–cor	0.600	60	0.083	0.248	+
	cor–max	0.685	60	0.068	0.205	+
Wien–Moravsky Jan	max–max	0.380	43	0.130	0.391	–
	max–cor	0.494	43	0.115	0.346	+
	cor–max	0.633	43	0.091	0.274	+

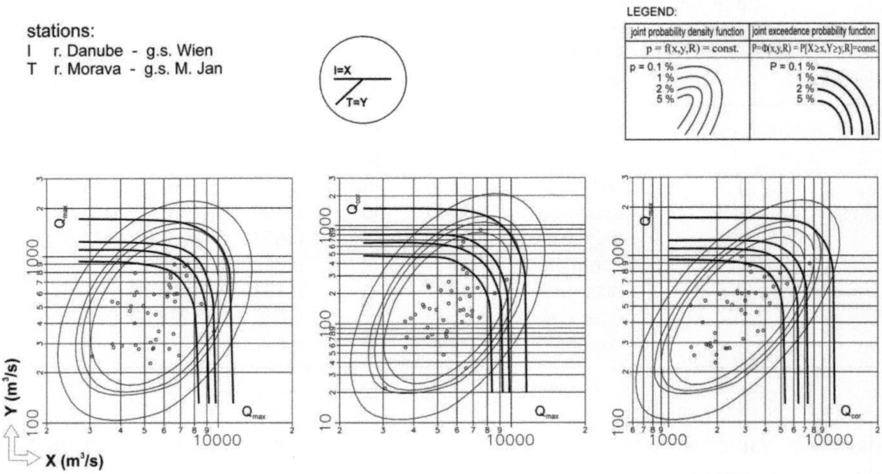

Fig. 6.6 Discharge coincidence of maximum annual flows

Table 6.10 Correlation of flood volumes of the Danube and Morava rivers

Stations	Combination of variables	*R*	*N*	*σ*	3σ	Significance
Bratislava–Wien	max–max	0.892	40	0.032	0.097	+
	max–cor	0.933	40	0.021	0.062	+
	cor–max	0.934	39	0.021	0.062	+
Bratislava–Moravsky Jan	max–max	0.420	53	0.113	0.339	+
	max–cor	0.328	44	0.135	0.404	–
	cor–max	0.664	48	0.081	0.242	+
Wien–Moravsky Jan	max–max	0.373	37	0.142	0.425	–
	max–cor	0.225	28	0.179	0.538	–
	cor–max	0.595	32	0.114	0.343	+

The results show statistically significant coincidence of flood volumes of the two examined rivers at Bratislava and Vienna. However, the maximum flood volume of the Danube River at Bratislava does not significantly correlate with the corresponding flood wave volume of the Morava River at Moravsky Jan. In addition, coincidence of the flood wave volumes of the Danube River at Vienna and the Morava River at Moravsky Jan is insignificant for two combinations of variables. For these two stations, only variable combinations related to the maximum annual flood wave volume at Moravsky Jan, and the corresponding flood wave volume at Vienna, can be regarded as having a statistically significant coincidence.

6.4.4 The Danube River (Bezdan–Bogojevo) – Drava River (Donji Miholjac) Sector

Results of analyses of flood wave peak discharges of the Danube and Drava rivers are presented in Table 6.11. As shown, almost all variable combinations are characterized by statistically significant correlation.

Only the variable combination of the maximum annual discharge of the Danube River (at two locations – Bezdan and Bogojevo) and the corresponding peak flow discharge of the Drava River (at Donji Miholjac) have insignificant correlation.

Comparable results were obtained when flood volume coincidence of the examined rivers/locations was analyzed (see Table 6.12). Statistically significant correlation characterizes all variable combinations related to flood volume of the Danube River (at Bogojevo and Bezdan). However, maximum flood wave volume of the Drava River at Donji Miholjac does not seem to coincide with the corresponding flood wave volumes of the Danube River (at either Bogojevo or Bezdan).

Table 6.11 Correlation of the maximum discharges of the Danube and Drava rivers

Stations	Combination of variables	R	N	σ	3σ	Significance
Bogojevo–Bezdan	max–max	0.917	61	0.020	0.061	+
	max–cor	0.934	61	0.016	0.049	+
	cor–max	0.899	61	0.025	0.074	+
Bogojevo–Donji Miholjac	max–max	0.563	54	0.093	0.279	+
	max–cor	0.354	54	0.119	0.357	–
	cor–max	0.618	54	0.084	0.252	+
Bezdan–Donji Miholjac	max–max	0.382	54	0.116	0.349	+
	max–cor	0.142	54	0.133	0.400	–
	cor–max	0.541	54	0.096	0.289	+

Table 6.12 Correlation of flood volumes of the Danube and Drava rivers

Stations	Combination of variables	*R*	*N*	σ	3σ	Significance
Bogojevo–Bezdan	max–max	0.948	53	0.014	0.042	+
	max–cor	0.958	53	0.011	0.034	+
	cor–max	0.948	53	0.014	0.042	+
Bogojevo–Donji Miholjac	max–max	0.500	49	0.107	0.322	+
	max–cor	0.482	43	0.117	0.351	+
	cor–max	0.339	44	0.133	0.400	–
Bezdan–Donji Miholjac	max–max	0.416	48	0.119	0.358	+
	max–cor	0.419	44	0.124	0.373	+
	cor–max	0.253	42	0.144	0.433	–

6.4.5 The Danube River (Bogojevo–Slankamen) – Tisa River (Senta) Sector

Results of analyses of flood discharge coincidence of the Danube and Tisa rivers are presented in Table 6.13. The results indicate that correlation coefficients are statistically significant for virtually all variable combinations. Only the maximum annual discharge of the Danube River at Bogojevo shows no significant correlation either to the peak flow or to the corresponding discharge of the Tisa River at Senta.

When flood volume coincidence of the examined rivers/locations is analyzed, statistically significant correlation is found for almost all variable combinations (see Table 6.14). One exception is the combination of maximum flood wave volumes of the Danube River at Bogojevo and the Tisa River at Senta.

Table 6.13 Correlation of maximum discharges of the Danube and Tisa rivers

Stations	Combination of variables	*R*	*N*	σ	3σ	Significance
Slankamen–Bogojevo	max–max	0.858	62	0.034	0.101	+
	max–cor	0.816	62	0.043	0.128	+
	cor–max	0.876	62	0.030	0.089	+
Slankamen–Senta	max–max	0.558	62	0.087	0.262	+
	max–cor	0.434	62	0.103	0.309	+
	cor–max	0.669	62	0.070	0.211	+
Bogojevo–Senta	max–max	0.279	63	0.116	0.349	–
	max–cor	0.198	63	0.121	0.363	–
	cor–max	0.366	63	0.109	0.327	+

Table 6.14 Correlation of flood volumes of the Danube and Tisa rivers

Stations	Combination of variables	R	N	σ	3σ	Significance
Slankamen–Bogojevo	max–max	0.800	57	0.048	0.143	+
	max–cor	0.869	57	0.032	0.097	+
	cor–max	0.920	57	0.020	0.061	+
Slankamen–Senta	max–max	0.685	51	0.074	0.223	+
	max–cor	0.583	46	0.097	0.292	+
	cor–max	0.839	51	0.042	0.125	+
Bogojevo–Senta	max–max	0.342	49	0.126	0.378	–
	max–cor	0.663	40	0.089	0.266	+
	cor–max	0.646	46	0.086	0.258	+

6.4.6 The Danube River (Slankamen–Pancevo) – Sava River (Sremska Mitrovica) Sector

The main results of the analyses of flood discharge coincidence of the Danube and Sava rivers are presented in Table 6.15 and illustrated in Fig. 6.7.

On the basis of these results, it can be concluded that the Danube River flood flows at Pancevo coincide with both upstream stations: the Danube River at Slankamen and the Sava River at Sremska Mitrovica. However, the discharge of the Danube River at the upstream station (Slankamen) shows no significant correlation with the Sava River flows at Sremska Mitrovica in either variable combination, except for max–max discharges.

When flood volume coincidence at these points is analyzed, statistically significant correlation is found for almost all variable combinations (see Table 6.16). Nevertheless, the maximum flood volume of the Danube River realized at both stations (Slankamen and Pancevo) does not significantly coincide with the maximum volume of the tributary (the Sava River at Sremska Mitrovica).

Table 6.15 Correlation of maximum discharges of the Danube and Sava rivers

Stations	Combination of variables	R	N	σ	3σ	Significance
Pancevo–Slankamen	max–max	0.797	62	0.046	0.139	+
	max–cor	0.843	62	0.037	0.110	+
	cor–max	0.776	62	0.051	0.152	+
Pancevo–Sr. Mitrovica	max–max	0.625	63	0.077	0.231	+
	max–cor	0.424	63	0.103	0.310	+
	cor–max	0.498	63	0.095	0.284	+
Slankamen–Sr. Mitrovica	max–max	0.426	62	0.104	0.312	+
	max–cor	0.204	62	0.122	0.365	–
	cor–max	0.250	62	0.119	0.357	–

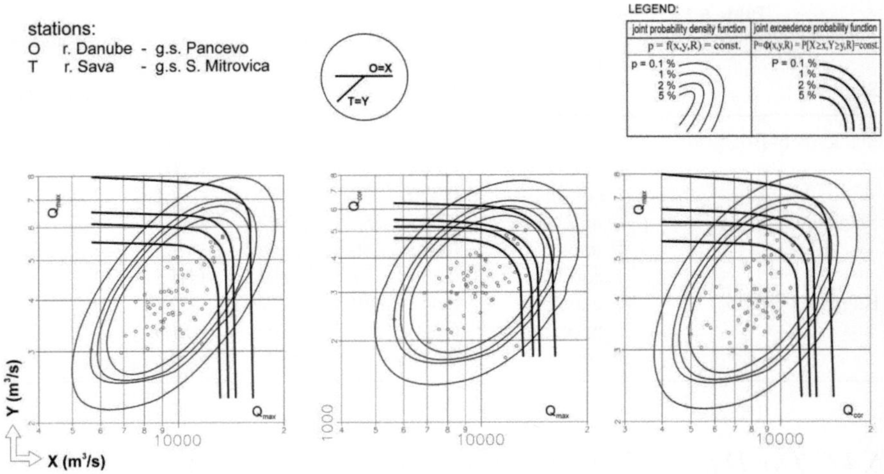

Fig. 6.7 Discharge coincidence of maximum annual flows

Table 6.16 Correlation of flood volumes of the Danube and Sava rivers

Stations	Combination of variables	R	N	σ	3σ	Significance
Pancevo–Slankamen	max–max	0.902	55	0.025	0.076	+
	max–cor	0.896	55	0.026	0.079	+
	cor–max	0.863	54	0.035	0.104	+
Pancevo–Sr. Mitrovica	max–max	0.362	39	0.139	0.417	–
	max–cor	0.659	39	0.091	0.272	+
	cor–max	0.803	39	0.057	0.170	+
Slankamen–Sr. Mitrovica	max–max	0.368	38	0.140	0.421	–
	max–cor	0.486	33	0.133	0.399	+
	cor–max	0.448	33	0.139	0.417	+

6.4.7 The Danube River (Pancevo–Veliko Gradiste) – Velika Morava River (Ljubicevski Most) Sector

Analyses of flood discharge coincidence of the Danube and Velika Morava rivers reveal that, in view of statistical significance, the examined events coincide in a similar manner as in the case of the Danube and Sava rivers. Namely, in all variable combinations of the Danube River (at both Veliko Gradiste and Pancevo), the Danube River (at Veliko Gradiste) and the Velika Morava River (at Ljubicevski Most), the correlation is statistically significant. A lack of correlation is evident between Pancevo (the Danube River) and Ljubicevski Most (the Morava River), except for the max–max variable combination. The results of these calculations are shown in Table 6.17.

Table 6.17 Correlation of maximum discharges of the Danube and Velika Morava rivers

Stations	Combination of variables	*R*	*N*	σ	3σ	Significance
Vel. Gradiste–Pancevo	max–max	0.957	63	0.011	0.032	+
	max–cor	0.905	63	0.023	0.069	+
	cor–max	0.964	63	0.009	0.027	+
Vel. Gradiste–Ljub. Most	max–max	0.596	61	0.083	0.248	+
	max–cor	0.369	61	0.111	0.332	+
	cor–max	0.437	61	0.104	0.311	+
Pancevo–Ljub. Most	max–max	0.574	61	0.086	0.258	+
	max–cor	0.336	61	0.114	0.341	–
	cor–max	0.325	61	0.115	0.344	–

Table 6.18 Correlation of flood volumes of the Danube and Velika Morava rivers

Stations	Combination of variables	*R*	*N*	σ	3σ	Significance
Vel. Gradiste–Pancevo	max–max	0.970	55	0.008	0.024	+
	max–cor	0.970	55	0.008	0.024	+
	cor–max	0.970	55	0.008	0.024	+
Vel. Gradiste–Ljub. Most	max–max	0.642	45	0.088	0.263	+
	max–cor	0.445	43	0.122	0.367	+
	cor–max	0.227	42	0.146	0.439	–
Pancevo–Ljub. Most	Max–max	0.600	45	0.095	0.286	+
	max–cor	0.522	43	0.111	0.333	+
	cor–max	0.201	39	0.154	0.461	–

When flood volume coincidence of the Danube and Morava rivers is analyzed, an obvious lack of correlation is seen between the maximum flood volume of the Morava River at Ljubicevski Most and the simultaneous flood wave volume of the Danube River (at both Veliko Gradiste and Pancevo). The results are presented in Table 6.18.

6.4.8 *The Danube River (Vadu Oii–Ceatal Izmail) – Siret River (Lungoci) Sector*

Analyses of flood wave peak discharge coincidence of the Danube and Siret rivers were based on data collected in Romania. The results are shown in Table 6.19.

The results indicate that statistically significant flood peak coincidence is characteristic of all variable combinations related to Vadu Oii and Ceatal Izmail (on the Danube River). As far as coincidence of the Danube and Siret rivers is concerned, significant correlation is present only between the maximum annual values realized on the Danube River (at both Ceatal Izmail and Vadu Oii) and the corresponding

Table 6.19 Correlation of maximum discharges of the Danube and Siret rivers

Stations	Combination of variables	*R*	*N*	σ	3σ	Significance
C. Izmail–Vadu Oii	max–max	0.942	59	0.015	0.044	+
	max–cor	0.930	59	0.018	0.053	+
	cor–max	0.887	59	0.028	0.083	+
C. Izmail–Lungoci	max–max	0.390	40	0.134	0.402	–
	max–cor	0.659	40	0.089	0.268	+
	cor–max	0.309	40	0.143	0.429	–
Vadu Oii–Lungoci	max–max	0.357	40	0.138	0.414	–
	max–cor	0.590	40	0.103	0.309	+
	cor–max	0.179	40	0.153	0.459	–

Table 6.20 Correlation of flood volumes of the Danube and Siret rivers

Stations	Combination of variables	*R*	*N*	σ	3σ	Significance
Ceatal Izmail–Vadu Oii	max–max	0.765	51	0.058	0.174	+
	max–cor	0.966	50	0.009	0.028	+
	cor–max	0.965	50	0.010	0.029	+
Ceatal Izmail–Lungoci	max–max	0.309	34	0.155	0.465	–
	max–cor	0.158	24	0.199	0.597	–
	cor–max	0.075	27	0.191	0.574	–
Vadu Oii–Lungoci	max–max	0.313	34	0.155	0.464	–
	max–cor	0.246	27	0.181	0.542	–
	cor–max	0.177	27	0.186	0.559	–

peak discharges of the Siret River at Lungoci. The remaining variable combinations do not exhibit statistically significant correlation.

In the case of flood volume coincidence of the Danube and Siret rivers (see Table 6.20), statistically significant correlation features all variable combinations related to flood volume of the Danube River itself (at Ceatal Izmail and Vadu Oii). A lack of correlation is obvious for all other variable combinations of the flood volume of the Danube River and its tributary – the Siret River.

6.4.9 The Danube River (Vadu Oii–Ceatal Izmail) – Prut River (Tchernovtsy) Sector

As in the case discussed in Section 6.4.8, analyses of flood coincidence were based on the data collected in Romania (for the Danube River) and in Ukraine (for the Prut River). The analyses were performed to assess the coincidence between the Danube River and the Prut River only. The coincidence involving flood waves on the Danube River itself (at Ceatal Izmail and Vadu Oii) was discussed in Section 6.4.8.

Table 6.21 Correlation of maximum discharges of the Danube and Prut rivers

Stations	Combination of variables	R	N	σ	3σ	Significance
Ceatal Izmail–Tchernovtsy	max–max	0.074	39	0.159	0.478	–
	max–cor	0.425	39	0.131	0.394	+
	cor–max	0.091	39	0.159	0.476	–
Vadu Oii–Tchernovtsy	max–max	0.152	39	0.156	0.469	–
	max–cor	0.321	39	0.144	0.431	–
	cor–max	0.084	39	0.159	0.477	–

Table 6.22 Correlation of flood volumes of the Danube and Prut rivers

Stations	Combination of variables	R	N	σ	3σ	Significance
Ceatal Izmail–Tchernovtsy	max–max	0.225	32	0.168	0.503	–
	max–cor	0.367	14	0.231	0.694	–
	cor–max	0.373	21	0.188	0.563	–
Vadu Oii–Tchernovtsy	max–max	0.213	33	0.166	0.498	–
	max–cor	0.710	17	0.120	0.361	+
	cor–max	0.383	21	0.186	0.559	–

Flood peak discharge results are shown in Table 6.21. On the basis of the values of correlation coefficients, a statistically insignificant coincidence of flood peaks is characteristic of most variable combinations related to the Danube River and its tributary – the Prut River. The only exceptions are the maximum flood peak of the Danube River (at Ceatal Izmail) and the corresponding flood peak of the Prut River (at Tchernovtsy).

Flood volume coincidence of the Danube and Prut rivers shows similar results to those seen above (Table 6.22). In almost all variable combinations, coincidences are statistically insignificant. A distinct conclusion can be drawn only for variable combinations involving maximum annual values of the Danube River at Vadu Oii, and the corresponding values of the Prut River at Tchernovtsy, where correlation coefficients appear significant.

6.5 Recommendation for the Use of Coincidence Analyses

The previously discussed model can be used for various purposes, including:

(a) Defining the design flood flow of the main river and its tributary where hydrological monitoring data are available for the confluence;
(b) Defining maximum flood flow levels in the absence of adequate hydrological monitoring data;

(c) Analyses necessary to investigate capacity enhancement of two adjacent streams;
(d) Assessment of the statistical significance of flood flow coincidence for the most interesting flood features – for observed and/or simulated flood hydrographs.

The ensuing text provides a brief outline of the theoretical approach, the justification of the analyses, and the use of results in the assessment of flood flow characteristics at a river confluence.

6.5.1 Results of the Implementation of Flood Flow Analysis Aimed at Determining Design Levels at a Confluence, Using Sufficient Hydrological Monitoring Data

A river confluence can be regarded as having sufficient hydrological observations if all required hydrological monitoring data are available for the inflow profiles (i.e., for both the main stream and its tributary), and for the outflow profiles (downstream profiles on the main river). For these types of analyses, the following data are required:

1. A series of recorded maximum annual discharge levels at all examined profiles,
2. The calculated results of flood flow coincidence (as described in Chapter 5), for the following combinations of variables:

   - maximum annual flows on the main river – maximum annual flow on the tributary
   - maximum annual flows on the main river – corresponding (simultaneous) flow on the tributary,
   - maximum annual flows on the tributary – simultaneous (corresponding) flow on the main river.

The design flood levels within the zone of confluence are obtained through hydraulic modelling, using design flood discharge and boundary conditions defined through the described procedure. The design flood discharge is selected as follows:

1. On the downstream reach of the main river, design levels are defined through probabilistic modelling of maximum annual flow at the gauging station downstream from the tributary, $QOUTmax,p$, for a chosen exceedance probability, p.
2. On the main river upstream from the tributary, within the zone of hydraulic interaction, the design flood level appears as the enveloping line obtained through hydraulic modelling of the flood level, using the boundary conditions stemming from coincidence analyses of the following combinations of variables:

   - the maximum annual discharge on the main river downstream from the confluence for the desired exceedance probability, p, and the simultaneous,

corresponding discharge of the main river at the upstream gauging station for the same probability (*Q*OUTmax;*Q*INcor1)*p*,
- the corresponding flow of the main river at the downstream gauge and the maximum annual discharge of the main river at the gauge located downstream from the confluence for the chosen exceedance probability, obtained through coincidence analyses (*Q*INmax;*Q*OUTcor1)*p*.

3. On the tributary, within the zone of influence of the main river the design flood level is an enveloping curve, which should be defined through hydraulic modelling using the following combinations of variables:

   - the maximum annual flow on the main river downstream from the tributary for a selected exceedance probability, *p*, and the simultaneous discharge on the tributary, upstream from the mouth, for the same probability of coincidence (*Q*OUTmax;*q*TRcor2)*p*,
   - the corresponding flow of the receiving river at the downstream gauge, simultaneous to the maximum annual discharge on the tributary, for a selected exceedance probability value, based on coincidence analyses for the same probability (*q*TRmax;*Q*OUTcor2)*p*.

4. On the main river outside the zone affected by the tributary, the design water level is obtained from probabilistic analyses of the maximum annual discharge at the upstream gauge, *Q*INmax,*p*.
5. On the tributary upstream from the zone of influence of the main river, the design flood level is defined using the maximum annual discharge, at the inflow gauge, *q*TRmax,*p*, for the selected exceedance probability value, *p*.
6. A schematic outline of the combinations used to define the design flood level, within the zone of interaction of the main river and its tributary, is presented in Fig. 6.8. The level of protection corresponds to the selected exceedance probability.

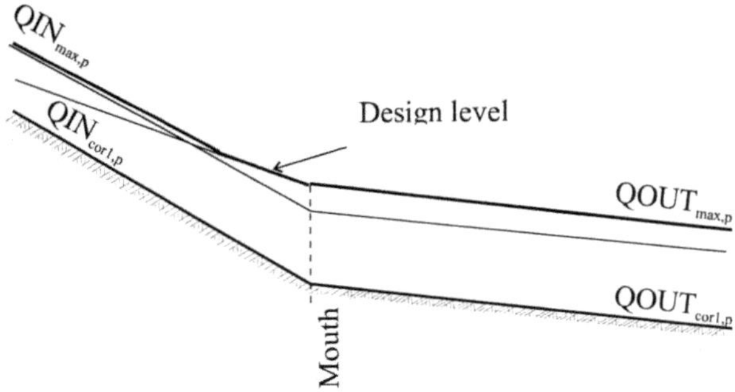

Fig. 6.8 Schematic outline of the design flood level within the zone of interaction of the main river and its tributary

Numerical indicators of the selection of flood flow values, which were used to define the design flood levels of the Danube River in the examined reaches, are shown in Table 6.23. All calculations were based on coincidence analyses outlined in Chapter 5, and the assumption was made that the required level of flood protection is a 100-year return period.

6.5.2 Results of the Implementation of Flood Flow Analysis Aimed at Determining Design Levels at the Confluence Using Sufficient Hydrological Monitoring Data

A river confluence is regarded as having inadequate hydrological data if observations are missing at one of three considered input–output stations. As such, all necessary statistical coincidence analyses should be performed on a series of available data, as described in Section 6.5.1.

The procedure to be followed in this case is fully analogous to the previously discussed case. The missing data are to be replaced in the following manner:

(a) Missing data for the outgoing profile (downstream gauge) of the main river:

- Maximum annual flow, QOUTmax,p, is replaced by the highest value of the sums of the corresponding discharge values (QINmax;qTRmax)p, associated with a selected exceedance probability, p,
- The corresponding discharge, QOUTcor1, is replaced by the largest sum of discharge pairs associated with a given coincidence probability, p [i.e., (QINmax;qTRcor2)p],
- The corresponding discharge, QOUTcor2, is replaced by the largest sum of discharge pairs associated with a given coincidence probability, p [i.e., (qTRmax; QINcor2)].

(b) Missing data for the inflow profile (upstream gauging stations) on the main river:

- Maximum annual discharge, QINmax,p, is replaced by the difference of a pair of discharges (QOUTmax – qTRmax)p of the most probable point along the coincidence line related to the variable combination (QOUTmax;qTRmax)p;
- Discharge QINcor1 can be substituted by the difference (QOUTmax – qTRcor1)p obtained along the coincidence line (QOUTmax;qTRcor1)p of the same probability, p;
- Discharge QINcor2 is replaced by the difference (qTRmax – QOUTcor2)p associated with the coincidence of variable combinations (qTRmax;QOUTcor2)p, for the same probability p.

Table 6.23 Selection of design floods on the Danube River and its tributary for purposes of defining water levels for a 100-year return period

		Design floods ($m^3 s^{-1}$)						
		Along main river				Along tributaries		
		Downstream	Upstream from the mouth			Upstream from the mouth		
		From the mouth	Within zone of mutual influence	Outside zone of mutual influence		Within zone of mutual influence	Outside zone of mutual influence	
No sector	River/station	QOUTmax,1%	QOUTcor1,%	QINcor1,1%	QINmax,1%	QOUTcor2,1%	qTRcor2,1%	qTRmax,1%
1	Danube (Hofkirchen–Achleiten)/Inn (Passau Ingling)	7,561	4,416	1,106	3,522	1,220	4,421	5,964
2	Danube (Linz–Krems)/Enns (Steyr)	10,525	3,002	4,686	8,347	802	800	3,112
3	Danube (Vienna–Bratislava)/Morava (Moravsky Jan)	10,324	6,191	6,450	9,402	4,922	20	1,518
4	Danube (Bezdan–Bogojevo)/Drava (Donji Miholjac)	8,596	6,510	6,821	7,487	2,171	902	2,282
5	Danube (Bogojevo–Slankamen)/Tisa (Senta)	10,574	6,296	5,953	8,596	3,200	350	3,787
6	Danube (Slankamen–Pančevo)/Sremska (Mitrovica)	14,600	5,924	7,700	10,574	3,900	2,238	6,597
7	Danube (Pančevo–V. Gradište)/V. Morava (Ljubičevski Most)	15,559	8,507	11,082	14,600	7,880	90	2,564
8	Danube (Vadu Oii–Izmail) Siret (Lungoci)	15,887	13,520	14,162	16,124	200	324	3,462
9	Danube (Vadu Oii–Izmail) Prut (Tchernovtsy)	15,887	13,520	14,162	16,124	5,931	10	4,372

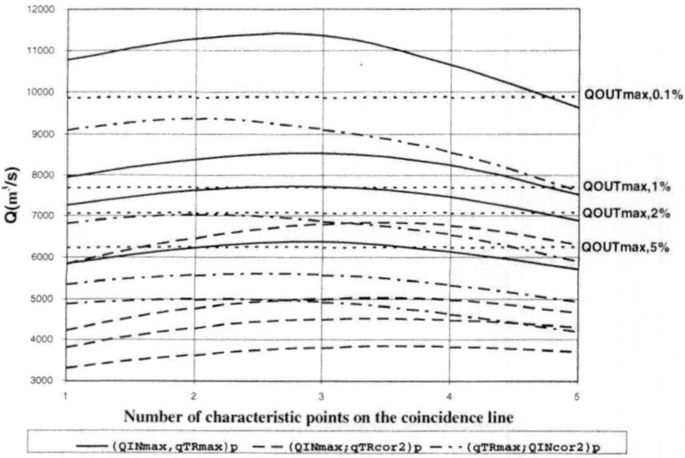

Fig. 6.9 Parallel representation of theoretical (statistical) and estimated flood flows of the Danube River (estimates based on the substitution of missing data with values obtained through coincidence analyses) at the Achleiten outflow gauge

(c) The insufficiency of data on the tributary can be overcome in the following way:

- Maximum annual discharge, *q*TRmax,*p*, can be replaced by the difference (*Q*OUTmax – *Q*INmax)*p* obtained from the coincidence curve (*Q*OUTmax; *Q*IN=max)*p* for the same probability value, *p*;
- Simultaneous discharge, *q*TRcor1, can be substituted by the difference (*Q*OUTmax – *Q*INcor1)*p*, taken from the coincidence line (*Q*OUTmax;*Q*INcor1)*p*, for the same value of probability;
- Simultaneous corresponding flow, *q*TRcor2, should be substituted with the value of the difference (*Q*INmax – *Q*OUTcor1)*p*, taken from the coincidence line (*Q*INmax;*Q*OUTcor1)*p*, for the same probability, *p*.

For purposes of illustration, Figs. 6.9 and 6.10 show the results of calculations for design flood levels obtained in cases where data regarding the outflow profile on the main river – the Danube River (for three chosen characteristic reaches) were missing.

Numerical values that justify the selected substitutions for the missing maximum annual discharge data at all inflow–outflow gauges on the Danube River and its tributary, for all related sectors, are shown in Table 6.24 for an exceedance probability of 1%. The results lead to the conclusion that errors introduced by using the indicated substitutions fall within a relatively wide range.

The most reliable estimates are obtained when the maximum annual discharge at the outflow gauge is substituted. Errors range between –7 and +13%, with an average of about 4%, which suggests a very good estimate. Somewhat less reliable estimates result when substitutions are made for the inflow gauge of the main river. The average error was about –8%, and varied within a range from –1 to –24%. The

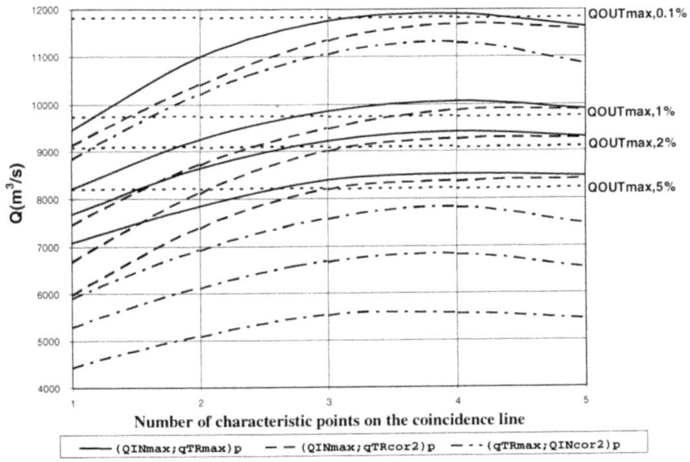

Fig. 6.10 Parallel representation of theoretical (statistical) and estimated flood flows of the Danube River (estimates based on the substitution of missing data with values obtained through coincidence analyses) at the Bratislava outflow gauge

largest errors are associated with the substitution of the maximum annual discharge of tributaries. Calculated errors deviate from –16 to –31% (with the average being about –23%). Evidently, floods are underestimated, which can be expected given that the estimates were derived from data recorded on the main river, which controls a much larger drainage area. Nevertheless, such substitution of missing data can be used successfully.

6.5.3 Analyses Related to Capacity Enhancement of Two Adjacent Streams

In some situations, flood protection from two adjacent streams is required. Designed protection measures can be implemented independently from one another. However, such a solution tends to be quite expensive, and is sometimes not feasible due to constraints imposed by existing development or other natural conditions. A more complex approach consists of integrated consideration of the necessary measures, taking into account flood coincidence of the studied rivers. In the ensuing text, a brief review of an integrated approach is presented.

The basic assumption is that reasonable natural conditions and technical possibilities to realize a floodway or other flood relief channel with all the necessary structures on it exist. The purpose of such a channel is to allow for joint functioning of the two rivers: a part of the flood water of one stream can be directed into the other, and vice-versa.

River channel capacities are generally analyzed in two ways. Upstream from the connecting flood relief channel, river channel capacities are designed to provide

Table 6.24 Numerical indicators of goodness of substitution of missing maximum annual flood data for a 100-year return period

No sector	River/station	*Q*OUTmax,1% ($m^3\ s^{-1}$)			*Q*INmax,1% ($m^3\ s^{-1}$)			*q*TRmax,1% ($m^3\ s^{-1}$)		
		Observed	Replaced	%	Observed	Replaced	%	Observed	Replaced	%
1	Danube (Hofkirchen–Achleiten)/Inn (Passau Ingling)	7,561	8,542	113	3,522	2,690	76	5,964	4,464	75
2	Danube (Linz–Krems)/Enns (Steyr)	10,525	9,813	93	8,347	7,827	94	3,112	2,159	69
3	Danube (Vienna–Bratislava)/Morava (Moravsky Jan)	10,324	10,064	98	9,402	9,313	99	1,518	1,192	79
4	Danube (Bezdan–Bogojevo)/Drava (Donji Miholjac)	8,596	8,992	105	7,487	7,276	97	2,282	1,909	84
5	Danube (Bogojevo–Slankamen)/Tisa (Senta)	10,574	11,075	105	8,596	8,185	95	3,787	3,053	81
6	Danube (Slankamen–Pancevo)/Sremska (Mitrovica)	14,600	15,720	108	10,574	9,439	89	6,597	5,577	84
7	Danube (Pancevo–V. Gradiste) V. Morava (Ljub–cevski Most)	15,559	15,629	100	14,600	13,773	94	2,564	1,894	74
8	Danube (Vadu Oii–Izmail) Siret (Lungoci)	15,887	16,915	106	16,124	14,631	91	3,462	5,570	71
9	Danube (Vadu Oii–Izmail) Prut (Tchernovtsy)	15,887	16,223	106	16,124	15,122	94	4,372		
	Average	104			92			77		

the required level of protection against their own flood water. Sections of the river downstream from the connecting flood relief channel are designed to handle the total flood discharge equal to the maximum sum of the simultaneous flow estimated from the flood coincidence curve for the required level of protection (i.e., at a 1% exceedance probability level).

If we assume that protection from two rivers, River *A* and River *B*, against floods of return period $T = 1/p$ is required (p is the flood exceedance probability), then we can propose to alleviate the flood hazard at minimum cost using a connecting flood relief channel between the two rivers. That channel would be used to provide flow in both directions. The design flood should then be selected as follows:

- The design flood at River *A* upstream from the connecting channel *AB*, is the maximum flood flow of River *A* for the required exceedance probability, i.e. $QA\text{max},p$;
- The design flood at River *B* upstream from the connecting channel *AB*, is the maximum flood flow of River *B* for the required exceedance probability, i.e. $QB\text{max},p$;
- The design capacity of both streams, *A* and *B*, downstream from the connecting channel, $Q\text{d}$, is equal to the maximum sum of flows obtained from the flood coincidence curve of the required exceedance probability of maximum annual flood flows [variable combination ($QA\text{max},p;QB\text{max},p$)]. The distribution of flows among the two rivers is defined based on natural, technical and economic conditions;
- The capacity of the connecting channel, $Q\text{c}$, is equal to the maximum difference of the total design flow of the downstream reaches *A* and *B*, and the discharges $QA\text{max},p$ and $QB\text{max},p$, i.e. $Q\text{c} = \sup\{Q\text{d} - QA\text{max},p;\ Q\text{d} - QB\text{max},p\}$.

A decrease in the design flood, $Q\text{d}$, as compared to the sum $QA\text{max},p + QB\text{max},p$, depends on the degree of correlation between maximum annual flows observed at Rivers *A* and *B*. If the correlation coefficient is $R = 1$, there is no decrease in the design flow, which then becomes the sum of maximum annual floods at the studied rivers ($Q\text{d} = QA\text{max},p + QB\text{max},p$). A maximum design flood decrease is achieved when there is no coincidence of flood flows at the considered streams ($R = 0$). In this case, the simultaneous occurrence of the maximum annual flows at both rivers, *A* and *B*, is equal to the product of their uni-variate probabilities. The practical meaning of this is that the probability of the sum of maximum annual discharges, each with a return period of 100 years, is 10,000 years. Experience shows, as described in Chapter 5, that correlation coefficients range between 0 and 1 (i.e., $0 < R < 1$).

Therefore, a decrease in the design capacity of a flood protection system is larger when the coefficient of correlation is smaller, and vice-versa.

6.5.4 Assessment of the Statistical Significance of Flood Flow Coincidence for the Most Interesting Flood Features of Recorded and/or Simulated Flood Hydrographs

Graphical representation of flood coincidence analyses is basically used to assess statistical significance of the flood magnitude at the main river and at its tributaries. It can be applied to assess the significance of both realized and expected floods.

The assessment of flood flow significance, for any combination of variables, is performed using the following procedure: A pair of flood flow values is plotted into the coordinate system. Then, the pair is compared to the coincidence lines associated with the exceedance probability value. The exceedance probability is determined using interpolation techniques (either linear or logarithmic). The reciprocal value of the exceedance probability is a flood return period, or its statistical significance for every desired flood wave element.

6.6 Concluding Remarks

Coincidence analysis of the most important flood hydrograph features of the Danube River and its tributaries confirmed that flood wave genesis is very complex within the Danube Basin. In particular, this complexity is evident at the confluence points. Evaluation of the characteristics of flood hydrographs, their interaction and concurrency, play an important role in the design of hydraulic structures within the flood defence system, both from a system development cost perspective and a degree of protection provided perspective.

Results obtained through coincidence analyses serve multiple important functions. First, they can be used for an evaluation of the statistical significance of the coincidence of various flood wave characteristics, and then to provide a comprehensive assessment of the flood situation on the Danube River and its tributaries. The practical implication of the results is that, when the coincidence of floods on the main river and its tributaries is not significant, a less expensive flood protection system can be developed based on a conventional uni-variate approach to flood analysis. Naturally, the system must nevertheless provide the required degree of safety. Second, the approach proposed here provides an opportunity for the determination of the design parameters of the optimal combinations of the examined variables with respect to both system development cost and the safety of hydraulic structures. Too, the methodology can be used to calculate the design water level at the confluence when hydrologic records are missing for one gauging station on the main river (either upstream or downstream from a tributary).

In addition, the analyses presented can prove to be advantageous under circumstances where a flat area should be protected from two adjacent streams. The protection system, based on a conventional flood estimation procedure for each river (uni-variate analyses), is often expensive and difficult to develop since the spatial constraints are usually very rigid and the required degree of protection high. The proposed methodology produces results which allow for an integrated examination of flood protection from both rivers, taking into account the coincidence of flood waves. The basic assumption associated with this concept is that there is the possibility of constructing (one or more) connecting channels between two adjacent streams with or without gates. If the coincidence is significant, a part of the flood flow can be directed from one river to the other, and vice-versa.

The approach can be used to conduct an overall evaluation of flood flow characteristics. In addition, upon a comprehensive examination of various flood hydrograph parameters (peak discharge, flood volume and flood duration with either the same or a different probability), the "design hydrograph" can be defined at each input/output profile (gauging stations). In that way the most "unfavourable" shape of the design flood can be identified for purposes of developing a flood protection system within the confluence area.

The analyses performed reveal that, within the Danube River Basin, the most significant coincidence can generally be expected for the following flood hydrograph elements:

(I) Flood discharge

- The maximum annual discharge of the main river considered with the maximum annual discharge of either the main river or its tributary, shows significant coincidence in 73% of cases;
- The maximum annual discharge of the main river when compared with the simultaneous (corresponding) discharge of either the main river or its tributary, demonstrates significant coincidence in 69% of cases;
- The main river discharge which corresponds to a maximum annual discharge of either the main river or its tributary shows significant coincidence in 69% of cases.

(II) Flood volume

- The maximum annual flood volume on the main river considered in conjunction with the maximum annual flood volume of either the main river or its tributary, shows significant coincidence in 65% of cases;
- The maximum annual flood volume on the main river, when compared with the volume of simultaneous (corresponding) flood of either the main river or its tributary, shows significant coincidence in 77% of cases;
- The main river flood wave volume which occurs simultaneously with the maximum annual flood volume of either the main river or its tributary shows significant coincidence in 65% of cases.

(III) Flood wave duration

- The maximum annual flood wave duration on the main river, in comparison with the maximum annual flood wave duration of either the main river or its tributary, shows significant coincidence in 58% of cases;

- The maximum annual flood wave duration on the main river, when compared with the duration of the simultaneous (corresponding) flood wave of either the main river or its tributary, shows significant coincidence in 77% of cases;
- The duration of the main river flood wave which occurs simultaneously with the maximum annual flood wave of either the main river or its tributary, shows significant coincidence in 65% of cases.

(IV) Concurrent peak discharges

- The coincidence between the maximum annual discharge at the main river downstream profile and time lag between the annual flood peak occurrence on either the main river or its tributary were found to be statistically insignificant at all examined confluences.

Investigations pertaining to particular flood characteristics lead to the following conclusions:

(A) All evaluated combinations involving maximum annual peak discharge, flood volume, and flood duration, revealed statistically significant coincidence;
(B) When analyses were performed for the main river and its tributary, the results could be summarized as follows:

(I) Flood discharge

- Peak annual discharge of the main river – peak annual discharge of the tributary (significant coincidence was observed in 78% of analyzed cases);
- Maximum annual discharge of the main river – simultaneous discharge of the tributary (78% of events);
- Discharge of the main river occurring simultaneously with the peak annual discharge of the tributary (67% of considered events).

(II) Flood wave volume

- The maximum annual flood volume of the receiving river – the maximum annual flow of the tributary (67% of cases),
- The maximum annual flood volume of the main river – the volume of the simultaneous flood wave of the tributary (67%);
- The flood wave volume of the main river corresponding to the maximum annual volume of the flood of the tributary's flood wave (observed in 56% of cases).

(III) Flood wave duration

- The maximum annual duration of the flood wave on the main river – the maximum annual duration of the flood wave on the tributary (56% of cases);

- The maximum annual duration of the flood wave on the main river – the duration of the corresponding flood wave on the tributary (67% of events);
- The duration of the main river flood wave corresponding to the maximum annual flood duration of the tributary (67% of cases).

(C) An examination of flood events at the upstream profile of the main river and at its tributary produced the following results:

(I) Flood discharge

- The peak annual discharge of the main river – the peak annual discharge of the tributary (significant coincidence was present in 44% of cases);
- The maximum annual discharge of the main river – the simultaneous (corresponding) discharge of the tributary (33% of cases);
- The discharge of the main river occurring simultaneously with the maximum annual discharge of the tributary (44% of considered cases).

(II) Flood wave volume

- The maximum annual flood volume of the main river – the maximum annual flow of the tributary (33% of cases),
- The maximum annual flood volume of the main river – the volume of the simultaneous flood wave of the tributary (56%);
- The flood wave volume of the main river corresponding to the maximum annual volume of the tributary flood wave (44%).

(III) Flood wave duration

- The maximum annual duration of the flood wave on the main river – the maximum annual duration of the flood wave on the tributary (22% of cases);
- The maximum annual duration of the flood wave on the main river – the duration of the simultaneous (corresponding) flood wave on the tributary (67%);
- The duration of the main river flood wave corresponding to the maximum annual flood duration of the tributary (33%).

The outlined results suggest that the most significant coincidence on the Danube River and its tributaries can be expected when flood events are examined at the upstream and downstream profiles of the main river. Somewhat less significant coincidence is associated with a flood event realized on the main river's downstream profile and the tributary. The coincidence of flood waves occurring at the main river's upstream profile and the tributary is in most cases statistically insignificant.

Taking this into account, the conclusion can be reached that the results obtained in this study have enormous practical value. They can be used to determine input parameters for the design of flood protection structures on key sectors along the

Danube River (i.e. in the vicinity of tributaries). The implementation of these results will lead to the development of an optimally sized flood protection system, which in addition ensures higher safety levels for the Danube's lower reaches.

Finally, it should be reiterated that this study should give an incentive to professionals and government authorities dealing with flood protection to pay appropriate attention to flood coincidence along the Danube River. The use of the outlined approach can greatly reduce spending of precious flood protection funds, while at the same time providing the required level of safety to flood-prone areas.

References

Abramowitz M, Stegun AI (1972) Handbook of mathematical functions with formulas, graphs, and mathematical tables. Dover Publications, Inc., New York, NY

Jovanovic S (1981) Primena metoda matematicke statistike u hidrologiji. Gradjevinski fakultet, Beograd

Yevjevich V (1972) Probability and statistics in hydrology. Water Resources Publications, Fort Collins, CO

Chapter 7
Basin-Wide Water Balance in the Danube River Basin

Pavel Petrovič, Katarína Mravcová, Ladislav Holko, Zdeněk Kostka, and Pavol Miklánek

Abstract This paper briefly describes the result of mutual work of the IHP UNESCO National Committees of the Danube countries and experts and scientists nominated by their countries. It represents part of the second phase of the co-operation of Danube countries within the frame of the IHP UNESCO in the field of hydrology. This project in fact represents a continuation, improvement and enlargement of the first Danube Monograph (Stančík et al. 1988, Hydrology of the River Danube, Publishing House Príroda, Bratislava). The main objectives of the work were estimation of water balance components on the basis of mathematical modelling and preparation of maps of water balance components for the Danube Basin. Maps of the mean annual precipitation, actual evapotranspiration and runoff were prepared. Difficulties in data collection in individual countries resulted in the obligatory period for data analysis of 1961–1990.

The first part of this work presents the administrative background and circumstances which favourably helped to intensify work on the project since 2002.

The second part deals with assessment of the general water balance of the Danube River Basin. In terms of working maps, the USGS (US Geological Survey) HYDRO1k digital terrain model was chosen. The Danube area was split into 109 balance regions. The intention was to apply the water balance model WatBal to each such region to calculate evapotranspiration for the balance regions as the main output.

The third part deals with monthly data collection and processing, WatBal model methodology and with model tuning. Calculation of water balance was performed for 109 balance regions, but tuning was successful for 84 balance regions only. Results are given both in a tabular and map form.

P. Petrovič (✉)
Water Research Institute, Bratislava, Slovakia

Pavel Petrovič and Katarína Mravcová are both from the NCs of IHP UNESCO of the Danube Countries

M. Brilly (ed.), *Hydrological Processes of the Danube River Basin*,
DOI 10.1007/978-90-481-3423-6_7,

Keywords Mean annual runoff · Danube water balance · Balance regions · Basin mean elevation · Mean annual precipitation

7.1 Introduction

The Danube River is the second longest river in Europe and flows through – or forms a part of the borders of – ten countries (http://en.wikipedia.org/wiki/Danube). The length of the main Danube channel is 2,860 km. The total drained area (inclusive parts of nine countries/areas with Danube tributaries only) covers 817,000 km^2, which represents 8% of the total European land surface. Therefore, one can see that this area is of interest both historically and in relation to the components of everyday life. This work deals with the water balance of this area.

The Danube River Basin is home to 81 million people with a wide range of cultures, languages and historical backgrounds (http://www.icpdr.org/icpdr-pages/river_basin.htm).

7.1.1 Some Remarks on the History of the Danube Water Balance

Co-operation in the field of hydrology on the official level (supplementary to the regular exchange of data for discharge forecast in the frame of the World Meteorological Organisation) began in 1972 as a co-operation among a group of states headed by UNESCO (Germany, Austria, Federal Republic of Yugoslavia) and a group of states (Czechoslovakia, Hungary, Soviet Union, Bulgaria, Romania) under the umbrella of the scientific working group for hydrology of the Danube (Navigation) Commission in Budapest. This first phase of co-operation, which was later unified into the International Hydrological Programme produced a very interesting result – the Hydrological Monograph of the Danube River (Stančík et al. 1988) – a representative publication in four languages giving an overview of tabular processing and spatial distribution of the main hydrological elements (precipitation, superficial runoff depth and actual superficial evapotranspiration) in the whole Danube River Basin. The methodology at that time was based on the national contribution in the form of maps of the balance elements. The role of the co-ordinator was first to provide guidance in relation to methodological questions and secondly to bring together the national input in the form of isolines of individual balance elements into "tailor-made" maps of water balance elements.

It was determined that this regional co-operation in the frame of the IHP UNESCO was of great significance and it was decided that the topic would be studied further and follow-up volumes to the Danube Hydrological Monograph would be published.

In the 1990s, a new initiative – at that time under the leadership of Dr. Oskar Behr (TU Vienna) – started to assemble a new version of the water balance for the whole Danube River Basin. The first methodological proposals were presented at the working meeting of the Chairmen of the NCs (National Committees) for the IHP UNESCO of the Danube Countries in Lednice (Czech Republic). Unfortunately in the Autumn of 1998 Dr. Oskar Behr ended his management of the co-operation due to serious illness and at the working meeting of the Chairmen of the NCs for the IHP UNESCO in Bratislava (May 1999) Slovakia was asked to take over responsibility for the whole subproject "5.3".

After studying related material, the Slovak NC IHP UNESCO prepared a detailed proposal of project development and authorised the Water Research Institute (WRI) in Bratislava to take the role of co-ordinator.

The Water Research Institute in Bratislava took into consideration the situation within the Danube Basin, where tasks in progress related to the co-operation within the frame of the IHP UNESCO were almost at the level of "voluntary" participation, and first focused its attention on increasing the level of importance in Danube River decision bodies.

The result was that after the IHP UNESCO obtained observer status at the International Commission of the Protection of the Danube River (ICPDR) our project "Basin-Wide Water Balance in the Danube River Basin" was recognised to be of great interest to the secretariat of the ICPDR in relation to the Joint Action Programme (JAP).

7.1.2 Some Administrative Milestones

The Plenary Session of the ICPDR in 2001 agreed the supporting resolution, asking

(a) Contracting Parties to ensure the availability of hydrometeorological data on air temperature, air humidity, precipitation and runoff for the agreed period of water balance evaluation to the national experts from the IHP Water Balance Working Group;
(b) The River Basin Management Expert Group to specify, in co-operation with the IHP Water Balance Working Group, the selections (sub-basins) as units for regional water balance assessment. This should include the provision of the dividing lines for sub-basins;
(c) The Contracting Parties to fully support (technically and financially) the work of national experts needed for completion of national contributions to the water balance (as stated under (a) and (b) above).

This was followed by an IHP UNESCO Meeting: *International Hydrological Programme of UNESCO – Regional Meeting of the IHP National Committees of the Electoral Groups I and II of UNESCO; 18 and 19 February 2002, Berlin/Germany*, where support was gained and a recommendation was agreed, which included a request addressed to the Secretariat (of UNESCO) and UVO ROSTE (UNESCO Venice Office – Regional Office for Science and Technology for Europe) to assist in the launching and implementation of the Water Balance of the Danube River Basin project in (since) 2002.

Thanks to the above-mentioned favourable atmosphere and to

(a) conclusions of the 15th working ordinary Danube IHP UNESCO chairmen group meeting in Deggendorf (May 2001);
(b) resolutions of the 4th Plenary Session of the ICPDR in Vienna on 29–30 November, 2001
(c) recommendation of the Regional Meeting of the IHP National Committees of the Electoral Groups I and II of UNESCO in Berlin on the 18 and 19 February, 2002
(d) the aim to join further work on the project "Water Balance of the Danube River Basin" with the present phase of the ICPDR JAP
(e) the possibility and willingness of the UVO ROSTE to support this project also financially it was possible to come to the agreement on the Project development intensification as much as possible.

Each Danube country in the CEE (Central and Eastern Europe) region with an economy in transition verified the proposed time schedule and work extent, and approved requirements for co-financing of the necessary part of the work on the project from international sources.

This resulted in UVO ROSTE – after supporting expert meetings in Zagreb in 2002 and in Sofia in 2003 – promising the amount of 80,000 USD for the period 2003–2004, to

(a) partially assist in the participation at planned meetings of the experts and steering committee, and
(b) partially support the international part of the work (co-ordination, methodology and development of guidelines, and assessment of the interim report and drafting of the final report).

In consonance with the expert group meeting in Sofia in 2003 requests for support were obtained from Bosnia and Herzegovina, Moldova, Romania, Serbia and Monte Negro, Slovakia and Ukraine. Details are given in the project final report (Petrovič et al.: Hydrological Monograph of the Danube River Basin).

Since the Sofia meeting regular Steering Committee meetings have been organised at least once a year. At the second meeting (March 2004) in Bratislava an improved methodology for the WatBal model application (WRI Bratislava) and a project based on GIS processing and mapping of the water balance elements (Institute of Hydrology SAS) were presented and accepted.

In consonance with the ICPDR recommendation, we have created contacts with the ICPDR GIS Expert Subgroup (GIS ESG); their observer took part in our Project Steering Committee meeting in March 2004 in Bratislava. We also contacted the EU JRC Ispra (Dr. R. Hiederer) which helped us with our GIS approach in the creation of draft water divides with their catchment-based information system output – Main European drainage basins on a 1:1,000,000 scale (designated WSEU1M).

Final delineation of balance regions in terms of areal extent was optionally defined preferably within the range 5,000–10,000 km^2 based on paper maps at the preferred scale of 1:50,000. Balance regions should take into consideration also the Water Framework Directive (WFD) (Directive 2000/60/EC). A GIS map of selected balance regions is included on a CD attached to the *project final report* (PFR) (Petrovič et al.: Hydrological Monograph of the Danube River Basin). The final printed version was prepared at a scale of 1:2,000,000. A list of the balance regions is presented in Table 7.1.

Table 7.1 List of balance regions

Region	River	Profile	Country	Area (km^2)	Min Elev (m)	Mean Elev (m)	Max Elev (m)	WatBal tuning level
1	Donau	Neu-Uln Bad Held	GE	7,586.4	465	749	2,401	2
2	Donau	Donauwoerth	GE	7,471.6	395	535	875	1
3	Donau	Kelheim	GE/AT	7,897.9	337	745	2,752	2
4	Naab	Heitzenhofen	GE	5,414.8	334	497	933	2
5	Donau	Schwabelweis	GE	6,977.0	325	515	1,322	2
6	Isar mit Muehlbaeche	Platting	GE/AT	8,922.0	316	742	2,557	1
7	Inn	Kajetansbrucke	SW/AT	2,156.8	967	2,302	3,652	2
8	Inn	Oberaudorf	AT	7,542.7	464	1,782	3,565	1
9	Salzach	Burghausen	GE/AT	6,627.7	352	1,279	3,369	2
10	Inn	Schaerding	GE/AT	9,192.8	300	616	2,082	2
11	Donau	Achleiten	GE/AT	6,632.4	314	449	1,277	2
12	Traun	Wels	AT	3,505.0	309	842	2,658	2
13	Enns	Steyr (Ortskai)	AT	6,066.2	284	2,302	3,652	2
14	Donau	Kienstock / Krems	AT	9,759.4	194	528	1,741	2
15	Donau	Wien	AT	6,057.7	154	432	1,548	2
16	Dyje	Ladna	CZ/AT	11,881.6	159	427	792	1
17	Morava	Straznice	CZ	9,119.4	164	414	1,469	2
18	Morava	Mor.Sv.Jan	CZ/SK	3,059.0	146	245	878	2
19	Morava	Devin	SK/AT	5,118.9	132	264	1,160	2
20	Vah	Zilina-Budatin	SK	5,684.3	332	855	2,197	2
21	Vah	Sala	SK	5,505.1	109	451	1,371	2
22	Mosoni-Duna	Abda	AT/HU	6,951.6	106	293	1,948	2
23	Raab	Kormend	AT/HU	4,671.5	184	469	1,698	2
24	Raba	Gyor	AT/HU	5,679.1	106	205	792	5
25	Nitra	Nove Zamky	SK	4,048.2	109	328	1,234	1
26	Danube	Komarno	SK/HU	6,734.2	104	152	591	2
27	Hron	Brehy	SK	3,814.3	194	673	1,768	2
28	Ipel	Ipelsky Sokolec	SK/HU	4,831.6	115	313	1,005	1
29	Duna	Nagymaros	SK/HU	3,964.1	99	214	852	5
30	Duna	Dombori	HU	8,700.3	83	136	689	5
31	Sió	Simontornya	HU	10,207.8	91	178	540	1
32	Duna	Mohács	HU	6,638.4	79	163	553	5
33	Mur	Spielfeld	AT	9,523.2	244	1,061	2,742	1

Table 7.1 (continued)

Region	River	Profile	Country	Area (km²)	Min Elev (m)	Mean Elev (m)	Max Elev (m)	WatBal tuning level
34	Drau	Villach	AT	5,581.8	486	1,663	3,365	2
35	Drau	Drau Grenze	AT	6,424.1	340	975	2,361	2
36	Drava	Botovo (incl. Mursko Sred.)	SI/HU/HR	9,671.8	114	337	1,936	2
37	Drava	Donji Miholjac	HU/HR	6,105.2	88	138	520	6
38	Danube	Bogojevo	HU/HR/SM	7,508.2	78	126	827	6
39	Sajó	Felsőzsolca	SK/HU	6,439.3	107	365	1,394	2
40	Hernád	Gesztely	SK	5,312.2	108	507	1,623	1
41	Bodrog	Streda n.B.	SK/UKR	8,557.9	91	352	1,439	2
42	Tisza	Tivadar	UKR/HU/RO	12,646.1	105	641	2,000	2
43	Somes	Dej	RO	8,816.9	232	648	2,303	1
44	Szamos	Csenger	HU/RO	9,024.1	113	237	1,839	2
45	Tisza	Záhony	HU/UKR	5,339.2	98	248	1,465	5
46	Tisza	Tiszapalkonya	SK/HU	6,051.3	87	145	809	5
47	Tisza	Szolnok	HU	11,143.2	78	156	921	5
48	Sebes-Körös	Körösladány	HU/RO	9,218.0	80	270	1,747	1
49	Hármas-Körös	Gyoma	HU/RO	10,843.4	78	274	1,669	5
50	Hármas-Körös	Kunszentmárton	HU	6,922.9	76	95	161	5
51	Tisza	Ocna Mures	RO	9,913.5	253	703	2,100	2
52	Mures	Mihalt	RO	6,221.6	227	532	1,800	2
53	Tarnava Mare	Hoghiz	RO	7,143.9	443	797	1,792	2
54	Olt	Cornet	RO	6,689.0	312	718	2,544	2
55	Olt	Branisca	RO	8,276.5	173	708	2,519	2
56	Maros	Makó	HU/RO	5,753.3	79	204	1,172	5
57	Maros	Szeged	HU	7,913.5	73	94	200	5
58	Stari Begej & Plovni Begej	Hetin-Srpski Itebej	SM/RO	4,226.1	76	180	1,254	2
59	Danube	Slankamen	HU/HR/SM/RO	15,703.7	70	95	491	5
60	Sava	Litija	SI	4,819.3	230	761	2,440	2
61	Sava	Catez	SI	5,263.6	137	498	1,984	1
62	Sava	Zagreb (incl. Catez)	SI/HR	2,276.2	112	275	886	2
63	Kolpa	Sisinec	SI/HR/BiH	7,535.1	95	440	1,348	6
64	Una	Kostajnica	HR/BiH	9,127.3	104	610	1,677	2
65	Sava	Jasenovac	HR/BIH	10,026.4	87	174	865	2
66	Vrbas	Delibasino selo	BiH	4,218.5	141	843	1,970	2
67	Bosna	Modrica	BiH	10,421.6	99	684	1,937	2
68	Sava	Zupanja	HR/BiH	9,018.6	84	203	941	2
69	Drina	Bastasi	BiH/SM	3,090.2	425	1,406	2,296	6
70	Drina	Bajina Bašta	BiH/SM	11,970.5	212	1,069	2,331	2
71	Sava	Sremska Mitrovica	HR/BiH/SM	10,572.3	72	286	1,409	2
72	Kolubara	Draževac	SM	3,575.5	71	286	1,188	2
73	Tamiš	Jaša Tomić	SM/RO	5,294.5	74	418	2,089	2
74	Brzava-Moravica-Karaš-Nera	Virtual 74	SM/RO	6,100.8	7	279	1,400	2

Table 7.1 (continued)

Region	River	Profile	Country	Area (km^2)	Min Elev (m)	Mean Elev (m)	Max Elev (m)	WatBal tuning level
75	Danube	Pančevo	SM/RO	7,003.8	67	101	574	5
76	Ibar	Lopatnica Lakat	SM	7,552.1	225	891	2,178	2
77	Zapadna Morava	Jasika	SM	7,029.0	139	561	1,723	2
78	Južna Morava	Korvin Grad	SM	9,515.5	188	680	1,840	2
79	Južna Morava	Mojsinje	SM/BG	6,008.7	137	675	1,895	2
80	Velika Morava	Ljubičevski Most	SM	7,160.6	73	337	1,724	2
81	Danube	Prahovo	SM/RO	12,864.3	29	330	2,015	2
82	Timok	Tamnič	SM/BG	4,207.5	59	498	1,788	2
83	Danube	Lom	SM/RO/BG	4,331.0	23	176	1,169	6
84	Jiu	Podari	RO	9,254.9	67	456	2,519	2
85	Ogosta	Mizia	BG	3,141.5	27	400	1,792	2
86	Iskar	Orjahovo	BG	8,286.0	31	686	2,665	2
87	Dunare	Corabia	RO/BG	9,107.7	26	155	2,036	5
88	Olt	Stoenesti	RO	8,855.9	60	491	2,535	2
89	Calmatui-Vedea-Teleor	Virtual_89	RO	5,385.2	15	151	600	2
90	Osam	Izgrev	BG	2,177.3	41	430	1,967	2
91	Jantra	Karantzi	BG	6,877.2	32	420	2,282	2
92	Rusenski Lom	Bojichen	BG	2,876.2	34	273	483	1
93	Dunare	Giurgiu/Russe	RO/BG	11,221.7	12	176	2,050	5
94	Arges	Virtual_94	RO	12,090.0	25	394	2,544	2
95	Dunare	Silistra	BG	8,577.9	14	105	447	5
96	Ialomita	Slobozia	RO	9,206.6	19	365	2,505	2
97	The rest of the Danube	Below Silistra	BG	4,863.2	7	243	442	7
98	Dunare	Vadu Oii	RO	8,380.7	9	51	63	7
99	Buzau	Racovita	RO	5,111.1	40	533	1,954	2
100	Siret	Lespezi	UKR/RO	6,053.8	213	455	1,470	2
101	Siret	Dragesti	RO	6,036.0	165	550	1,857	2
102	Siret	Racatau	RO	7,724.8	122	860	2,279	2
103	Barlad	Tecuci	RO	6,801.5	31	219	525	2
104	Siret	Lungoci	RO	9,688.2	12	526	1,777	2
105	Dunare	Galati	RO	9,664.1	6	72	1,406	7
106	Prut	Chernivtsi	UKR	6,798.7	100	513	1,848	2
107	Prut	Ungheni	UKR/RO/MD	8,793.5	31	183	453	1
108	Jijia	Chiperesti	UKR/RO	5,575.7	30	155	587	2
109	Danube	Ceatal Izmail	UKR/RO/MD	13,483.4	5	104	375	6

As far as the GIS component is concerned, we were in contact with the ICPDR – RBM/GIS and we also used the free USGS Digital Elevation Model (DEM) with the only difference being that we used as our source the data set HYDRO1K, which has equal area pixels, so that the area of any particular selected region is given just by the amount of pixels inside the region frontier (inside the chosen polygon).

7.2 WatBal Methodology and Results

The first edition of the Danube River Basin Hydrological Monograph was based completely on national contributions dealing with maps of all studied water balance elements: precipitation, actual evapotranspiration and runoff depth. The role of the co-ordinator of that study was mainly in recommendation and regulation of methods used in particular countries for assessment and mapping of selected water balance elements. The main problem was to connect particular isolines as they crossed state borders, where the aim was to prepare tailor-made maps of all studied elements (precipitation, actual evapotranspiration and runoff depth).

The first study in its "internal" version was published in both Russian and German. At that time it was also agreed that a representative quadrilingual version should be organised by the Water Research Institute in Bratislava. This task was realised in 1988 with significant UNESCO support for the printing of this book and four maps (now already out of stock) (Stančík et al. 1988).

As political changes in Europe began to play a serious role, the interest in a mutual water management approach increased and there were moves to prepare a second edition of the Monograph with extended periods of observation and more sophisticated methods.

The methodology and results of the present study are based on contemporarily available technology and data processing tools. This section consists of three parts. The first deals with the GIS base and tools for improved input data preparation for further study and processing. The second part describes data assembly. The third part of the study provides an evaluation of water balance modelling and tuning for selected balance regions using a modified mathematical model of water balance working with lumped parameters for a balance region. The model WatBal, recommended by the US EPA (Yates 1994, Yates and Strzepek 1994) for evaluation of possible climate change impact in river basins in CEE countries in the 1990s was selected. The WatBal model was used by the author for evaluation of possible climate change in the Nitra River Basin (Petrovič 2002, 1998a, b).

Further in the PFR (Petrovič et al.: Hydrological Monograph of the Danube River Basin) water balance evaluation in grid or isoline form is based on extrapolation of water balance elements from single (lumped) values for balance regions for spatial presentation on maps of water balance for each balance region, or indeed for tailor-made maps covering the whole Danube River Basin. This part of the work was performed by the Institute of Hydrology of the Slovak Academy of Sciences.

7.2.1 GIS Technology in Modelling of Water Balance

A final version of this part of the water balance processing studies was presented at the XXIInd Danube Conference in Brno and also at the Conference "The Danube and Europe: Integrated space applications in the Danube Basin" (23–25 June 2004, Mamaia, Romania) (Petrovič and Bad'urová 2004). An electronic copy of the paper is on a CD attached to the PFR (Petrovič et al.: Hydrological Monograph of the Danube River Basin).

7.2.1.1 Common Geographical Co-ordinate System

It was agreed that for the whole Danube Basin the co-ordinate system of the USGS defined for the Hydro1k DEM for European space will be used:

Projection used: Lambert Azimuthal Equal Area

Units = meters
Pixel size = 1,000 m
Radius of sphere of influence = 6,370,997 m
Longitude of origin = 200,000 E
Latitude of origin = 550,000 N
False Easting = 0.0
False Northing = 0.0

It is optional that for future use the final GIS version can be transformed into the ETRS 89 system.

7.2.1.2 Base Map: Digital Elevation Model

The Digital Elevation Model HYDRO1k was freely available from the Internet. Its resolution: a pixel size of 1 × 1 km. A visualisation of this DEM for the Danube Basin can be seen in Fig. 7.1.

Usage: estimation of basin-weighted areal mean, minimal and maximal elevation, vertical gradient of particular hydrometeorological elements and verification of the areal extent for each balance region.

This seems to be sufficient for such a purpose with respect to meteorological input data precision and values of vertical gradients of respective elements versus vertical simplification of the area into the grid map.

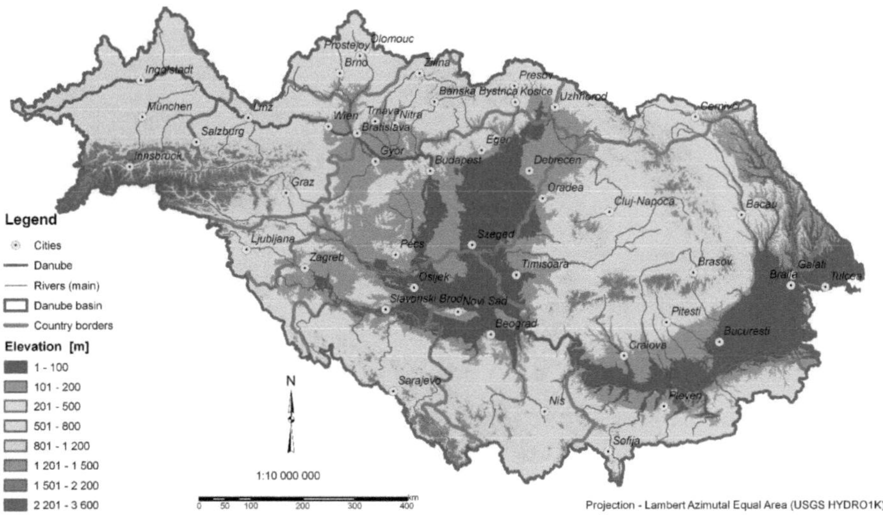

Fig. 7.1 Digital elevation model and the Danube Basin boundary

7.2.1.3 Assessment of Water Dividing Lines and River Net

Construction of water divides was a task for each participating country itself (through IHP UNESCO NCs and national hydrological services). The expected precision was a reference scale of 1:50,000 or better.

The water dividing lines form "boundaries" of balance regions. Therefore, the balance regions should be natural runoff areas (sub-basins or groups of sub-basins) from our point of view with quasi-homogeneous characteristics of vertical gradient of basic hydrometeorological elements. The regions (some of them transboundary) must fully cover the whole territory of the Danube Basin and they should take into consideration the definition of "water districts", in the sense of Art. 2 – Definition, par. 15 of the WFD 2000/60/EC agreed on 23 October 2000 (Directive 2000/60/EC). In order to be used in WatBal, the balance regions represent polygons delineating drainage areas above selected closing profiles with runoff measurement. The region area should be between 5,000 and 10,000 km^2.

The resulting map of selected balance regions can be seen in Fig. 7.2.

The GIS presentation of the river net was finally developed on the basis of the JRC Ispra Italy data used in maps. Dr. R. Hiederer (prepared by Ms. Katalin Bodis) kindly gave the authors the recent version for dissemination and use free of charge within our project. Some gaps in the system have been filled in by the author.

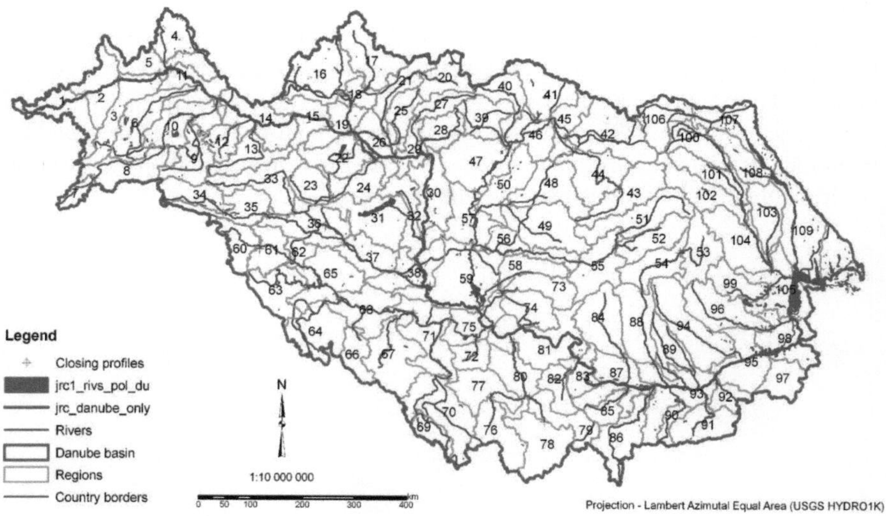

Fig. 7.2 Map of balance regions

7.2.2 WatBal Model Input Data Assembly

7.2.2.1 Input Data in General

Water balance is evaluated for each (sub) basin of the Danube River area. For estimation of all monthly mean input data it is necessary to have basin area and

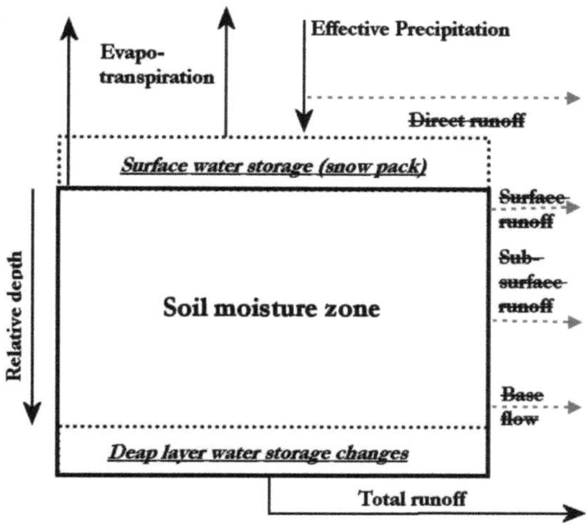

Fig. 7.3 Structure in modified WatBal model (*dashed rectangles* – new introduced components; *dashed gray arrows* – all components of runoff are accumulated in total runoff only)

area/elevation distribution; meteorological data on precipitation (including snowfall period), air temperature (for evaluation of the solid and liquid part of precipitation) and air humidity for potential and actual evapotranspiration estimation – natural water consumption demand. Measured discharge in relation to the closing profile (or difference between inflow and outflow) for each balance region is also needed. A set of methodological supporting EXCEL-files was given for dissemination to all participating countries and the latest version of this methodology is included in the CD attached to the PFR (Petrovič et al.: Hydrological Monograph of the Danube River Basin) (see Fig. 7.3).

7.2.2.2 Basin Characteristics

Water dividing lines of the selected river basin related to the closing profile in suitable discharge stations need to be estimated from "water management" or similar maps at a scale 1:50,000 or better. The Lambert Azimuthal Equal Area co-ordinate system was used according to the USGS definition given in Section 7.2.1.1.

The amount (count) of pixels of the USGS DEM lying inside the selected river basin area is assumed to be the total area of the selected balance region.

Furthermore, the "statistics" function in ArcView used for each selected region gives the mean elevation, elevation range in a basin, and finally areal extent, as can be seen in Fig. 7.4.

7.2.2.3 Precipitation Data

Areal and temporal variability of precipitation is relatively high. For areal mean monthly precipitation estimation, as many precipitation stations with complete data sets as possible should be used. In our pilot basin (the Nitra River Basin) and its close surroundings, we selected 37 precipitation stations, which were originally prepared on the basis of long-term (1901–1970) precipitation series evaluation by

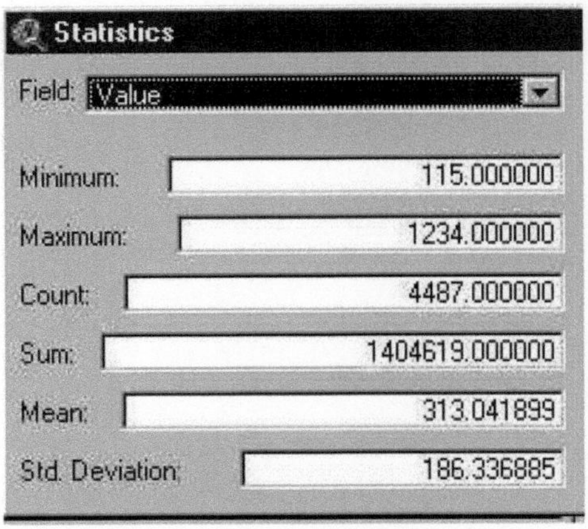

Fig. 7.4 Application of the statistics function in ArcView (v. 3.2) for the Nitra River Basin

Šamaj and Valovič (1978). These data series were verified and extended up to the year 2000 in co-operation with P. Faško from the Slovak Hydrometeorological Institute.

The number of precipitation stations used in our study was limited by the amount obtained from participating countries. An overview of existing information is given in Table 7.2 .

Areal monthly precipitation totals could be obtained by different methods. The most precise method would be evaluation of areal precipitation from monthly maps of isohyets (drawn and processed e.g. by GIS tools or manually by the expert), but this is an extremely time-consuming approach. For orographic conditions in Slovakia, the monthly precipitation data seem to have an approximately linear increase of precipitation with elevation (Petrovič 1972), except for some lee localities. The situation in the Austrian Alps can be different, as shown in Appen-

Table 7.2 Amount of stations used in precipitation map assembly

Country	Number of precipitation stations	Approximate area (GIS) in the Danube Basin (km^2)	Density of P stations (km^2/station)
AT	687	80,338	117
BG	32	47,926	1,498
BiH	95	37,710	397
CZ	56	21,627	386
GE	557	55,828	100
HR	19	33,710	1,774
HU	82	92,759	1,131
MD	7	12,650	1,807
RO	74	230,739	3,118
SI	14	16,154	1,154
SK	204	46,678	229
SM	55	88,394	1,607
UA	17	30,759	1,809
CH	2	1,782	891
SUM	1,901	797,054	

dix 4 of the PFR (Kling and Nachtnebel 2004). Anyway, our assumption concerning a linear increase of precipitation was applied to the whole Danube River Basin.

Areal precipitation totals for mean elevation of the analysed (sub) basin were used, and all the processing was done in an EXCEL worksheet by use of the TREND internal function in the recommended way using as the muster the file NIin6190.xls, sheet RRNitMM which is on the CD attached to the PFR (Petrovič et al.: Hydrological Monograph of the Danube River Basin).

7.2.2.4 Meteorological Data Related to the Basin Mean Elevation

For recalculation of air temperature and air humidity for the vertical "gravity" point (weighted mean elevation, obtained in ArcView – see above – h_w) at least three meteorological stations have to be used.

Not all of them must lie directly in the studied (sub) basin, but all of them should sufficiently represent the character of the landscape, its orientation to cardinal points and as far as it is possible also the elevation range. A regression approach is used; this means that for each region at least three meteorological stations need to be defined.

Evaluation of air temperature in relation to mean elevation is based on the linear decrease of temperature with the increase in elevation. The calculation is performed for each month of the 30-year period separately in EXCEL employing the TREND function using parameters giving the non-zero "b" coefficient of linear regression. The respective EXCEL file is included in the CD attached to the PFR [2].

Estimation of air humidity is more complicated. On the basis of the equation of vertical distribution of vapour pressure, decrease of vapour pressure e with elevation growth has to be exponential (nearly linear for log e). Comparison of results obtained by linear regression of relative air humidity with elevation and by linear regression of log e led practically to the same results. The ratio of the linear estimation of relative air humidity and the relative humidity obtained from exponential regression lies for the Nitra River Basin in the range from 0.9706 to 0.9994 with an average equal to 0.9958. Obtained results for the Nitra River Basin allow us to use linear regression of relative air humidity directly; individual error will not exceed the acceptable value. The respective EXCEL file is included in the CD attached to the PFR.

Potential evapotranspiration (PET), in principle, is estimated according to Budyko–Zubenokova (Kuz'min 1976). The reason for choosing the Budyko–Zubenokova method is the fact, that in principle only air temperature and air humidity is needed. All other methods recommended for example by FAO need more input data sets. Precision of the obtained PET data seems to be acceptable (Petrovič 1980).

Selection of the appropriate region for evaluation can be done on the basis of the long-term areal means of air temperature and air humidity from Fig. 7.5 according to Konstantinov (Konstantinov et al. 1971).

A set of nomograms prepared for different geobotanical zones (an example for the region "forest-step" is shown in Fig. 7.6) was digitised and incorporated initially

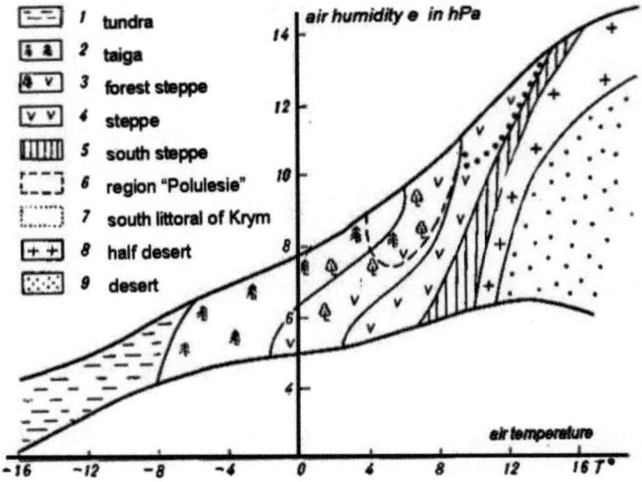

Fig. 7.5 Geobotanical balance regions according to Konstantinov

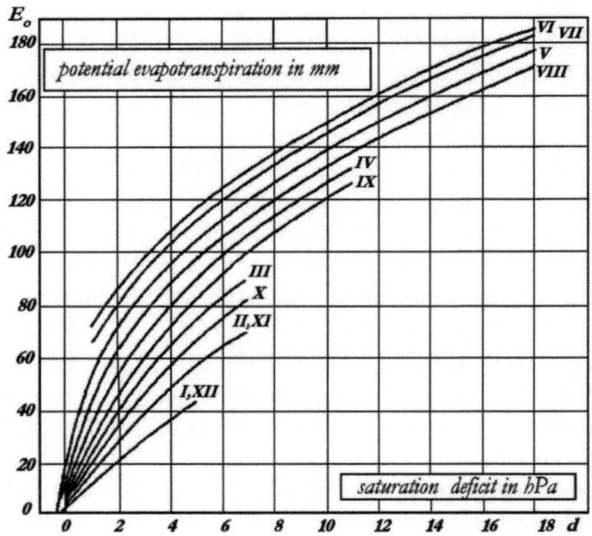

Fig. 7.6 Potential evapotranspiration for "forest-step"

into the program in Fortran77 and later into the EXCEL worksheet. The value of PET is obtained with the help of the "LOOKUP TABLE" in rows representing a given month as an interpolation between columns giving PET values in the first step for the computed integer part of the saturation deficit, in the second step for the computed real fraction of the saturated deficit. In our case the EXCEL file allows selection of one of three digitised (from nine in total) geobotanical regions named in Russian literature:

- mixed forest (deciduous and coniferous)
- deciduous forest
- forest-step

while for our Slovak territory, the latter seems to be optimal up to the elevation of 500 m a.s.l.

The file for PET estimation was prepared for dissemination to participating countries and is included in the attached CD in the PFR (Petrovič et al.: Hydrological Monograph of the Danube River Basin).

7.2.3 *WatBal Model Tuning*

7.2.3.1 Model Parameters and Variables Used

A detailed description of the model methodology is given in the PFR (Petrovič et al.: Hydrological Monograph of the Danube River Basin).

In principle for water balance evaluation a basic equation is valid, in our case in the form:

$$\mathrm{PRE} = \mathrm{AET} + \mathrm{FLOW} + \mathrm{DSM} + \mathrm{DDWS} \tag{7.1}$$

Where PRE is precipitation areal mean total for a given time step
AET is the actual evapotranspiration sum according to Budyko–Zubenokova

$$\mathrm{AET} = \frac{\mathrm{PET}}{2 * \mathrm{WCRIT}} * \left(\mathrm{SM1} + \mathrm{SM1} * \frac{1 - \frac{\mathrm{PET}}{2*\mathrm{WCRIT}}}{1 + \frac{\mathrm{PET}}{2*\mathrm{WCRIT}}} \right) \tag{7.2}$$

and

AET = min(AET,PET)

PET is potential evapotranspiration according to Budyko–Zubenokova (methodological EXEL file is included in the attached CD to the PFR [2])

WCRIT – value of “critical” soil moisture content (above the WP) threshold, for the actual soil moisture (WACT) lower than this value the actual evapotranspiration (AET) is less than the potential one (PET) and it can be expressed in the following form:

$$\mathrm{AET} = \mathrm{PET} * \mathrm{WACT}/\mathrm{WCRIT} \tag{7.3}$$

and an increase in this parameter means a decrease of actual ET (the opposite also being true).

In the final version of the model we have expected possible values of WCRIT in the range of 50–150% of SFFC, this condition does not need to be obligatorily applied in the WatBal model version 3b and the final version included in the CD attached to the PFR (Petrovič et al.: Hydrological Monograph of the Danube River Basin).

FLOW (in the model FLOWM) is runoff depth
DSM is a change (difference SM2-SM1) in soil moisture content estimated for the actual time step, where $SM1_{t+1}=SM2_t$
DDWS is a change (difference WDELTM2 – WDELTM1) in "deep water storage" depth; and all variables in the equation are in the same units – mm per time step, in our case the time step is one month.

Within the model run some supporting parameters (representing in a certain sense boundary conditions or thresholds levels) are needed; these parameters are also used in the model tuning process.

ATRAIN – is air temperature threshold; if the air temperature is higher, all precipitation in the basin is in liquid phase (must be positive);
ATSNOW – is air temperature threshold, if the air temperature is lower, all precipitation in the basin is in solid phase (must be negative); the condition ATRAIN $\geq$ ATSNOW must be valid;
MDGFAC – is monthly snowmelt degree factor (potential depth of snow water, which can melt thanks to each positive degree of air temperature TEMP above zero);
PET is potential evapotranspiration according to Budyko–Zubenokova (methodological EXEL file is included in the CD attached to the PFR [2])
PREL – is liquid portion of monthly precipitation, where

$$\text{PREL} = \text{PRE} + \text{SNMELT} - \text{SOLPRE}$$

PRIESKO – represents a "fast seepage coefficient" and gives the portion of liquid precipitation, which enters (infiltrates into) the soil layer at the beginning of the time step and plays a role in the AET evaluation in a given time step; it must be in the interval <0; 1> (for tuning a help construction is used);
SFFC – is a soil moisture of full field capacity, it represents saturated water content above the wilting point (WP) in the soil layer (the implicitly assumed soil layer has a thickness of 1 m);
SNMELT – is the portion of snow which melts in a given time step as a function of ATEMP and MDGFAC
SOLPRE – is the monthly portion of precipitation which is stored at the surface as solid precipitation, where

for (ATEMP <= ATSNOW)SOLPRE=PRE
for (T>ATRAIN)SOLPRE=0
for (ATSNOW<=ATEMP and ATEMP<=ATRAIN)

$$\text{SOLPRE} = \text{PREC} * \left(1 - \frac{\text{ATEMP} - \text{ATSNOW}}{\text{ATRAIN} - \text{ATSNOW}}\right) \tag{7.4}$$

WACT – is an actual soil moisture value (SM1 and SM2 at the boundary of the time step, or the mean of both values) above the wilting point (WP) in a given time step; printed just for information, in a model run, where

$SM1_t = SM2_{t-1}$ is the soil moisture content at the beginning of a given time step and

SM2 is the soil moisture content at the end of a given time step, where

for SM1 – AET <= WSFFC SM2 = SM1 – AET
for SM1 – AET > WSFFC SM2 = WSFFC and

$$\text{WDEEP}_t = \text{WDEEP}_t - 1 + \text{SM1} - \text{AET} - \text{WSFFC} \tag{7.5}$$

WSFFC – is the Water content in soil by the Full Field Capacity
WSURF – is the water layer stored in a given month at the soil surface, where

$$\text{WSURF}_t = \text{WSURF}_{t-1} + \text{SOLPRE} - \text{SNMELT} \tag{7.6}$$

Finally, the starting sheet for WatBal tuning (Fig. 7.7) shows some of above-mentioned "hidden" internal parameters used at the beginning and end of each time step and some "starting and ending" values for different variables (e.g. starting soil moisture SM – in cell G3 and/or snow pack water content WSURF in cell K3 in the model file).

7.2.3.2 Tuning Process

In this section, the tuning of required parameters is described. The EXCEL version "3ab" of the WatBal model was used in the participating countries. The version "3b" was used in the WRI in the final fine calibration of results. Details of the calibration process are available to the involved experts on request or from the CD attached to the PFR (Petrovič et al.: Hydrological Monograph of the Danube River Basin).

B9 =0,5*C7+C7*(ABS(B8)/(1+ABS(B8)))

	A	B	C	D	E	F	G	H
1	**Nove Zamky**							
2	VARIABLES - STARTING DATA						SM	WDELTM
3			START	-1,4235	5,889		104,53	0
4			END	-0,1025	4,452		104,53	0,0
5	OPTIMISATION PARAMETERS				in gray			
6		WCRIT	WSFFC	ATSNOW	ATRAIN	ATMIX	MDGFAC	PRIESK
7		***90,3124***	114,80887	-0,102521	4,452435	4,55496	74,791	0,45000
8		-0,4018						0,81818
9		90,3124						

Fig. 7.7 Tuning parameters of WatBal in version V3b

A balance region is taken as a "virtual point/virtual hydrometeorological station" located in the reference elevation of the basin.

As precipitation input or the areal mean of precipitation is taken, this means precipitation is estimated from the GIS approach or data values computed for the reference elevation value from the linear gradient of precipitation totals with the elevation. Estimation of areal precipitation is based on as many stations as possible in the basin (and nearest surroundings), it is in principle presented in Section 7.2.2.3.

The runoff depth is taken as the volume of discharge in the closing profile divided by the total basin area and duration of the time step. Details of the tuning procedure are described in the PFR (Petrovič et al.: Hydrological Monograph of the Danube River Basin) in the attached CD.

The model for PET was initially re-written from the FORTRAN 77 version into EXCEL. It is based on two components (see Section 7.2.2.4) – the air temperature and humidity (methodological support in the file NIin6190.xls) and separate estimation of PET on the basis of air temperature and air humidity using the Budyko–Zubenokova method (file WB-POTET.xls).

Air temperature data is further required directly in the file for the WatBal model (WB-MODELV3b.xls) for decisions on the liquid and/or solid part of precipitation and on snowmelt, the snow pack melting process and snow water equivalent in each given time step.

This step is performed with a set of target tasks:

Values of ATSNOW and ATRAIN can be judged by evaluation of basic elevation range and considering dynamic meteorology equations. This means that ATRAIN is the temperature in the basin weighted average elevation at which the temperature at the highest point of the water dividing line is equal to zero. At a temperature equal to or higher than ATRAIN, precipitation in the whole watershed is in the form of rain.

On the other hand, the ATSNOW represents the temperature in the basin-weighted average elevation at which the temperature at the lowest point of the water dividing line is equal to zero. At a temperature equal to or lower than ATSNOW, precipitation in the whole watershed is snowfall. For this first judgement, the vertical temperature gradient can be taken as a critical adiabatic gradient –0.65°C/100 m. In the model version V3b by filling in the elevation range of the balance region in cells "TIMESEQ!L394:L396" the starting boundary values of ATSNOW and ATRAIN can be seen in cells D3:E3 – see Fig. 7.7. Vertical air temperature inversions are not taken into consideration in this study.

Optimisation of the model through tuning is performed in EXCEL with the use of the SOLVER "ad-ins" procedure. We used for tuning of the model the optimisation parameters WCRIT, WSFFC, ATSNOW, ATRAIN, MGDFAC and PRIESKO (calculated with the help of cell H8 proper setting), as can be seen in Fig. 7.7.

Snowmelt and mixed precipitation for different elevations in the catchments for the air temperature between ATSNOW and ATRAIN are considered, too. The snowmelt monthly degree factor MDGFAC (*M*ONTH*D*E*GFAC*) was in the daily time step application for East Slovakia around 4.0. Practical "tuning" of the model by method trial and error has shown that for a monthly step another value (somewhere between 10 and 120) has to be used. A detailed explanation is not easy and variance of obtained values has a certain relation to the vertical extent of a basin. In the case where the value of MDGFAC is out of the mentioned range offered for solution by the SOLVER, it cannot be accepted.

SFFC and WCRIT should be the same at the beginning and at the end, which represents the basic assumption, that any decrease of WACT below the maximum possible available soil moisture at the SFFC causes a proportional decrease of actual evapotranspiration according to Eq. (7.2). This can be achieved by optimising the value of PRIESKO, if other values are in a converging range.

The period of data processing was enlarged from that initially selected (1951–1980) to the period of 1951–2000 in which the binding period agreed by all participants and used in computation and map assessment was the 1961–1990 period.

All the computation is done in a DO LOOP, where the aim is to achieve the same final actual soil moisture SM, as the starting one (this is simply done by setting the new starting soil moisture content equal to the "old" ending soil moisture content before the next optimisation of the WatBal). Similarly, the starting and final surface moisture storage WSURF representing the snow pack water content should be as far as is possible the same.

On the other hand, philosophically we can consider the representativeness of the chosen 30-year period for the WatBal application and reflect on the "identity" of the initial Autumn in the year before the analysis starts and the Autumn in the latest year of processing. We are talking about the representative period of observation, but such a period is representative just by definition.

Practical tuning showed that model tuning has a convergent solution for quasi-real values of SFFC between 70 mm (light sandy soils) and 300 mm (heavy soils). In cases of extreme values of difference between precipitation and potential evapotranspiration, the parameter WCRIT can be used independently on the SFFC (but must stay positive).

The results of modelling are actual evapotranspiration, soil moisture and some other "helpful" characteristics and they are computed within the model run for each monthly time step.

It is clear, that in the assessment of mean monthly water balance the resulting actual evapotranspiration must have the same yearly total. The influence of choosing SFFC and tuning PRIESKO is propagated in variation of the yearly course of monthly evapotranspiration totals, as is presented in Fig. 7.8; nevertheless, the annual course of these elements was not the main aim of this project.

The obtained results show that evapotranspiration values at sufficient soil moisture for lower WCRIT are higher in the first half of the year than when there is insufficient soil moisture, on the other hand a higher WCRIT causes a certain delay in soil moisture use in relation to evapotranspiration needs.

By tuning the mutual compensation (relation) of WCRIT's and PRIESKO's influence on the error minimising process a relation between these two parameters was found, as can be seen from Fig. 7.9.

Anyway, for the final "fine" tuning with version V3b the difference between SFFC and WCRIT can be accepted.

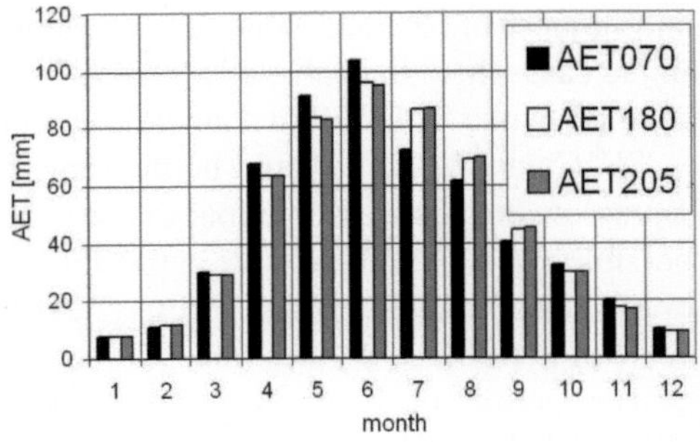

Fig. 7.8 Resulting monthly mean areal evapotranspiration AETxxx obtained for different values of starting critical soil moisture content WCRIT (xxx in mm) with optimised PRIESKO coefficient

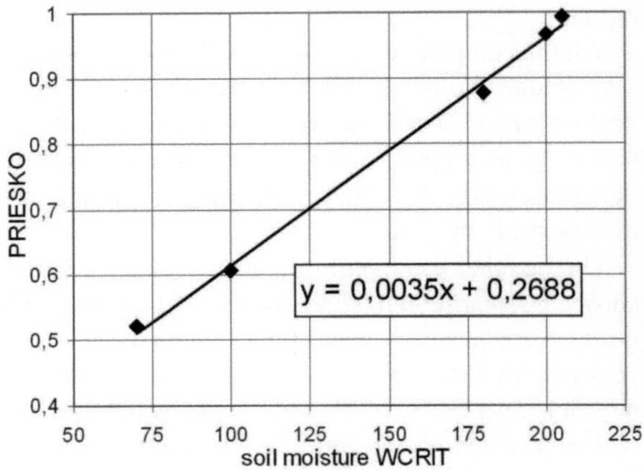

Fig. 7.9 Optimised values of PRIESKO for chosen WCRIT values for mean long-term water balance

Tuning is performed in the EXCEL worksheet, where input columns represent air temperature TEMP, precipitation PRE, potential evapotranspiration PET and observed runoff depth (discharge expressed in mm/month) FLOWM.

It is recommended to set initial values as follows:

Parameters for the first run of tuning
value WCRIT (with the help of the value) in B8 = 1
value WSFFC in C7 = 200
value MDGFAC in G7= 80
Parameter for tuning infiltration coefficient
PRIESKO in H8 = 1

Obtaining (in a DO LOOP) the minimum difference between soil moisture content SM (G3:G4) and surface (solid phase) water storage WSURF (K3:K4) at the beginning and at the end of all time steps is the first criterion for model tuning. In our situation it was not fully possible to gain a closed cycle of elements for fulfilling all the main criteria – the final water balance error (BDELTA) equal to zero, the mean of deep water storage (representing in a certain sense the groundwater aquifer) WDELTM (cell H390 in the model file) equal to zero and the final deep water storage WDELTM (cell H372 in the model file) equal to zero. Because of this circumstance, the tuning in different couples of criteria was performed in EXCEL using the internal "Add-Ins" function SOLVER. A detailed description is outside the frame of this report; some information is present in the PFR (Petrovič et al.: Hydrological Monograph of the Danube River Basin) (Chapter 9.2; Appendix 3). More details are included for dissemination together with the WatBal model files set in the CD attached to the PFR.

Within previous tasks in the WRI Bratislava, we have already tried a different approach to rainfall–runoff modelling, and model tuning was based on minimising the differences between measured and modelled runoff depth. For example simulations based on antecedent precipitation index (API) and the non-linear regression approach have shown some interesting results.

However, the climate change impact scenario application in non-linear models caused significant problems. The present version of the model uses measured runoff depth and for tuning just minimisation of the difference in the balance equation and setting the long term mean of deviation in deep soil water storage as close as possible to zero [see Appendix 3 in the PFR (Petrovič et al.: Hydrological Monograph of the Danube River Basin)].

A sample of the model results processed with a monthly time step during the 30-year period (360 time steps) for one balance region can be seen in Table 7.3.

In this Table TEMP is mean air temperature (°C), PRE is precipitation total; PET and AET are potential and actual evapotranspiration monthly mean and finally FLOWMM is the runoff depth; all water balance elements are in mm. All the available results (final tables) of the WatBal application are available in a separate

Table 7.3 Example of the WatBal model results: Monthly course of long-term mean values of balance elements up to the profile Žilina

Month	TEMP	PRE	PET	AET	FLOWMM
1	−5.06	55.89	9.39	5.54	41.08
2	−3.39	52.82	14.03	8.99	42.54
3	0.09	53.21	29.53	24.06	48.80
4	5.08	65.34	57.17	45.74	60.24
5	10.09	99.29	85.83	65.14	57.72
6	13.06	113.46	102.94	72.90	51.85
7	14.44	101.16	99.58	64.44	41.26
8	13.90	94.82	81.98	53.55	37.95
9	10.53	73.93	54.60	35.30	28.45
10	6.12	66.57	37.62	23.44	34.02
11	0.94	76.64	18.02	11.81	30.73
12	−3.38	69.96	9.66	6.19	38.92
Mean or Total	5.2	923.1	600.4	417.1	513.5
Maximum	16.69	237.50	115.46	99.34	150.84
Mean	5.20	76.92	50.03	34.76	42.80
Minimum	−10.87	5.52	6.23	2.84	14.47

EXCEL file and in hard copy, they are printed as Annex 4 in the PFR (Petrovič et al.: Hydrological Monograph of the Danube River Basin).

7.3 Maps of Balance Elements

This section explains the preparation of the final maps of mean annual precipitation (Fig. 7.10), actual evapotranspiration and runoff (1961–1990) for the Danube River Basin.

7.3.1 Methodology

The maps were prepared based on data provided by participating countries and their resolution is 1 km. The precipitation map was constructed by interpolation of measured rain-gauge data. Kriging with extended drift was used for interpolation. The map of actual evapotranspiration was based on data from WatBal simulations performed for the sub-basins. Actual evapotranspiration simulated by WatBal for each sub-basin was attributed to the centre of gravity of the sub-basin. The map was constructed by interpolation among the centres of gravity by means of kriging with extended drift. The same approach was used to construct the map of mean annual runoff.

7.3.2 Map of Mean Annual Precipitation

The precipitation map is based on data from 1901 stations (rain gauges). The numbers of gauges from particular Danube countries and their densities are shown in

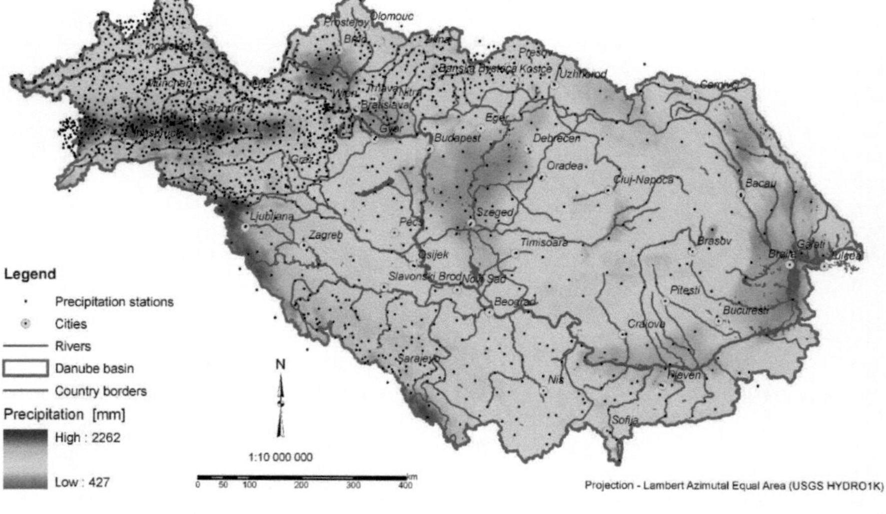

Fig. 7.10 Map of mean annual precipitation 1961–1990

Fig. 7.11 Location of precipitation stations available for interpolation of the precipitation map

Table 7.2. The spatial distribution of precipitation stations is shown in Fig. 7.11. Geostatistical analysis of the data revealed two correlation lengths – about 100 km and about 240 km. The semivariogram was fitted for the first one. The interpolated precipitation map is shown in Fig. 7.13. Compared to other published regional maps, which were based on much more data, precipitation in mountain ranges (e.g. the Alps, the Western Carpathians) in our map is underestimated.

A comparison of measured precipitation with precipitation extracted from the interpolated map is presented in Fig. 7.12. It should be noted that measured precipitation represents a point value, while the extracted precipitation represents the area of 1 km^2, which is the resolution of the precipitation map. The two values are therefore a priori different. Further differences between measured and extracted precipitation may be caused by the differences between the true elevation of the station that was used in interpolation and the elevation provided by the specific grid of the digital elevation model. Nevertheless, Fig. 7.12 shows that for the majority of the

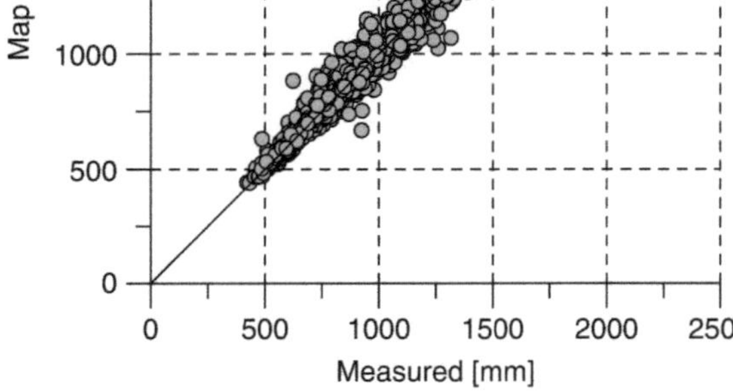

Fig. 7.12 Comparison of measured precipitation with precipitation extracted from a precipitation map; the diagonal represents the 1:1 line

points both values were comparable. Most of the differences among measured and interpolated precipitation were within the interval of ±10%.

Mean differences (and their standard deviations) in precipitation for the Danube countries calculated as the averages for gauges situated in a particular country are: Austria +2% (5%), Bulgaria +3% (4%), Bosnia and Herzegovina +2% (6%), Croatia –1% (2%), Czech Republic +1% (7%), Germany +1% (6%), Hungary +0% (2%), Moldavia 0% (3%), Romania 0% (4%), Slovenia 0% (3%), Slovakia +1% (1%), Serbia and Montenegro +1% (5%), Ukraine 0% (3%).

7.3.3 Map of Mean Annual Actual Evapotranspiration

Mean annual actual evapotranspiration (ETa) for each sub-basin was simulated by WatBal. The ETa values were attributed to the centres of gravity of the sub-basins. An evapotranspiration map was then created by interpolation among the centres of gravity using kriging with extended drift. The map of spatial distribution of mean annual evapotranspiration is shown in Fig. 7.13. Figure 7.14 shows the comparison of ETa from WatBal with sub-basin actual evapotranspiration extracted from the interpolated map. It reveals that the sub-basin values extracted from the map are rather different from those calculated by WatBal. The mean difference for all the sub-basins is about 3%, which means that the evapotranspiration from the map for the Danube Basin is 3% higher than that from WatBal. The differences are within the interval ±10% for about 56% of the sub-basins (Table 7.4). The total number of processed sub-basins is 101 instead of 109 listed in Table 7.1 (the above-mentioned number of water balance regions), because for some sub-basins all the WatBal values were not calculated due to problems with the water balance equation.

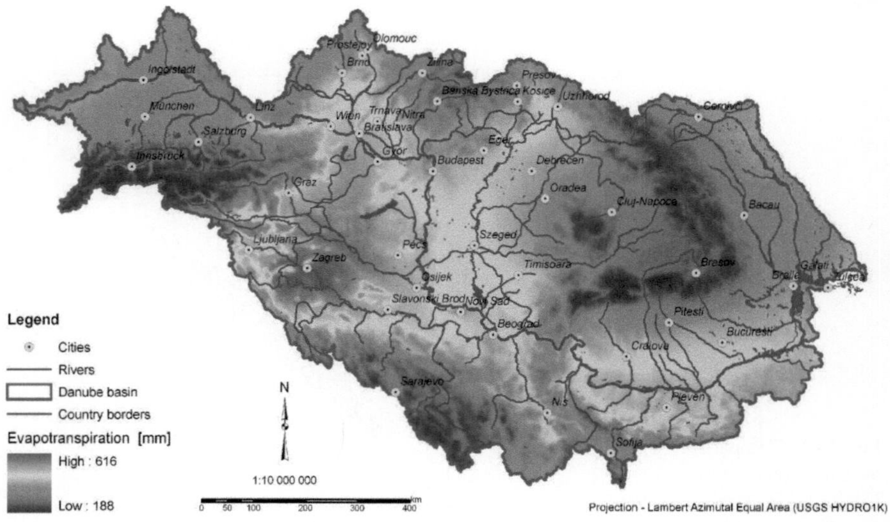

Fig. 7.13 Map of mean annual actual evapotranspiration 1961–1990

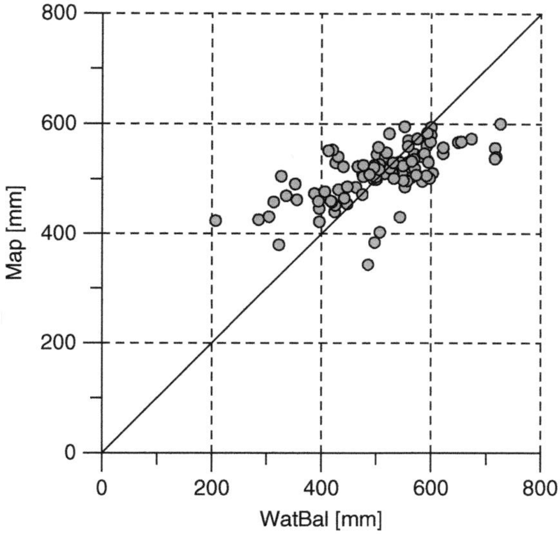

Fig. 7.14 Comparison of sub-basin actual evapotranspiration simulated by WatBal and extracted from the interpolated map; the diagonal represents the 1:1 line

Table 7.4 Comparison of evapotranspiration and runoff extracted from interpolated maps with WatBal (evapotranspiration) and measure values (runoff); negative differences mean that the values from the map are smaller

Differences (%)	Evapotranspiration No. of sub-basins	Runoff No. of sub-basins
Higher than –40%	0	3
–30 to –40%	1	3
–20 to –30%	6	9
–10 to –20%	16	13
0 to –10%	30	14
0–10%	27	14
10–20%	8	20
20–30%	4	9
30–40%	4	5
Higher than 40%	5	11

The spatial distribution of the differences exceeding 10% is shown in Fig. 7.15. It can be seen that many such differences occur in mountain sub-basins.

7.3.4 Map of Mean Annual Runoff

The map of mean annual runoff was constructed in the same way as the map of mean annual evapotranspiration. Measured runoff values were attributed to the centre of gravity of each sub-basin and kriging with extended drift was used to interpolate the map. The interpolated map of mean annual runoff is shown in Fig. 7.16. Figure 7.17 shows the correlation between measured runoff and runoff extracted from the interpolated map. The mean difference for all sub-basins was 8%, which means that the

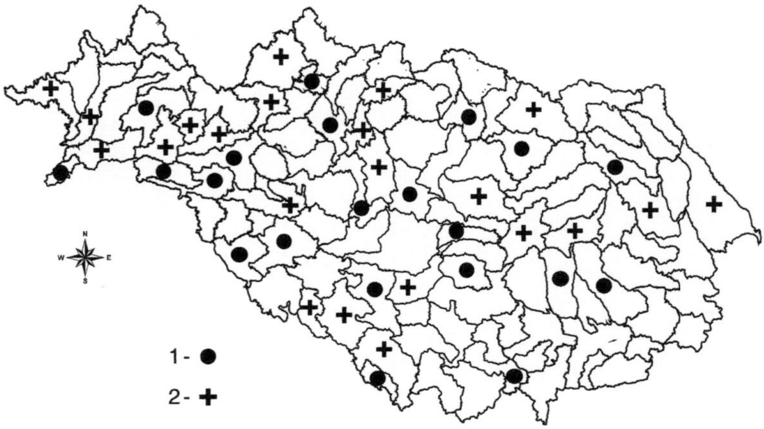

Fig. 7.15 Sub-basins of the Danube Basin and spatial distribution of differences between ETa (from WatBal) and mean annual evapotranspiration extracted from the interpolated map; 1-values from the map are lower than ETa by more than 10%; 2-values from the map are higher than ETa by more than 10%

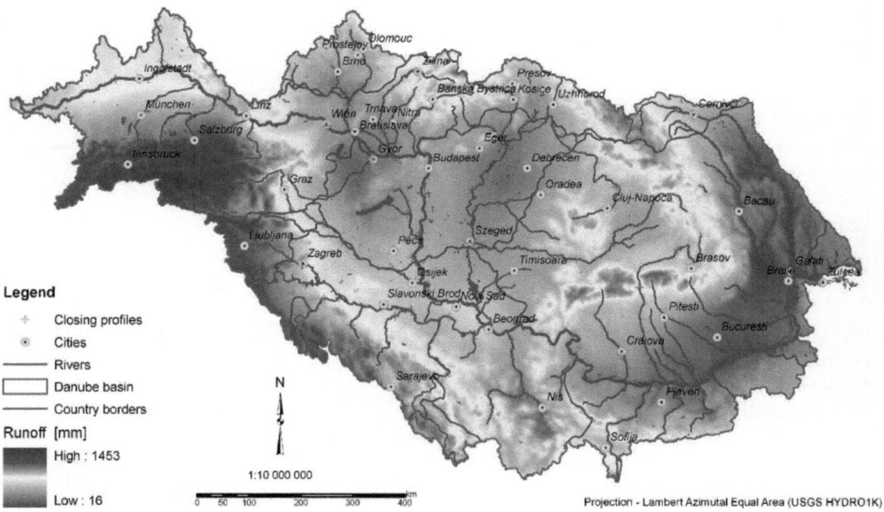

Fig. 7.16 Map of mean annual runoff 1961–1990

map overestimated measured runoff for the whole Danube Basin by 8%. Table 7.4 indicates that small differences among measured and interpolated runoff, i.e. the ones from interval ±10% existed just for about one fourth of the catchments. The differences were higher than ±20% for about 40% of sub-basins.

The spatial distribution of the differences exceeding 20% is shown in Fig. 7.18. It can be seen that such differences are usually connected with sub-basins situated at lower elevations.

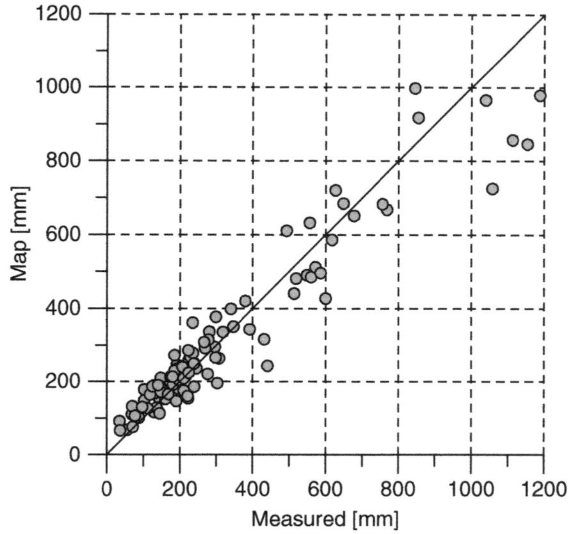

Fig. 7.17 Comparison of measured runoff with the runoff extracted from the interpolated map (Fig. 7.18); the diagonal represents the 1:1 line

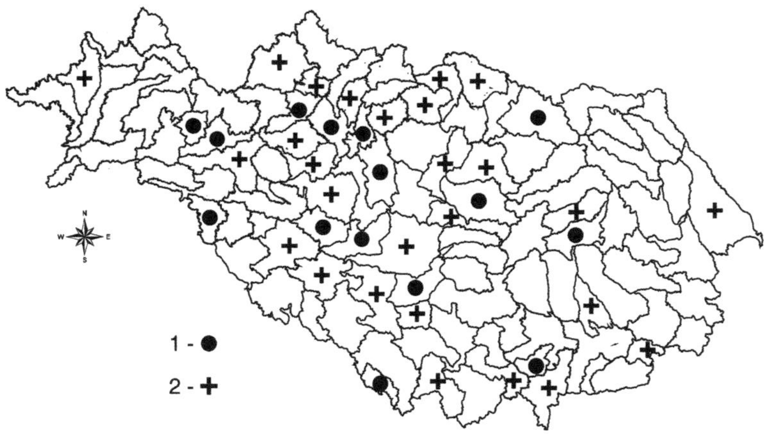

Fig. 7.18 Sub-basins of the Danube Basin and spatial distribution of differences between measured runoff and runoff extracted from the interpolated map; 1-values from the map are lower than measured runoff by more than 20%; 2-values from the map are higher than measured runoff by more than 20%

7.3.5 GIS Analysis Conclusions

In the Section 7.3 it was analysed the approach and performed the application of methodology for the Danube Basin.

A summary of final results related to individual countries is shown in Table 7.5. It can be seen from this table, that the total area of the Danube Basin included into processing was only 797,054 km^2. This is because "small" parts of the Danube

Table 7.5 Selected characteristics of the Danube countries; statistics of precipitation, actual evapotranspiration and runoff were extracted from maps, explanation is given in Section 7.3

	Area in the Danube Basin		Precipitation (mm)	Actual evapo-transpiration (mm)	Runoff depth (mm)
Country	km^2	%	Mean	Mean	Mean
CH	1,782	0.22	1,093.6	343.0	916.6
D	55,828	7.00	969.8	507.4	449.9
AT	80,338	10.08	1,040.2	482.6	603.5
CZ	21,627	2.71	694.0	524.7	196.4
SK	46,678	5.86	716.1	516.5	234.9
HU	92,759	11.64	599.2	559.2	133.4
SI	16,154	2.03	1,308.5	562.3	659.3
HR	33,710	4.23	935.9	582.1	395.0
SM	88,394	11.09	758.8	509.9	252.7
BiH	37,710	4.73	1,052.2	499.9	501.0
RO	230,739	28.95	676.0	493.2	198.5
BG	47,926	6.01	669.1	540.4	164.0
UA	30,759	3.86	832.9	496.9	297.2
MD	12,650	1.59	579.8	523.5	74.0
Sum or mean	797,054	100.00	784.6	513.5	292.2

Basin in some countries (Italy, Poland, Macedonia, Albania) and part of the Danube Delta below the profile Ceatal Izmail were not included into Table 7.5.

7.4 Open Topics

The application of the WatBal model was not possible for some balance regions. The reasons for this vary.

Up to the end of 2005, almost no data were obtained from transboundary regions for Bulgaria/Romania.

A further reason is, that in some balance regions between the up and down stream profiles (inflow and outflow) the resulting *discharge gives negative values* for some time steps (predominantly in the summer time), and for some balance regions (e.g. Nagymaros in Hungary and Kostajnica in Serbia and Montenegro) even a negative long-term mean was found. This can only be to a limited extent explained by the very intensive river bank infiltration which is enriching the horizon of the shallow ground water recipient and this water is consumed by evapotranspiration thanks to the capillary rise from the ground water level to the layer of the active root zone of existing vegetation. To this group belong practically all the Hungarian regions for which the EXCEL structure for tuning was prepared by Hungary. A similar situation occurred in Serbia and Montenegro.

The present version of the model does not allow evaluation of such a situation without further model modification. The co-ordinator tried to perform such a model change and calculate water balance for such cases, where the "missing" discharge is formally considered as an additional amount of precipitation. With full incorporation of such infiltrated water into the "effective precipitation" it could occur that such a sum of precipitation is higher than the sum of potential evapotranspiration and resting (positive) runoff depth. This means that part of the problem is really caused by errors in estimation of input elements.

From the aspect of data quality it was not possible to wait for mutual unification of method for precipitation corrections in all the Danube countries, therefore non-corrected precipitation was used. A similar situation also occurs with some discharge measurements performed by methods not unified by the trans-boundary (bilateral) river commissions and this can significantly contribute to discrepancies in interbasins in which discharge "contribution" is calculated as a difference between discharge and all inflows, measured in different countries.

In a few cases, we also had a situation, where mean runoff from the region was greater than the mean precipitation total. Such a situation is acceptable in a karst region when the surface water divide is different from the subsurface one. Participants of the project in Serbia solved the situation by approximate estimation of the water divide for the ground water aquifer.

Finally yet importantly, it is necessary to make some theoretical remarks on the limits of method used that was discussed by the Austrian participants in the PFR [Appendix 4 (Kling and Nachtnebel 2004)] and by a Slovak reviewer of the final draft [M. Lapin, personal communication] related to precipitation in mountainous areas:

The mean systematic error of precipitation total measurements ranges from about 15% in the lowland to more than 25% in the mountains. This may explain some discrepancies in the obtained results. Determination of PET by the Budyko–Zubenokova method is reliable in the case of terrain with simple topography (in the lowland preferably), but not in high mountains with steep slopes. It seems that the PET in the mountains also in relation to Fig. 7.14 needs to be verified.

Finally, within the Danube River Basin, there are also areas like the delta or similar to the delta, for which the surface runoff is due to the very flat top of soil which is practically not measurable. In such cases, the region is very often full of irrigation and amelioration canals and water balance here should be estimated in a different way.

Output tables from the application of the WatBal model (for basins where calculation was possible) are included in the PFR (Petrovič et al.: Hydrological Monograph of the Danube River Basin).

Discussion of the results of the final draft took place at the NCs IHP UNESCO chairmen meeting in Passau in June 2005 concluding in agreement with publication of this material (Petrovič et al.: Hydrological Monograph of the Danube River Basin).

7.5 Conclusions and Recommendations

7.5.1 *Achievements*

A harmonised and unified methodology in consonance with the provisions of the WFD was used for general basin-wide water balance estimation.

The water balance covers 92% of the Danube Catchment area and it is the first comprehensive water balance estimation of this extent ever prepared for the Danube River Basin.

The assembly of hydrological and meteorological data for this project finished in December 2005 and the collected data can be used for further expected studies in the Danube River Basin.

The WatBal model was applied in 109 sub-basins of the Danube Catchment (optionally with the catchment area between 5,000 and 10,000 km^2) related to selected discharge stations with long-term observation series. Training in the use of the WatBal model for delegates from all participating countries was performed in Sofia in June 2003. The methodology of the WatBal model application is included in the CD attached to the PFR [2]; the experience of trained national experts is available for further work with the WatBal model.

Two-day expert training on the method of spatial (GIS) distribution of results provided by the tuned WatBal model for a balance region was performed in Bratislava in December 2004.

Maps of balance elements (precipitation, evaporation and runoff depths) at the scale 1:2,000,000 were prepared by the Institute of Hydrology of the Slovak Academy of Sciences using the WatBal modelling results.

7.5.2 *Conclusions*

The project final report (PFR) (Petrovič et al.: Hydrological Monograph of the Danube River Basin) is the result of a collaboration of National Committees of IHP UNESCO within the Danube River Basin and it represents theme 5.3 of the co-operation in the field of hydrology within the Danube Basin and is made available to all involved partners (CD – attached to the PFR is available on request from the Slovak National Committee for IHP UNESCO).

At the same time, the report fulfils the provision of the ICPDR Joint Action Programme (Paragraph 3.13, p. 26) stating "Contracting Parties shall establish on a basis of harmonised methodology the general water balance of the Danube River Basin. As an input for this purpose, the Contracting Parties to the extent necessary shall provide connecting data, which are sufficiently comparable through the application of the harmonised methodology. On the same data base water balances can also be compiled for the main tributaries of the Danube River".

7.5.3 Recommendations and Outlook

For further optimal use of the assembled data within the co-operation of the Danube countries in the frame of the IHP UNESCO, it would be reasonable to continue with further data processing. The presently collected data fulfils the basic requirements for a detailed study of the annual course of water balance. Moreover, through the application of climate change scenarios to meteorological elements and consecutive impacts on runoff regime, water management and water requirements in relation to vegetation crop can be studied.

Acknowledgments This project was supported by the UNESCO Regional Bureau for Science and Culture in Europe (BRESCE), Italy and by the Slovak Science and Technology Agency – contract No. APVT-99-018202.

Full report "Hydrological Monograph of the Danube River Basin." was published by the Water Research Institute Bratislava, Slovakia (ISBN: 80-89062-49-0; EAN: 9788089062492) (Petrovič et al.: Hydrological Monograph of the Danube River Basin).

The authors would like to express their thanks to the International Commission for the Protection of the Danube Basin (ICPDR), to the EU Joint Research Centre (JRC) Ispra (Italy) for their significant help with the processing of this project, and further to the USGS for allowing the use of their digital terrain model.

Personally, we would like to thank Mr. Dr. Philippe Pypaert for his significant organisational and financial management of this project.

References

Directive 2000/60/EC of the European Parliament and of the Council of 23 October 2000 establishing a framework for Community action in the field of water policy. Official Journal of the European Communities L 327, 22 December 2000

Kling H, Nachtnebel HP (2004) "WatBal modelling Austrian balance regions" Material BOKU

Konstantinov AP, Astachova NI, Levenko AA (1971) Metody rascheta ispareniya s sel'skochozyajstvennych polej (Methods of the evaporation from agricultural fields estimation – 126 pp) GIMIZ Leningrad

Kuz'min PP (Red., 1976) Rekomendacii po raschetu ispareniya s poverkhnosti sushi. Utverzhdeny GUGMS (Recommendation for soil evapotranspiration. Agreed by the Main Administration of the Hydrometeorological Service) Published by: Min. melioracii i vodnogo khozyajstva SSSR. Gidrometeoizdat, Leningrad

Petrovič Š (1972) Zmena zrážok s výškou na Slovensku (Variation in rainfall in Slovakia as influenced by height). In: Proceedings of the Vplyv klimatických prvkov na hydrologické procesy, HMÚ Bratislava, pp 23–32

Petrovič P (1980) Výpočet potenciálnej evapotranspirácie modifikovaným programom Organizácie pre výživu poľnohospodárstvo – FAO (Computation of potential evapotranspiration using the modified programme of the FAO). In: Vodohospodársky časopis SAV, 28 (1980), 6:609–619

Petrovič P (1998a) Climate change impact on hydrological regime for two profiles in the Nitra river basin. In: Proceedings of the 19th conference of the Danube Countries on hydrological forecasting and hydrological bases of water management, Osijek, Croatia, pp 117–123

Petrovič P (1998b) Possible climate impacts on the water resources of the Danube river basin, case study: Subbasin of the Nitra river. In: Proceedings of the 2nd international conference on climate and water, vol 2, Espoo, Finland, 17–20 August 1998, Edita Ltd, Helsinki, pp 981–990

Petrovič P (2002) The Danube Basin water balance – case study: The Nitra river basin. In: Proceedings of the 21st conference of the Danubian Countries on the hydrological forecasting and hydrological bases of water management. Bucharest, 2–6 September 2002
Petrovič P, Baďurová K (2004) GIS technology use in basin-wide water balance estimation of the Danube river basin. In: Proceedings of the Conference "The Danube and Europe: integrated space applications in the Danube Basin", Mamaia, Romania, 23–25 June 2004 and also In: Proceedings of the 22nd conference of Danubian countries on the hydrological forecasting and hydrological bases of water management, Brno, Czech Republic, 30 August–2 September 2004
Petrovič S et al. Hydrological Monograph of the Danube River Basin. Published by the Water Research Institute Bratislava, Slovakia (ISBN: 80-89062-49-0; EAN: 9788089062492) also referred to as the Project Final Report (PFR)
Šamaj F, Valovič Š (1978) Dlhodobé priemery úhrnu zrážok na Slovensku za obdobie 1901–1970 (Long-term precipitation means in Slovakia for the period of 1901–1970). In: Zborník prác Hydrometeorologického ústavu v Bratislave, zv. 14. Alfa, Bratislava
Stančík A, Jovanovič S [Chairmen of the authors team], et al. (1988) Hydrology of the River Danube. Published by the Publishing House Príroda in Bratislava for Ministerstvo lesného a vodného hospodárstva SSR and Výskumný ústav vodného hospodárstva v Bratislave. 272 pp + 4 maps. HRD 064-209-88
Yates D (1994) WatBal – an integrated water balance model of river basin runoff. IIASA WP-94-64
Yates D, Strzepek K (1994) Comparison of models for climate change assessment of river basin runoff. IIASA WP-94-45, Laxenburg, July 1994 http://en.wikipedia.org/wiki/Danube http://www.icpdr.org/icpdr-pages/river_basin.htm

Chapter 8
Thermal and Ice Regimes of the Danube River and Its Tributaries

Alžbeta Stančíková

Abstract The aim of this study was to present an overview of the Danube thermal regime, its change and tendencies, with respect to time and along the river channel, and to elucidate these changes in relation to natural factors, whether geographic, morphologic, climatic, or anthropogenic, in the river channels or in the basins. The ice regime is closely linked to navigation, hydro energy, and to protection against floods. Therefore, another aim of this study is to compose the mean and extreme characteristics of the appearance and duration of ice phenomena and ice cover, and to examine their relationships with meteorological data.

Data were processed in relation to air and water temperatures, water discharge, and characteristics of the ice phenomena and freeze-up, from the whole observation period, and particularly for the period 1956–1985. These data are processed and arranged in easy-to-survey tables; some are also in graphical form. The achieved results should be of use in project assessments for active interventions in the river, for example for cooling in energetics and industry, and in solution of ecological or other problems.

Keywords Temperatures of the water · Thermal regime of the river · Ice phenomena · Freeze-up · Ice regime of the water

8.1 Introduction

Knowledge of the River Danube's thermal and ice regime are of extreme importance for countries through which it flows. The Danube countries develop along this river expressway important national water and economic activities, depending directly upon the state of these regimes. On the other hand, these activities can change

A. Stančíková (✉)
Water Research Institute, Bratislava, Slovakia
e-mail: alzbeta.stancik@stonline.sk

M. Brilly (ed.), *Hydrological Processes of the Danube River Basin*,
DOI 10.1007/978-90-481-3423-6_8, © Springer Science+Business Media B.V. 2010

the Danube's existing thermal and ice regime character, and that of its important tributaries.

Regional co-operation of the Danube countries within the IHP UNESCO has formed preconditions for elaboration of theme II of the Hydrological Monograph (IHP Nationalkomitees der Donauländer 1986) "Thermal and Ice Regime of Danube and of its Main Tributaries" (Regionale Zusammenarbeit der Donauländer im Rahmen IHP der UNESCO 1993). In its elaboration experts from Germany (Bayer 1989), Austria (Technische Universität Wien 1989), former Czechoslovakia (Forschungsanstalt für Wasserwitschaft 1989), Hungary (VITUKI 1989), former Yugoslavia (Federal Institut for Hydrometeorology 1990), Bulgaria (BNK po MGP 1990), and Romania (Institut meteorogii i gidrologii 1990) took part. In the study data from publications of the Danube Commission (Donaukommission 1967, 1987) were also used. Data were processed in relation to air and water temperatures, water discharge, and characteristics of the ice phenomena and freeze-up, from the whole observation period, and particularly for the period 1956–1985 (in hydrological years). Observation/measurement times do not differ much in particular Danube states.

8.1.1 Aims, Amount, and Documentation of the Input Data

The aim of this study was to present an overview of the Danube thermal regime, its change and tendencies, with respect to time and along the river channel, and to elucidate these changes in relation to natural factors, whether geographic, morphologic, climatic, or anthropogenic, in the river channels or in the basins. The achieved results should be of use in project assessments for active interventions in the river, for example for cooling in energetics and industry, and in solution of ecological or other problems. The ice regime is closely linked to navigation, hydro energy, and to protection against floods. Therefore, another aim of this study is to compose the mean and extreme characteristics of the appearance and duration of ice phenomena and ice cover, and to examine their relationships with meteorological data.

For the formation of the thermal regime, many factors take part. From the hydrological aspect, it is first of all discharge, water diversions and inflows, groundwater inflows, and for all of those, their temperature and magnitude. Meteorological factors assume a role through air temperature and through the precipitation amount and temperature. Morphological factors influence temperature distribution in the river channel cross-section and along its longitudinal profile. Of particular importance are the anthropogenic interventions in the catchments. These include, for example, corrections of the river channels, storage of water in reservoirs, its utilization and control, warm wastewater releases from industry and hydropower plants.

On the basis of the approved methodology (Forschungsanstalt für Wasserwitschaft 1988), it was decided that for characterization of the thermal regime, representative data on water temperature and on related factors from hydrological and the nearest suitable meteorological stations, for the whole period of observations, up to the year 1985, would be used. The database of these input data consists of:

- the mean annual, maximum and minimum daily temperature of water (t_o) and of air (t_a),
- the mean annual water discharge (Q).

The fact that the yearly mean values are used should be taken into account. Because of the size of the analyzed region (catchment area 817,000 km^2, length of the river channels 2,857 km), more detailed data, such as for example monthly or seasonal, have not been used.

Several factors influence ice regime formation in rivers. Those are primarily climatic and hydrological factors in the river and in its catchment; together with these factors also morphological characteristics of the river channel, and anthropogenic interventions are of importance. The ice regime is characterized by several forms of ice phenomena, depending upon conditions prevailing on the particular river location. In this study, two of these forms are evaluated, viz.

- ice phenomena in general (IP), i.e. the surface ice, including border (shore) ice, frazil ice, anchor (bottom) ice, ice sludge, and ice run,
- freeze-up (FU) of a given river section, irrespective of the fact, whether it is an ice dam, or an ice jam, spreading out in the upstream direction.

Ice phenomena appear immediately in relation to temperature of the air and water. The impact of discharge is of minor significance, as is precipitation. However, "warm rain" and snow pack can play a major role, locally.

The following data were used to characterize the river Danube ice regime:

- observation dates, terms, and particular ice regime data from individual gauging stations, viz.:
  - date of begin and end of ice phenomena (BIP, EIP),
  - date of begin and end of freeze-up (BFU, EFU),
  - total duration and longest non-interrupted duration of the IP,
  - total and longest non-interrupted duration of the freeze-up,
  - water level (stage) during duration of the IP and FU, particularly its mean and maximum values,
  - mean and maximum discharge during duration of the IP,
- long-term characteristics of the Danube ice regime, viz.:
  - period of observations, number of years with the IP and FU,
  - the earliest and latest date of BIP and EIP, the earliest and latest date of BFU and EFU,

- thermal characteristics of winters with occurrences of IP and FU, viz.:
  - sum of negative temperatures up to BIP and BFU $\sum t_a^-$,
  - mean air temperature in the first day of occurrence of IP and FU,
  - sum of positive temperatures up to EIP and EFU $\sum t_a^+$,
  - number of days with negative mean daily temperature t_a up to BIP,
  - mean water temperature in the first day before occurrence of IP and FU.

These data are processed and arranged in the easy-to-survey tables. Some of them also in a graphical form, such as the ice regime in critical winter periods, for example 1955/1956, 1962/1963, 1963/1964, and others.

8.1.2 Anthropogenic Impacts Relevant for the Thermal and Ice Regimes

In addition to basic hydrological and meteorological data for elaboration of this study, of great importance are also data on human activities along the Danube, its tributaries, and in the whole Danube catchment. Regulations in the river channel were take into account, particulary power station construction, thermal, and industrial waste releases, as they were presented by experts from the co-operating Danube countries.

8.1.2.1 River Danube Section in Germany

The first water gauging station incorporated for processing into this study is Ingolstadt (2,458.3 km). Upstream of the gauging station, up to the outlet of the right-hand tributary Lech, over a river stretch of approximately 40 km, a cascade of four water structures has been constructed, which began operation between 1964 and 1971. In all, on the river stretch from Ulm (2,588.0 km) up to Ingolstadt – some 130 km, 15 water structures were constructed from 1952–1984. Forty kilometres downstream of Ingolstadt the Rhine-Main-Danube Canal enters the Danube. From there on, up to Kachlet (the first water structure built on the Danube in 1927), another four water structures were built over a stretch of ca. 190 km between 1977–1985. On the German–Austrian border the water structure Jochenstein has been in operation since 1954.

Construction of the thermal power plants on this river section began in the year 1965. The following are conventional plants on the Danube: Ingolstadt (2,451.4 km, built in 1965), Irsching (2,445.9 km, 1969), and Pleiting (2,256.1 km, 1969). On the Isar – a right-hand tributary to the Danube – two Nuclear power stations (Isar I and Isar II) have been in operation since 1976 and 1988, respectively.

8.1.2.2 River Danube Section in Austria

The Danube flows through Austria over a total length of ca. 350 km. From this stretch, 21 km between the structures of Kachlet and Jochenstein, are common with

Germany, and downstream a stretch of 8 km is common with Slovakia. Up to 1985, a stretch of 250 km from this section was under backwater influence, upstream of eight water structures, which began operation between the years 1954 and 1985.

From the point of view of the ice regime, not without importance are also the water structures on the Danube tributaries of the Alps in Germany and Austria (Lech, Isar, Inn, and Enns).

The waste heat in this section, originates from the Linz and Vienna regions. Already in 1971 the waste heat from Linz together with that of the Traun thermal input was capable of melting almost 370,000 m^3 of ice daily. This energy output corresponds to that of the nuclear power station Zwentendorf, which has not yet began operation.

8.1.2.3 River Danube Section in Slovakia and the Common Slovak–Hungarian Section

An 8-km river section downstream of the Morava outlet is common between Slovakia and Austria. The following 30 km downstream is the Slovak national section. The rest of the 142 km stretch (downstream to the outlet of the River Ipel') is the common Slovak–Hungarian section. So far, there is only one water structure on this whole river stretch, but it has been in operation to short a time to assess its impact. In the examined period (up to 1985), this structure did not exist. Warm wastewaters are produced by the conurbation regions of Bratislava, Komárno and Štúrovo, however, they are only of local impact.

8.1.2.4 River Danube Section in Hungary

From the outlet of the River Ipel' downstream to the Hungarian–former Yugoslavian border (275 km), the Danube river channel is not barred by a water structure. River channel regulation works were constructed from 1958–1975 over a length of 67 km between Foeldvár and the border, which influenced ice formation predominantly in this section. Wastewater outlets influence the Danube water temperature mainly in the Budapest region and in the area of the nuclear power station Paks.

8.1.2.5 River Danube Section in Romania

The Danube flows through Romania over a length of 1,024 km. Of this length; 229 km is a common section with former Yugoslavia, ca. 472 km with Bulgaria, and some 80 km with Ukraine and Moldavia. The Djerdap I river system – for navigation and water power production – has a decisive impact upon the heat and ice regime. It was constructed as a common Yugoslav–Romanian structure over 942.25 km between the years 1964–1971. Later Djerdap II was constructed over 863.0 km; it was finished in 1984. In addition, there are dams and reservoirs on the tributaries Jiu, Olt, Arges, Ialomita and Siret, as well as thermal power plants located close to the Danube. Also there are also important artificial interventions on other tributaries.

8.1.2.6 The Rest of the Danube River Sections

An overview of the technical interventions affecting the Danube river sections and its tributaries, influencing their thermal and ice regime in Bulgaria and Ukraine, has not been so far available.

8.2 Thermal Regime

8.2.1 Method and Terms of Observations

Hydrographic services in the Danube countries, performing among others also air and water temperature measurements, assume approximately the same daily temperature time course and amplitudes. Service regulations in individual countries state the observers exact location, method and term of observation. There is little difference in these terms and observation instruments. Temperature is measured by an immersed thermometer with a reading accuracy of 0.1°C. Measurements are performed once a day between 7.00 and 8.00 a.m., in the river cross-section so that the immersed instrument is located in a homogeneous water environment. Together with water temperature, also some meteorological elements (e.g. precipitation, fog), and other phenomena (ice) are observed.

To these general data some of the co-operating organizations add:

- In the German river section, from the beginning of the observation period (1929), all of the water temperature observations have been performed as described earlier, and thus this observation series is from this aspect homogeneous. Some of the newer stations are equipped with resistance thermometers for continuous registration, enabling registration of the daily temperature course.
- On the Austrian territory, the older observations are slightly different to those mentioned above; also some observation sites have been changed. The observation data series are long – also from the 1980s. For these reasons, and because of intensive anthropogenic interferences, these long data series are not often homogeneous.
- On the Slovak territory, temperature measurements are made at 7.00 a.m., only in forecasting stations at 6.00 a.m.; selection of these stations was verified by an extensive study on daily term observations for selected stations.
- On the Hungarian river section, measurements are taken also at 7.00 a.m., together with reading of the water gauge; instruments, method, and accuracy correspond to the above standards.
- In Romania, the water temperature is measured at 8.00 a.m., on tributaries at 7.00 and 17.00; a single measurement on the Danube is considered satisfactory, because of the small daily temperature fluctuation, and because the large water volume and water surface area are exposed to heat exchange with the atmosphere.

Air temperature is measured in three daily intervals, from which the mean value is calculated. In the Danube countries, these terms are uniform.

8.2.2 Station Network

Thirty-one stations on the Danube and 14 stations on its tributaries were used for analysis of the thermal regime between Ingolstadt and Braila (river section length of 2,288.6 km). This equates to a mean for the river section of 74 km per station. However, this figure changes along the river. Up to the Inn outlet it is 30 km, between the Inn and Morava outlets ca. 113 km, up to the Ipel' outlet 57 km, between Ipel' and Dráva outlets 65 km, from there on to Braila even 135 km. On the upper Danube, the network is denser. It was an effort to best characterize the changes downstream of tributaries, as well as impacts from human intervention.

On the German river section, data from eight stations were processed on the River Danube and one on the Isar tributary. The oldest observations are from Schwabelweis where they began in 1929, which represents a 57-year data series up to 1985. The shortest data series are from Oberdorf where they began in 1973 (13 years).

Observations have been carried out at four stations on the Austrian Danube since 1898; however, they have been processed only since 1901, representing an 85-year data series. However, the observation series in Engelhartszell was interrupted between 1941 and 1947, so it is considered inhomogeneous.

In the Slovak Danube section, observations are from three stations. The oldest are from Bratislava, since 1925 (61 years), Komárno, since 1947 (39 years), and from Gabčíkovo, since 1952 (34 years). Four stations on the tributaries (Morava, Hron, Váh, and Ipel') have observations since 1962, 1963, 1962 and 1947 (series from 23 to 39 years long), respectively.

In Hungary there are two stations with observations since 1945, one since 1950, and one since 1956, which represents series of 41, 36, and 30-year long observations.

On the former Yugoslavian territory, the Danube water temperature is observed at four stations, established between 1946 and 1953, so they have observations 40–33 years long. The tributaries Tisa, Drava, Sava, and V. Morava have 40-year long observation series, since the year 1946.

Five Bulgarian stations have observations since 1941, which represents a 45-year long series.

In Romania temperatures on the Danube have been observed at three stations since 1954, thus for 32 years. The tributaries Jiu and Olt have data since 1963 (23 years), Arges since 1955 (31 years), Ialomita since 1967 (19 years), and on Siret since 1960 (26 years).

An overview of the hydrological stations with the most suitable meteorological stations is presented in Table 8.1.

Table 8.1 Basic data on hydrological and climatic stations on the Danube and on selected tributaries

River	Hydrological station	Location (km)	Observed from t_o	Observed from Q	Climatological station	Observation t_a
1	2	3	4	5	6	7
Danube	Ingolstadt	2,458.3	1932	1901	Regensburg	1947
Danube	Oberndorf	2,397.4	1973	1926	Regensburg	1947
Danube	Schwabelweis	2,376.2	1929	1901	Regensburg	1947
Danube	Straubing	2,321.3	1972	–	Regensburg	1947
Danube	Deggendorf	2,284.6	1933	–	Passau	1948
Danube	Vilshofen	2,249.4	1971	1901	Passau	1948
Danube	Passau-Kachlet	2,225.0	1934	–	Passau	1948
Danube	Achleiten (Jochenstein)	2,223.1	1969	1901	Passau	1948
Danube	Engelhartszell	2,200.7	1901	–	Rannwerk	1949
Danube	Linz	2,135.2	1901	1893	Linz MZA	1898
Danube	Stein-Krems	2,003.5	1901	1893	Krems MZA	1898
Danube	Vienna	1,929.1	1901	1893	Wien-H. Warte	1901
Morava	Moravský Ján	67.6	1962	1901	Malacky	1965
Danube	Bratislava	1,868.8	1925	1901	Bratislava	1931
Danube	Gabčíkovo	1,819.6	1952	–	Gabčíkovo	1952
Danube	Komárno	1,767.1	1947	1901	Hurbanovo	1931
Váh	Šaľa	58.5	1963	1963	Žihárec	1963
Hron	Brehy	102.5	1962	1962	N. Baňa (Žiar)	1962
Ipeľ	Ipeľ. Sokolec	42.9	1947	–	Želiezovce	1951
Danube	Budapest	1,646.5	1945	1923	Budapest	1871
Danube	Dunaujváros	1,580.6	1945	1924	Budapest	1871
Danube	Paks	1,531.3	1945	–	Kaloča	1861
Danube	Mohacs	1,446.9	1950	1918	Kaloča	1861
Danube	Bezdan	1,425.5	1946	1946	Sombor	1946
Danube	D. Miholjac	74.6	1946	1946	D. Miholjac	1956

Table 8.1 (continued)

River	Hydrological station	Location (km)	Observed from t_o	Observed from Q	Climatological station	Observation t_a
1	2	3	4	5	6	7
Danube	Bogojevo	1,367.4	1946	1946	Osijek	1946
Tisa	Senta	122.0	1946	1946	Senta	1946
Sava	S. Mitrovica	136.0	1946	1946	S. Mitrovica	1946
Danube	Smeredevo	1,116.2	1946	1946	Smeredevo	1946
V. Morava	Lj. Most	39.9	1946	1946	Lj. Most	1954
Danube	V. Gradište	1,059.8	1953	1949	V. Gradište	1949
Danube	Drobeta-T. Severin	931.0	1948	1932	T. Severin	1932
Danube	Novo Selo	833.6	1946	1937	Novo Selo	1948
Danube	Lom	743.3	1937	1921	Lom	1892
Jiu	Zăvalu	7.4	1963–1969 1973–1985	1963	Beket	1963
Olt	Izbiceni	6.0	1962–1969 1973–1985	1960	Turnu Magurele	1962
Danube	Svistov	554.3	1937	1917	Svistov	1894
Danube	Ruse	495.6	1937	1908	Ruse	1866
Danube	Giurgiu	493.5	1948	1932	Giurgiu	1932
Argeş	Budeşti	31.0	1955–1969 1973–1985	1951	Giurgiu	1932
Danube	Silistra	375.5	1943	1941	Silistra	1892
Ialomiţa	Slobozia	77.0	1962	1951	Merkulesti	1962
Danube	Braila	169.7	1948	1932	Braila (Galaţi)	1932–1960 1961–1985
Siret	Lungoci	74.0	1951–1969 1972–1985	1951	Braila (Galaţi)	1951–1960 1961–1985

8.2.3 Thermal Regime Characteristics

The Danube catchment is in accordance with its geological and geographical structure divided into three parts (IHP Nationalkomitees der Donauländer 1986):

- The upper part extends from the river spring in the Schwarzwald up to the so-called Devín Gate. Decisive factors in this part of the catchment are the Alpine foreland, and the prevailing cold Alpine tributaries.
- The central part extends from the Devín Gate to the Iron Gate. In terms of area it is the greatest catchment part. From the north, it is fringed by the Carpathian range, from the Small Carpathians down to the Danube breach through the Southern Carpathians, to the Balkan range. To the west and south it is limited by the Carnic, Julian and Dinaric Alps.
- The lower catchment part is formed by the Lower Danube lowland, with catchments of the Siret and Prut tributaries, and the surrounding land.

Waters from the high mountainous regions, mountain plains, hollows, and lowlands contribute to the Danube. Therefore, the river character is variable and changeable, with characteristics of a mountain river type, as well as a lowland river. Characteristic and deciding factors, influencing particular Danube catchment parts, are also a great influence upon its thermal regime.

To evaluate the Danube thermal regime, the mean annual water temperatures were used, taken for the hydrological year. Time course data were processed for the selected period 1956 to 1985 (out of this, 30-year period, and also from shorter periods, i.e. 1956–1965, 1966–1975, and 1976–1985), and presented in tables and graphs. The Danube thermal regime is not characterized only by temporal thermal courses, but also by its courses along the river channel, by comparison between particular river sections, and by its relationships to particular hydrological and meteorological quantities, mainly to air temperature and to water discharge. To express these characteristics and relations, the basic series were further processed, mainly by statistical methods. The results were:

- exceedance curves of the selected stations,
- correlations between the air and water temperatures on one hand, and water temperature and discharge on the other hand,
- smoothing of the selected observation series by the weighted moving averages method, and
- trends in development of the selected quantities.

The exceedance curves of water temperature, air temperature and discharge, were set up as theoretical curves with binomial distribution type. Their basic statistical characteristics t_o – mean, C_s and C_v, are presented with particular data series in Tables 8.2, 8.3 and 8.4, for stations in the upper, central, and lower Danube. The C_v values are small, indicating relatively well equalized data series. The asymmetry coefficients C_s, except for three curves (Ingolstadt, Drobeta-Turnu Severin, and Lj.

Most) show a negative asymmetry. It holds, that the C_s values fulfil the condition $C_s > 2C_v$ (with the exceptions of the Danube at Deggendorf and of the Drava at D. Miholjac), but their calculation from a series of only 30 elements, can include a high random error.

In the following will be presented the temperature characteristic of stations, for particular Danube sections.

Table 8.2 Mean annual t_o values and statistical characteristics of selected stations of the upper Danube

Year	Ingolstadt	Deggendorf	Passau	Engelhartszell	Vienna
	°C				
1	2	3	4	5	6
1956	8.9	9.0	9.6	8.4	9.0
1957	9.3	9.8	10.4	9.5	9.8
1958	9.2	9.7	10.4	9.0	9.7
1959	9.8	10.4	11.0	9.6	10.0
1960	9.6	10.0	10.4	9.5	9.9
1961	10.4	10.5	11.1	9.8	10.4
1962	9.3	9.4	10.0	8.7	9.0
1963	8.8	9.2	9.8	9.0	9.5
1964	9.8	10.2	10.9	9.4	9.7
1965	8.6	8.9	9.6	8.3	8.7
1966	9.6	10.1	10.8	9.3	9.7
1967	9.7	10.2	10.9	9.3	9.8
1968	9.5	10.2	11.1	9.4	9.9
1969	10.0	10.3	11.1	9.6	10.0
1970	9.2	9.4	9.8	8.7	9.0
1971	10.7	10.8	11.3	9.9	10.4
1972	10.4	10.9	11.3	9.6	9.8
1973	10.4	10.7	11.1	9.3	9.7
1974	10.2	11.0	11.1	9.6	10.1
1975	10.7	10.9	11.3	9.4	9.7
1976	11.2	11.1	11.6	9.9	10.1
1977	10.9	10.7	11.3	9.7	10.0
1978	9.8	9.8	10.7	9.3	9.5
1979	9.6	9.9	10.9	9.4	9.4
1980	10.0	9.7	10.6	9.0	8.0
1981	10.0	9.9	11.0	9.5	9.5
1982	10.2	10.3	11.1	9.9	10.1
1983	11.2	10.9	11.9	9.9	10.4
1984	–	9.4	10.4	9.3	9.7
1985	8.5	11.4	10.7	9.4	9.7
Mean	9.9	10.1	10.7	9.3	9.7
C_s	0.187	−0.097	−0.523	−0.877	−0.571
C_v	0.070	0.064	0.054	0.045	0.046
Change t_o	0.33	0.34	0.41	0.23	0.21
Change t_a	0.13	0.09	0.09	−0.08	0.11

8.2.3.1 The Upper Danube

The Danube water temperature t_o from Ingolstadt to Jochenstein in mean yearly values, is always higher than the air temperature t_a. The absolute values of t_o are between 8.6 and 12.1°C, and those of t_a between 6.6 and 8.9°C. The temperature difference between air and water, in a given year can be from 0.9 to 3.5 K.

Table 8.3 Mean annual t_o values and statistical characteristics in the central Danube selected stations

Year	Bratislava	Komárno	Budapest	Mohacs	Bogojevo	Smeredevo
	°C					
1	2	3	4	5	6	7
1956	9.1	9.4	9.5	10.0	9.7	10.5
1957	9.8	10.1	10.5	11.2	11.5	12.2
1958	9.8	10.2	10.5	11.2	11.4	12.3
1959	10.0	10.4	10.7	11.8	11.4	12.4
1960	10.3	10.3	10.8	11.3	11.4	12.4
1961	10.3	10.5	11.1	11.9	11.7	13.0
1962	9.3	9.3	9.7	10.4	10.8	11.7
1963	9.5	9.9	10.5	11.0	11.1	12.0
1964	9.8	10.0	10.4	10.9	11.2	11.9
1965	9.0	9.1	9.4	10.0	10.1	11.1
1966	10.0	10.3	10.8	11.2	11.8	12.2
1967	10.0	10.2	10.8	11.4	11.6	12.3
1968	9.9	10.1	10.9	11.7	11.3	12.3
1969	9.9	10.5	10.7	11.5	11.2	11.9
1970	9.0	9.5	9.9	10.6	10.6	11.5
1971	10.1	11.1	11.2	12.0	11.6	12.0
1972	10.0	11.0	11.0	11.8	11.6	12.1
1973	9.8	10.3	10.4	11.4	11.4	12.0
1974	10.2	10.6	11.0	11.9	11.6	12.3
1975	10.4	10.5	11.0	12.0	11.6	12.5
1976	10.5	10.4	11.2	12.3	11.8	–
1977	9.8	10.2	11.9	12.2	11.6	12.7
1978	9.5	9.8	10.5	11.6	10.8	11.7
1979	9.7	10.0	10.6	10.9	11.3	12.4
1980	9.2	9.3	9.8	10.1	10.4	11.3
1981	10.0	9.9	10.9	11.1	11.6	12.2
1982	10.2	10.2	11.0	11.5	11.9	12.5
1983	10.4	10.7	11.2	11.6	12.0	13.1
1984	9.5	9.3	10.5	11.0	11.3	11.8
1985	9.9	10.3	10.3	10.7	11.1	11.7
Mean	9.8	10.1	10.6	11.3	11.3	12.1
C_s	−0.492	−0.298	−0.414	−0.478	−1.307	−0.741
C_v	0.042	0.049	0.052	0.056	0.047	0.044
Change t_o (K)	0.03	−0.08	0.01	0.14	0.19	−0.11
Change t_a (K)	−0.01	−0.05	−0.35	−0.21	−0.28	−0.18

In the Austrian stations, t_o values are not that much higher than those of t_a. Values t_o move from 8.3 up to 10.4°C, and t_a from 6.5 to 11.1°C. In comparison with the German station, also the t_o–t_a difference has changed. It is in Engelhartszell 0.0–2.2, and in Vienna only 0.0–1.0.

In all stations of the upper Danube, the mean annual water temperature has a rising trend, expressible higher than that of the air temperature. The highest rise is

Table 8.4 Mean annual t_o values and statistical characteristics in the lower Danube selected stations

Year	Drobeta-Turnu Severin	Lom	Svistov	Silistra	Braila
	°C				
1	2	3	4	5	6
1956	11.5	11.3	11.4	11.6	11.7
1957	11.9	12.0	12.4	12.7	12.5
1958	11.8	12.1	12.6	12.8	12.6
1959	11.8	12.1	12.3	12.6	12.4
1960	11.9	12.6	12.7	13.0	12.9
1961	12.6	13.1	13.2	13.3	13.3
1962	11.6	12.3	12.1	12.5	12.6
1963	11.8	12.3	12.6	12.8	12.7
1964	11.7	12.0	12.3	12.4	12.3
1965	10.9	11.5	11.9	12.0	11.9
1966	12.0	12.6	13.1	13.3	12.7
1967	12.1	12.5	12.9	13.2	12.9
1968	12.2	13.0	12.7	12.9	12.7
1969	12.3	12.2	12.2	12.3	12.1
1970	11.5	12.1	12.2	12.4	12.5
1971	12.0	12.6	12.7	13.0	12.7
1972	11.5	11.9	12.6	12.7	12.0
1973	11.8	12.1	12.4	12.3	10.8
1974	11.6	12.1	12.2	12.6	12.2
1975	12.6	12.8	12.9	13.0	12.4
1976	12.0	11.9	12.1	12.4	11.8
1977	13.3	12.9	13.1	13.1	11.6
1978	11.4	11.8	12.0	12.0	12.3
1979	12.6	12.5	12.7	12.8	12.8
1980	10.7	11.5	12.0	12.0	11.8
1981	11.4	12.3	12.6	12.9	12.1
1982	12.4	12.5	13.0	13.1	12.2
1983	11.6	12.3	13.4	13.4	12.2
1984	11.7	12.2	13.0	12.9	12.9
1985	11.0	11.3	12.8	12.4	11.8
Mean	11.8	12.2	12.5	12.7	12.3
C	0.334	−0.160	−0.292	−0.460	−0.692
C_v	0.046	0.038	0.036	0.035	0.041
Change t_o (K)	0.00	−0.01	0.11	0.12	−0.16
Change t_a (K)	0.00	−0.04	0.00	0.13	−0.19

in Passau, the lowest in Vienna. As for t_a, there is a decrease in Engelhartszell, and a negligible rise in other stations (Table 8.2 – changes).

8.2.3.2 Central Danube

There is a similar tendency as in Austria, in the river sections between Bratislava and Budapest. There were years and periods with water temperature higher than that of the air, but also the other way round. The water temperature changed from 9.0 to 11.9°C, and of the air from 8.4 to 12.1°C. Temperature amplitudes are 1.5–2.5 K for t_o and 1.9–2.2 K for t_a. Temperature differences of t_o and t_a values are up to 1.3 K in a year.

From Budapest to Gradište, again t_o (8.9–12.3°C) is expressively higher than t_a (from 6.0 to 11.2°C), and also the differences are higher (up to 5.0 K), as compared with the previous river section. They even have had a tendency to increase since 1960.

On the lower lying central Danube stations, the t_o values (from 9.4 to 13.1°C) remain higher than t_a (from 9.3 to 12.6°C).

In this Danube river section, the long-term development trends are different. The water temperature, except at Komárno and Smederevo, rises from 0.01 to 0.19 K. Contrary to this, the air temperature shows a decreasing trend, from 0.01 to 0.35 K. Again, changes are more expressive in the lower stations of the central Danube.

8.2.3.3 Lower Danube

Also in this Danube River section, the water temperatures exceed those of the air, by ca. 1.5 K. Exceptional are some years, for example in Braila, where the value is up to 3.0 K. Absolute mean yearly t_o values are from 10.7 to 13.5°C, and t_a from 9.2 to 13.4°C. The t_o amplitude is 1.8 to 2.6 K, that of t_a is from 2.0 to 2.9 K.

Neither on the lower Danube, are the temperature development trends unique. Without change are both, t_o and t_a, in the Drobeta-Turnu Severin station. A decreasing trend is shown for both temperatures in stations Lom and Braila, and a rising one, in stations Svistov and Silistra.

8.2.3.4 Tributaries

On the upper Danube, the cold Alpine tributaries have a decisive impact upon its thermal regime. Only temperature data were available on the River Isar from the Landshut station. Temperatures changed from 8.4 to 11.0°C, with amplitude of 2.6 K. For air temperature, no data were submitted.

In the central Danube, data were processed, from the left-hand tributaries Morava, Váh, Ipel' and Tisa, and from the right-hand ones Drava, Sava, and V(elika)

Morava. From each of these tributaries, data were available only from a single station, and the distances from the outlets to the Danube are 40 and more kilometres. Therefore, it is not possible to reliably assess the tributaries impact upon the Danube's thermal regime.

The tributaries Morava (its outlet at 1,880 Danube km), Váh (1,760 km), Hron (1,716 km), and Ipel' (1,708 km), originate in mountainous regions; this manifests itself also upon their thermal regime. In general, the Morava is considered the warmest tributary in the Devín–Štúrovo Danube section. Mean annual water temperature was within the interval 9.4–11.4°C. However, its thermal regime is influenced by a thermal power plant with through-flow cooling.

The longest observation is on the Ipel', from station Ipel'ský Sokolec. The mean annual t_o value changed from 7.9 to 10.9°C (amplitude 3.0 K) and t_a from 8.0 to 10.4°C (amplitude 2.4 K). Water temperature was approximately in half of the examined period higher than the air temperature. The long-term temperature trend was decreasing, but 10-times higher for water than that for air. The left-hand Tisa tributary has mean annual temperatures within the range from 10.8 to 13.2°C (amplitude 2.4 K). The right-hand Drava has temperatures from 9.5 to 12.0°C (amplitude 2.5 K), and V. Morava from 11.6 to 13.9°C (amplitude 2.3 K).

Of the lower Danube tributaries, Tisa at the Senta station (122.0 km above its outlet to the Danube) was, according to its water temperature exceedance curve, as warm, as the Danube even at Drobeta-Turnu Severin (285.0 km downstream from Tisa at Senta). Similar were the temperatures of V. Morava.

Water and air temperature trends have in individual stations a different character. While both on the Ipel' have a decreasing trend, on the Drava t_o has a rising and t_a a decreasing one, on the Tisa t_o is without change and t_a slightly decreases. For V. Morava only the t_o trend is known, which decreases.

From the Romanian Danube tributaries, the temperature regime of rivers Jiu, Olt, Arges, Ialomita and Siret was processed. Data series from these stations are not complete; therefore, they were not processed statistically.

An overview of the selected Danube tributaries temperatures, for the processed hydrological characteristics, and for the long-term t_o and t_a development trends, is given in Table 8.5.

8.2.4 Thermal Regime Changes Observed After Anthropogenic Interventions into the River

When examining correlation relationships between the water and the air, we assumed a linear correlation. An overview of correlation coefficients for selected stations of the whole evaluated period 1956–1985, as well as for the decennia within this period, is presented in Table 8.6. The comparisons for particular stations indicate which changes in relations between t_o and t_a occurred in these decennia, and thus connections with anthropogenic interventions in these periods can be looked for.

Table 8.5 Mean annual t_o values and statistical characteristics for selected stations of the Danube tributaries

Year	Landshut Isar	Ipel'. Sokolec Ipel'	D. Miholjac Drava	Senta Tisa	Lj. Most V. Morava
	°C				
1	2	3	4	5	6
1956	9.0	9.0	11.0	11.3	11.9
1957	9.7	9.8	10.9	11.7	13.3
1958	10.0	9.8	10.2	12.0	13.0
1959	10.3	9.8	10.6	12.2	12.7
1960	10.1	10.9	10.8	12.0	13.3
1961	10.6	10.7	11.9	12.7	13.1
1962	9.7	9.5	9.5	11.3	12.2
1963	9.8	10.2	10.4	12.3	12.8
1964	10.2	9.9	10.7	11.5	12.0
1965	8.4	9.4	10.8	11.0	11.9
1966	10.1	10.5	11.7	12.2	13.9
1967	10.2	9.7	11.6	12.5	12.8
1968	10.1	10.0	11.5	12.7	12.8
1969	10.7	9.8	11.2	11.6	12.2
1970	9.5	9.8	10.2	11.4	12.1
1971	10.8	9.7	11.3	12.3	12.4
1972	10.9	9.9	10.3	12.6	12.9
1973	10.1	9.2	10.8	11.8	11.9
1974	9.9	9.3	11.2	11.8	12.1
1975	10.1	9.4	10.7	12.6	12.1
1976	10.5	7.6	11.1	11.7	11.6
1977	10.0	8.0	10.6	12.3	13.5
1978	9.6	7.9	10.6	11.1	11.6
1979	9.6	8.6	10.9	12.3	12.9
1980	9.7	9.0	10.2	10.8	11.6
1981	10.2	8.1	11.0	12.2	12.1
1982	10.5	8.9	11.6	12.2	12.6
1983	11.0	9.9	12.0	12.2	12.8
1984	9.9	9.8	11.0	11.8	12.2
1985	10.1	10.0	10.8	11.6	12.2
Mean	10.0	9.5	10.9	11.9	12.5
C_s	−0.750	−0.681	0.098	−0.370	0.412
C_v	0.054	0.085	0.053	0.043	0.048
Change t_o (K)	–	−0.41	0.09	0.00	−0.21
Change t_a (K)	–	−0.04	−0.16	−0.04	–

To supplement the information achieved from this type of data processing, air and water temperature courses along the river Danube (with 50% certainty, from 1956–1985) were constructed. They are shown in Fig. 8.1.

To arrive at a picture of the time series of the examined elements (t_o and t_a), a moving average method was used for linearization, with nine binomial expansion

Table 8.6 Overview of correlation coefficients for selected Danube stations

	r_{xy}			
Station	1956-1985	1956-1965	1966-1975	1976-1985
Ingolstadt	0.68	−0.45	0.19	−0.64
Deggendorf	0.55	0.88	0.23	0.34
Passau	0.72	0.85	0.38	0.83
Engelhartszell	0.65	0.74	0.06	0.90
Vienna	0.88	0.72	0.30	0.74
Bratislava	0.78	0.91	0.60	0.71
Komárno	0.63	0.86	0.08	0.60
Budapest	0.46	0.89	0.47	0.72
Mohacs	0.48	0.82	0.45	0.41
Bogojevo	0.70	0.77	0.69	0.81
Smeredevo	0.67	0.84	0.52	0.89
Drobeta-Turnu Severin	0.32	0.38	0.12	0.37
Lom	0.74	0.75	0.66	0.75
Svistov	0.65	0.80	0.70	0.69
Silistra	0.78	0.84	0.7	0.84
Braila	0.62	0.73	0.53	0.49

elements. This evaluation was supplemented by calculation of trends according to relations

$$t_o = B_o T + A_o$$

$$t_a = B_a T + A_a,$$

where A_o, A_a, B_o, and B_a are the trend line parameters, and T is the order number of the year, beginning with the first year of observations. These trend equations for selected stations are given in Table 8.7.

From this information, it can be stated, in relation to natural and anthropogenic interventions, that

- Between Ingolstadt and Passau, water temperature rose at various intensities, while the air temperature decreased, or did not change. This river section has the highest concentration of water structures.
- Below the cold Inn river outlet, t_o dropped greatly, while t_a increased.
- From Engelhartszell, where the t_o value is lower than at Ingolstadt, the temperature up to Dunaújváros rose to a level close to that at Passau, in accordance with the air temperature. An exception was the t_a drop in the section between Vienna and Bratislava.
- Between Dunaújváros and Bezdan, t_o first dropped up to Mohács (as well as t_a), then it rose in spite of the fact that t_a did not change, similarly as between Mohács and Bezdan, where t_o decreased.

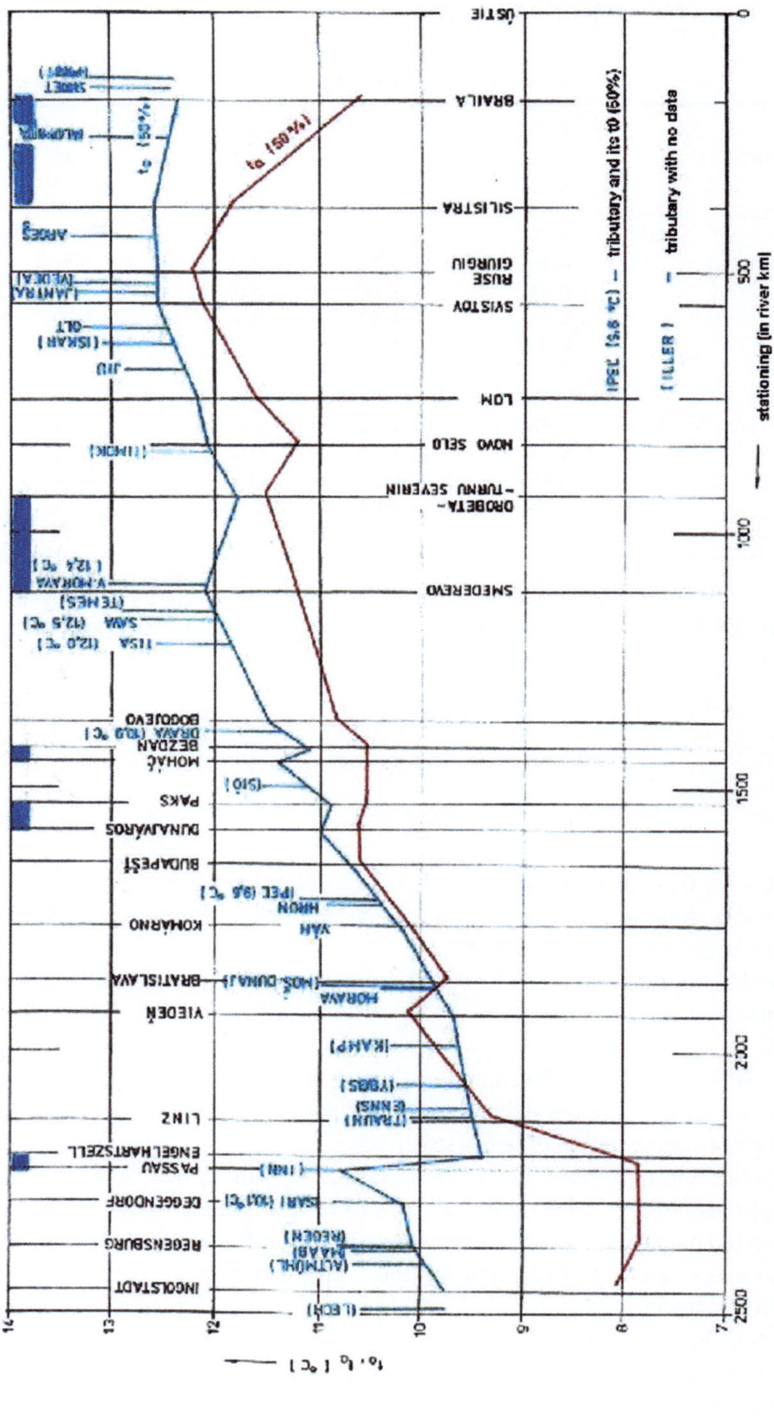

Fig. 8.1 Courses of the air (t_a) and the water (t_o) temperatures, with a certainty of 50% along the Danube for the 1956–1985 period

Table 8.7 Overview of the trend relationships for the selected Danube stations and for its tributaries

River	Hydrological station	$t_o = B_oT + A_o$	$t_a = B_aT + A_a$
1	2	3	4
Danube	Ingolstadt	0.03300 T + 7.49	0.01300 T + 7.12
Danube	Deggendorf	0.03400 T + 7.76	0.00960 T + 7.17
Danube	Passau	0.04110 T + 7.87	0.00900 T + 7.26
Danube	Engelhartszell	0.02300 T + 7.73	0.00076 T + 8.28
Danube	Wien	0.02100 T + 9.23	0.01100 T + 8.95
Danube	Bratislava	0.00280 T + 9.61	–0.00090 T + 9.81
Danube	Komárno	–0.00800 T + 10.70	–0.00530 T + 10.38
Danube	Budapest	0.00130 T + 10.59	–0.03500 T + 13.13
Danube	Mohacs	0.00700 T + 10.45	–0.02100 T + 11.84
Danube	Bogojevo	0.01900 T + 9.92	–0.02800 T + 12.78
Danube	Smeredevo	–0.01100 T + 12.95	–0.01800 T + 12.51
Danube	Drobeta-Turnu Severin	11.79	11.41
Danube	Lom	–0.00100 T + 12.32	–0.00400 T + 11.84
Danube	Svistov	0.01100 T + 11.83	12.04
Danube	Silistra	0.01200 T + 11.85	0.01300 T + 10.86
Danube	Braila	0.01600 T + 13.44	0.01900 T + 11.91
Ipel'	Ipel'ský Sokolec	–0.04100 T + 12.39	–0.00420 T + 9.80
Drava	Donji Miholjac	0.00880 T + 10.28	–0.01600 T + 11.84
Tisa	Senta	11.93	–0.00310 T + 10.74
V. Morava	Ljubičeskij Most	–0.02100 T + 13.99	–

- From Bezdan to Smederevo, both temperatures rose; from Smederevo to Novo Selo they had opposite tendencies.
- From Novo Selo to Ruse both temperatures rose, an impact from interventions on numerous tributaries took place.
- In a further section to Silistra t_o still rose slightly, and then to Braila decreased by ca. 0.3 K; but t_a decreased from Ruse by as much as 1.7 K. Contributing to this regime were the tributaries Arges (warmer than the Danube), and Ialomita and Siret (both much colder than the Danube).

8.3 Ice Regime

The Danube has been used for a long time, mainly for navigation. Industrial development in individual Danube countries needed a more rational and multipurpose use of its potential. This concerned mainly water energy, irrigation possibility, and cooling for industry and power production. On the other hand, the Danube represents also a great danger during floods, particularly those caused by ice. The catastrophic flood in 1956 incited the Danube countries to begin a common ice regime observation programme. In the Danube Commission (DC) Secretariat, two analytical overviews were elaborated.

In the first, "About the Danube Ice Regime", published in 1959, an analysis is presented, of the ice regime relations of the Danube navigation section from Engelhartszell to Sulina (2,200 km), based on the 1900/1901–1955/1956 data, i.e. 56 years of observations. In its four chapters are:

- the ice regime characteristics of this section,
- relations between the ice regime and the air temperature,
- relationship between the ice regime and the river channel morphology,
- methods of the "fight with ice".

The second publication "Report on the River Danube Ice Regime" (Donaukommission 1967), published in 1967, was devoted to the river section ice regime between Regensburg and Sulina (2,376 km), and was based on 60 years of observations, beginning with the winter of 1900/1901. It contains information on conditions of the origin and formation of the ice phenomena (IP) and of freeze-up (FU), input data and their processing methods, the ice regime characteristics, impacts of the water structures, and methods of "fighting" IP and FU.

These publications present an extensive source material on the Danube ice regime up to the year 1960. In 1973 knowledge on the Danube ice regime was supplemented by a further volume "Ice Regime I", which apart from observations, processing methods, and short-term forecasting, presented also the long-term IP forecast, relationships of their occurrence to the atmospheric circulation above Europe, and air temperature characteristics forecasts. In addition to that, since 1953, a yearbook has been published by the DC called the "Hydrological yearbook of the River Danube" (Donaukommission 1987), where in addition to statistical data processing of water levels, discharges, water and air temperature, and bed-load sediment regime in the main water level stations, also the ice regime characteristics are indicated.

The aim of theme II "Thermal and ice regime of the Danube, and its main tributaries" of the Danube Hydrological Monograph was to extend and supplement data and analyses of the former publications up to the year 1985, i.e. by a further 25 years of observations. In fact, during this last period, many important interventions into the Danube thermal and ice regime have been realized.

8.3.1 Ice Phenomena Observations

Ice phenomena observations on the Danube began in the twentieth century. However, it did not reach the observation level of other hydrological elements, because of the strong influence of the observer's subjectivity. Observations are made daily during the morning water stages and temperature observations. In addition to type of IP, other characteristics are also registered, for example approximate width of the border ice, intensity of the ice run, and in selected stations daily, or in other time intervals, also the thickness of the ice cover.

8.3.2 Network of Observation Stations

For processing of the ice phenomena characteristics of the Danube and its tributaries, data were used from 55 hydrological stations. Their selection is shown in Table 8.8. In addition to stationing (km), there is indicated the number of years with IP and FU, up to winter 1984/1985.

Stations, water structures, thermal and nuclear power stations (set into operation up to 1985), are schematically presented in Fig. 8.2.

Table 8.8 Basic data of stations on the Danube and its tributaries, with IP and FU observations

River	Hydrological station	Stationing (km)	Observation (years)		
			Total	Only IP	Only FU
1	2	3	4	5	6
Danube	1 Ingolstadt	2,458.3	85	58	11
Danube	2 Neustadt	2,432.3	80	59	19
Danube	3 Kelheim	2,414.8	85	59	16
Danube	5 Niederwinzer	2,381.7	77	58	12
Danube	6 Regensburg	2,379.3	85	57	12
Danube	7 Schwabelweis	2,376.2	70	50	11
Danube	8 Straubing	2,321.3	85	70	20
Danube	9 Pfelling	2,305.5	56	39	14
Danube	10 Deggendorf	2,284.6	85	79	19
Danube	11 Hofkirchen	2,256.9	60	46	25
Danube	12 Vilshofen	2,249.4	84	70	35
Danube	14 Passau Ilzstadt	2,225.3	85	60	8
Danube	15 Achleiten-Jochens.	2,223.1	85	56	6
Danube	16 Engelhartszell	2,200.7	91	67	4
Danube	17 Linz	2,135.2	91	69	1
Danube	18 Stein-Krems	2,003.5	91	70	1
Danube	19 Vienna	1,929.1	91	66	7
Morava	20 Moravský Ján	67.6	60	50	19
Danube	21 Bratislava	1,868.8	85	63	13
Danube	22 Palkovičovo	1,810.0	55	42	7
Danube	24 Komárno	1,767.1	85	69	17
Váh	25 Šal'a	58.5	50	47	30
Hron	26 Brehy	102.5	34	30	18
Ipel'	27 Ipel'. Sokolec	42.9	60	56	38
Danube	28 Budapest	1,646.5	85	77	25
Danube	29 Dunaujváros	1,580.6	40	36	7
Danube	30 Paks	1,531.3	40	36	10
Danube	31 Mohacs	1,446.9	84	78	42
Danube	32 Bezdan	1,425.5	88	70	38
Drava	33 D. Miholjac	74.6	88	54	29
Danube	34 Bogojevo	1,367.4	88	66	24

Table 8.8 (continued)

River	Hydrological station	Stationing (km)	Observation (years)		
			Total	Only IP	Only FU
1	2	3	4	5	6
Danube	35 Novi Sad	1,255.1	88	66	24
Tisa	36 Senta	122.0	74	68	48
Danube	37 Zemun	1,172.9	88	63	24
Sava	38 S. Mitrovica	136.0	82	42	20
Danube	40 Smeredevo	1,116.2	88	58	24
V. Morava	41 Lj. Most	39.9	60	46	25
Danube	42 V. Gradište	1,059.8	56	43	19
Danube	43 Drobeta-T. Severin	931.0	53	33	1
Danube	44 Novo Selo	833.6	44	32	3
Danube	45 Lom	743.3	44	33	5
Jiu	46 Zăvalu	7.4	34	31	18
Olt	47 Izbiceni	6.0	34	31	18
Danube	48 Svistov	554.3	44	31	10
Danube	49 Ruse	495.6	44	35	11
Danube	50 Giurgiu	493.5	44	32	11
Argeş	51 Budeşti	31.0	34	34	11
Danube	52 Silistra	375.5	44	35	15
Ialomiţa	53 Slobozia	77.0	34	34	22
Danube	54 Braila	169.7	53	41	20
Siret	55 Lungoci	74.0	34	33	25

Fig. 8.2. (continued) Scheme of the ice observation stations, constructed water structures and power stations

Mark on map	Type of structure	Name	Mark on map	Type of structure	Name
A	ws	Ingolstadt	N	ws	Wallsee
B	TPS	Grossmehring	O	ws	Ybbs-Persenbeug
C	TPS	Irsching	P	ws	Melk
D	ws	Bad Abbach	R	ws	Altenwörth
E	ws	Regensburg	S	ws	Greifenstein
F	ws	Geisling	T	NPS	Paks
G	ws	Straubing	U	ws	Djerdap I.
H	TPS	Pleiting	V	ws	Djerdap II.
I	ws	Kachlet			
J	ws	Jochenstein	ws	Water structure	
K	ws	Aschach	TPS	Thermal power station	
L	ws	Ottensheim	NPS	Nuclear power station	
M	ws	Abwinden Asten	Station numbers on map correspond to numbers in Table 8.8		

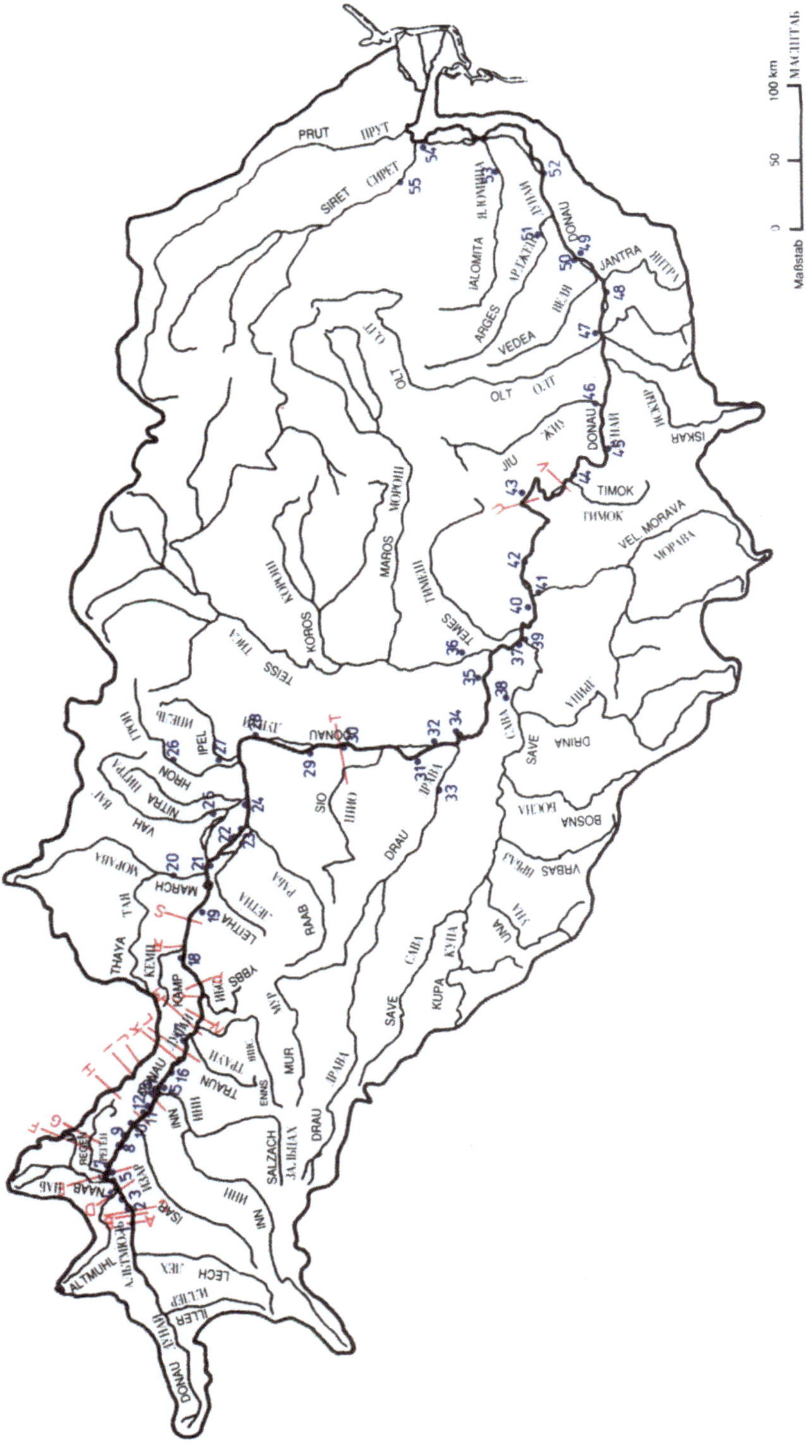

Fig. 8.2 (continued)

8.3.3 Ice Regime Characteristics

In conformity with the approved methodology (Forschungsanstalt für Wasserwitschaft 1988) for ice regime characterization, the following data were used: dates of the IP begin (BIP), of the IP end (EIP), dates of the FU begin (BFU), of the FU end (EFU), and of their duration (DIP, DFU). For their evaluation, the extreme dates and the extreme durations were chosen in particular stations:

- the earliest begin of BIP and of BFU,
- the latest end of EIP and of EFU,
- the longest duration of DIP and of DFU,

which are presented in Tables 8.9 and 8.10 for the Danube, and in Tables 8.11 and 8.12 for the tributaries.

From the point of view of anthropogenic interventions into the river, similarly as for the thermal regime, the data are not homogeneous. On the basis of the overview in Tables 8.9 and 8.10, it was possible to generally state changes in characteristic dates and durations, as follows:

- At the upper and central Danube stations, for the whole observation period, the earliest BIP falls on the second November half. Exceptions are stations with shorter observation periods (Schwabelweis, Pfelling, Hofkirchen, Palkovičovo, and stations between Dunaújváros and Drobeta-Turnu Severin). The latest BIP is from February middle to its end; for the lower Danube stations the later BIP falls to middle of December, and the latest one to the second half of February.
- The earliest ice departure (EIP) starts from the end of November to the end of December (exceptions are stations Giurgiu and Braila with EIP at the beginning of January), the latest EIP occurs in the whole river from 10.03 to 31.03.
- The mean IP duration was 14–36 days; from the point of view of industrial utilization and navigation, more significant are values of the undisturbed IP duration, which are between 17 and 91 days.
- Apart from the processing of the whole observation period, also shorter periods of data were processed, namely 40-years 1945–1985, 30-years 1955–1985, 20-years 1965–1985, and the selected strong winters 1953/1954, 1955/1956, 1962/1963 and 1984/1985. In the 40-year period, the first IP and FU occur in December, with the exception of Ingolstadt, Regensburg, Vienna, further on from Dunaújváros to Paks, in Novi Sad, Smederevo and Drobeta-T. Severin. Ice departure and FU end falls to March; IP and FU duration is similar as for the whole period. In the 30-year period, the first IP are in December, FU in December to January; in the river section between Engelhartszell and Bratislava and from Drobeta-T. Severin to Svistov, FU did not occur at all; EIP and EFU fell to March. In the 20-year period, large changes are recorded in dates as well as in IP and FU durations; first IP was in December, the last in February on the upper Danube, on the central and lower Danube up to the middle of March; FU on the upper and central Danube started

Table 8.9 Long-term Danube IP characteristics

	Ice Phenomena (IP)						
	Begin of IP		End of IP		Duration		
						Longest duration	
					Mean	Total	Uninterr.
Hydrological station	Earliest	Latest	Earliest	Latest	Days		
1	2	3	4	5	6	7	8
Ingolstadt	19.11	26.02	30.11	31.03	16	80	29
Neustadt	19.11	24.02	01.12	13.03	16	88	19
Kelheim	19.11	21.02	30.11	13.03	18	88	36
Niederwinzer	27.11	29.02	29.11	14.03	18	70	44
Regensburg	27.11	25.02	02.12	14.03	18	69	31
Schwabelweis	04.12	24.02	17.12	15.03	18	80	31
Straubing	19.11	21.02	02.12	19.03	20	79	27
Pfelling	03.12	09.02	05.12	14.03	20	87	17
Deggendorf	19.11	19.02	02.12	19.03	23	89	35
Hofkirchen	05.12	26.02	14.12	14.03	18	89	22
Vilshofen	19.11	26.02	02.12	15.03	22	94	35
Passau Ilzstadt	19.11	12.02	02.12	19.03	15	78	22
Achleiten	19.11	15.02	02.12	11.03	19	83	39
Engelhartszell	26.11	12.02	27.11	10.03	17	66	63
Linz	27.11	12.02	02.12	10.03	21	79	59
Stein-Krems	29.11	13.02	30.11	13.03	16	70	56
Vienna	28.11	13.02	28.11	15.03	19	68	68
Bratislava	17.11	13.02	02.12	22.03	24	83	72
Palkovičovo	03.12	21.02	29.12	17.03	22	91	91
Komárno	18.11	13.02	03.12	22.03	27	90	90
Budapest	17.11	10.02	01.12	24.03	32	93	40
Dunaujváros	01.11	21.02	04.12	20.03	27	93	81
Paks	01.11	23.02	13.12	20.03	27	93	82
Mohacs	16.11	26.02	05.12	27.03	36	95	84
Bezdan	01.12	27.02	06.12	25.03	14	45	30
Bogojevo	01.12	27.02	05.12	26.03	19	60	34
Novi Sad	01.12	28.02	04.12	25.03	23	63	49
Zemun	06.12	23.02	07.12	25.03	14	67	43
Smeredevo	01.12	09.02	13.12	28.03	17	35	26
V. Gradište	06.12	05.03	31.12	21.03	16	36	27
Drobeta-T. Severin	13.12	09.02	28.12	16.03	25	76	76
Novo Selo	13.12	23.02	25.12	15.03	32	84	84
Lom	11.12	01.03	25.12	15.03	31	83	83
Svistov	10.12	22.02	26.12	20.03	31	90	90
Ruse	09.12	22.02	26.12	20.03	31	89	89
Giurgiu	10.12	22.02	08.01	18.03	28	88	32
Silistra	12.12	22.02	28.12	26.03	33	91	91
Braila	11.12	07.02	14.01	25.03	36	91	37

Table 8.10 Long-term Danube FU characteristics

	Freeze-up (FU)						
	Begin of FU		End of FU		Duration		
						Longest duration	
					Mean	Total	Uninterr.
Hydrological station	Earliest	Latest	Earliest	Latest	Days		
1	2	3	4	5	6	7	8
Ingolstadt	14.12	06.02	17.12	10.03	18	51	51
Neustadt	08.12	17.02	13.12	11.03	21	75	75
Kelheim	09.12	17.02	18.12	12.03	18	44	42
Niederwinzer	18.12	12.02	25.12	12.03	19	38	38
Regensburg	16.12	12.02	27.12	13.03	23	38	38
Schwabelweis	16.12	17.02	27.12	12.03	22	40	38
Straubing	07.12	10.02	17.12	18.03	30	63	59
Pfelling	20.12	10.02	18.01	13.03	35	73	73
Deggendorf	14.12	02.02	17.12	18.03	34	76	76
Hofkirchen	09.12	14.02	29.12	12.03	20	73	59
Vilshofen	07.12	03.02	17.12	15.03	21	85	76
Passau Ilzstadt	25.12	14.02	05.01	04.03	17	32	32
Achleiten	28.12	03.02	06.01	07.03	18	50	36
Engelhartszell	20.12	10.03	23.12	10.03	6	18	18
Linz	20.01	20.01	16.02	16.02	2	2	1
Stein-Krems	15.02	15.02	13.03	13.03	27	27	27
Vienna	26.12	18.02	05.01	15.03	29	43	43
Bratislava	15.12	18.02	19.12	16.03	29	67	60
Palkovičovo	22.12	13.02	09.01	11.03	33	79	79
Komárno	14.12	18.02	21.12	20.03	27	78	78
Budapest	17.12	16.02	05.01	19.03	27	83	83
Dunaujváros	05.01	09.02	19.01	13.03	30	55	55
Paks	05.01	07.02	17.01	14.03	32	61	61
Mohacs	09.12	16.02	30.12	20.03	33	77	77
Bezdan	08.12	11.02	27.12	20.03	33	85	85
Bogojevo	16.12	17.02	21.12	19.03	27	68	68
Novi Sad	22.12	13.02	27.12	15.03	24	64	64
Zemun	11.12	17.02	13.01	21.03	19	67	61
Smeredevo	16.12	11.02	29.12	16.03	24	65	65
V. Gradište	26.12	08.02	07.01	13.03	26	68	68
Drobeta-T. Severin	14.01	14.01	28.01	28.01	16	16	16
Novo Selo	09.01	29.01	13.02	08.03	42	55	55
Lom	01.01	06.02	12.01	09.03	31	55	55
Svistov	29.12	30.01	14.01	13.03	27	67	67
Ruse	26.12	04.02	21.01	13.03	34	66	66
Giurgiu	26.12	01.02	11.01	16.03	33	69	69
Silistra	22.12	22.02	21.01	20.03	35	76	76
Braila	19.12	15.02	27.01	21.03	40	83	83

Table 8.11 Long-term IP characteristics on selected tributaries

	Ice phenomena (IP)						
	Begin of IP		End of IP		Duration		
						Longest duration	
					Mean	Total	Unintrr.
Hydrological station	Earliest	Latest	Earliest	Latest	Days		
1	2	3	4	5	6	7	8
Moravský Ján	01.12	11.02	12.12	22.03	31	97	97
Šal'a	23.11	20.02	19.12	23.03	38	94	94
Brehy	17.11	01.02	17.01	17.03	38	93	88
Ipel'ský Sokolec	16.11	14.02	20.01	26.03	52	107	103
Donji Miholjac	01.12	13.03	16.01	13.03	12	50	24
Senta	01.12	02.03	07.12	22.03	13	39	33
Sremska Mitrovica	14.12	11.02	24.12	15.03	4	19	14
Ljubičeskij Most	02.12	03.02	16.12	20.03	11	53	47
Zăvalu	12.11	15.02	11.12	22.03	33	100	55
Isbicen	22.11	15.02	02.01	23.03	40	102	48
Budeşti	10.11	10.02	07.12	24.03	43	102	53
Slobozia	19.11	24.01	21.01	28.03	46	108	55
Lungoci	09.11	18.02	06.02	31.03	63	113	55

in January and beginning of February, while on the lower Danube in January, and the FU departure in March.

The situation on the tributaries is demonstrated in Tables 8.11 and 8.12:

- tributaries originating in mountainous regions had earlier IP arrival; the earliest from 9.11 to 23.11, the others in the first half of December (from 1.12 to 14.12); the latest arrivals were shifted to the end of January up to the middle of March;
- the earliest IP departure was from 11.12 to 21.01, and on Siret up to 6.02; the latest ice departure on all tributaries was in the second half of March;
- freeze-up occurred the earliest in December (from 1.12 to 25.12) and the latest from the beginning of February to the beginning of March;
- end of FU occurred from the middle of December up to the middle of January at the earliest, at the latest then from 10.3 to 24.3;
- the lowest IP duration values were on Sava – 4 days, duration 11–13 days occurred on Drava, Tisa and V. Morava, the other tributaries had a mean duration of 31–46 days, the longest was on Ipel' – 52 days, and on Siret – 63 days;
- the longest total duration was 19 up to 113 days;
- the longest uninterrupted duration was 47–55 days, but on Sava it was only 14 days, while on Morava, Váh, Hron and Ipel' 88 up to 103 days;

Table 8.12 Long-term FU characteristics on selected tributaries

	Freeze-up (FU) (ZÁ)						
	Begin of FU		End of FU		Duration		
						Longest duration	
					Mean	Total	Uninterr.
Hydrological station	Earliest	Latest	Earliest	Latest	Days		
1	2	3	4	5	6	7	8
Moravský Ján	01.12	02.02	17.12	19.03	36	72	72
Šal'a	04.12	22.02	03.01	21.03	33	89	89
Brehy	12.12	10.02	15.12	11.03	26	79	79
Ipel'ský Sokolec	01.12	24.02	14.12	20.03	37	84	84
Donji Miholjac	10.12	14.02	23.12	10.03	27	76	76
Senta	07.12	26.02	16.12	24.03	36	96	94
Sremska Mitrovica	16.12	15.02	16.01	13.03	9	60	60
Ljubičeskij Most	08.12	19.02	28.12	15.03	10	62	59
Zăvalu	18.12	05.02	02.01	11.03	19	48	40
Isbicen	25.12	21.02	05.01	19.03	26	81	81
Budeşti	12.12	09.03	24.12	12.03	41	77	77
Slobozia	05.12	11.03	21.12	18.03	30	78	75
Lungoci	11.12	08.03	13.01	14.03	21	73	70

- the mean FU duration was shortest on Sava and V. Morava (9–10 days), on the rest of the tributaries it was 19–41 days;
- the total FU duration was 48 (Jiu) up to 96 days (Tisa);
- the longest uninterrupted FU duration was 40–94 days.

In addition to these characteristics, relations between IP and FU were examined to the air and water temperatures, namely:

- air temperature on the first day of occurrence of IP and FU (t_a^{1IP}, t_a^{1FU}),
- negative temperatures sum up to BIP and BFU ($\sum -t_a^{BIP}$, $\sum -t_a^{BFU}$),
- positive temperatures sum up to EIP and EFU ($\sum +t_a^{EIP}$, $\sum +t_a^{EFU}$),
- water temperature on the first day of occurrence of IP and FU (t_o^{1IP}, t_o^{1FU}),
- number of days with $\sum -t_a^{BIP}$ and $\sum -t_a^{BFU}$.

These data were processed in tabular and graphical form. An overview of the characteristics for the central Danube station (Vienna, 30-year period), is given in Table 8.13.

According to the data in this table it is evident, that the Vienna industrial agglomeration, within the examined period of 30 years, positively influenced ice

Table 8.13 Temperature characteristics of winters with IP and FU for station Vienna

	Air temperature t_a						Number of days with	Water temp. 1 day with	
	$\Sigma{-}t_a$		In first day with		$\Sigma{+}t_a$				
	BIP	BFU	IP	FU	EIP	EFU		IP	FU
Winter	°C						Days	°C	
1	2	3	4	5	6	7	8	9	10
1955–1956	–35.9	–	–17.3	–	0.0	–	6	0.1	–
1956–1957	–17.5	–	–9.3	–	0.0	–	4	0.6	–
1957–1958	–20.9	–	–5.9	–	13.8	–	6	0.2	–
1958–1959	–40.5	–	–4.3	–	0.0	–	11	0.2	–
1959–1960	–40.4	–	–9.8	–	1.9	–	6	0.2	–
1960–1961	–24.6	–	–6.5	–	3.0	–	8	0.3	–
1961–1962	–15.1	–	–11.1	–	0.0	–	3	0.4	–
1962–1963	–15.9	–	–11.5	–	0.0	–	4	0.2	–
1963–1964	–17.5	–	–4.9	–	1.6	–	5	0.7	–
1964–1965	–	–	–	–	–	–	–	–	
1965–1966	–44.2	–	–6.4	–	0.0	–	8	0.0	–
1966–1967	–	–	–	–	–	–	–	–	
1967–1968	–30.4	–	–8.1	–	8.3	–	4	0.0	–
1968–1969	–50.3	–	–5.0	–	0.8	–	11	0.0	–
1969–1970	–19.8	–	–9.5	–	1.1	–	4	0.0	–
1970–1971	–23.0	–	–7.1	–	0.0	–	4	0.0	–
1971–1972	–9.0	–	–7.0	–	0.0	–	3	0.8	–
1972–1973	–21.8	–	–4.8	–	7.9	–	8	0.0	–
1973–1974	–	–	–	–	–	–	–	–	–
1974–1975	–	–	–	–	–	–	–	–	
1975–1976	–	–	–	–	–	–	–	–	–
1976–1977	–	–	–	–	–	–	–	–	
1977–1978	–	–	–	–	–	–	–	–	
1978–1979	–76.1	–	–1.4	–	2.8	–	16	0.2	–
1979–1980	–	–	–	–	–	–	–	–	
1980–1981	–	–	–	–	–	–	–	–	
1981–1982	–44.0	–	–10.9	–	0.0	–	6	0.6	–
1982–1983	–	–	–	–	–	–	–	–	
1983–1984	–	–	–	–	–	–	–	–	
1984–1985	–59.8	–	–16.3	–	1.4	–	14	0.2	–

formation – alleviated it; in this period freeze-up did not occur at all, and there was 11 years without IP (particularly in the period from 1972/1973 to 1984/1985 – out of 13 winters nine times). This in spite of the fact, that during the whole observation period from the 1900/1901 winter to the 1954/1955 one, i.e. for 55 years, IP did not occur in 14 years, and in the remaining 41 years, in addition to IP also FU occurred in 5 years.

8.3.4 Ice Regime Changes Due to Human Activity

During the last century the Danube *river section in Germany* was influenced by river channel corrections, regulation and construction of water structures. After regulation, the river length was shortened, the bifurcations removed and the river channel cross-section changed. Simultaneously, the meanders and islands were removed, which were obstacles for free ice departure. They were often responsible for the formation of ice jams and barriers, causing catastrophic ice floods. These regulations were performed over the Kelheim to Weltenburg straits, in the upstream direction. Observations confirmed that the ice problems diminished quantitatively. Numerical data were not available. The IP observations were made at the water level gauging stations only; data from the mid-sections were also not available.

The principal change for the ice regime was the construction of a cascade of water structures. During frosty days, ice cover on the surface of the reservoirs formed which subsequently, particularly when covered with snow, prevented further cooling of the water surface. Thus, the ice melt and following ice flow continued without problems, even during flood situations. Ice flow and ice run (of the frazil ice and ice mud) did not occur, because there were no conditions for its formation. An exception is the so-called secondary ice run, which occurs with a heavy frost during floods after the gates on weirs have been opened.

The release of waste heat from thermal power plants is of great importance for the Danube ice regime. It was determined that ice begins to form ca. 25 km downstream from the water release location, approximately 1 day later than at the other locations. However, this only occurs if the power plant produced a sufficient power output. After restoration of the steam power plants at Ingolstadt, Irsching and Pleinting from 1977–1978, no changes of the thermal regime were observed.

The river correction and regulation activities on the *Austrian* Danube *river section* influenced the river ice regime greatly. The Danube flows here through a 150-km long narrow strait, and through a 200-km long wide alluvial lowland mead. The 1830 ice flood has mobilized all efforts for improvement of the river channel. The professional literature confirms the positive impact upon the flood regime of the extensive river regulation activities, performed between 1850 and 1938. The construction of the water structures had a great influence. There were five of them, completed already by 1970 on the Austrian section of the River Danube. As a consequence of the systematic river regulation, the following changes of the ice regime were noted:

- Freeze-up occurrence probability decreased ($P = 0.675$ decreased to $P = 0.240$); FU occurred in this river section only when extended from the Slovak or Slovak–Hungarian river sections.
- Probability of BIP up to Bratislava was the same.
- Ice departure started first on the upper section; EIP in Bratislava, compared with Engelhartszell and Vienna, was 1 week later on average, and fell on the latter part of March.

- Analysis of the mud ice flow has shown that from the completion of the water structure Kachlet to the start of construction of the other structures, ice drift flow intensity decreased on average, to 46% in Aschach, to 36% in Linz, and to 54% in Ybbs; this water structure stopped the ice flow from the whole Bavarian Danube. Ice flow from the tributaries Inn and Enns stopped only after the construction of water structures on them had started.
- It is documented in relation to all characteristics, that not only the ice flow intensity, but also its quantity and quality has changed; only during extremely frosty winters, in the free river sections, was an ice regime observed, that was similar to before the operation of the water constructions. The probability of such extreme cases was very low.
- Long experience has shown, that on water structures, an isolated ice cover originates; it should also be mentioned, that in free river sections, the IP occurrence does not last as long as the FU on water structures.
- Finally it needs to be stated, that the Austrian Danube river regulation had a positive influence upon the ice regime of the Slovak river section; however, this does not mean, that the danger and threat of ice formation has been removed.

In the *Slovak* Danube *river section* between Bratislava and Komárno, anthropogenic interventions are represented only by the correction and regulation works and wastewater releases. They almost certainly influenced the ice regime positively. IP, which occurred every year between 1876 and 1900, were a consequence of missing river channel regulation structures, particularly those for low and medium discharges. Numerous meanders, sharp bends and gravel benches enhanced ice formation. In the first 30 years of the twentieth century, there were five winters without ice, and in the second 30 years there were four such winters. In the period 1961–1985, there were in Bratislava 12 winters without ice, in Komárno seven winters. It is assumed, that this is a consequence of the construction of Austrian water structures on the Danube, inflow of warm groundwater from the Danube infiltration zone, and of the wastewater outflows from industry and energy production. The Váh tributary also causes a difference in the number of days without ice between Bratislava and Komárno.

In the Danube *river section in Hungary*, the most dangerous was the one downstream of Dunaföldvár (ca. 1,560.0 km). In order to reduce this danger, river channel regulation activities had already begun in the nineteenth century. Up to 1945 (over a period of 80 years), between Dunaföldvár and Vukovar on 23 sites, embankment protections, directional structures and protection levees were constructed. After 1945, activities were performed in the Baja–Moháč section. From 1958–1975, in agreement with a unique channel regulation concept, the Danube was corrected over a total length of 67 km (between Dunaföldvár and the border with former Yugoslavia), and on the territory of former Yugoslavia in the section from 1,436.0–1,333.0 km. These activities decreased ice jam and barrier occurrence to an acceptable level and thus also the need for the use of icebreakers. The warm wastewaters released to the Danube, influenced the ice regime partly by increasing water temperature, and partly by changing water quality. This influences the physical processes of ice formation (increase of dissolved matter concentration slows down this process).

During the last 50 years (up to 1985), the number of days with FU in Budapest has decreased, and a similar situation is present in the other Hungarian Danube stations. To a significant extent, it shows up downstream of the nuclear power station Paks. In addition, here, except for local interventions, an impact can be documented from the Austrian Danube water structures. It is an objective fact that ice volume inflow to Hungary decreased.

According to relevant research (Institut meteorogii i gidrologii 1990), the anthropogenic interventions on the *Romanian* Danube *river section*, influence the ice regime more than the thermal one. The water structure reservoirs have created favourable conditions for ice formation (low velocities), compared with those on the free river sections, and hastened IP occurrence. Downstream of the reservoirs, the conditions worsened, partly because of the higher velocities and turbulence, and because of ice deposition in the reservoirs. Here also, warm wastewater releases contributed to a decrease of ice occurrence. For example, in the Jiu tributary, the warm wastewater ensured flow without ice, even in the coldest winters over a long river stretch downstream of the wastewater output.

8.4 Conclusion

The water temperature from Ingolstadt to Braila (2,288.6 km), from 1956–1985 has been raised by 0.5–4.0°C. Between particular water gauging stations, episodes occurred with temperature decrease (Tables 8.2, 8.3, and 8.4). This was a consequence of mutual climatic relations affecting each other and changing along the river. It was also influenced by the tributaries and the various anthropogenic interventions. Quantitative characteristics of the ice regime at individual stations, on the upper, central, and lower Danube are presented in Tables 8.2, 8.3 and 8.4, and for the tributaries in Table 8.5. An analysis of the ice regime in relation to causal factors, mainly to anthropogenic interventions, is given in Section 8.2.4.

On the Danube, two sorts of ice regime occur, namely the ice regime on a free river section, and also the one on a section with a submerged backwater. The long-term IP and FU characteristics were elaborated, based on data from selected stations on the upper, central, and lower Danube (Tables 8.9 and 8.10), and on some tributaries (Tables 8.11 and 8.12). The beginning, end and duration of IP and FU along the river, and relationships between the ice regime, climate changes and the anthropogenic interventions, are described in Section 8.3.4.

References

Bayer. Landesamt für Wasserwirtschaft: Projekt Nr. 2 "Temperatur- und Eisregime der Donau und ihrer wichtigsten Zubringer". I. Teil, München, September 1989; II. Teil, München, April 1990

BNK po MGP: Projekt No 2 "Termičeskij i ledovyj režim Dunaja i jego osnovnych pritokov". Sofija, February 1990

Donaukommission (1967) Studie über dasEisregime des Flusse Donau. Budapest
Donaukommission: Hydrologisches Jahrbuch des Flusses Donau. Jahrgänge 1953 bis 1987. Budapest
Federal Institut for Hydrometeorology: Projekt Nr.2 "Temperatur- und Eisregime der Donau und ihrer wichtigeren Zubringer". Beograd, July 1990
Forschungsanstalt für Wasserwitschaft (1988) Temperatur- und Eisregime der Donau und ihrer wichtigeren Zubringer. Methodik des Projektes Nr.2, Bratislava
Forschungsanstalt für Wasserwitschaft: Projekt Nr.2 "Temperatur- und Eisregime der Donau und ihrer wichtigeren Zubringer". I. Teil, Bratislava, December 1989
IHP Nationalkomitees der Donauländer (1986) Die Donau und ihr Einzugsgebiet – Eine hydrologische Monographie
Institut meteorogii i gidrologii: Projekt No 2 "Termičeskij i ledovyj režim Dunaja i jego osnovnych pritokov", Nacionalnyj otčet. Bucharest, October 1990
Regionale Zusammenarbeit der Donauländer im Rahmen IHP der UNESCO (1993): Temperatur- und Eisregime der Donau und ihrer witigeren Zubringer. Bratislava
Technische Universität Wien: Temperatur- und Eisregime der Donau und ihrer wichtigeren Zubringer, Österreichischer Beitrag. Wien, September 1989
VITUKI: Projekt Nr.2 "Temperatur- und Eisregime der Donau und ihrer wichtigeren Zubringer". I. Teil, Budapest, September 1989; II. Teil, October 1989

Chapter 9
Sediment Regime of the River Danube (1956–1985)

László Rákóczi

Abstract During the investigated period (1956–1985), the natural sediment regime of the River Danube has changed significantly due to various human interventions on the Danube itself and on its main tributaries. The total mass of suspended sediment transported to the Black Sea annually varies between 25 and 80 $\times$ 10^6 t. The sediment transport quickly rises where the bigger tributaries (Tisza, Sava, V. Morava) join the Danube within a relatively short reach. Before the construction of barrages and hydropower stations, the share of bedload in the total sediment transport reached about 20% on the German and Austrian Danube reach. On low-land sections this share drops to about 1% or less.

The most important human interventions affecting sediment transport include retaining and collecting the eroded soil and rock materials on the slopes of basins; damming the rivers for navigation and energy production purposes; dredging the bed material from the river channel for industrial use at a rate much higher than the bedload transporting potential of the river, etc. Because of the summarized effects of these anthropogenic factors, the suspended sediment and bedload transport show a decreasing tendency at almost every station investigated.

The presented results also show how the damming effect of the river barrages causes characteristic changes in the grain-size distribution of both the suspended sediment and the riverbed material. As human interventions have continued on the River Danube and in its basin since 1985, it is now high time to collate all available sediment data for the following 20-year period and compile a similar monographic summary (1986–2005) based on this data.

Keywords Artificial impacts · Changes of sediment transport in time and space · Effects of river barrages · Scarcity of bedload data · Gravel mining from the riverbed · Changes of grain-size composition of bed material

---

L. Rákóczi (✉)
Research Centre for Environment and Water (VITUKI), Budapest, Hungary
e-mail: lrakoczi@t-online.hu

M. Brilly (ed.), *Hydrological Processes of the Danube River Basin*,
DOI 10.1007/978-90-481-3423-6_9,

9.1 Introduction

The full text of this study was published in 1993 as follow-up volume No. 1 of the Hydrographical Monograph of the River Danube RC of DC (1986) in German and Russian (Hydrological Working Group of the Danubian Countries). The monograph was the first attempt to collate all available suspended sediment, bedload and bed material data for a 30-year period from the eight riparian countries of that time. The locations and names of the 20 sediment measuring stations supplying data for this monograph are indicated on the map of the Danube Basin (Fig. 9.1). During the 1956–1985 period, several significant anthropogenic measures took place across the Basin, therefore, the main aim of the work became to demonstrate the effects of the various natural and artificial impacts on the sediment regime of the River Danube.

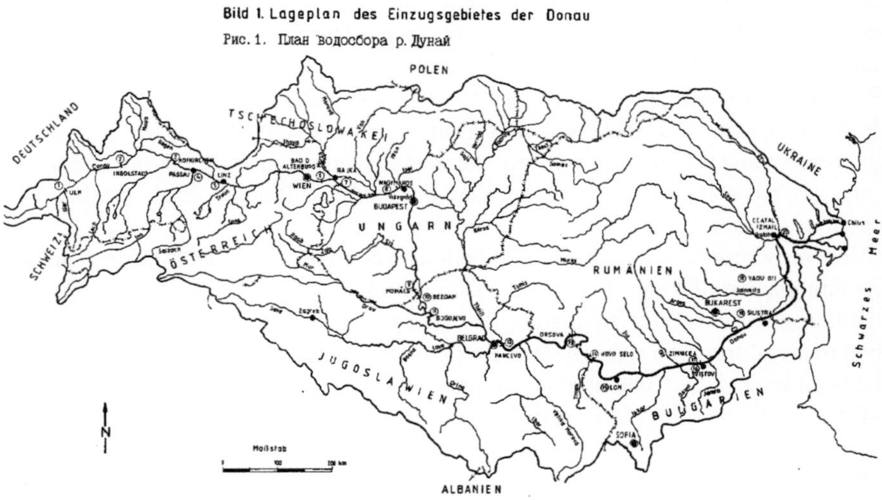

Fig. 9.1 Basin of the Danube River showing the sediment measuring stations (1956–1985)

9.2 Suspended Sediment Transport

After processing the received sediment data in a uniform way, the time series of the annually transported suspended sediment masses in tons together with the annual flow volumes were presented graphically for the sediment measuring stations (see e.g. in Fig. 9.2 for Passau, Linz and Bad Deutsch-Altenburg). Suspended sediment transport is significantly increased by the contribution of the River Inn, joining the Danube at Passau. Because of the sediment settling effect of three run-of-river hydropower stations constructed on the German Danube reach during the 1956–1974 time period, the difference in annual sediment transport at Passau and Linz has seemingly decreased after 1964 compared with the preceding years with moderate

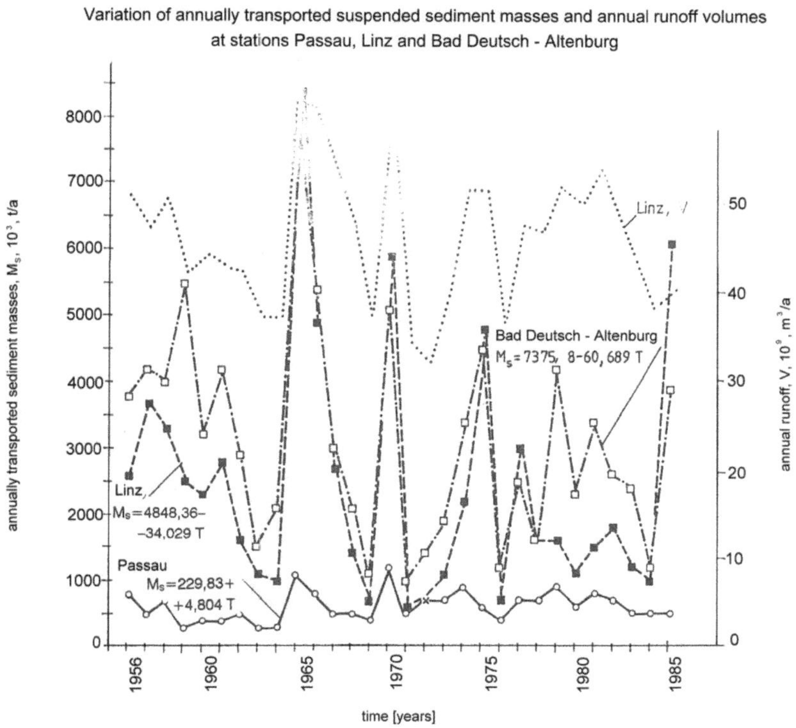

Fig. 9.2 Variation of annually transported suspended sediment masses and annual runoff volumes at stations Passau, Linz and Bad Deutsch-Altenburg

flood activity. Because of the erection of another six hydropower stations on the Austrian stretch, the sediment transport at Linz in 1965, 1970 and 1975 was greater than at Bad Deutsch-Altenburg.

The annual minimum, mean and maximum suspended sediment masses are also shown in tabulated form for each sediment-measuring station (Table 9.1). As can be seen, the ratios of maximum and minimum data are rather high at several stations, surpassing even 10 at Linz, Rajka and Orsova, presumably due to the sediment settling effects of upstream hydropower structures. The annual suspended sediment transport at a given station therefore can not be well characterized by the mean transport only, rather by all the three tabulated data, indicating also the range of variation. For example, at Ceatal Izmail, i.e. at the closest measuring station to the Black Sea estuary of the River Danube, the long-term mean suspended sediment transport was found to be 43 $\times$ 10^6 t/a, however, during the investigated 30-year period, the transport changed between 14.7 $\times$ 10^6 and 71 $\times$ 10^6 t/a.

The variation of suspended sediment transport along the River Danube is presented in Fig. 9.3 for three dry years with long low flow periods and without great flood waves. It demonstrates that since the early 1960s, suspended sediment transport has decreased significantly, mainly due to the settling effect of the

Table 9.1 Sediment measuring stations on the River Danube considered for investigation of suspended sediment and bedload transport

No.	Name of the station	Distance from the confluence in km	Country
1	Ulm	2,586.7	Germany
2	Ingolstadt	2,458.3	Germany
3	Hofkirchen	2,256.9	Germany
4	Passau	2,230.6	Germany
5	Linz	2,135.1	Austria
6	Bad Deutsch-Altenburg	1,887.1	Austria
7	Rajka	1,848.4	Hungary
8	Nagymaros	1,694.6	Hungary
9	Mohács	1,446.8	Hungary
10	Bezdan	1,425.5	Yugoslavia
11	Bogojevo	1,367.4	Yugoslavia
12	Pančevo	1,153.3	Yugoslavia
13	Orşova[a]	955.0	Romania
14	Novo Selo	833.6	Bulgaria
15	Lom	743.3	Bulgaria
16	Svistov	554.3	Bulgaria
17	Zimnicea	554.3	Romania
18	Silistra	375.5	Bulgaria
19	Vadu Oii	238.0	Romania
20	Ceatal Izmail	79.6	Romania

[a]Since 1971: Dobreta-Turnu Severin, 937 km.

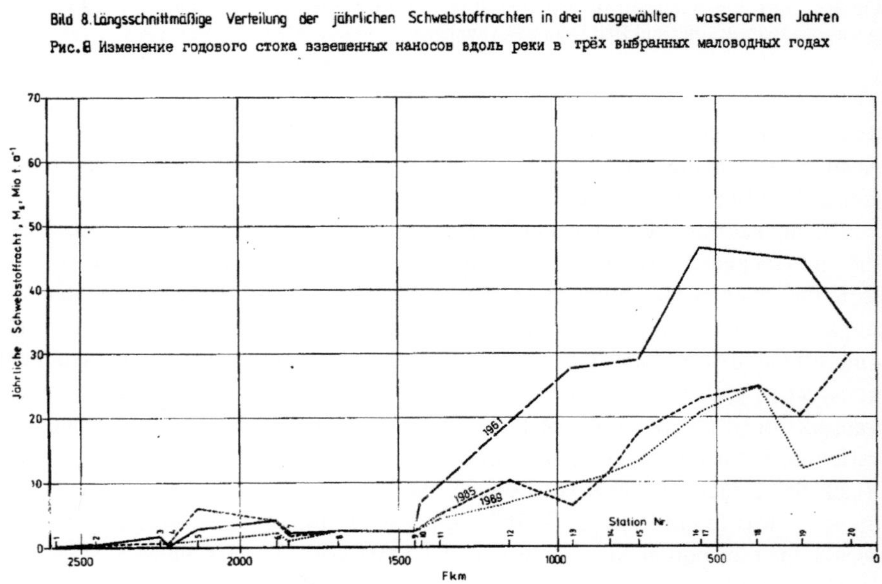

Fig. 9.3 Variation of annually transported suspended sediment masses in three selected dry years along the River Danube

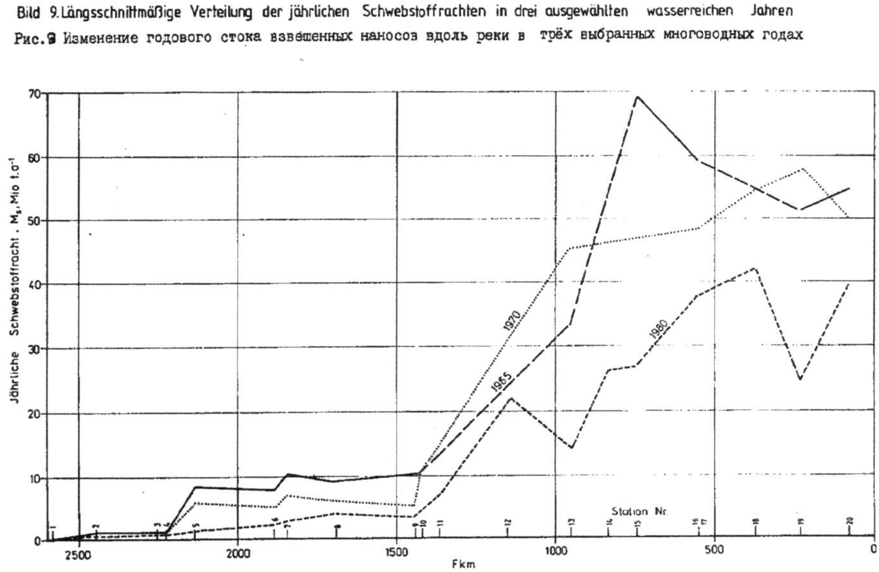

Fig. 9.4 Variation of annually transported suspended sediment masses in three selected wet years along the River Danube

hydropower stations on the upstream reaches. Another anthropogenic influence is the great reservoir of the Iron Gate I, indicated by the decreasing suspended sediment transport at Orsova station.

In Fig. 9.4 the longitudinal variation of annual sediment transport data are plotted for three wet years with great floods. The sediment contribution of the tributaries at high flow is clearly visible (e.g. Inn, Morava, V. Morava and Isker Rivers). The graph for 1980 shows the sediment settling capacity of the Iron Gate I reservoir even for flood waves. The significant increase of suspended sediment transport between Orsova and Novo Selo stations indicates a significant riverbed scouring activity downstream of the hydropower plant. Further downstream, the sediment contribution of the Siret River should be mentioned.

Table 9.2 presents the suspended sediment contribution of the important tributaries. Besides their names, the location of their confluences to the River Danube and the sizes of their basins, the annual minimum, mean and maximum suspended sediment masses are given for nine tributaries. As is indicated, the highest annual sediment yield is contributed by the V. Morava, Olt and Siret Rivers. The possible reasons for this are their relative large and easily erodible catchment areas and the low number of sediment retarding structures along them.

The mean diameter of the suspended sediment grains on the middle reach of the River Danube varies between 0.06 and 0.07 mm and before the operation of barrages generally decreased in a downstream direction. As an example, the effect of the reservoir Iron Gate I is presented in Fig. 9.5. As the solid curves No. 1 and 2 show before the damming, the sediment samples taken near the water surface and

Table 9.2 Suspended sediment contribution of important tributaries to the River Danube for the period 1956–1985

Name of the river	Confluence into the Danube (km)	Catchment area F (km^2)	Annual sediment transport (103 t × a^{-1}) Minimum	Mean	Maximum
Inn	2,225	26,130	716	2,819	6,461
Morava[a]	1,794	26,658	–	98	–
Drava	1,384	40,150	290	1,572	3,285
Tisza	1,215	157,220	1,323	4,964	12,260
Sava	1,171	95,712	2,003	5,480	13,035
V. Morava	1,103	37,444	1,323	6,904	24,457
Iskar	637	8,646	130	880	2,450
Olt	604	24,010	697	6,824	22,958
Siret	155	47,610	1,053	11,786	31,189

[a]Station: Moravský Jan (F = 9,882 km^2; period 1951–1974).

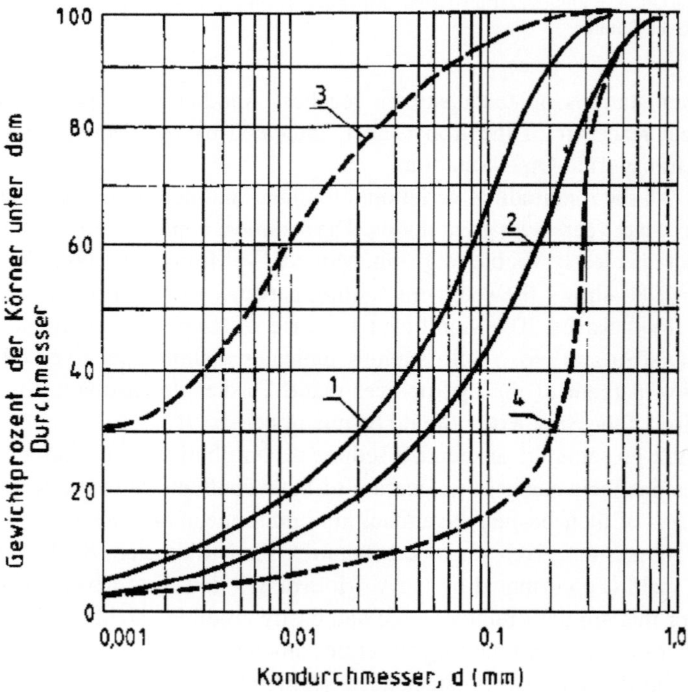

Fig. 9.5 Grain-size distribution of suspended sediment at Novo Selo before and after the construction of HPP Iron Gate I (after Pecinov 1984)

the riverbed at station Novo Selo, 110 km downstream of the barrage, had rather similar grain-size compositions with 0.06 and 0.14 mm mean diameters. Because of damming, however, the near-surface sample (No. 3) refined to 0.0067 mm while the near-bed sample (No. 4) became much coarser due to increased bottom scour: about 0.28 mm (Pecinov 1984). The decrease of mean diameter is more gradual further downstream on the Rumanian reach of the River Danube and drops below 0.02 mm approaching the sea (Bondar and State 1977).

9.3 Bedload Transport

The amount of available bedload data is unfortunately much less than that of the suspended sediment. One reason for this is that the share of bedload usually is low in the total sediment transport and therefore its measurement is neglected. As can be seen in Fig. 9.6 (Kresser and Lászlóffy 1964), before the construction of the large river barrages, along the German and Austrian Danube reaches there was a 20–30% share of bedload in the total sediment transport, however, in Hungary this dropped well below 10%, even close to 1%. Another point is that the bedload sampling is a rather difficult procedure, mainly because the movement takes place intermittently in time and in certain stripes of the riverbed only. Additionally, the samplers have several sources of error and the different countries apply various types of sampling equipment and measurement techniques.

Table 9.3 shows the annually transported mean bedload masses for 15 measuring stations. Because of the reasons mentioned above, the minimum and maximum values could be given for five stations only. On the German Danube stretch in the mid-1960s, the approximate mass of the annual bedload transport was about 111,000 t a^{-1} (at Kelheim), increasing to 800,000 t a^{-1} on the central Austrian reach (at Ybbs-Persenbeug). The mentioned drastic decrease of transport at the Hungarian stations (Nagymaros) within the investigation period was mainly due to the existence of an inland delta of the Danube River entering the flat plain. The loss of slope resulted in a large-scale deposition of coarse sediment at that location. The low bedload transport on the sand-bed Danube stretch at station Orsova indicates

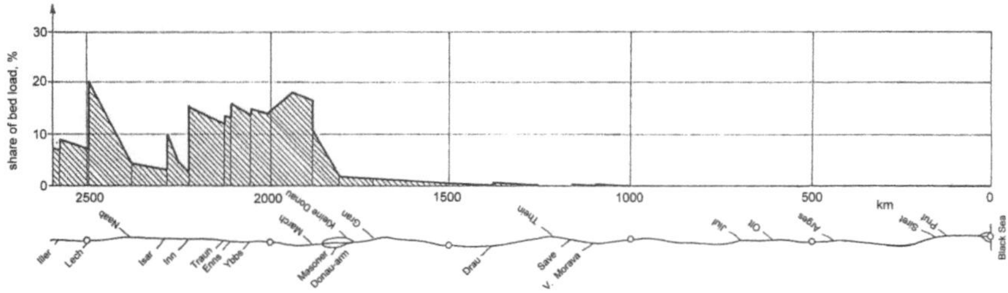

Fig. 9.6 Percentile share of bedload of the River Danube in the total sediment transport before 1960 (Kresser and Lászlóffy 1964)

Table 9.3 Annually transported bedload masses at 15 stations

		Annual		
		Minimum	Mean	Maximum
Name of the station	rkm	Mass (*G*b 106 kg a^{-a}) of the bedload sediment		
Donauwörth	2,509	–	13.0	–
Ingolstadt	2,459	–	94.0	–
Kelheim	2,404	–	111.0	–
Pfatter	2,348	–	100.0	–
Straubing	2,318	–	81.0	–
Metten	2,288	–	72.0	–
Jochenstein	2,205	–	540.0	–
Ybbs-Persenbeug	2,060	–	822.0	–
Bad Deutsch-Altenburg	1,887	–	1,080.0	–
Nagymaros	1,695	3.0	10.0	20.0
Mohács	1,447	9.0	15.0	28.0
Orsova	955	–	88.0	–
Zimnicea	554	112.0	493.0	1,354.0
Vadu Oii	238	54.0	141.0	358.0
Ceatal Izmail	80	17.0	73.0	279.0

the damming effect of the barrage Iron Gate I. Finally, the decreasing data at Vadu Oii and Ceatal Izmail are the result of increased sediment settling near and within the delta.

The effects in terms of decreasing sediment transport of river barrages and riverbed dredging activities are greater on the bedload than on the suspended sediment (Rákóczi 1990). The bedload transported to a river barrage is stopped completely at the upstream end of the backwater curve and can not be spilled out during floods from the reservoir as occurs frequently with the deposited suspended matter. Bauer (1965) has investigated the filling-up processes in reservoirs built before 1960 on the German Danube stretch. At the barrage Böfinger Halde erected in 1952, the average annual gravel deposition was found to be 19,000 m^3 a^{-1}. On the Austrian Danube reach, almost 200,000 m^3 a^{-1} filling-up was observed in the first 4 years of the Ybbs-Persenbeug reservoir. The volume of annual deposition depends on the length of the free-flowing river reaches between two successive barrages and on the flow regime (frequency of floods) of the River Danube. The deposited sediment was later excavated from every location of significant accumulation (Kobilka and Hauck 1982).

In 1968 an intensive gravel dredging campaign started along the Czechoslovak–Hungarian Danube stretch with annual volumes well surpassing the annual bedload transporting capacity of the river (max. 300,000 m^3 a^{-1}) and lasted more than a decade. In some years, the two countries even extracted more than 1×10^6 m^3 a^{-1} riverbed material.

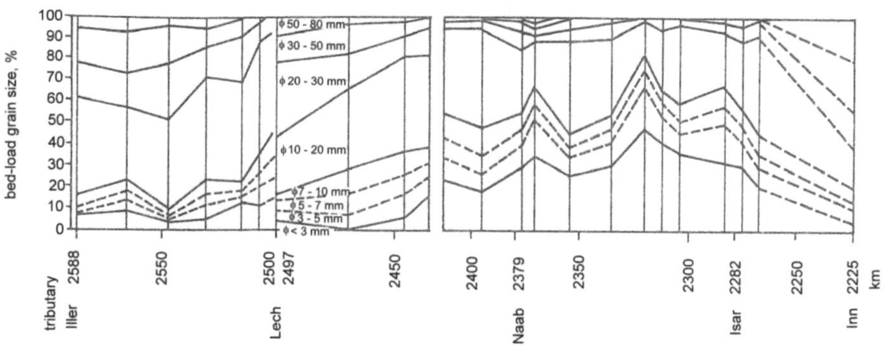

Fig. 9.7 Variation of bedload grain sizes along the German reach of the Danube River (Bauer 1965)

The variation of grain-size composition of bedload on the German Danube reach is shown in Fig. 9.7. Downstream of rkm 2,282 a gradual decrease of grain diameter can be observed. According to Schmutterer (1961), on the Austrian stretch the mean grain size of bedload decreases from 37 (rkm 2,092), to 22 mm (rkm 2,084) and to 13 mm at Bad Deutsch-Altenburg (rkm 1,889). On the upper part of the Hungarian Danube reach, Bogárdi (1971) found the mean diameter varied between 10 and 15 mm, however, on the middle part it was around 0.4–0.5 mm and at the lowest part only 0.3 mm. On the Rumanian/Bulgarian Danube reach, the barrage Iron Gate I caused a further refinement of the bedload grains (Fig. 9.8a, b). Though the range of this effect decreases in the downstream direction, it still remains visible even at C. Izmail.

9.4 Grain-Size Composition of Bed Material

Though the procedure of bed material sampling is easier than that of bedload sampling, some Danubian countries do not collect such data. The main channel of the River Danube is too wide, for the grain-size distribution of its cross-sections to be characterized by one sample only. Since aggrading, degrading and dynamically stable parts can simultaneously occur within any cross-section, five to ten samples are usually necessary to reveal the most important lateral variations. The composition of bed material might drastically change also in time, for example during and after the passage of significant floods.

Figure 9.9 presents the changes in grain-size distribution of Danube bed materials, between Aschach and Silistra (Rajnov et al. 1975). The gradual but not continuous decrease of grain diameter can be observed from the Austrian river reach to the mouth at the Black Sea. It can also be seen that the coarse gravel bed material ceases along the Hungarian reach and that downstream of Bezdan, the riverbed consists entirely of sand. Downstream of the barrage Iron Gate I, the bed material

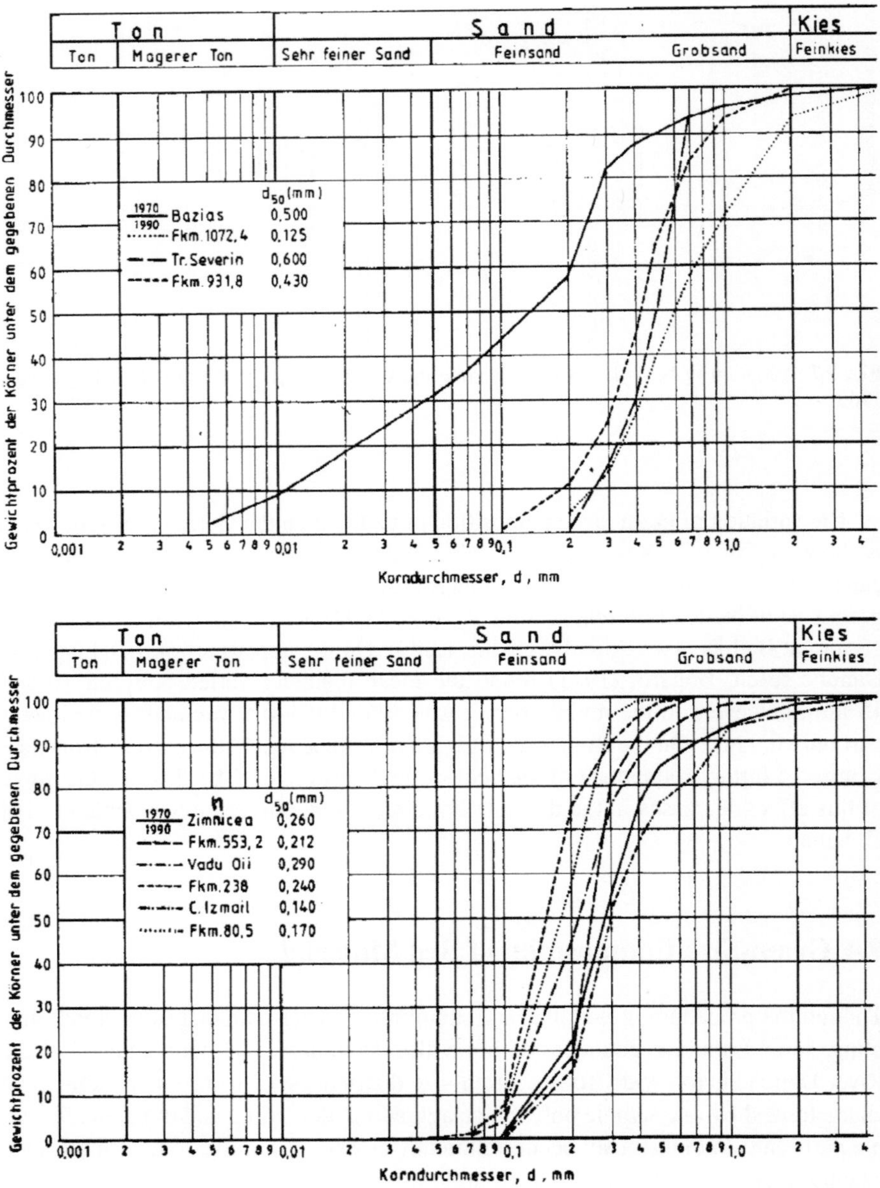

Fig. 9.8 Grain-size distribution of bedload along the Rumanian reach of the River Danube before and after the construction of HPP Iron Gate I

becomes gravely again for a relatively short, 35-km long reach, because of the high erosive power of flow leaving the dammed reservoir. Further downstream the sand fractions are predominant and about 180 km downstream of the barrage, the bed consists of fine sand like at Beograd.

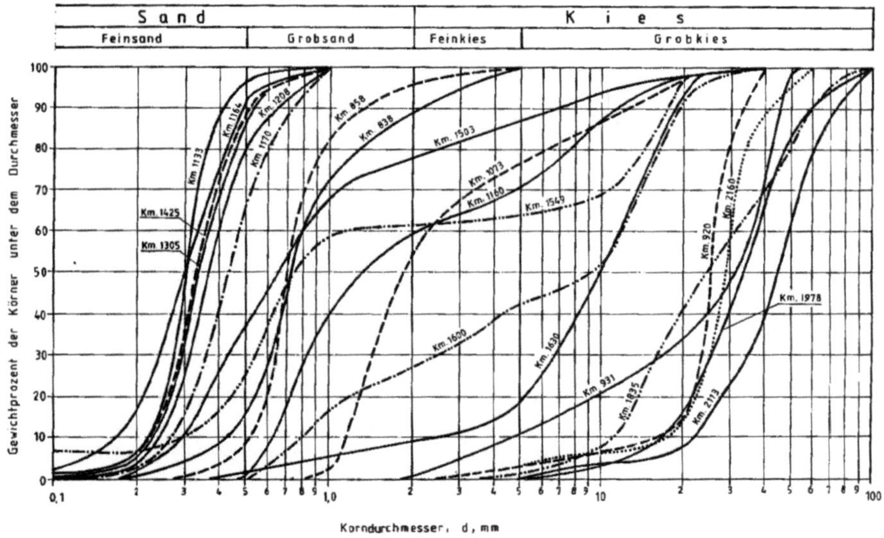

Fig. 9.9 Grain-size distribution of bed material of the River Danube between rkms 833 and 2,161 (Rajnov et al. 1975)

9.5 Conclusions

In the investigated period (1956–1985), the River Danube has transported annually 25–80 $\times$ 10^6 t suspended sediment to the Black Sea. The total mass of suspended sediment, passing any measuring section shows a 5- to 6-fold variation. The sediment transport quickly rises where the bigger tributaries (Tisza, Sava, V. Morava) join the Danube within a relatively short reach. Before the construction of barrages and hydropower stations, the share of bedload in the total sediment transport reached 20% on the German and Austrian Danube reach only. On other sections this share is about 1% or less.

Both the erosion over the river basins and the transport of generated sediment by streams are heavily influenced by human interventions. These include for example retaining and collecting the eroded materials on the slopes of basins, damming the rivers for navigation purposes and to harvest the energy of the flow, dredging the bed material from the river channel for the construction industry at a rate much higher than the re-supplying potential of the sediment transport, etc. Because of the summarized effects of these anthropogenic factors, the suspended sediment and bedload transport show a decreasing tendency at almost every station investigated.

Regarding the first results of the international research project "SEDBAL", initiated by the UNESCO/ISI, the collation of all available sediment data from the Danubian countries and the continuation of the presented monograph for the next 20-year period (1986–2005) is extremely urgent.

References

Bauer F (1965) Der Geschiebehaushalt der bayerischen Donau in Wandel wasserbaulicher Maßnahmen. Die Wasserwirtschaft 55. Jg. H. 4+5

Bogárdi J (1971) Feststofführung der Gewässer. Akademischer Verlag, Budapest

Bondar C, State I (1977) The Danube River contribution with water and silts to the Black Sea. Cercetari Mar IRCM 10:51–66

Kobilka J, Hauck HH (1982) Feststoffhaushalt in den Stauräumen de österreichischen Donaukraftwerke. Internationale Talsperrenkommission, 14. Talsperrenkongress

Kresser W, Lászlóffy W (1964) Hydrologie du Danube. La Houille Blanche 2:133–178

Pecinov D (1984) Änderungen in der granulometrischen Zusammensetzung der Schwebstoffe im Fluß Donau unter dem Einfluss der Hydrosystems "Eisernes Tor". In: Proceedings of the 12th Konferenz der Donauländer über hydrologische, Vorhersagen, Bratislava

Rajnov S et al (1975) Der Einfluss der Staustufe "Eisernes Tor" auf die Schwebstofführung der Donau. In: Proceedings of the 8th Konferenz der Donauländer über hydrologische, Vorhersagen, Regensburg

Rákóczi L (1990) Vorhersage von Flussbettveränderungen an der ungarischen Donaustrecke. In: Proceedings of the 15th Konferenz der Donauländer über hydrologische, Vorhersagen, Varna

RC of DC (Regional Co-operation of Danube Countries) (1986) Die Donau und Ihr Einzugsgebiet. Eine Hydrologische Monographie. Bayer. Landesamt f. Wasserwirtschaft, München

Schmutterer J (1961) Geschiebe- und Schwebstofführung der österreichischen Donau. Wasser und Abwasser. Jg

Chapter 10
Training of the Danube River Channel

Alžbeta Stančíková

Abstract This report provides an exhaustive account of all modifications made to the Danube River, from the spring at Donaueschingen, Germany, downstream to the river mouth at Sulina. One important feature of this report is that it provides a historical overview of how our predecessors influenced the evolution of the natural riverbed and how the watercourse was used and optimized for agriculture, industry, trouble-free navigation, and flood protection. Emphasis is placed also on environmental, ecological, recreational and land-management needs. This report also provides an account of previous projects, how they progressed over time, and their implications to the river. Attention is paid also to the achievements in navigation, flood protection, and the efficiency of the training projects in terms of sediment and ice transport.

Keywords Channel training · Flood protection · Flood levee · River bank · Revetments protecting

10.1 Introduction

The first systematic channel training projects on the Danube River can be traced back to the late nineteenth century and the early twentieth century. Depending on requirements, the training projects were implemented in several phases. Initially, the primary goal was to furnish flood protection for the adjacent residential and agricultural areas. Later, proposals were made to build water intake facilities, to drain the adjacent floodplains, to restore the disturbed water runoff, to eliminate formation of ice jams, to prevent extensive silting, and perhaps most importantly to upgrade navigation on the Danube River due to its important economic advantages. Raising

A. Stančíková (✉)
Water Research Institute, Bratislava, Slovakia
e-mail: alzbeta.stancik@stonline.sk

M. Brilly (ed.), *Hydrological Processes of the Danube River Basin*,
DOI 10.1007/978-90-481-3423-6_10,

levees, cutting off numerous braided channels, rectifying the main channel, and other modifications to the natural channel – made it possible to achieve the desired effects – although not always conducted in desired proportions. The major gravel-bed tributaries draining the Alps Mountains influenced, to some degree, the flow regime of the river as well. Training projects on these tributaries were considered important, especially in terms of the intents to modify the main channel.

The river responded to these alterations by tending to reach a new equilibrium condition, which was often accompanied with local manifestations of silting and incision of the riverbed. As a result, new issues came forth. The influence of natural obstructions to the flow (e.g. narrowed rocky reaches – gorges, etc.) had to be accounted for in the design of training structures. At some places, dredging was deployed. Later, heavy industrial dredging showed its adverse effects in many reaches of the river. This led to restrictions limiting the extent of dredging. Several flood control structures were installed on the main channel and its major tributaries. Backwater zones that were formed by the then-newly completed hydro schemes provided better conditions for navigation; yet a higher level of flood protection was achieved as well.

The construction of a cascade of reservoirs on the river gave rise to new concerns and created conflicts between environmental protection and land-use management. Incorporating these two interests into the design of training structures gave rise to an additional set of issues – including the need to rehabilitate the already modified reaches. As will be documented later in this report, many of the conflicts between the design of training structures and the requirements to maintain a prosperous environment have not been revealed yet, and probably will remain unresolved in the near future.

Training works on the Danube River and their anticipated achievements were the objective of the Project entitled "The Danube River Channel Training", which was implemented within the third phase of the regional co-operation of the Danube-basin countries under the auspices of the International Hydrological Programme of UNESCO (Stančíková et al. 1999).

This report provides an exhaustive account of all modifications made to the Danube River, from the spring at Donaueschingen, Germany, downstream to the river mouth at Sulina. One important feature of this report is that it provides a historical overview of how our predecessors influenced the evolution of the natural riverbed and how the watercourse was used and optimized for agriculture, industry, trouble-free navigation, and flood protection. Emphasis is placed also on environmental, ecological, recreational and land-management needs. This report also provides an account of previous projects, how they progressed over time, and their implications to the river. Attention is paid also to the achievements in navigation, flood protection, and the efficiency of the training projects in terms of sediment and ice transport.

This report is a compilation of practical and theoretical knowledge acquired prior to 1990 – under the assumption that all necessary background materials had been processed and made available by the individual co-operating countries before the time of its writing. Since the assumption of basin-wide data availability was never completely met, some disproportion in description of the individual reaches of the Danube River may exist.

Finally, this report stems from a project co-ordinated and partially conducted by the Water Management Institute in Bratislava, Slovakia, in collaboration with experts from Germany, Austria, Hungary, Croatia, Yugoslavia, Romania, Bulgaria and Ukraine.

10.2 Objectives and Methods

10.2.1 General

The goal of the first attempts to actively modify the channel morphology was primarily to render flood protection. However, these efforts were carried out mostly locally; and, over time, new problems inevitably had emerged. Therefore, a systematic approach had to be considered. The primary concept of the modifications was to stress the importance of the river in terms of navigation, with emphasis placed on it as a source of potable and industrial water. But protection against disastrous and devastating actions of the watercourse during floods was considered also. All of the above-sketched criteria were covered in the following three main objectives:

- flood protection for residential areas, crop lands, transportation and various industrial buildings;
- sufficient quantities of water for residents and industry,
- international navigation.

An ecological concern associated with the installation of training structures was strongly emphasized. Priorities of the project objectives were subject to change depending on when and where the channel training works were to be conducted.

In the first half of the twentieth century (after World War I), the "International Commission for the Danube River" (CID) was established. The Commission's primary objective was to unify all procedures applied for navigation along the entire length of the Danube River. In 1949, signing the "Beograd Convention" – (BC), the "Danube Commission" – (DC) – began to operate. The DC's mandate was to watch over the enforcement of the Beograd Convention on navigation on the Danube River, and in co-operation with all countries in the Danube Basin to co-ordinate the program targeted toward making the river fully navigable.

10.2.2 Types of Training Projects

The Danube River has a wild morphology both in its mountainous and lowland sections. This means that not only the basin, but the riverbed and the adjacent floodplain areas also differ considerably along the watercourse. In this respect, the Danube Basin is divided into three smaller portions:

- The Upper Danube, stretching from the spring to the confluence with the Morava River terminating at the Devín Gate, Slovakia (encompassing some 13% of the entire Danube Basin)
- The Middle Danube, draining the largest portion (58%) of the entire basin. This section extends from the Devín Gate downstream to the juncture between the Southern Carpathians and the Balkan Mountains at the Iron Gate;
- The Lower Danube, draining the Lower-Danube lowland with its surrounding foothills and mountains (29% of the entire basin).

Traditionally, in the Upper and Middle Danube, alterations of the channel are made for high water, mean water and low water.

River training for high water is, in many ways, associated primarily with flood protection. The goal of modifications for high water is to protect the cultivated area along the river against flooding and to affect the flood-induced changes in the riverbed.

River training for mean water is intended to establish steady-state conditions in the riverbed and to achieve an unrestricted flow of water, ice and sediments. Recent modifications to the channel usually resulted in better conditions for navigation. However, this kind of modification must respect the groundwater regime of the adjacent floodplains.

River training for low water is supposed to act ancillary to mean-water alterations. The purpose is to make river navigation smooth, and to ensure an unrestricted conveyance of ice.

As for alterations for high water, of special importance are the minimum pool elevation and navigable depths (hereafter referred to as MPE and ND). These are defined and regularly updated by the Danube Commission for every ten successive years, see "Rekomendaciji po ustanovleniju jedinogo metoda opredelenija nizkogo sudochodnogo urovnja vody na Dunaje". Water surface elevation profile is an important indicator of how the riverbed gradient evolves in time, and how the spatial characteristics of the channel change. This is an objective indicator of natural and human-made changes made to the channel.

The following three main categories of river projects are recognized: reinforcement of riverbanks, construction of flow-guiding structures positioned within the river channel, and other structures including protective levees.

Revetments protecting river banks against scour and weathering by the actions of water flow, ice, surge waves, and other factors, may be performed by means of vegetation planting (grasses, aquatic and riparian plants, willows, and other woody plants), or by non-vegetative means (cobbles, Gabion mattresses, impacted rock fills, filters with loose materials and geotextiles) or a combination of these methods.

The following flow-guiding structures are commonly used: groynes (deflecting, concentrating, repelling), either longitudinal or perpendicular to the bank.

Other types of flow-guiding structures are sills, drop structures and chutes. Sills are channel traversing structures made of stone, concrete, prefabricates or other suitable material. Drop structures are used to abruptly change the longitudinal gradient of the riverbed. In contrast to drop structures, chutes change the bed elevation gradually.

Flood levees are artificial embankments raised, revetted and designed to protect the adjacent area against flooding at times of high-water stages. When minor flooding of the adjacent floodplains is acceptable, levees are designed for a 100-year discharge ($Q < Q_{100}$) – and such levees are commonly known as "100-year levees". In some of the Danube Basin's countries the term "summer levee" is used to refer to a 100-year levee.

10.3 River Training for High Water

Floods on the Danube River often result from spring-time snowmelt and heavy atmospheric precipitations. Currently, spring-time floods are of minor severity; only at times of coinciding ice phenomena in the headwaters and summer-time heavy rainstorms (June–August). Tributaries including the Inn, Traun and Enns Rivers are important contributors to floods also.

Although the era of unsystematic approaches to flood control was followed by specifically prepared projects, the existing system of flood protection was already burdened by mistakes made in the past. Common mistakes made during the period of unsystematic alterations were:

- Inappropriate layout of levees – interconnecting levees constructed for local demands;
- Improper technology, construction material, material permeability, foundations, fill-material impaction etc.;
- Failure to conduct geological surveys;
- Insufficient dimensions of levees; and underestimating the importance of ancillary installations.

10.3.1 Levees on the Upper Danube

In the *Baden-Württemberg section* between Donaueschingen and Ulm, the primary purpose of the levees was to protect the municipalities against flooding. These levees were designed mostly for 100-year floods. In the winter season, floodwaters often overtopped the levees, inundating the valley floodplain and endangering the nearby built-up area. Some levees were designed even only for Q_5. At other places, floodwaters remained within the zone of protection without overtopping up to Q_{60}. Higher stages resulted in overflows, but the levees were still able to reduce the peaking Q_{100}. This somehow reduced the risk of flooding of the nearby municipalities.

Other important river reaches, for example the industrial area of Donautal at Ulm, were protected with levees designed for Q_{100}. In the Bavarian section of the Danube, two levels of flood protection were in use: "100-year levees" or "summer levees" as they are often referred to in the Danube region, and levees for higher flows. In the 1930s, the section between Ulm and Passau was protected by means of the old levees that were erected around residential areas and valuable agricultural lands, and new levees built on both sides of the channel with a total length of 182 km, with an additional 18 km of levees on the left side. Later, this system went through a series of innovations to form an enclosed network of levees with a total length of approximately 300 km, involving 48 water pumping stations, and 600 km of drainage canals.

Under these circumstances, in the 1940s, the total inundation areas down to the state border encompassed 36,140 ha of crop fields and 12,660 ha of natural floodplains and forests. The total expanse of the area protected by levees was 31,700 ha.

In the years that followed, construction of various hydro projects in the Bavarian section of the Danube meant a major change in the levees' layout and the reconstruction of the already installed training structures. Comparing the coverage of flooded areas between 1954 and 1985 and the 1930s suggests that a 51% reduction was achieved, i.e. 24,904 ha.

In the *Austrian portion* of the river, floods and ice jams affected especially the settlements in the Vienna Basin in the Tulln and Morava fields (Bitterer 1969). The city of Vienna with its suburbs immediately adjacent to the river experienced several major floods; although as early as 1830 a levee was erected to protect the suburb of Leopoldstadt. The disastrous flood of February 1862 affected the lower located districts of Vienna and Lower Austria with devastating consequences to both property and life. In 1869, a decision was made to alter the river channel in order to prevent floods in the future. Consequently, a series of still-existing flood control structures were built between the confluence of the Isper and Devín, Slovakia. These include flood control structures in the Krems region, between Stockerau and Langenzersdorf; in Vienna on the left-hand side riverbank between Langenzersdorf and Schönau; the system of levees downstream from Vienna to the confluence with the Morava River; the levees in the Southern Tulln Fields; and other flood control structures in Vienna and on the right-hand side of the Danube below Vienna. The entire system of flood control structures consists of a series of levees with a total length of almost 190 km and other ancillary installations and structures such as levee closures, weirs, cut-offs, training structures in terminating confluences, dewatering stations and protective structures built around ports.

10.3.2 Levees on the Middle Danube

The area *between the mouth of the Morava River and the confluence of the Ipel' River* was partially protected by a system of levees (in the Upper Rye Island around the year 1450; in the lower part of the Rye Island by 1670); but this system of levees was rather weak, low and susceptible to failures. An important milestone in the levee's outlay in this region was the flood of 1853. As the floodwaters receded and damage to property and life was tallied up, the levees were examined for failures and repaired along their entire length – the pattern of levees has been preserved up to the present time. After these modifications, the entire length of the levees in Slovakia's portion of Rye Island was 195 km, 59 km of levees were below the town of Komárno, some 23 km at Bratislava along the southern riverbank, and about 40 km on the northern tributaries that were affected by the backwater in the Danube channel. The flood protected area in Rye Island encompassed around 145,000 ha, while in other regions the protected area was only 40,000 ha (Szolgay et al., 1991).

Flood protection and the effectiveness of the river engineering projects in this region were inevitably related to the phenomenon of internal waters. It was at the beginning of the twentieth century that, apart from the protection against external waters, increased attention was paid to the protection against internal waters also. The importance of flood protection measures was somehow underestimated; which resulted in severe consequences during the floods of 1954 and 1965. During the flood of 1954, internal waters flooded an area of 10,000 ha. Therefore, new drainage canals were constructed, grout injection was applied at sites with weak levee foundations, and new levees were erected using modern technology. During the flood of 1965 the levees broke once again. 10,000 ha of land were under water at the town of Patince and around 55,000 ha in the lower part of Rye Island, which made this flood the biggest natural disaster in the modern history of Slovakia.

In 1967 a final proposal was introduced for construction of flood control structures; although not yet counting on the then-scheduled international hydro project between Gabčíkovo (Slovakia) and Nagymaros (Hungary) – (G-N).

With regard to the construction of the Gabčíkovo Hydro project, only very minor amendments to the original project were made. Since the floodwater regime was unlikely to be affected solely by the hydro plant at Gabčíkovo, no changes to the original concept of flood prevention relying on the original G-N hydro scheme were needed. What remained unresolved was the complex protection against floods in the area of Nagyamaros. Along the Danube River (Chl'aba–mouth of the Ipel' River, Štúrovo–mouth of the Hron River, Kravany–Iža–Komárno–Medveď ov) and further along the terminating tributaries, a newly completed and reconstructed levee system was already in place. Because of the intrinsic geomorphology of Rye Island, attention was paid to a simultaneously conducted reconstruction of the then separately operated drainage system into one unified irrigation–drainage system with new water pumping stations.

In the eighteenth century, the area *between Szob and Mohács* was protected only by local earth-fill dikes constructed to protect towns, castles and points of confluence with other smaller tributaries. Flood levees, as they are perceived today, were first erected in this region in the late eighteenth century and in the early nineteenth century. The first contiguous levees were built between Paks and Báta between 1820 and 1825 with a total length of 464 km. The development in the design of flood control structures in the early twentieth century was accelerated by the expanding agricultural lands, but especially due to the consequences of previous disastrous floods. Between 1876 and 1945, several river engineering projects were completed at Sárközi, Margitsziget and below Mohács. Although another series of floods surpassed all water levels that were encountered earlier; the number of levee failures decreased markedly as well as the expanses of inundated areas and damage. In the modern era of river training works, floods exceeding all historic maxima re-examined the strength of the protective structures several times. On the basis of these experiences, and considering the existing sources of construction materials, upgrading of the existing structures began to improve their flood protective potential. Many of these reconstruction works are still underway, with a total of 542 km of new levees along the Danube, protecting an area of 3,477 km^2.

The levees on the northern bank of the Danube River, Tisza River, Sava River, and Tamisa River and other smaller watercourses of the *Vojvodina region* (Serbian–Croatian and Serbian section of the Danube River stretching from the border to Hungary downstream to the mouth of the Timok River), were installed between the mid-nineteenth century and World War I (Katiš and Sretenivič 1990). During this period of time, some 50 levee failures were observed, flooding more than 300,000 ha of land. Almost 51% of the land was exposed to risk of flooding either directly or indirectly, with 173 municipalities being inundated. The flood of 1965 and the long-lasting elevated water stages on the Tisa River and Sava River underscored the importance of levee revetments not only on the Danube River, but on all major rivers in the Vojvodina region also.

In 1967 the "Program for Reconstruction of the Flood Protection System" was introduced with an objective to raise and strengthen the levees in this region. As a result, between 1965 and 1974, levees with a total length of 135.5 km were erected. In addition, some 72.0 km of levees were reconstructed between Pančevo and Dubovac. This project was coupled with the construction of Djerdap I, whose end-point of the backwater zone is observed upstream at the mouth of the Tisa River. The reconstruction required an upgrade of more than 472 km of levees along the watercourses in Vojvodina, of which 82 km of the levees were along the Danube River and the mouth of the Tisa River. This project also involved the construction of shorter levees along the Danube at Mladenov, Beočina and Sremski Karlovac.

The right-hand side riverbank, starting at the border to Hungary, runs farther downstream to Ilok, *Croatia*. Because of the construction of levees and the topographic characteristics of the area, this river reach may be divided into two sections; one, forming the borderline (rkm 1,433.0) and running downstream to the mouth of the Drava River (rkm 1,383.0) which is protected with a system of levees; and another one extending from the mouth of the Drava River down to Ilok (rkm 1,295.0) with raised banks. The first section runs along the state border – at Draž with a length of 6.8 km, Gomboš levee with a length of 2.3 km, and Zmajevac levee – at Kopačevo with a length of 32.7 km. The Drava–Danube levee, which is hooked to the Zmajevac–Kopačevo levee, is part of this system of levees also.

10.3.3 *Levees on the Lower Danube*

The natural channel of the Danube River between Griu (rkm 851.0) and its delta (Black Sea) extends in an inland direction as much as 2–20 km (Mittiade and Popa 1966). The total inundation area is around 880,000 ha with 43,500 ha on the right-hand side in Bulgaria, and 570,000 ha in Romania including a part of the delta. The remainder of the delta is located in Ukraine and Romania.

Similarly to other countries, *Romania's* economic interests supported the decision to build protective levees in the second half of the twentieth century, after World War II. At that time, old levees were reconstructed, while new levees were built. The intensity of the river engineering projects is documented by the expanses of flood protected areas. In 1940 the flood protected areas encompassed some 50,000 ha, whereas in 1964, this figure increased up to 375,000 ha. The total length of levees exceeded 1,000 km. At the same time, due to irrigation needs after World War II, detention facilities and water pumping stations were built, and these served as flood control structures as well.

The levees enclosing the inundation area affected the natural regime of flooding. The levees confined water within the river channel and thus excluded the surrounding areas from being flooded. This caused higher flood flows in the channel with prolonged flood durations. The annual average discharge increased by 1% at Zimnicea, at Vadu Oi by 3%, and at Tulcea by around 5%.

Bulgaria shares with Romania some 471.6 km of the Danube channel. Although Bulgaria's side of the river is at a higher elevation than that of Romania, the need to prevent flooding urged both countries to build levees on both sides of the river. These levees were built between 1930 and 1950, with a total length of 260.7 km and furnishing protection against floods for an area of 85,340 ha.

10.3.4 *Characteristics of the Levees and Their Intent*

The period of unsystematic construction of levees on the Danube was followed by an intense development of systematic levee design in the second half of the twentieth century, and this has continued up to the present time. Flood protection, especially construction of flood levees, requires that a set of complex questions need to be answered. These questions pertain to the geometric parameters of levees using knowledge on design flood discharges, stability and permeability of the levees, the foundation strata, levee layout, including methods of diverting internal waters.

10.3.4.1 Design Flows

The question "to what flood should a channel capacity be designed" is perhaps as old as the field of river engineering itself. Flood protection, generally speaking, cannot go that far so that every possible danger could be eliminated. In deciding to what degree protective measures are feasible, the principle of cost-effectiveness proportionality should be considered. Safety of the public is of utmost importance, whereas environmental and landscape issues and the costs of maintenance – as stated in German recommendations for a proper choice of design (Gewässerdirektion Donau/Bodensee 1994) – are accepted in all Danube countries.

"Design flood" is a term that is defined by the peak discharge, maximum water stage, duration and volume of the flood wave, and as such it is used to set flood control measures and design parameters. In most instances, the highest possible flood is not taken as the sole factor; rather, an "elevated risk" of flooding with its acceptable consequences is considered (acceptable risk).

In the *Baden-Württemberg* section of the Danube, the system of protective levees is interrupted and the levees are partially dimensioned for the summer-season or winter-season floods. At some locations, dimensions of the levees are able to prevent flooding for discharges up to $Q = 510\ m^3\ s^{-1}$, which corresponds roughly to Q_{100} and other places to "summer floods" with a discharge of $350\ m^3\ s^{-1}$. At other places, the current state of flood protection is designed only for a discharge between Q_{10} and Q_{25}, despite the fact that the channel capacity increased by lowering the channel bottom by about 1 m. The industrial area in Donautal is protected on its left side against the 100-year discharge $Q_{100} = 590\ m^3\ s^{-1}$. Downstream from the bridge at the Wiblingen hydro plant there is no levee on the left-hand side bank of the channel (overflowing occurs at Q_{20}), while the area adjacent to the right-hand side of the river is protected against flooding by a levee built around the hydro plant.

In the 188-km long *Bavarian* section between Neu-Ulm and Kelheim two levee systems were built with a total length of about 70 km. Currently, this section of the river experiences backwater effects brought about by a cascade of 16 dam projects. The levees that were erected around these damming systems improved flood protection for the surrounding area against the 100-year flood Q_{100}. The subsequent section between Kelheim and Passau (160 km) has an uninterrupted system of levees (from Regen to Vils). Nevertheless, these levees furnish only partial protection, considering the currently applicable criteria. The program of restoring the flood-damaged levees was aimed at strengthening these levees according to the applicable principles and within the scope of the existing flood protection status – i.e. for a 25-year flood. In the section of "permanent" protection, the levee system must withstand 100-year floods. In 1977 and 1994, a cascade of four hydro projects was constructed at – Bad Abbach, Regensburg, Geisling and Straubing. Their levees served as flood protection. These levees are designed to be able to convey 100-year floods Q_{100}. Those reaches of the Danube River where there is no backwater effect observed, i.e. below the Donauwörth, Vohburg and Straubing hydro projects, were step-by-step upgraded to meet the new requirements for flood protection up to Q_{100}.

Flood protection measures in the *Austrian* section of the Danube can be, considering the topography of the country, divided into three categories–lowland levee system, mountainous levee system and region Vienna (Bitterer 1969). Lowland levee system is Ybbs-Persenbeug-Melk, Tulln Fields, the area below Vienna down to Porta Hungarica at Hainburg, and on the right-hand side of the river the lowland between the base of the Hundsheim Hills and the state border to Slovakia. The objective of the flood protection was to protect property and life, but attention was also paid to the possible risk of sinking groundwater level. Moreover, the intention was to maintain natural flooding of the adjacent floodplain forests. These requirements predestined the overall layout of the levee system in this area. Lower levees, with crest heights corresponding to Q_{10}–Q_{20} are sufficient to protect the agricultural lands. The urbanized areas need, on the other hand, to be protected against 100-year floods. The mountainous levee system extends from the mouth of the Isper River downstream to Persenbeug and the Wachau region to Melk and Krems. At these locations, elevated flows are accompanied with high-water stages. This makes the construction of protective structures a challenging task. From Isper to Ybbs-Persenbeug, by the construction of the hydro scheme and raising the embankment 1.5 m above the backwater level, the entire area is protected against extreme floods. For the capital city of Vienna, different design criteria have been adopted. If the left-hand side area or the lower suburbs of the city were flooded, this would have disastrous consequences. All flood barriers within the city limits were built for water stages corresponding to 9.32 m at the Vienna Reichsbrücke gauge station. Later, the left-hand side levee was raised to an elevation equal to water levels during floods with a discharge of 12,000 m3 s^{-1}. The floods of 1954 and 1965 revealed that the body of the levee is not strong enough to withstand prolonged discharges above 10,000 m3 s^{-1}. According to Prof. Kresser (Kresser 1986), the maximum widthstandable discharge for the levees in Vienna is 13,500 m^3 s^{-1}; however, due to the retention volumes above Vienna this value should be 14,000 m^3 s^{-1}. The new protective scheme already accepts the latter discharge.

Water co-operatives that were building the levees in the *Slovak and Slovak–Hungarian* section of the river (between Devín and Szob) could not adhere to, for a long time, the unified principles for construction of levees and other flood protection structures due to unavailable reference data on floodwater profiles. Moreover, such reference water profiles changed after every flood event. Although, the first initiatives to reconstruct the levees trace back to the late nineteenth century, only after the severe flood of 1954 more attention was paid to this issue. After 1954, based on hydrological surveys, plans for alterations of the then-existing levees were proposed. A similar situation occurred in 1965 after a failure of the levee at Čičov. With respect to the proposal to build the G-N hydro scheme, the complex set of flood-preventing measures was accompanied with reconstruction of levees. As a result, the newly constructed flood protection system was upgraded not only along the main channel, but also along the major terminating tributaries. For levees, the 100-year discharge Q_{100} was chosen to represent the design discharge.

In *Hungary*, the common practice prior to 1970 was that the largest flood recorded over the last period was used as the design discharge. This practice meant difficulties in terms of evaluating the safety provided by the flood control system. Therefore, in 1973, several amendments to legislative standards were made; introducing a new term: "design flood" defined by three parameters – discharge, flood wave duration expressed in days, and a safe water stage. Adopting these rules in practice means the design 100-year flood (without ice phenomena) for the Danube River, excluding the section between Esztergom and the southern border of Hungary where the standard is the envelope curve of flood water stages with ice phenomena, and excluding the cities of Györ and Budapest, where the calculated 1,000-year flood is applied. The safety freeboard on all watercourses in Hungary is one metre above the water surface profile, except for sites with a prescribed height of the freeboard. Duration of safe floods, expressed in days, has to be considered in levee design parameters. As an acceptable measure is flood wave duration with a 1% probability of exceedance.

Even below *Mohács*, the levees were designed and constructed for the highest discharge in the previous period of record. Higher discharges caused levees to fail, and, as a result, led to subsequent raising and reinforcing of levees. The flood of 1965 became a reference design flood, which can be understood as the Q_{100}. After 1965 the levees were revised and reconstructed for the last time. Left-hand side levees on the Danube were raised by 0.9 m and their freeboard above the level of the design flood was 1.2 m.

For the right-hand side levees of the Danube in *Croatia* as the design, discharge was set to the Q_{100} which corresponded to the flood of 1965. The freeboard of the levee system in the section state border–Draž exceeds the height of the flood levels in 1965 by 1.2 m; for the Gomboš levee the freeboard is 1.0 m; and in the case of the Zmajevac–Kopačevo levee the crown of the levee is by 1.0 m higher than the water stage during the 1965 flood. Some authors claim that levees with such a freeboard should be sufficient even for 1,000-year floods $Q_{1,000}$.

In 1972, the Djerdap I Power Scheme was built at *river kilometre 942.95*. The flood protection system relies on the levees within the power scheme and the levee system extending from the scheme downstream to the Nera confluence (constructed mostly on the left-hand side and, where necessary, on the right-hand side at reaches with low banks). The levee system of the Djerdap II power scheme (rkm 863.0, completed in 1984) provides flood protection for the area between the two hydro schemes. These levees were designed and constructed to withstand Q_{100} with a safe freeboard.

Along the borderline separating *Romania from Bulgaria and further along the Romanian section* of the Danube the levees are designed for Q_{100}. In limited cases the levees on the Bulgarian side are designed for $Q_{1,000}$.

10.3.4.2 Geometric Parameters of Levees

As time progressed, the period of unsystematically conducted flood control was soon followed by systematic design and parameterization of levees, with proper choice of construction materials and construction requirements on subsidiary flood control structures (drainage channels, withdrawal stations, etc.).

The need to improve the stability of the levees resulted in widening of the levee crests, less-steep landside slopes of the levees, construction of impermeable walls within the levees, revetments, drainage trenches on landside slopes of levees, etc.

The development in levee geometry follows the empirical experience in each of the considered countries and the state of theoretical knowledge.

10.3.5 Discussion on Training Projects for High Water

The effectiveness of the training works should be perceived in how floodwaters can be safely conveyed from the floodplain, but the mechanical vigour of the levees against water pressure, wave actions and ice phenomena should be considered also. Since the late twentieth century, the levees were raised after every major flood event (e.g. in 1853, 1876, 1890, 1899 and particularly in 1954 and 1965). The concept of high-water protection in this period was not definite neither complex, especially at the beginning, and its consequences were not always apparent. Over time, local flood control structures were interconnected to form a long system of levees; but this meant a serious problem in terms of its impact on hydrology, hydraulics, channel morphology and environmental issues. For instance:

- Restricted natural inundation of lowlands, increased water stages and elevated risk of spill foods,
- Intensified sediment process in the downstream sections, continually raising the elevation of the inundation area, and silting of the main recipient and its side branches; all resulting in higher high-water level;
- Large-scale deterioration of the regime of internal waters compared to the original conditions;
- Increased likelihood of levee failures and flooding of the already protected agricultural lands;
- Impaired biodiversity of fauna and flora, if this is dependent on the water regime of the Danube (surface and subsurface water).

These processes were partially influenced by the training works aimed at improving the navigability on the river (alterations for mean water, and later for low water). The above-outlined training works were not conducted in a desirable sequence; hence, the expected water-management effects were not achieved.

The protection system for internal and external water is continuously monitored in each of the Danube countries. The purpose of the monitoring is to provide information on the state of the systems, but emphasis is placed on its effectiveness at times of extreme phenomena; creating an operationally manageable control of the flood protection system.

The operational flood protection system is defined by domestic regulations for flood protection, operating plans of hydro projects – both devised for various levels of emergency and subject to international co-ordination and cross-examination.

10.4 River Training for Mean Water

The Danube River was subjected to systematically conducted training works in the early nineteenth century. Most of these river training projects were accomplished on the cusp of the nineteenth and twentieth century, but there are still several sections where training works have not been completed yet. This resulted in a shorter length of the watercourse with a steeper gradient, causing higher flow velocities and an intensified scour. To maintain the flow capacity of the channel, extensive excavation works were deployed at places where transitory sediment depositions were observed. These interventions are classified as alterations for mean water, and are expected to create steady-state conditions in the riverbed and to secure an unrestricted passage for water, ice and sediments.

Riverbed stabilization for establishing equilibrium conditions in the riverbed is usually carried out in one of the following manners:

- Creating stream cross-section to allow mean-water flows to be discharged within a specific channel width; the channel has to be equipped with flow-guiding structures. This method was successfully used in Germany – and therefore it is now being referred to as the "German method".
- Another option is to use various cross-sections placed between inflection points in the bend (cut-offs, sinusoidal bends with alternating straight sections); this method was used first by Girardone on the Rhine, and came to be known as the "French method".

10.4.1 Locations and Particulars of the Training Works

10.4.1.1 River Training on the Upper Danube

Major river training works in the *Baden-Württemberg* section were conducted between 1820 and 1890. Numerous meanders were cut off and the Danube was artificially shortened to approximately one fourth of its original length (about 118.27 km), of which some 43.2 km (from Beuron to Scheer) ran through its natural floodplain. Another 25.7 km (from Scheer to Riedlingen) were straightened and the river resembled more a canal then a natural watercourse. The total length of the modified channel in this section was 9,504 m, consisting of chute cut-off, revetments and a weir at the end of the section. In another 28.2-km long section between Riedlingen and Munderkingen a cut-off with a length of 650 m was made plus some other alterations. Eleven cut-offs and other alterations were made in the subsequent downstream section. An 11-km long section between Öpfingen and Ulm was rectified with cut-offs. However, these alterations were not purely intended for mean water. Their purpose was to provide flood protection during times of high-water stages and to modify the channel to allow construction of bridges. As a result, a new riverbed was formed to serve as a safe passage for 100-year floods. An adequate dewatering of extremely wet and swampy areas at Donaurieden and Eisingen was achieved also. As reported in the literature (Bauer 1965), in 1965, restoration of the riverbed was sought at numerous places in this section. Remedial measures that were deemed appropriate at that time were later incorporated into a project entitled: "Integriertes Donauprogram des Landes Baden-Württemberg (IDP)". Limited funding slowed the implementation of the project.

In the *Bavarian* section the Danube was still a "wild" river at the beginning of the nineteenth century (DVWK Merkblätter 1989). However, some improvement was seen in this section in terms of navigation needs. The objective was to straighten the channel as much as possible, to make the riverbed cut deeper and thus to create a confined channel. A series of cut-offs followed in the first third of the twentieth century; although being performed only locally. These cut-offs began to deteriorate soon. As river steamboat navigation increased, it was necessary to enhance the navigability of the waterway. By 1836, the first training works were underway, with first accomplishments achieved before 1867. Even here, as was the case for many Bavarian sections, wooden groynes were replaced with longitudinal guiding structures made of rock.

Between the confluence with the Iller and the state border the river channel was, in terms of alterations for mean water, divided into two sections. In the first section (91.0 km), the length of the channel was shortened by 27 km (22% of the length), the riverbanks were reinforced with stone revetments or boulders, and the already existing structures were maintained. Between 1906 and 1911 groynes were used to modify the channel for mean water. Groynes with the combined effect of channel straightening and dredging caused the riverbed to cut deeper by 50–70 cm. Because of the undesired bed incision, the groynes were not maintained and fell apart.

In the second section (approx. 84.0 km), in 1876, a more elaborated training of the river was completed. The channel was shortened and straightened. After 1876, maintenance works were conducted and the existing structures were rewetted. In 1909, works in this section were completed. Except for minor maintenance no additional major measures were needed. After straightening the channel, intense scouring in the riverbed caused huge masses of bedload to become problematic. The effect of backwater in the gorge at Weltenburg caused silting in this section, which was eliminated by dredging. It did not take long and every remedial action needed to be followed by additional training works.

The third section (with a length of 213.3 km) was part of the proposed waterway connecting the Danube to the Main and Rhine Rivers. This section underwent modifications also. Until the late 1930s, there were about 90 km of training structures and some 10 km of other auxiliary structures. Difficulties arose for high-capacity navigation between Pleiting and Passau – the so-called "Kachlet section". At first, controlled blasting was deployed; but later the backwater induced by a 9-m high dam of the Kachlet hydro project influenced this section.

Until the mid-nineteenth century, *Austria's section of the Danube* was basically left unchanged. Only a few local flood protection structures existed. It was only after 1850 that the first systematically prepared projects were proposed. The objective was to create a single channel for mean water. Side branches had to be cut off, sharp bends were smoothed by cut-offs and groynes, and riverbanks were rewetted. However, this was still not adequate for navigation corridors of sufficient width that were greatly needed because of increasing waterway traffic. Modifications for low-water conditions were needed at some places too, but mostly at the site of the cut-off at Vienna below Fischamend.

10.4.1.2 River Training on the Middle Danube

Training projects for mean water *between the mouth of the Morava River and the Ipel' River* trace back to the second half of the nineteenth century. The objective of these projects was to create a navigation corridor with a depth of 20 dm and an unrestricted passage of ice-sheets.

In the Rajka–Gönyü section, alterations to the channel for mean water were conducted at the end of the nineteenth century, being an important step in water management in this section. But the need for a navigable corridor was not completely satisfied. Following the General Plan of 1936 proposing training works in this section, a decision was made to use design water stage (water stage CID 1936), for which, at that time, was taken the lowest navigable water stage. The channel between Devín and Gabčíkovo had to be modified to achieve widths up to 300 m, in the Gabčíkovo–Medved'ov section 325 m, between Medved'ov and Vének 380 m, between Vének and Komárom 420 m, and 450 m between Komárno and Radvaň. The navigation depth was subject to international negotiations. The goal was to achieve a 25 dm depth below the CID-1936 water stage. Revetments were used to stabilize banks, and cut-offs, longitudinal dikes and banquettes were installed as well. It did not take long and in the mid-twentieth century, it became obvious that implementing these training works will not be sufficient. The riverbed was not stabilized, conditions for a navigable channel were not created, and the training structures were not functional.

In compliance with the then-applicable directives and following the goals defined in the "General Project for Training Works on the Danube River" for this section, in the 1970s, the braided channel was confined into a single channel by constructing numerous dams in the side branches, detention dikes of the first and second line, flow-guiding structures, bank revetments and by altering the network of side branches. It was expected that the Gabčíkovo-Nagymaros project would substantially alter the water regime and the transport of ice and bedload in this section of the river. Moreover, this section was to be utilized for its hydroenergetic potential and navigation capacity. As for the reservoir located upstream of Dunakiliti (rkm 1,842), the installed river structures were used for mean-water conditions; therefore it was recommended that only necessary maintenance of the riverbed should be conducted before commissioning the hydro scheme. Below rkm 1,816, downstream to the end of the outlet canal (rkm 1,811) a transitory section was proposed as the point where the original channel would rejoin the deepened section of the Nagymaros impoundment.

One of the recommendations of the "General Project for River Training" was that the Gönyü–Szob section (Csoma et al. 1975) should be modified for low water so that a thalweg mean-water condition could be created. Another recommendation was to narrow the section between rkm 1,728 and 1,722, by installing a longitudinal structure and several "T" structures. According to one study (Szolgay et al., 1981), it was preferable that, in the future, these structures remain in place, and are maintained and monitored for their efficiency – possibly extending beyond the completion of the planned hydro schemes.

The proposed extent of the training works involved installation of revetments, longitudinal flow-guiding structures and branch closures that were to be adjusted to the water stage defined by the Danube Commission (+3.0 m above this stage in sections upstream of rkm 1,740; a continuous transition from +3.0 to +2.5 m further to rkm 1,730; and +2.5 m between rkm 1,730 and 1,708).

The river training works in the section *between Szob and the southern border of Hungary* (approx. rkm 1,433) were conducted in several phases. In the period 1871–1910 the primary intent was to build revetments and guiding structures. In the Szob–Dunaföldvár section there were 12.8 km of flow-guiding structures and 42.5 km of revetments, out of which some 21.5 km were within the city limits of Budapest; in the section between Dunaföldvár and the state border there were 4.1 km of guiding structures, 57.1 km of revetments, 14 closures and 34 "T" structures.

In the 1930s began the second phase of reconstructions and construction of additional control structures; but now using groynes. In the first section, prior to 1956, there were already installed 1.15 km of guiding structures; 19.75 km of revetments; six closures; nine groynes – and a 20.5-km long stretch of revetments in the second section; 24 closures, four "T" structures and two groynes.

Problems concerning the water runoff, sediment, bedload regime and navigation gave rise to another phase of river training works. Between 1960 and 1976 regulations in the first section involved some 8.05 km of guiding structures; 31.2 km of bank revetments including a 9.0-km long floodwall in Budapest; one closure, and one "T" structure, and 42 groynes. The second section was supplemented with 1.02 km of guiding structures; 10.0 km of revetments; 56 side-branch closures, and four "T" structures. In the 1970s, based on an evaluation, it was concluded (A Duna áltolános szabályozási terve Szob-Dunaföldvár közöt 1978, Alsodunavölgyi Vízügyi Igazgatóság 1978, VITUKI 1976) that narrowing the channel for mean water significantly improved navigability and transport of ice; although such a narrowed channel began to deepen at some places, but this was expected to happen.

Training works carried out between *rkm 1,433 and 1,295* can be dated back to the nineteenth century (Brnič-Levada et al. 1977). In 1894, a general plan for river training was devised and inspired by the perplexity of the channel structure (inappropriate dimensions of bends in the channel below Mohács, at Batina etc., numerous islands and side branches supporting the formation of ice jams, an enhanced debris and bedload deposition and impaired floodwater runoff). Navigation through this section began to be dangerous, while the stability of riverbanks and levees was jeopardized also. The primary objective was to rejoin the braided channel of the Danube into a single channel with a uniform width for normal flow conditions and with appropriate riverbed morphology. All of the river training works were completed prior to 1914. Later, between 1914 and 1960, additional effort was devoted to maintain the existing condition of the channel.

The great ice-jam flood of 1935 created many concerns in this section. In 1963, in accord with the provisions of the Danube Regulation Act (International Convention for the Danube River), a draft of the "Joint Directive for Sustaining the Inland Waterways" was proposed in the former Yugoslavia. In 1970, an agreement on collaboration in this section of the river was signed. The aim of the agreement was to improve navigability and undisturbed transport of ice in order to reduce the risk of ice-jam floods.

From Ilok to Romania's border (the mouth of the Nera) the length of the Danube is about 218.1 km. Modifications conducted at troublesome sites in terms of navigation and floodwaters and ice transit were focused on constructing riverbank revetments in order to protect towns and ports, cut-offs to eliminate sharp bends, flow-guiding structures, side-branch closures and spur-dikes (Chmelár).

From the confluence of the Nera River downstream to the Iron Gate the Danube runs through natural rocky barriers-cataracts. In this 117-km long section the river could expand in its width, but channel deepening was possible. The narrow riverbed, hardpoints, sudden changes in channel width, specific flow velocity conditions, qualified this river reach as an extraordinarily dangerous section for navigation. The first training works were conducted in the middle of the nineteenth century. Despite several plans to construct canals and navigation locks in this section, such an endeavour was never accomplished. Therefore, an over-land route was proposed to provide a detour around the cataract section. This route ran through a cliff starting at Baziasu and ending at Orsova. This route was commissioned in 1837. This period witnessed the first river training works aimed at deepening the channels in rocky sills. All projects were completed by the end of the nineteenth century. The navigability in this section of the river improved substantially. Still there were other problems to be solved. A set of fundamentally complex solutions was adopted by Romania and the former Yugoslavia only in 1964; when pre-construction works for the hydroelectric and navigation system of Djerdap I at rkm 943.0 were already initiated. This hydro scheme went into operation in 1971. In 1984, Djerdap II at rkm 863.0 complemented Djerdap I.

10.4.1.3 River Training on the Lower Danube

Downstream the Iron Gate, except for the Danube Delta , no substantial alteration works were conducted. Several modifications were conducted on the right-hand side riverbank (Bulgaria) with a total length of 77.7 km. These included vegetation planting, stabilization of riverbanks with quarry stone, embankment walls in several towns and protection of ports, and 21 groynes. The need for river training works in this section was not as profound as on the Upper and Middle Danube. The river discharge in this portion of the river is rather high; for example at Orsova the discharge during low-water stages is around 2,000 $m^3 s^{-1}$ and during high-water stages 16,000 $m^3 s^{-1}$. The riverbed slope is rather moderate here, with flow velocities ranging from 0.2 to 0.8 $m s^{-1}$. High discharge and low flow velocities require wider channel widths. Since the flow regime in this reach is stable, the ratio between channel width and depth is favourable. Navigation depth at times of low-water stages only very rarely drops below 2.5 m; fords emerge only during periods of extreme droughts. But at some places the minimum depth of 2.0 m for navigation is not secured. For better navigability, the ford sections are excavated.

The Danube Delta deserves special attention when it comes to river training works. The delta begins 8 km upstream of Tulcea, where one of its branches turns to the northeast – the Chilija branch. This is the longest and largest branch of the Danube River. The Chilija branch has a length of 117 km and, in 1910, 72% of the flow in the Danube was discharged through this branch, compared to today's 59.9%. Historical river gauges are located along this branch also. This branch creates a natural borderline between Romania and Ukraine. Geometrical parameters of its channel favour the navigation of seagoing ships to the harbour at Izmail. The second branch of the Danube flows south-easterly and at Tulcea it splits into two smaller branches – the Sulina branch and St. Georghi branch. The St. Georghi branch is considered to be the oldest and was originally a large navigable branch with a harbour at St. Georghi located in its mouth. The length of this branch is 109 km with numerous meanders, discharging some 20% of the main-channel discharge. This branch ceased to be important in terms of navigation in the early nineteenth century, when its mouth was heavily silted. The smallest branch, perhaps for navigation the most important one, is the Sulina branch with a port at Sulina. Prior to the training works in this section it had a length of 83.8 km and diverted only 7% of the main-channel discharge. This section was subject to intensive river training.

After the Crimean War and upon signing the Peace Treaty of Paris in 1856, the CED (Commission Europeenne du Danube) commission was established. Increased attention was paid to the Sulina branch. In 1858 two funnel-shaped guiding dikes were built in its mouth; with a length of 1,412 m in the north and 915 m in the south (NC IHP Romania 1996). After completion, the depth of the branch in the mouth increased to 5 m. Later, when the southern dike was prolonged, water depths in this branch increased to 6 m. In the Sulina branch alone, the water depth was only 2.5 m in 1958, and therefore river training was justified. Wrecks of abandoned vessels were disposed of, revetments were installed along the channel banks, and dredging was deployed. After straightening the bends in the branch the entire length was shortened to 63.0 km. As a result of the dredging works, navigability in this section was improved by achieving a navigable depth of 6 m in the branch and its mouth. In the middle of the twentieth century the Sulina Branch resembled more an artificial canal than a natural watercourse (Töry 1954). Natural material was used in all structures. 334 groynes and bank revetment was built prior to 1911. Dredging played an important role also. While in 1911 the navigation depth in the Sulina Branch upstream of Braila (rkm 169.0) was 6.0 m, after 1914 this depth increased to 7.32 m as a result of channel modifications. But difficulties appeared in maintaining this depth. After World War I the International Commission convened by the CED issued a report stating that the water depths at the mouth were decreasing. As the causal agent was identified the increased discharge in the branch and wind conditions in the gulf – favouring aeolian sediment depositions within the mouth. A possible solution was to prolong the guiding dikes, which, at first, were equipped with openings. Later, however, these openings were closed in order to promote their effect. Dredging contributed also to the incision of the thalweg, now with a mean depth of 7.32 m. Maintenance of the thalweg required extensive dredging. However, difficulties continue up to the present time, and in most cases relief is achieved only by dredging.

The training works described in this chapter were not oriented solely on modifications for normal flow conditions. The aim was to ensure navigability under conditions of low-water stages also.

10.4.2 Design Parameters for Training Works

If the channel modification has to ensure an unrestricted flow of mean water, ice and bedload, including a safe navigation on the river, the parameters of the riverbed have to be appropriately designed. Apart from a correctly chosen bankfull discharge, geometric dimensions of the riverbed, i.e. the required bed width and depth, are important also. These parameters along with the morphology of the riverbed define the hydraulic parameters, i.e. friction, gradient/slope and flow velocity (n, i and v).

The principles and definitions used in proposing design parameters were not unified prior to the actual implementation of river training works. For this reason, we encounter two different interpretations of the term design discharge. While in the sections of the Upper Danube as the design discharge is taken the average value of discharge over a chosen period of record; on the Middle Danube, particularly in the section between the Devín Gate and Mohács, the design discharge is related to bankfull water stages. Bankfull discharge, however, usually exceeds the average discharge. The crest/height of control structures in the middle section was derived from the lowest navigable stages plus 1–3 m up to the natural bank height. The design discharge was determined retrospectively in relation to the lowest navigable water stage. Unfortunately, since not all necessary background material was available at the time of writing this report, only an incomplete review of design parameters used in the training works for mean water are provided here.

In the *Baden-Württemberg* section of the Danube, numerous hydroelectric plants have been constructed on diverting canals. Discharge capacity of hydroelectric turbines usually equals the average discharge Qavg. In the natural channel parallel to the diverting canals, only average low flows are maintained. In case of restoration measures this discharge flows through the new canal. In the original channel gravel bars are formed that have to be removed in approximately 10-year intervals. The reason is the long-prevailing low flows in these channels, even at times of high-water stages and floods in the main channel. This, on the other hand, promotes an intensive growth of vegetation preventing the gravel bars from migrating. On the contrary, gravel bars hinder the flow of water and increase maintenance costs. The proposed parameters of training works at selected reaches are listed in Table 10.1.

For the *Bavarian section* of the Danube, which is a part of the Rhine–Main–Danube waterway (R-M-D), the proposed design parameters for selected gauges are listed in Table 10.1. These are modern-day data processed as of 1985 for river sections between points of terminating tributaries (Iller, Lech, Altmühl, Naab, Regen, Isar and Inn) and take into account the presence of 16 weirs, hydro plants, six hydro schemes that were commissioned in the 1950s, including newly constructed hydro schemes Vohburg and Straubing.

Table 10.1 Selected river training parameters for mean water

Profile/river reach	From – to/backwater (km)	Length (km)	Station	Stationing (km)	Q (m^3 s^{-1})	B (m)	I (%)
1	2	3	4	5	6	7	8
Beuron	2,710.7–2,710.2	0.50	Beuron	2,717.7	11.0	–	0.78
Hundersingen	2,664.3–2,660.6	3.7	Hundersingen	2,662.4	24.6	–	1.05
Berg–Őpfingen	2,613.1–2,607.0	6.1			38.0	–	0.75
Donaustetter–Iller mouth	2,595.0–2,588.0	7.0			50.0	–	–
Iller mouth–border	2,588.0–2,580.0	8.0			120.0	–	–
Donauwörth–Lech mouth	2,508.1–2,497.0	86.0	Donauwörth	2,508.1	190.0		
Lech mouth–Altmühl mouth	2,497.0–2,411.0	26.0	Kelheim	2,414.8	300–330	120	–
Altmühl mouth–Naab mouth	2,411.0–2,385.0	26.0	Oberndorf	2,397.0	350		0.34
Isar mouth–Vilshofen	2,282.0–2,250.0	32.0	Hofkirchen	2,256.9	637.0	175	0.30
Inn mouth–border	2,223.0–2,203.0	20.0	Achleiten	2,223.1	675.0	–	–
Ottensheim–Wilhering	2,146.73	16.0	Linz	2,135.20	1,475.0	320–410	
Melk	2,038.20	22.5	Kienstock	2,015.2	1,882.0	320–410	–
Greifenstein	1,949.18	31.0	Wien	1,929.1	1,916.0	320–410	–
Devín–Medveďov	1,880.0–1,805.0	75.0	Bratislava	1,868.8	3,000.0	300–320	–
Kližská Nemá–Komárno	1,792.0–1,767.0	25.0			3,350.0	420	–
Ipeľ mouth–Nagymaros	1,708.0–1,692.0	16.0	Nagymaros	1,694.6	3,600.0	–	0.08
Fajsz–border	1,508.0–1,433.0	75.0	Mohács	1,446.9	3,700.0	–	0.05
Vukovar–Ilok	1,374.0–1,301.0	73.0	Bogojevo	1,367.0	2,900.0	400	–

Q – design discharge, B – width of the stream, I – slope.

A similar situation can be seen in the *Austrian section* of the Danube, where several hydro scheme projects began to emerge similarly as in Bavaria in the 1950s. These projects created a backwater area within the 350-km long reach in Austria. Only the section between the facilities at Altenwörth and Melk (approx. 28 km) and below the Freudenau hydro project to the state border to Slovakia (approx. 40 km) remained without the backwater effect. Design parameters for the riverbed for mean water have been defined for chosen gauge stations and hydro facilities, see Table 10.1.

Up until 1992, the 1,000-km long section between the water structures Freudenau and Djerdap I remained unimpounded, perhaps only with necessary modifications done for mean water and low water conditions.

Design discharge for the section *between Gönyü and Szob* was defined based on research outcomes (DVWK Merkblätter 1989, Szolgay et al., 1975). Recommendations were issued for the whole section of the Danube with the following figures proposed for the individual sections – 3,000 m^3 s^{-1} for the section between the Devín Gate and Medveďov; 3,250 m^3 s^{-1} for the Medveďov–Kližská Nemá (Gönyü) section; 3,350 m^3 s^{-1} for Kližská Nemá–Komárom; and 3,600 m^3 s^{-1} for the section between Komárom and Nagymaros.

A review of the implemented training projects for mean water on the *Hungarian section* of the Danube was made in the early 1960s (A Duna áltolános szabályozási terve Szob-Dunaföldvár közöt 1978, Alsodunavölgyi Vízügyi Igazgatóság 1978). Design discharge, channel roughness coefficients and slope conditions were identified for mean water conditions. The relationship between water stages corresponding to the design discharge and the height of riverbanks were examined also. Design parameters of the channel for mean water in Slovakia, the Slovak–Hungarian section and in Croatia's section of the Danube are summarized in Table 10.1.

10.4.3 Discussions on Mean-Water Training

Alterations for mean water were achieved by constructing cut-offs in large meanders, which resulted in an altered channel. Using longitudinal structures and transversal dikes, the flow was concentrated to a single channel. Riverbanks needed to be stabilized as well. The consequences were that many lateral branches were separated from the main channel at times when water stages in the branch system dropped below those in the main channel.

These interventions into the river channel in the Upper and Middle sections usually resulted in shortening of the watercourse, increased bed slopes, increased current velocities, local bed incision and at some places even occasional silting. Although the modifications for mean water met the requirements, however, additional modifications for low water had to be made on major sections of the watercourse particularly because of navigation needs.

Training for mean water on the Danube and its tributaries was investigated in terms of its impact on flooding (Oberste Baubehörde im Staatsministerium des

Innern 1927). In 1927, emphasis was placed on revealing how *the Bavarian section* was being affected. As stated in the report, a substantial improvement was observed in the movement of floodwaters; which was attributed to the improved transport of bedload and ice.

Kresser (Kresser 1986) conducted a study on the *Austrian section* of the Danube, concluding that the Danube River in this section gradually turned from being a "wide" river into a fairly "predictable" watercourse; which, as he explained, was achieved by the launch of several hydro schemes in the upper reaches – improving navigability – yet still with some weak points. The first hydro schemes were constructed along the Austrian section in the early 1960s. The effect of training for mean water was gradually overcome by new hydraulic conditions established in the newly created impoundments in the headwaters and tail waters of the hydro schemes.

Establishing a single bank line, constructing weirs and cascades in the lateral channels in the section *between Rajka and Gönyü* meant a positive change in flow conditions (Szolgay et al., 1978). Mean and low water training measures were adjusted for the height of spillways, which positively influenced the riverbed forming processes, and transport of bedload became uniform. The overall effect of the training initiatives was promoted by simultaneously performed training for mean water and low water – resulting in acceptable flow conditions at the project sites.

In the *Szob–Mohács section* (Chmelár 1994), investigations carried out in the 1970s (A Duna áltolános szabályozási terve Szob-Dunaföldvár közöt 1978) revealed that the earlier alterations for mean water and interventions in the low-water channel significantly improved the transit of ice and navigation, despite the fact that narrowing the channel for low flows was accompanied by locally occurring scours.

In summary, alterations for mean water substantially improved the runoff of mean- and low waters and hence the transport capacity of the main channel. An improvement was achieved in feeding the side branches and their flow regime, runoff of high waters, and in many respects better navigability in the main channel.

10.5 River Training for Low Water

Alterations for low water are additional to alterations for mean water to upgrade the efficiency of the training projects for mean water and to create favourable conditions for an unrestricted transport of ice and better navigation conditions at times of low-water stages. Naturally, the Middle as well as the Lower sections of the Danube were predominantly utilized for navigation, but even here, several reaches were identified as inappropriate or even dangerous for navigation due to the Danube cataracts. This fact was considered in the forthcoming plans for implementation of training projects to improve the navigation conditions that would meet the requirements of cargo vessels passing the European inland waterways.

10.5.1 Training Accomplishments and Project Features

10.5.1.1 The Upper Danube

In the *Baden-Württemberg section* of the Danube, river training initiatives that were introduced in the mid-nineteenth century and the early twentieth century caused the riverbed to deepen. It was then necessary to hinder the process of bed incision and reverse it wherever possible. The problem was relieved by constructing channel-traversing dikes. These measures increased sedimentation rates resulting in higher elevations of the riverbed, which subsequently resulted in an increase in water levels. These works were conducted between Scheer and Riedlingen, whereas the highest effectiveness of the works was achieved at Bloching.

The *Bavarian section* of the Danube was, between 1850 and 1920, altered for mean water. The alteration measures, however, did not bring the desired effect with respect to navigation and engineering design. Therefore, further initiatives were aimed at constructing hydro plants. Backwater areas were expected to increase the depth of the thalweg. In the most troublesome section – the gorge near Passau – the first hydro project was built in 1927 – Kachlet. After World War II, in 1953, at the lower end of the section below Ulm the Böfingen Halde project was completed. In the 1960–1970 decade, the Danube River was backwatered by the operation of ten projects; while in the 1980s an additional four hydro schemes went into operation.

Increased attention was paid to the waterway connecting Regensburg with Vilshofen, where several obstructions impeded the entrance of vessels into the Regensburg Port; while at the same time this section of the river was an important part of the planned international navigation link (Ertl 1985). Most of the projects were completed before 1965. A 2-m depth for the thalweg was achieved along the whole section. On longer sections, particularly between Regensburg and Straubing, the thalweg was only about 40–60 m wide. Although the alterations for low water did improve, to some degree, navigability of the waterway linking Regensburg and Vilshofen, complete satisfaction was never achieved. In the late 1980s, the Bad Abbach hydro plant at rkm 2,401.05 and the Regensburg hydro plant at rkm 2,381.30 were completed, and it was expected that their operation would resolve the problems downstream to Regensburg. Even before the project was completed upstream of Regensburg, it became clear that problems with navigability on this section will continue to exist and only one-way navigation will be possible.

In relation to navigation and water-management requirements, engineering design, and environmental concerns, priority was given only to those solutions that relied on the operation of the Geisling project that is located in the middle of this section, and the Straubing project at the downstream end of the section. Because of the approximately equal length of the impounded reach, both of these operating plants elevate the low-water levels by more than six metres.

Alterations for low water on the *Austrian section* of the Danube were initiated in the late nineteenth century. Restrictions for navigation were imposed by the Schlögen Loop, Grein Strudel and Aschach Gorge (Michalke 1969). Difficulties occurred also in the Wachau area, in Lower Austria and the Vienna cut-off. Groynes constructed mostly in groups were predominantly used as control structures, with the assistance of dredging and sills.

Most of the alteration works for low water were accomplished prior to the outbreak of World War II. After 1945, only damages caused by the war were eliminated and actions facilitating navigation through fords were carried out.

After training for low water, there remained places on the Austrian section that made navigation through this section problematic. It would be too laborious and costly to maintain the navigability of these sites; therefore, several hydro schemes were constructed in this section also. In 1958 the hydro scheme Ybbs-Persenbeug went into operation, and the Schlögen Loop and Grein Strudel were inundated in 1964 for the sake of the Aschach project; whereupon another series of hydro projects followed soon thereafter (seven facilities). Navigation conditions in the ford sections improved, but new concerns emerged as to the altered sediment transport regime in this section.

10.5.1.2 The Middle Danube

The first attempts to conduct systematic alterations for low water between *the confluence of the Morava River and the Ipel' River* were made in the early twentieth century. During periods of low flows, water was concentrated in the main channel by means of longitudinal structures and closures installed in lateral branches. Since the anticipated results were never completely achieved, a system of groynes had to be introduced in order to achieve the desired depths in the thalweg. Dredging was deployed at sites where the erosive actions of groynes and strong currents needed to be intensified. For convenience, the entire section is divided into three distinct subsections:

- The Devín–Bratislava section, which is a cross-boundary section separating Slovakia from Austria, has long been considered a problematic section with respect to navigation. With the need to create a canal suitable for navigation and to prevent ice-jam floods, training for low water required construction of 48 alternating groynes on both sides of the canal, several left-side branches needed to be cut off, and longitudinal structures were built.

  A project framework for river engineering in this section was introduced in 1972 (Rahmenplan, Fassung 1972, Österreich-Czechoslovak Rep., Grenzstrecke). In 1989, a conclusion was made that the section extending from rkm 1,880.0 to 1,876.0 maintained its original morphology; and downstream rkm 1,876.0 the earlier proposed initiative for improving navigability at Wolfsthal was not achieved. At that time, navigability in the section between rkm 1,876.0 and 1,874.8 was considered as a most troublesome task, mainly because the area was susceptible to the emergence of fords. As this section was located within the area of influence of backwater induced by the Gabčíkovo hydro scheme, conditions for a favourable thalweg became more complicated. After placing the Gabčíkovo hydro scheme into operation in 1992, channel parameters for low-water conditions changed slightly.
- After World War I, there was no maintenance work carried out on the section between Bratislava and Gönyü. The situation changed only after signing the so-called “Pacte du Danube” Treaty, according to which, Czechoslovakia and

Hungary were committed to mutually co-ordinate their river training projects on the Danube River. Because of the training works for mean water and partially for low water also, the depths at ideal inflection points were determined at –3 and –6 m below the low water reference height. The next phase of river training was conducted in accordance with the General Plan for river training on both sides of the river. The initiative to build a hydro scheme between Gabčíkovo and Nagymaros introduced additional changes in terms of river training in this section of the river. In the transitional period, the thalweg needed to be maintained by dredging. Extensive dredging was not always necessarily accompanied by improvements in the regime of water stages and navigability.

In 1992, the dam at Čunovo, Slovakia, became reality. In this respect, the question of navigability of the Danube River had to be discussed, as suggested by the recommendations of the DC; with attention to be paid to the section below the end of the outlet canal of the Gabčíkovo power plant. A project framework was proposed for training works to be carried out in the section between the end of the outlet canal and Szob for alternative thalweg depths. Now, the riverbed below this point is highly unstable. It is expected that degradation in this section will continue to exist even in the future.

- In the section between Gönyü and the Ipeľ confluence there are of concern for navigation rock sills at rkm 1,734; a narrow navigation corridor at the right-hand side bank between rkm 1,734 and 1,732, the Boží kopec-Obid section; other critical sites are at rkm 1,711, the narrow reach at the end of Helemba Island and the ford between rkm 1,714 and 1,713.

River training for mean water on the Danube *between Szob and Mohács* immensely improved navigability through this section. Yet, there were still other places that restricted navigation upstream Budapest at rkm 1,722, downstream at the ford at Vác (rkm 1,679) and at Göde (rkm 1,669). The reach below Budapest was more favourable for navigation, for even low-water stages in this section provided sufficient depths for navigation (2 m). After 1950, complementary river training projects were carried out; groyne structures were constructed and industrial dredging was deployed (A Duna áltolános szabályozási terve Szob-Dunaföldvár közöt 1978), which profoundly affected the morphology of the riverbed between Nagymaros and Budapest. In addition to the above-mentioned training works, more excavation projects were carried out in the period between 1970 and 1987 to improve navigability. Between 1970 and 1990, complementary repair works were carried out; in the period 1990–1991 between rkm 1,567 and 1,565, and between 1994 and 1996 between rkm 1,616 and 1,615.

Alterations for low water on the Danube River below Mohács downstream to the confluence of the Nera were a part of the initiative to create a single channel for mean water. These works are described in Section 10.4.1.3.

10.5.1.3 The Lower Danube

Alterations to the channel of the Danube River to improve navigation downstream of the confluence of the Nera cannot be characterized as alterations for low water. The system of groynes was installed only along the Sulina branch. A brief description is given in Section 10.4.1.3.

10.5.2 Design Criteria for River Training and Recommendations of the DC

Design criteria for low-water alterations on the Danube are determined by a fictive reference water level. Such a fictive water level for alterations and navigation was first introduced in 1936, commonly known as the "CID" water level. After 1957 this level has been determined in accordance with the requirements described in (Danube Commission 1992), a document issued by the DC in Budapest. Later, these regulations were subject to revisions and innovations. Now, the regulation of 1995 is in force (Danube Commission 1995). The water level is determined by statistically evaluating discharge time-series for the past 40 years (with excluded periods with ice phenomena) at reselected gauges on the Danube. Discharges obtained from the rating curves at the control gauge stations, with a 94% probability of exceedance, are considered obligatory for navigation once approved by the DC, and are applicable for 10 years.

Minimum pool elevation and navigable depths were determined accordingly for 1956–1965 (MPE and ND-DC'57), and for 1966–1975 (MPE and ND-DC'66). In the early 1970s, quantification of consequences resulting from river training on several sections of the Danube was used in preparing documents for the plans to carry out alterations to the river channel. Increased attention was paid to dredging and its impact on deformation of the riverbed. Subsequently the MPE was adopted and ND-DC' 76 was determined from the period 1931–1970. Intense dredging in the riverbed (industrial dredging and dredging aimed at modifying the riverbed) soon caused these water stages to lose their applicability. Therefore, the 52nd meeting of the DC on April 21, 1994, adopted new criteria and background materials for the calculation of the water stages, now using the 30-year period 1961–1991 (Danube Commission 1995). With respect to navigation on the Danube, the executive board of the DC proposed a document in 1969, defining minimal widths, depths and radius of curvature of the watercourse in order to maintain it navigable. In 1988 these recommendations were amended and extended to involve another section: the Kelheim–Vienna section, and modified for the Georgievskij Ceatal–Sulina section.

10.5.3 Effectiveness of River Training Works for Navigation and Other Purposes

The impact of river training is to be viewed from two perspectives: first, how it affects the migration of the riverbed – which consequences are manifested in changes of water stages; and secondly, the geometry and winding of the watercourse – which has to be accounted for in the navigation corridors.

The need to sustain the navigability of the *Bavarian section* of the Danube was accomplished by alterations for low water and by straightening the watercourse. The overall effect of the alterations was examined by comparing the original conditions with measurements made in certain time intervals after completion of these river training projects (Bauer 1965). For the original presumably natural conditions for navigability, data was used from the periods prior to 1826. Although imposing several limitations for water management, the natural course of the Danube was in a state of equilibrium with respect to bedload transport; but in terms of its morphology, the river appeared to be stable. The channel's capacity to export bedload equals the bedload import.

The introduction of the first training structures (chute cut-offs) was accompanied by some changes in the watercourse. The river attempted to maintain its original gradient. Export of bedload exceeded the import. The altered watercourse was no longer in equilibrium with the movement of bedload, which caused an intense incision of the bed. The deep cutting of the riverbed on the upper Bavarian section was inevitably accompanied by silting in the lower navigable sections of the Danube.

As for the navigation corridors, priority was given to the Kelheim–Vilshofen section that is a part of the R-M-D waterway. After training for low water, the channel's depths were sufficient for navigation; however, the width of the channel was still insufficient especially in sections downstream of Regensburg. This situation was relieved by constructing the Bad Abbach and Regensburg hydro plants, and Geisling and Straubing hydro plants.

After completion of the major river training projects, the *Austrian section* between 1893 and 1952 experienced changes in relation to its previous low-water conditions. Between Engelhartszel and Hainburg (a more than 300 km long stretch) a series of alternating silted and incised sections could be observed, at least until the late 1930s. After placing the Jochenstein project into operation, the Austrian section of the Danube experienced several changes becoming a waterway with several hydroelectric plants. This made it possible to stabilize the riverbed in the modified sections; however, at the same time, increased retention of bedload material was observed along with intensified erosion below the most downstream hydro scheme. For the river section in the Viennese Basin below Hainburg this meant an abrupt change in the incision of the riverbed. This effect became even more profound after the Greifenstein hydro scheme went into operation, because the import of bedload from upper sections was completely hindered. Upon completing the Freudenau project, the effect of bed incision was amplified. Below Hainburg, bed incision became a severe problem. This is the so-called "backward" erosion that will cease only after the Čunovo reservoir is in full operation.

Training for low water manifested its effects differently along the *Devín–Szob section*. In 1982, a new longitudinal profile of the Danube was formed in the section between Devín and Rajka due to the intense dredging and subsequent bed erosion; which led to increased gradients of the bed upstream the site of dredging, and lower slopes at downstream sites where dredging of bed material was carried out. The most obvious changes showed their effects in declined MPE and ND-DC by 1.7 and 0.8 m compared to the water level in 1976, respectively. Changed geometric parameters induced changes in other hydraulic parameters also; the regime of bed-load transport was changed in such a manner upstream the site of dredging that the transit of depositing material was more intense compared to the downstream section – where the phenomenon of "hungry" water was observed. In the next period, water levels at times of low flows tended to drop along the entire section. Only 30% of the investigated sites complied with the criteria imposed on the international navigation waterway, as defined in the recommendations of the DC. As for the Rajka–Gönyü section, a conclusion could be drawn in 1977 that the training projects and random low flows had alleviated, at least to some extent, the previous silting of the riverbed. Navigation widths for the 25 dm depth below the MPE and ND-DC improved, or at least did not worsen. Evaluation reports of 1991 concluded that the linkage between the water stages only confirms the ongoing trend of the riverbed – i.e. its deep cutting. For the Gönyü-mouth–Ipel' section a conclusion was made that the observed changes are tremendous in their magnitude. Through the intense dredging, navigable depth was formed at many places, which was projected to be achieved only after canalization of the river, i.e. 35 dm below the MPE and ND-DC.

After the training interventions and especially the intense dredging *between Szob and the southern border of Hungary*, it was found that navigation width was ensured at the depth of 25 dm in all but a 9 km reach, which constitutes 3.2% of the entire length of the section. For 30 dm below the MPE and ND-DC'76, the navigation width is not achieved over a length of 41 km, which is 14.8% of the entire length. For a depth of 35 dm, the required width of 180 m is not achieved along the length of 98.6 km, which is 33% of the entire length of the section.

In the period between 1970 and 1975, this section was subject to large-scale dredging, which caused the riverbed to sink by one metre on average. Surveys revealed that, between 1959 and 1984, low-flow water profiles had declined in the section between Szob and Paks; and this situation continued in the following years.

The results of a survey evaluating the effects of the training works conducted on the Danube *between the southern border of Hungary and Vukovar*, and how these influenced the navigation corridor, were published in (Petkovic, Varga).

A study reviewing the character and efficiency of the implemented river training projects in the section *between Bazias and Sulina* was based on channel training in the navigation way over the period 1962–1990 (NC IHP Romania 1996). The first part of the section extending from Bazias to Kazan involves an area of major cataracts. Since the first training projects conducted in the nineteenth century, the training initiatives had to be constantly supplemented with additional works until a principal solution was achieved by proposing construction of the Djerdap I hydro project. Comparing the state of the riverbed as it was in 1962 with 1990 one can see a prevailing trend of silting. Below Djerdap II downstream to the entrance to the Sulina branch, the bed elevation in 1990 appears to be well above the elevation of 1962. The prevailing trend of silting is not in accord with the theory of the so-called "hungry water". Cross-comparisons and measurements suggest that during periods of low flow, the pulsation regime causes the water levels to fluctuate within a range of 2.5 m and velocities to fluctuate within a range of

0.5 m s^{-1}, the consequences of which are reflected in a declined bedload transport. This fact can be explained (NC IHP Romania 1996) by comparing the long-term discharge regime of the Danube River prior to the construction of major hydro schemes and the drier period afterwards. This resulted in lower current velocities and a lower turbulence, which had an impact on the morphological processes in the riverbed.

In the remainder *of the Sulina branch*, the elevation of the bed underwent several changes over the period of investigation as well. In 1962, the riverbed was elevated compared to that of 1990, suggesting the presence of scour. This condition may be related to the scour (wash out) in the delta during low-flow periods. As for the navigation way parameters after completion of the hydro project, the upstream section offered better conditions for navigation. Obstacles to free navigation in the lower section, in the form of fords, are eliminated by dredging.

10.5.4 Summary

Conditions for navigation (navigation way depths, widths, curvature radii and fords) are regularly maintained in the DC's databases. In 1993 conditions for navigation were evaluated for the period 1981–1990 (Danube Commission 1992). Restricted conditions for navigation, in terms of navigable depths, over the investigated decade were found particularly in the Vienna–Mohács section. The most critical reach was found between Devín and Szob, where the required 25 dm depth of the thalweg was not reached in 141–202 days in a year. As for the Devín–Sap subsection, the impoundment at Čunovo, in 1992, and the navigation canal at Gabčíkovo relieved this undesirable situation. For the reaches below Szap where the outlet canal rejoins the Danube (in case the Nagymaros hydro scheme was not completed) it was necessary to continue to modify the watercourse to ensure favourable thalweg proportions in the same manner as at other critical sites in this section.

10.6 Other Training Works

As can be implied from the previous sections, construction of hydro schemes remains an essential part of river engineering. The Upper Danube is intensively utilized by several hydro schemes constructed in the Austrian and German sections. On the Lower Danube, Djerdap I was completed in 1977 and followed by Djerdap II in 1984. The Middle Danube was for some reason slower to exploit the energetic potential of the river and has long been viewed as the most troublesome section for navigation. The construction of hydro schemes and their relation to the problems that were encountered during the channel training projects are briefly described in a separate section on design parameters and the purpose of hydro schemes. Navigation canals and important waterways linking the Danube River with other rivers are also described.

10.6.1 Basic Characteristics and the Purpose of Hydro Schemes

In the Baden-Württemberg section of the Danube, numerous hydro schemes were proposed to serve as a source of water for municipal and industrial facilities, and at the same time, for generation of electricity. In the Donaueschingen–Öpfingen section weirs dominate with 1–5 segments with a width of 4–102 m and effective heights of 1.7–3.8 m. Powerhouses have been constructed at some of these weirs, equipped with 1–3 turbines capable of operating with discharges between 2.4 and 27.0 $m^3 s^{-1}$. Three hydroelectric plants are in operation in the Öpfingen–Böfingen Halde section – namely the Öpfingen, Donaustetten and Wiblingen hydropower plants. These plants operate in a run-of-river mode and are placed within diversion canals. Only the Böfingen Halde plant at the border to Bavaria is a true run-of-river power plant.

In Bavaria's section of the Danube, only the Jochenstein hydro plant is currently in operation, a joint project between Austria and Germany, equipped with 16 weirs with powerhouses; and there are another six hydroenergetic-navigation facilities. The storage volume of their impoundments is 1.3–6.5 million m^3; design discharge at the individual plants is 150–500 $m^3 s^{-1}$ with a generation capacity of 7.35–23.70 MW. The last of the series of power plants, i.e. the power plant at Vohnburg, has four weir segments with a width of 80 m, and 9.7 million m^3 storage volume. Its discharge capacity and generation capacity are comparable to the 15 plants located upstream. Navigation locks are operated at Bad Abbach, Regensburg, Geisling, Straubing, Kachlet and Jochenstein; since this section (below Kelheim) is part of the international R-M-D waterway. Their weirs have four to six segments with a total length of 96–175 m. Storage volume of the impoundments ranges from 5.3 to 44.6 million m^3. The design discharge is 230–2,050 $m^3 s^{-1}$ and output power is 6.4–132.0 MW.

The series of hydro plants operating within the Bavarian section is followed by several hydro schemes in Austria. Nine hydro schemes used for navigational and hydroenergetic purposes are in operation between Jochenstein and Vienna, including the Freudenau hydro project. Austria's hydro plants are equipped with weirs comprising four to six segments and with a total length of 24 m (Ybbs-Persenbeug – 30 m). These plants are able to work with flows in the range of 250–510 $m^3 s^{-1}$, producing maximum output power of 168–328 MW. Each of the plants has two navigation locks that are 24 m wide. Their length is 230 m; only the locks of the Freudenau plant are longer – 275 m.

In the *Middle Danube section*, only the plant at Čunovo and the Gabčíkovo diversion canal are currently in operation (Lahoda et al. 1993). The impoundment at Čunovo is equipped with a weir feeding an inundation area. This weir has twenty 24-m wide segments. An integral part of this weir is a powerhouse with five turbines rated for 360 $m^3 s^{-1}$ with 52-MW output power. Below this powerhouse is installed another weir with three 24-m wide segments with a discharge capacity of 3,300 $m^3 s^{-1}$. An ancillary navigation lock is located on the left side of the weir. During elevated water stages this weir is used as an additional weir segment. The purpose of this weir is to create an alternative navigation corridor through the original channel of the Danube. Another weir is located at a chute cut-off at rkm 1,851.75. This weir has four 18-m wide segments. The Gabčíkovo hydro scheme consists of a power plant and several navigation locks. The power plant is equipped with eight turbines with 90-MW output power. The navigation locks are 34 m wide and 275 m long. The tail water goes through an 8.2-km long outlet canal rejoining the Danube at Szap (rkm 1,811.0).

In the *Lower Danube section* two run-of-river hydro plants are in operation: Djerdap I and Djerdap II – the result of a joint project between Romania and Yugoslavia. Djerdap I (1964–1971) consists of a mid-channel positioned weir with 14 segments, and two lateral powerhouses equipped with six turbines rated for 178 MW and flow rates up to 4,250 m^3 s^{-1}. Two locks, 34 m wide and 310 m long, are placed parallel to the riverbanks; with a minimum depth of 4.5 m above the sill. The impoundment covers an area of 133 km^2 with a storage value of 1,350–2,031 million m^3. The backwater effect induced by this hydro scheme can be seen up at the confluence of the Tisza River (rkm 1,222). In December of 1984, another hydro scheme, Djerdap II, was constructed some 80 km downstream from the first one. The powerhouse was built in the middle of the transect, equipped with 20 turbines rated for 540 MW and flow rates up to 6,800 m^3 s^{-1}. A weir with four segments is attached to the power plant. Navigation through this section of the river is secured by two lock (34 m wide and 310 m long). The impoundment encompasses an area of 63.25 km^2 and provides a storage volume of 425 million m^3.

If we envision hydro schemes as channel training structures, it must be said that hydro schemes serve multiple purposes, inasmuch as they can actively affect the runoff regime, and change the morphology and hydraulic parameters of a watercourse. The major and most obvious benefits of channel training works are optimized conditions of water runoff (flood protection and artificial increasing of low flows), elimination of freeze-up phenomena, and optimal navigation depths during low-flow periods. On the other hand, in respect to bedload transport, training structures may create undesired flow conditions also. This is apparent especially at the end-point of backwater areas where large quantities of sediment deposits may accumulate; while the so-called "hungry water" phenomenon may emerge below the dammed sections.

10.6.2 Danube Waterways

The idea of connecting the major European rivers to improve navigation and water management is actually very old. Three important canal systems are currently present on the Danube River: the waterway connecting the Danube with the Rhine and Main rivers (R-M-D), the link Danube–Tisza–Danube (DTD), and finally the Danube (Crna Voda)–Black Sea (Konstanca) waterway.

10.6.2.1 The Rhine–Main–Danube Waterway

The idea to connect the Rhine, Main and the Danube was accomplished by constructing a 117-km long navigation canal starting at Bamberg on the Main and ending at Kelheim on the Danube. This is the "cordial" section of the 3,500-km long waterway to which more than 15 European countries have a direct approach. The altitude of the waterway, with its highest point at 496.0 m above the North Sea, makes this waterway the highest one among all waterways in Europe. The almost 1,200-year-old idea, that was first introduced during the reign of the king Charles the Great (793 AD), became reality through this waterway. The very first attempt to link these rivers was the construction of the Fossa Carolina Trench. Although the concept of inland waterways and construction methods changed over time, the original idea became reality by the commissioning of the R-M-D waterway on September 25, 1992.

This waterway involves a 64-km long canal between the navigation locks at Eibach and the bifurcation at Altmühl below the navigation lock at Dietfurt; and 34 km from Altmühl between Dietfurt and its mouth to the Danube. The elevation difference between water levels at the Norimberg Port and the upper impoundment at Hipoltstein is 93.5 m; while between the upper impoundment and the mouth at Altmühl the elevation difference is 67.8 m. The project involved the construction of one navigation lock with a rise of 19.5 m; three locks with a rise of 24.7 m; and three locks with a rise of 8.4 m, while the latter one has a navigation chamber, three 15-m wide weirs, and turbines capable of operating in reversed pumping mode.

Navigation locks of the Main–Danube juncture are fed with water taken from the Danube. Pumping facilities were installed in five plants – Kelheim, Riedenburg, Dietfurt, Berching and Bachhausen – to convey water to navigation locks and to increase water stages in the Regnitz River during low-flow periods. The water pumps that are installed at the hydro plants at Riedenburg and Kelheim can operate in two modes – as pumps or turbines.

10.6.2.2 Navigation and Irrigation System of the Danube–Tisza–Danube Rivers

The need to ensure sufficient quantities of water for irrigation of the southern portions of Vojvodina (Bačka and Banat), and to drain the lowest parts of this area at times of prevailing high-water stages on the local streams, forced the local water managers to find solutions to these needs. Water from the Danube, Tisza, Begeja and Tamiša Rivers could not have been utilized without altering their flow regime. In the seventeenth and eighteenth century, the navigation canal Begej and a canal connecting the Danube and Tisza were constructed, later followed by new flood levees erected along the Danube and Tisza with a total length of 1,300 km. In addition to these projects, irrigation-drainage canals, pumping stations and other facilities were constructed – to form the complex navigation-hydro melioration system as it is known today. This system involves 20 canals with 24 weirs, five auxiliary safety weirs, 16 navigation locks and six pumping stations. This system is able to drain an area of 760,000 ha, and irrigate some 500,000 ha of land. The total length of its canals is 960 km, of which some 664 km may be used for up-bound and down-bound navigation (Osnovnij projekt Velikog).

New navigation locks in the DTD system have thalwegs complying with west European standards for navigation in canals. The length of its locks is 85 m, with a width of 12 m and the minimum depth over the sill is 3 m.

The DTD water management system is one of the most important and complex systems implemented not only in the Danube Basin. The water regime of the canals does not substantially affect the flow regime of the Danube; since it is controllable and incoming or outgoing water is negligible compared to the discharge of the Danube. For example, the weir at Bezdan discharges 60 $m^3\ s^{-1}$ and in reversed operation the flow rate is 230 $m^3\ s^{-1}$, while the annual average discharge of the Danube at Bezdan is 2,360 $m^3\ s^{-1}$.

10.6.2.3 Danube–Black Sea Waterway

The problems encountered with navigation in the Danube Delta gave rise to a project with a goal to interconnect the town of Cerna Voda and Constanta. This canal made access to the Black Sea shorter by 240 km than the one leading through the Sulina branch. Construction works were initiated in 1949 to connect the port at Cerna Voda with the port at Medea, but in 1953, the project was interrupted. In 1975, the initiative was renewed and a more complex solution was proposed. The new alternative involved the connection of Cerna Voda with Porta Alba and downstream along the original canal to the Medea port, or through the new canal to the port at Constanta – Southern Agigea. The result of the intensive construction works was a 64.4-km long canal, which went into operation in 1984. Its width at the bottom is 91 m with a depth of 7 m. Navigation locks are positioned at the inlet and outlet of the canal; one on the Danube at Cerna Voda and the other one at Agigei and Medei. The locks operate in pairs and their width in chambers is 25 m, and 310 m in length with a 7.5 m depth over the sill. The dimensions of the canal enable navigation for river-going vessels with lengths of 296 m and sea-going vessels with an uplift of 3,000 t. This canal serves not only as an inland link connecting important industrial centres, but its operation actively supports agriculture and serves as a source of water for irrigation of the steppe area of Dobrudža. The water regime of this canal affects the runoff regime of the Danube only very slightly, because its average flow is only 60 $m^3\ s^{-1}$ while the annual average flow of the Danube at the gauge at Silistra is 6,800 $m^3\ s^{-1}$.

10.7 Conclusions

On the *Upper Danube*, with its almost 1,000 km and covering about one third of the entire length of the Danube, most of the training works for high water, low water and mean water were completed before the 1920s. However, some of the training works especially those for low water were carried out even later. But there still remained several sections with restricted or even hindered navigation. Among these sections were the gorge at Weltenburg, narrow and shallow reaches between Ulm, Regensburg, Vilshofen and Passau, the Schlögen Tangle, Aschach Gorge, Grein Struden, the area near Wachau, and several reaches in Lower Austria.

This was the beginning of the era of hydro projects. During times of alterations for low and mean water, the goal of the first hydro projects was to provide water for the local industry and municipal communities. Twenty-three hydro schemes are in operation in the Baden-Württemberg section between Neudingen and Ulm. Because of the low crest, and height of the training structures in these sections (1.70–3.85 m) the zone of backwater is rather short causing low-water stages in the diversion canals. The altering effect of these structures is only local which is not always favourable. In the Bavarian section between Ulm and Jochensteinom at the border to Austria, 22 hydro projects were constructed. The first hydro project – Kachlet, which was completed as early as 1927 helped to relieve many of the problems encountered in the gorge between Passau and Vilshofen. Between 1953 and 1994, many of the navigation problems were eliminated by these hydro schemes. Six of these hydro schemes are equipped with navigation locks. These hydro schemes are part of the waterway connecting Rhine–Main–Danube, and comply with the international standards for navigation. In the first phase, nine hydro stations were constructed in this section, while four of them were commissioned before 1960. Later, seven hydro schemes were constructed. Another two projects followed after 1990. In this section, the reach between Straubing and Vilshofen remains problematic. Construction of two hydro schemes in this section is also being considered.

In the period from 1959 to 1997, ten hydro projects were completed and placed into operation in the section between Jochenstein and Devín. Backwaters induced by these projects gradually began to affect the dangerous sections such as the Schlögen Tangle, Aschach Gorge and Grein Struden. The remaining sections without the backwater effect are the area near Wachau, between Krems and Schönhübel (altered during the period 1984–1988); and finally the reach between Freudenau and Devín where the backwater effect from Čunovo is not observed.

For now, consensus between water managers and ecologists has been reached. The need to establish a symbiosis between the environment and human interventions becomes a priority. This requires costly investments into the design of river engineering projects and long-term monitoring of the environment. Quantification of environmental impacts based on long-term surveys remains important for the design of a hydro scheme. This delays the implementation of hydro projects.

The *Middle Danube*, which is also one third of the entire length of the Danube, is problematic in terms of navigation. Alterations for mean and low water were conducted in the late nineteenth century. Many of these structures were disrupted or nonfunctional already at the beginning of the twentieth century due to riverbed-forming processes. In the upper reach between Devín and Szob, groynes were reinstalled and repaired between 1963 and 1974. During the preparatory phase for the construction of the G-N hydro scheme, a decision was made to modify the critical reaches within this section. For the remainder of reaches only necessary maintenance was recommended. Through commissioning of the plants at Gabčíkovo and Čunovo, in 1992, some of the critical sites were "relieved" indeed. However, the section below Szap became even more problematic due to the failure to complete the hydro project at Nagymaros.

The structures that were built in the nineteenth century in the section Szob–southern Hungarian border were renewed in the 1930s. Later, between 1960 and 1976, between the southern border and the mouth of the Nera River the training works of the late nineteenth century and early twentieth century were renewed and supplemented in a non-uniform manner. More was done after 1963 and 1970, i.e. after signing the Danube International Convention in 1963. Several reaches were altered either uniformly along their entire length, or only locally at sites where the risk of ice jams and hindered navigation were observed. The last portion of this section, from the mouth of the Nera River to the Iron Gate, is an area of numerous cataracts. Only after the completion of the Djerdap I hydro project was relief achieved. The backwater zone of this project reaches back to the mouth of the Tessa River, which substantially improves navigation through this rocky section.

While in the upper third of the Upper Danube (below Ulm) the flow regime and transport of ice is improved by a cascade of 31 hydro projects, the Middle Danube with its length comparable to that of the Upper Danube has only two hydro projects. This fact supports the formation of critical sites that must be systematically maintained and altered. In the second half of the nineteenth century, the Lower Danube was altered for high water by constructing levees. Renewal and construction of additional levees continued in the 1950s. Only necessary dredging and rewetting works were conducted in ports and urban areas near the watercourse. Alteration for low and mean water was carried out only in the Sulina branch in the late nineteenth century. These structures are still subject to reconstruction.

Certainly, increased attention must be paid to the following issues:

- monitoring of the riverbed geometry, morphological parameters of the surrounding area and their relation to biotic parameters,
- updating of minimum pool elevations and navigation depths for records shorter then 10 years;
  and
- examination of flood protection measures. This will create a database of hot environmental issues and training works on the watercourse and the area adjacent to it.

References

A Duna áltolános szabályozási terve Szob-Dunaföldvár közöt (1978) Baja

Alsodunavölgyi Vízügyi Igazgatóság (1978) A Dunaföldvár-déli országhatár közötti Dunaszakasz áltolános szabályozási terve. Baja

Bauer F (1965) Der Geschiebehaushalt der bayerischen Donau im Wandel wasserbaulicher Manaßhmen. Die Wasserwirtschaft, 55(4/5):8–10

Bitterer F (1969) Der Hochwasserschutz im Donauabschnitt zwischen den Mündung der Isper und der March. Österr.Wasserwirtschaft 21(1/2):34–47

Brnič-Levada D et al (1977) Regulierung des Donaubettes. Teilprojekt 6.2. Nationalbericht. Kroatien, Osijek

Chmelár V (1994) Dunaj historický a dnešný. ELECTRA Publishing House, Žilina, p 224
Csoma J et al (1975) Podklady a zásady potrebné pre vypracovanie generálneho projektu úpravy Dunaja v km 1792 až 1708. Správa VÚV-VITUKI, Bratislava-Budapest
DVWK Merkblätter (1989) Wahl des Bemessungshochwassers Nr.209. Hamburg-Berlín, pp 1–6
Danube Commission (1992) Informacija o soderžaniji sudochodnogo farvatera I o perekatach na Dunaje ot Ulma do Suliny. Budapest
Danube Commission (1995) Novyj rasčet nizkich sudochodnych i reguljacionnych urovnej vody po osnovnym vodomernym postam na Dunaje. Budapešť, p 85
Ertl W (1985) Die Donau zwischen Kelheim und Jochenstein-Teil der Main-Donau Wasserstraβe. München
Gewässerdirektion Donau/Bodensee (1994). Projekt 6.2. Donau in Baden-Württemberg. Ulm
Katiš B, Sretenivič M (1990) Hidroenergetski i plovidbeni sistem Djerdap I. Vodoprivreda 22:1–2
Kresser W (1986) Vorhersage von Flussbettänderung in der österreichischen Donaustrecke. 12. Konferenz der Donauländer, Belgrade
Lahoda F et al (1993) Vodné dielo Gabčíkovo. História a riešenie. Vodohosp. výstavba, Bratislava
Michalke K (1969) Regulierungsarbeiten und Schiffahrtsverhältnisse im Aschacher Kachlet. Österreichische Wasserwirtschaft 21(1/2):23–27
Mittiade N, Popa G (1966) Zatopljajemaja zona Dunaja i jejo aktual'nyj vid. Gidrotechnika, Vodnoje chozjajstvo, Meteorologija, II, 9:470–475. Bucharest
NC IHP Romania (1996) Regulirovanije rusla reky Dunaj. Nacional'nyj otčet Ruminiji. Bucharest
Oberste Baubehörde im Staatsministerium des Innern (1927) Der Hochwasserschutz an der Donau in Bayern. München
Osnovnij projekt Velikog – kanala Dunav – Tisa – Dunav. Belgrade
Petkovic S, Varga S Smernice i principy projektovanja regulacije Dunava na sektoru Bezdan. Vukovar, Belgrade
Stančíková A et al (1999) Regulierung des Donaubettes. Eine hydrologische Monographie, Folgeband V., Teilbericht 3. Regionale Zusammenarbeit der Donauländer in Rahmen des IHP UNESCO. Bratislava
Szolgay J, Capeková Z, Holubová K (1991) Riešenie špecifických problémov úpravy, využitia a splavnosti Dunaja. Záver. správa VÚVH, Bratislava
Szolgay J, Náther B, Mišút O (1975) Štúdia hydrologicko – hydrotechnickej schémy pre biotechnický projekt úpravy medzihrádzového priestoru Dunaja od km 1842.0 po km 1810.0. Final report, VÚVH, Bratislava
Szolgay J, Stančíková A, Náther B, Csoma J, Laczay I (1978) Vyhodnoternie vplyvu uskutočnených úprav na vývoj koryta Dunaja v section Rajka – Gönyü. Final report, VÚVH and VITUKI, Bratislava–Budapešť
Szolgay J, Stančíková A, Náther B, Csoma J, Laczay I (1981) Zásady pre úpravu Dunaja v section Gönyü – Szob do obdobia prevádzky SVD G – N. Final report, VÚVH and VITUKI, Bratislava–Budapešť
Töry K (1954) A Duna és szabályozása. Budapešť
VITUKI (1976) Alapadatok és alapelvek a Duna Szob – déli határ közöti szakasza általános szabályozási tervéhez. Budapešť

Chapter 11
The Fords of the Danube

László Goda sr.

Abstract The paper deals with the fords of the riverbed of the navigable Danube reach between Regensburg and Sulina, on the basis of relevant data collected and published by the Danube Commission from 1961 to 1990. First the notion of ford and the related characteristics are defined. Then the sources of ford data and the methods of their processing are described. In the main section, a survey is offered of the ford conditions along the Danube stretch considered, including the morphological factors favouring ford genesis and the frequency of ford occurrence. The hydrological regime of the period investigated is analyzed. The distribution of ford-days both in time and along the river is evaluated. Finally, the measures taken by the Danube Countries and the River Authorities for maintaining the waterway are surveyed. An additional investigation is directed to the impacts of dredging of the riverbed, both for the purposes of river training and the construction industry.

Keywords Danube Catchment · Ford · Danube Commission · Waterway maintenance · Dredging · River training

11.1 Introduction

When investigating the anthropogenic efforts aiming at the improvement of riverbed conditions of the Danube (RZD 1999), various characteristic periods can be distinguished within the last two centuries. After the flood protection measures taken during the second half of the nineteenth century, mean water regulations were carried out and continued also in the twentieth century. During the first half of the latter, however, low-water regulation works were the most typical ones, aiming at the mitigation or termination of navigation obstacles. The latter measures were

L. Goda sr. (✉)
Research Institute for Environment and Water (VITUKI), Budapest H-1453, Hungary
e-mail: godalaszlo@gmail.hu

M. Brilly (ed.), *Hydrological Processes of the Danube River Basin*,
DOI 10.1007/978-90-481-3423-6_11, © Springer Science+Business Media B.V. 2010

extended on different shorter or longer stretches of the river, consisting at the beginning mostly of riverbed dredging, and later of river canalization and of erection barrages.

Any ford (or sandbank or shoal) constitutes an important characteristic for the effectiveness of the measures taken for low-water regulation as well as of the dredging carried out to this purpose. From the behaviour of individual fords and of whole river stretches riddled with them, from the number of ford-days and from the time development of ford depths one can conclude if the measures taken will have a long-lasting effect, or whether further alterations have still to be expected. One has also to find out if the various reaches of the river possess the capacity for building and maintaining fords, depending first on local conditions (slope, curvatures, sediment transport (RZD 1993), riverbed material, etc.). One of the consequences of all this is that along the various river reaches certain fords gradually sink in, others rise, others disappear or come into being, while the entirety of the parameters characterising the given river reach may even remain unchanged.

In the second half of the twentieth century, due to the resulting effects of a number of various impact factors, generally an increasing sinking tendency of the riverbed of the Danube could be observed. This tendency can partly be explained by the fact that as a consequence of the establishment of a series of river barrages, rapidly following each other on the German and Austrian reach of the Danube and some of its tributaries, sediment transport by the Danube was practically discontinued, as well as by the fact that at the same time formerly unthinkable quantities of riverbed material were dredged both for the purposes of river training and gravel exploitation.

The general deepening of the riverbed has caused also remarkable changes in the ford conditions of the Danube. As a consequence of low-water regulation and gravel exploitation, a number of fords have disappeared; while at the same time new fords have also arisen, especially along the stretches between subsequent dredged sites. At certain points, as a consequence of the silting up of new river training works, even old fords were revitalized, formerly considered nonexisting over several decades. As a consequence of the erection of barrages and the intensive dredging of the riverbed, unprecedented alterations of the low-water riverbed took place during the last 3–4 decades, whose repetition in the future is also improbable. This circumstance underlines the timeliness of thoroughly investigating the fords as typical formations of the low-water riverbed.

The arrangement and analysis of ford data result also in useful information for river training and flood protection. By quantitatively analyzing the ford conditions and comparing them with the river training measures taken along the given stretch, experts may recognize important relationships. Conclusions can be drawn regarding the efficiency of the so-far adopted methods of low-water training (including dredging and stone works) and eventually decisions may be taken to give priority to other methods. From the point of view of flood defence, however, it is a very important experience that on the Danube the period of ice cover often coincides with that of low discharges, when unfavourable ford conditions may considerably hamper ice transport, thus causing ice clogging.

The ford conditions of the Danube are continuously changing. The typical ford data of the various years are collected in the publications of the Danube Commission (Comission du Danube 1961–78a, 1979–90c). In the framework of the project (RZD 1996), whose main results are presented in this paper, a database was established, which satisfactorily and reliably characterizes the ford situation of the period 1961–1990, offering a reliable comparative basis also for future conditions.

The data presented in this paper do of course describe only the past conditions along the canalized river stretches. At the same time, one has to take into account that downstream of the canalized Danube stretch between Regensburg and Gabčíkovo (rkm 2,377–1,852) there is still an uncanalized, 628-km long reach, prone to ford generation, down until the upper limit of the impoundment of Iron Gate I. On the other hand, also downstream the barrage of Iron Gate II (rkm 943), no significant changes of the former ford conditions were experienced as yet. Thus, it is seen that along two-thirds of the total length of the Danube, the fords may still be significant obstacles to navigation.

11.2 Definition of the Notion "Ford"

The word *ford* (or sandbank or shoal) marks a river reach with a very limited water depth. A ford can be either permanent or changeable. The permanent fords are situated in the rocky breakthrough stretches of the rivers.

A changeable ford is a special element of the bottom orography of the riverbed. It comes into being by settling of the transported solid matter and consists of a hump of the river bottom, closing an acute angle with the stream line.

From the *genetic point of view*, the changeable ford can be considered as a solitary sandbank, developing itself, in a meandering river, along the transition reaches, between the subsequent curvatures, in a direction opposite to the flow. Also the rock banks situated in the riverbed are considered as fords, although being independent from the settling of solid matter.

From the *point of view of the navigation*, the notion "ford" is being interpreted in a significantly wider sense. Practically each shorter or longer river reach is considered as a ford, along which the water depth is smaller than desired navigation depth within a certain width. Such a condition may arise everywhere if the stream velocity drops. Such a dropping may be due to

- slope reduction
- widening of the riverbed
- in- and/or outlet of side arms, tributaries and canals
- sedimentation of solid matter input by tributaries
- natural or anthropogenic impoundment
- objects situated in the riverbed, increasing its roughness (ship- and bridge-wrecks, etc.)

A ford above which passes the line of maximum depths without any significant interruption is called a *good ford*, another, above which this line is interrupted several times, is a *bad ford*.

The notions adopted in the present investigation are defined as follows:

- *Ford depth*: water depth above the highest point of the river bottom.
- *Navigation* or *regulation water stage*: a water stage determined for a given cross section of the river, at which the minimum water depth within the prescribed width is secured.
- *Navigation depth deficit*: the difference between the prescribed minimum water depth and the ford depth.
- *Duration of the ford period*: total number of days within a given calendar unit (month, year) with ford depths lower than the minimum navigation depth.
- *Ford frequency*: number of fords existing along a unit of river length (100, 1,000 km).
- *Total number of ford-days*: annual sum of days, during which the ford depth does not reach the minimum navigation depth on one or more fords of the given river reach. This parameter is higher than the longest duration of the various fords of the given river reach.
- *Top ford* or *limiting ford*: that ford which along a given river reach hinders navigation during one or more days, showing the smallest ford depth of that river reach.

11.3 The Data on the Fords

The data of the fords of the navigable reach of the Danube are collected in the publications of the Comission du Danube (1961–78a), Comission du Danube (1961–78b), Comission du Danube (1979–90c). The observations of ford parameters include, according to the recommendations for hydrometeorologic observations and measurements (Comission du Danube 1970d), all periods without drifting ice and ice cover, during which the water depth above the given ford is smaller than the prescribed *navigation water depth*. The latter is, on the reaches without impoundment, 20-dm upstream of Vienna and 25 dm between Vienna and Brăila (Comission du Danube 1966, Comission du Danube 1988).

The information gathered on fords include the following data: location of ford (rkm), length of ford, width of waterway along that length, daily data of water depth measured above the ford, daily water stages read on the next gauge to the ford, and eventually also the daily discharges in that gauging cross section. In the first third of our investigation period (1961–1970), water depths were mostly measured by sounding. Since the early 1980s, ultrasound and other acoustic depth measurements have generally been adopted.

Since the interpretation of ford data would not have been suitable without taking into account the dredging data related to the given ford or to the given river reach, also the relevant dredging data published in the volumes (Comission du Danube 1961–78b) and (Comission du Danube 1979–90c) had to be utilized.

In the annual reports of the Danube Commission, the data of 250 fords and gorges of the Danube reach between Regensburg and Sulina were published for the period 1961–1990.

Complete, systematic ford observations were carried out down to the location Brăila (rkm 170). Along the Danube reach under maritime impact, between Brăila and Sulina, the waterway is being maintained by the "River Direction of the Lower Danube".

From these available data, those of 236 fords were utilized in the present work. The data of gorges without depth information were not taken into account, because they are not suitable for composing time series. Further, also the data of those fords were neglected, where the duration of fords did not surpass altogether 1–2 days during the investigation period of 30 years.

As the final result of data arrangement and processing, the following information was made available for each ford:

- serial number of the ford (starting with No. 1 at Regensburg and ending with No. 236 at Sulina);
- name of the ford;
- location of the ford (with up- and downstream limit rkm, taken from annual reports);
- month and year;
- number of ford-months and ford-years during the period 1961–1990;
- total sum of ford-days;
- a so-called prominence characteristic of the given ford, i.e., the number of months during the investigated 30 years, in which the given ford used to be the top ford of the given river reach or of the whole Danube;
- the total quantities of dredged material during 30 years.

11.4 General Description of the Ford Conditions of the Danube

11.4.1 Morphological Preconditions for the Genesis of Fords

On every non-impounded stretch of the Danube there are a certain number of fords, above which the water depth would strongly decrease during the low discharge periods. This may lead to limitations or even temporary suspension of navigation. In order to improve the conditions of the top fords which basically determine the navigation conditions on the different stretches of the river, various ambitious works of river training and dredging are regularly carried out. Experience shows, however, that a final solution of this problem can only be expected from river regulation by impoundments, i.e. by river canalization.

11.4.1.1 The Upper Danube

The Danube stretch between Regensburg and Devín (rkm 2,379–1,880) is a river of primarily alpine character. This statement holds both for the type of river valley and for the structure of the riverbed and the runoff regime. The river valley is generally narrow and bordered by steep slopes, while the narrow stretches are occasionally interrupted by smaller basins. Along the cataract stretches, there are often rocky banks, so-called "Kachlets" under the water. On the somewhat wider bottoms of the basins, situated between narrower stretches, the river is meandering, while the width of its bed increases, thus favouring the development of fords. The major tributaries, hardly less significant than the Danube itself, are transporting large sediment masses, thus considerably contributing to the genesis of mouth-fords (e.g., the estuaries of the Isar (Fig. 11.1), the Alte Ohe, the Schwechat, etc.). Until 1990, the number of fords significantly decreased due to the fact that by today about 50% of the considered stretch of the Upper Danube has been impounded.

Favourable preconditions for ford genesis can also be recognized on the number of fords observed between 1961 and 1990. On the Danube stretch in question, a total number of 84 fords have been registered (Fig. 11.2). The number of fords per 100 km was particularly high on the reach between rkm 2,300 and 2,200 (Fig. 11.3), where the average distance between subsequent fords used to be 2–3 km. An extraordinarily high ford frequency was observed also on the stretch rkm 2,360–2,330 (15 fords). A similar frequency has characterized also the 20-km long stretch just downstream the Isar mouth (rkm 2,281), a pure consequence of the sediment supply arriving with the Isar itself. Along the further part of the Upper Danube, the inclination to ford building slightly decreases resulting in a frequency below 15 fords per 100 km.

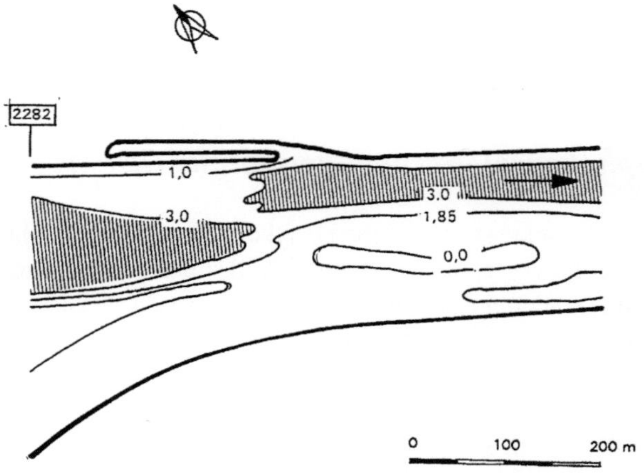

Fig. 11.1 Layout of the ford near to the mouth of the Isar River (rkm 2,280)

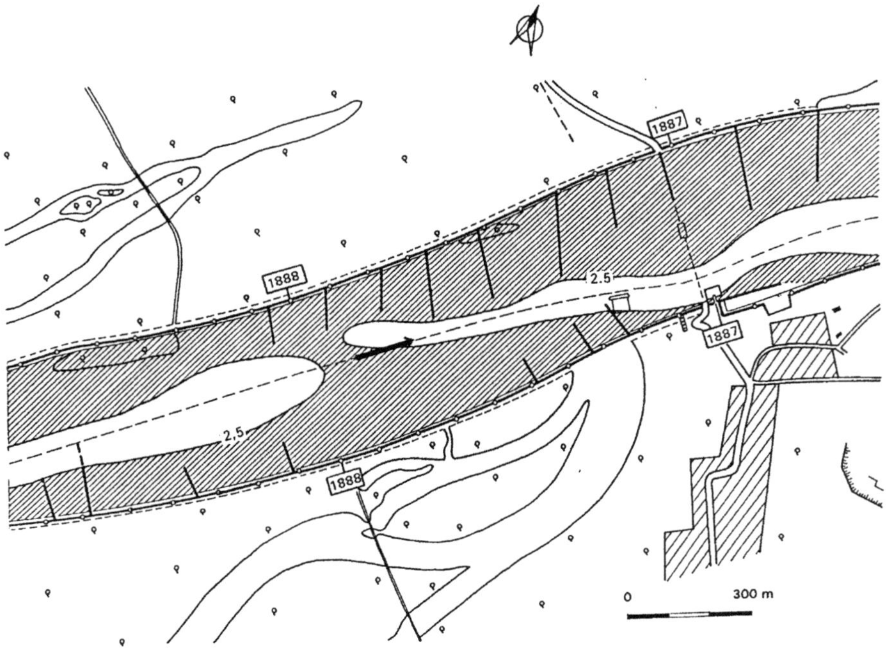

Fig. 11.2 Layout of the ford Deutsch-Altenburg (rkm 1954)

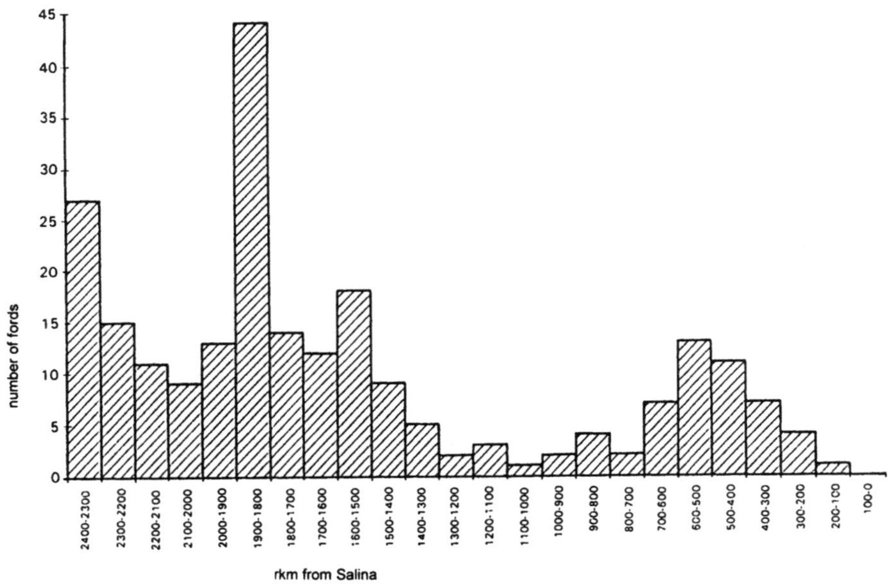

Fig. 11.3 Frequency of ford occurrence on the Danube stretch between Regensburg and Sulina between 1961 and 1990

11.4.1.2 The Central Danube

The relatively short Danube reach between Devín and Gönyű (rkm 1,880–1,791) is characterized by extraordinary morphological features, creating particularly difficult problems both for river training and navigation. Here the Danube is flowing through a very large talus fan – which might also be considered an "inland delta" – built by the river itself, just upstream the sudden slope reduction at the Gönyű section (from 35–40 to 8–10 cm km^{-1}). Before the establishment of barrages along the German and Austrian Danube, the estimated solid matter transport in this cross section used to be 400–600,000 t year^{-1} (RZD 1986).

The river, meandering through the talus, produced over the centuries innumerable side branches and islands, thus providing most favourable preconditions for the genesis of a great number of permanent fords. The out- and inlet of each side branch as well as each island provides the possibility of ford creation (Fig. 11.4). Along this stretch, with its 34 fords, one can observe the highest ford frequency of the whole Danube (Fig. 11.3). A particularly strong ford genesis took place on the reach rkm 1,810–1,880, where alone eight fords were registered.

Between Gönyű und Turnu Severin (rkm 1,791–931), the Danube can be described as a "lowland river", as regards both the characteristics of its valley and its runoff regime. The exceptions are only those short stretches, where the Danube breaks through mountain ridges, showing an alpine character once again. Such breakthrough stretches are the gorge of Visegrád as well as the Danube stretches up-

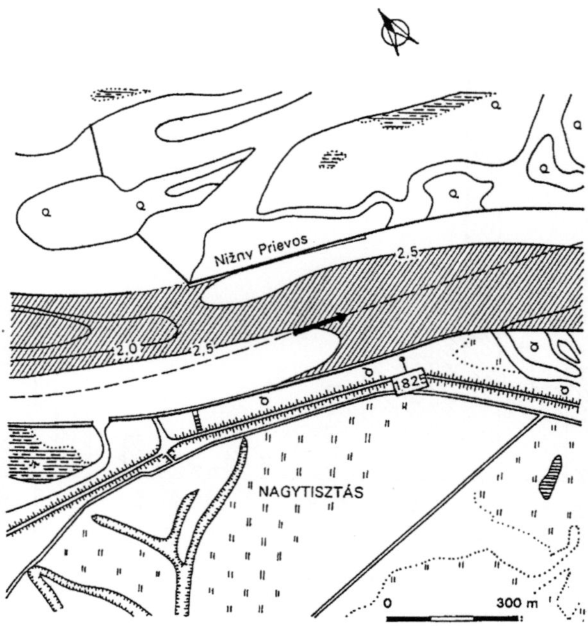

Fig. 11.4 Layout of the ford Dunaremete (rkm 1,825)

and downstream the Iron Gate. The slope of the water surface is 8–10 cm km^{-1} at Gönyű, gradually decreasing until the cross section of Moldova-Veche (rkm 1,048) down to 5 cm km^{-1}.

On the lowland stretches of the river, its wide valley is accompanied by alluvial terraces, criss-crossed by side branches. In the breakthrough stretches, the valley becomes as narrow as 0.6–2.5 km, the slopes of the side walls are steep and the riverbed is rocky.

A great part of the Central Danube is meandering, but both the length of the straight reaches and the radius of the curvatures are here generally quite a lot greater than along the Upper Danube.

As for the ford genesis, distinction has to be made between the lowland stretches and the breakthrough reaches. On the flatland stretches, the driving force of ford genesis is first of all the inclination to generate side branches, as particularly well displayed on the Danube reach between Paks and Mohács (rkm 1,532–1,477). In the surroundings of Apatin (rkm 1,401), navigation is heavily hampered by the fords situated along the inflexion stretches between the narrow curvatures. Along the lowest part of the Central Danube, it is the large sediment masses supplied by the great tributaries (Drava – rkm 1,384, Tisa – rkm 1,215, Sava – rkm 1,171 and Velika Morava – rkm 1,103) which create particularly favourable preconditions for ford genesis.

As for the two breakthrough stretches, in one of them, the "Iron Gate", navigability has been well secured both by the regulatory works carried out in the first half of the nineteenth century and by the erection of the two barrages in 1972. In the surroundings of the Visegrád Pass, however, the rocky fords are increasingly bulging, due to gradual deepening of the neighbouring river stretches.

During the period 1961–1990, altogether 54 fords were registered along the Central Danube. Their majority (39 fords) were observed along the Danube stretch between the section Szob and the mouth of the tributary Drava, with an average frequency of 9–17 fords/100 km. Downstream the mouth of the Drava, their frequency was below five.

11.4.1.3 Lower Danube

The Lower Danube is a lowland river with wide wash-lands. The inclination to ford genesis is decreasing, the radius of the curvatures increases considerably and there are even rather long straight stretches. The slope of the water surface keeps decreasing gradually. At Turnu Severin, it is 5 cm km^{-1}, and finally, at the mouth into the Black Sea, only 1 cm km^{-1}.

A typical feature of the Lower Danube is its inclination towards generating river islands (Fig. 11.5). Between Turnu Severin and Brăila (rkm 931–170) there are altogether 194 such islands, enclosed by 59 river branches whose total length amounts to about 66% of the length of 761 km of the Danube reach in question. Along the remaining reaches of the Lower Danube, there are no islands.

The genesis, the development and the disappearance of river islands is a manifestation of the laws of stream processes. A hump, consisting of sand and gravel, is moved in the stream direction, occasionally taking a rest, during low discharge

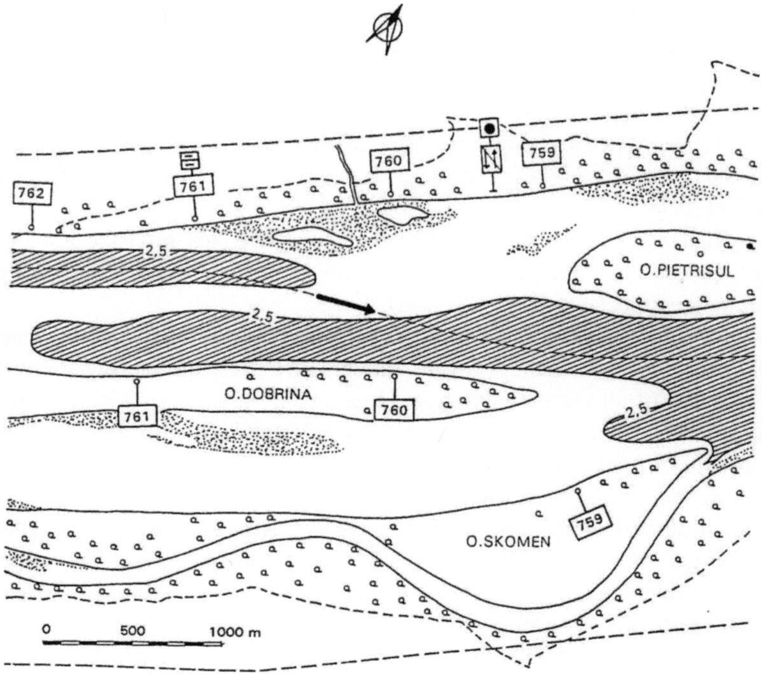

Fig. 11.5 Layout of the surroundings of Dobrina Island (rkm 761)

periods. During low-water stages, the hump towers slightly above the water, and is stabilized by vegetation thus forming an island. Along the river reaches dispersed with islands, the slope of the water surface is 3–7 cm km^{-1}: at the lower limit value the sediment transport of the Danube is suspended and at the upper limit value even the formerly deposited sediment will be swept away.

The widths of the riverbed vary between 1,000 and 1,500 m. In the lower domain, the conditions for island genesis are not particularly favourable.

During major floods, the islands are inundated in such a way that the settled sediment causes a further increase of the island's height. According to available measurements, the major islands have risen up by 0.6–3.3 m between 1908 and 1966.

Along the Lower Danube, so far practically no river training measures have been undertaken, save for occasional riverbank protections and dredging for the sake of navigation.

At the upper and lower ends of the islands, as well as in the secondary ramifications along them, stream velocity decreases, leading to the genesis of new fords and sandbanks. As a consequence, the main river branch will often be transferred, while the water depth in the former waterway decreases rapidly, occasionally as much as below 10 dm. In such cases, of course, also the trace of the waterway has to be transferred (Fig. 11.6).

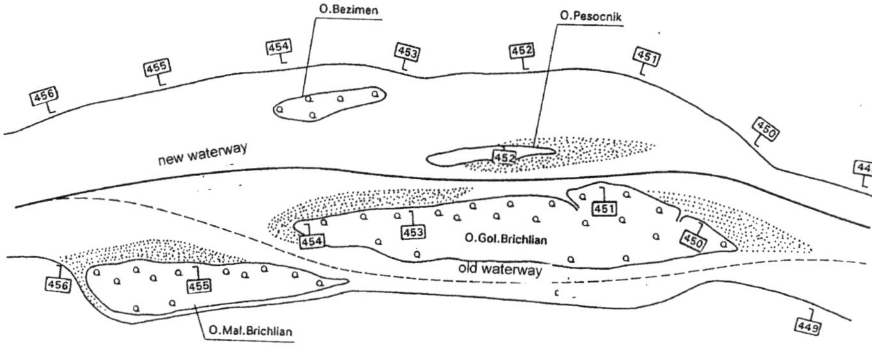

Fig. 11.6 Transfer of the waterway along the Danube stretch rkm 456–449

As one group of objects causing fords, also the wrecks of ships stranded or destroyed during the Second World War have to be mentioned. Between 1961 and 1975, more than 100 wrecks were hoisted from the Lower Danube, indicating the gravity of this phenomenon. A particularly great number of wrecks were found in the surroundings of the back-inlet of the side branch Gogos (rkm 863–859), as documented also in the navigation map published in 1967 (Fig. 11.7).

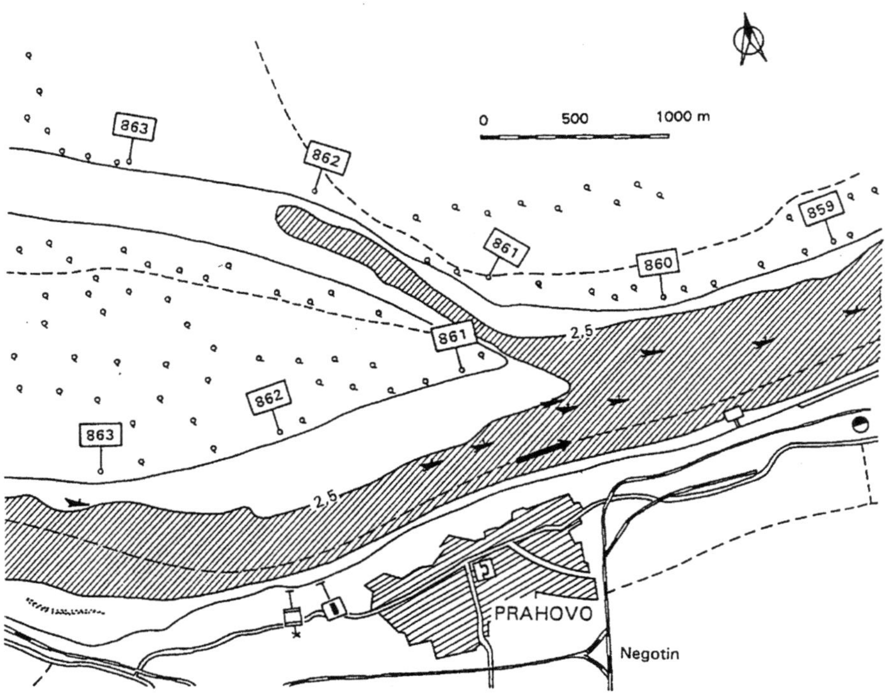

Fig. 11.7 Submerged ships on the Danube stretch from rkm 863 to 859

The number of fords on the Lower Danube was 51. Their frequency referred to 100 km was particularly high along the reach rkm 600–400 (Fig. 11.3). Here the fords appear often in pairs (rkms 859 and 858, 761 and 759, 632 and 631, 568 and 564, 538 and 537, 505 and 504, 345 and 344).

11.4.2 Impact of the Runoff Regime on the Fords

Any ford is the result of the mutual effects between the riverbed and the water and sediment masses moving therein, thus it has a close relationship also with the runoff regime. In this connection, the following two questions may be of particular importance:

- Impact of the actual runoff regime of a given ford
- Impacts of the annual and multi-annual runoff regime on the entirety of the fords of a particular river reach.

The first of the above questions cannot be answered satisfactorily on the basis of the data set available, since for the majority of the fords there are no continuous observations, but, at the best, only 2–3 months long series of measurements, limited to the low-water periods. On the basis of these series, the following conclusions can be drawn, without taking into account the statements of the references (Stănescu 1967, Nicolae 1970):

- In the case of (fix) fords in rocks, the actual water depth above the ford depends only on the water stage
- In the case of fords consisting of sand and gravel, two options are possible, depending on the local factors (width of riverbed, slope of water surface, outlet of side branches, etc.):
- whenever the water stage (and with it, the area of stream cross section) decreases, the local resistance against stream increases, thus the ford will be eroded, the decreasing of water depth above the ford stops, or it may even increase during a long lasting low-water period;
- whenever flood is decreasing, so does the stream velocity, thus the transported solid material will be settled, particularly in wide riverbeds. As a consequence, the ford keeps developing. The ford depth decreases, while the main stream abandons, due to the increased resistance, the ford, transferring itself to the opposite side of the riverbed or the island.

11.4.3 Characteristic Parameters of the Ford Conditions

In order to characterize the ford conditions of the whole Danube and its various partial reaches, the following indices were calculated by utilizing the data set available:

- Number of fords and frequency of their occurrence
- Time duration of ford periods and their distribution within the year
- Total number of ford-days, i.e., the number of days, on which navigation was hindered at least by one ford along the Danube reach in question (this value is in general higher than the duration of any of the individual fords)
- The minimum ford depth

The annual number of fords of the Danube between Regensburg and Sulina reached during the period 1961–1990 its maximum in 1973, when 87 fords were registered, and its minimum in 1981, with 13 fords (Fig. 11.8). During this period, there was no year without any fords. The annual mean number was ca. 50 fords, both for the whole investigation period and its first and second half. The frequency of ford occurrence along the Danube is displayed in Fig. 11.9. It can be seen that more than half of the 236 fords registered, namely 139 fords, occurred only 1–5 times during the 30 years. The most frequently – in 27 out of 30 years – registered ford was that of Medvedov (rkm 1,806).

The time series of the total number of ford-days corresponds more or less to the course of ford-days (Fig. 11.11). The maximum of this index, 281 days, was registered in 1971, its minimum, 31 days, in the year 1981. The multi-annual average of the annual duration of ford-days was 145 days.

The occurrence of ford periods is to be expected, in accordance with the runoff regime, mostly during the autumn and winter months. The data referring to the whole Danube (Fig. 11.10) show that ca. one quarter of the annual ford-days is to be expected in late winter (January–March). After this period, until the end of July, ford-days occur most infrequently. In August and September, the low-water character of the runoff regime increases, as a consequence of which one quarter of the annual ford-days fall onto October. The ford periods of November and December are

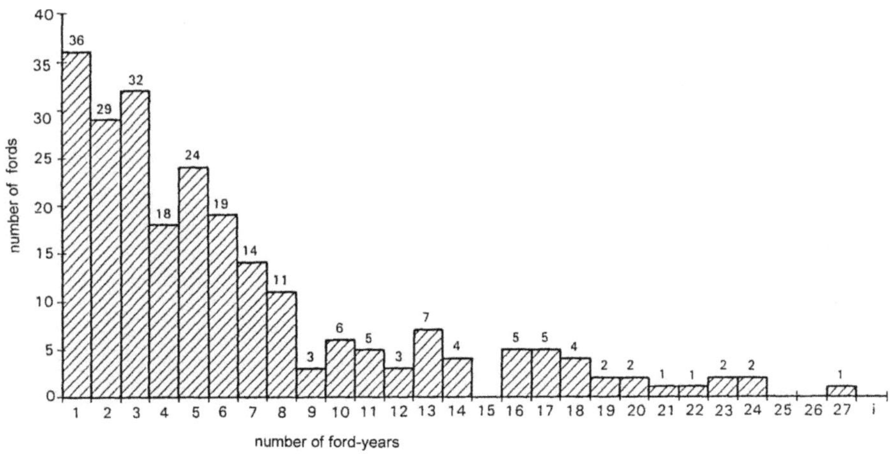

Fig. 11.8 Characteristics of ford conditions of the Danube stretch between Regensburg and Sulina during 1961–1990

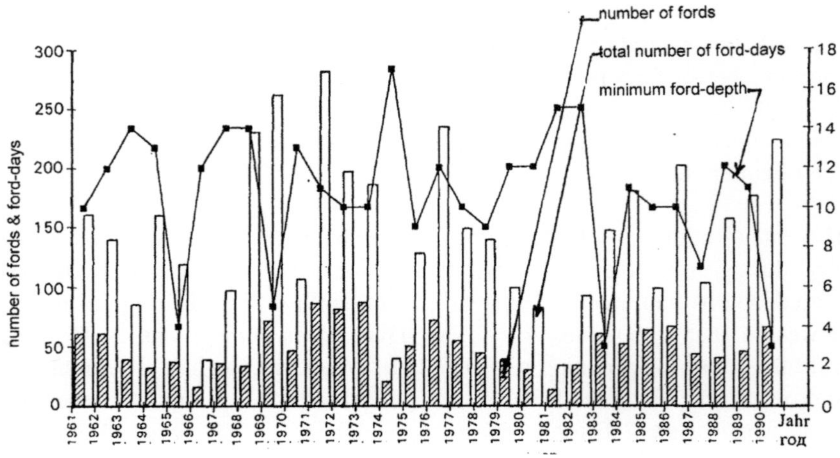

Fig. 11.9 Number of fords that occurred in years out of 30 years on the Danube between Regensburg and Sulina

still significant, although with a diminishing tendency, within the annual distribution of ford-days.

In comparison with the statements made above, referring to the whole navigable reach of the Danube, typical differences do characterize the various partial reaches of the river. Along the German reach, the distribution within the year is somewhat smoother. Here the share of January (18%) even surpasses those of the fall months. On the Austrian reach, the largest number of fords are registered in October.

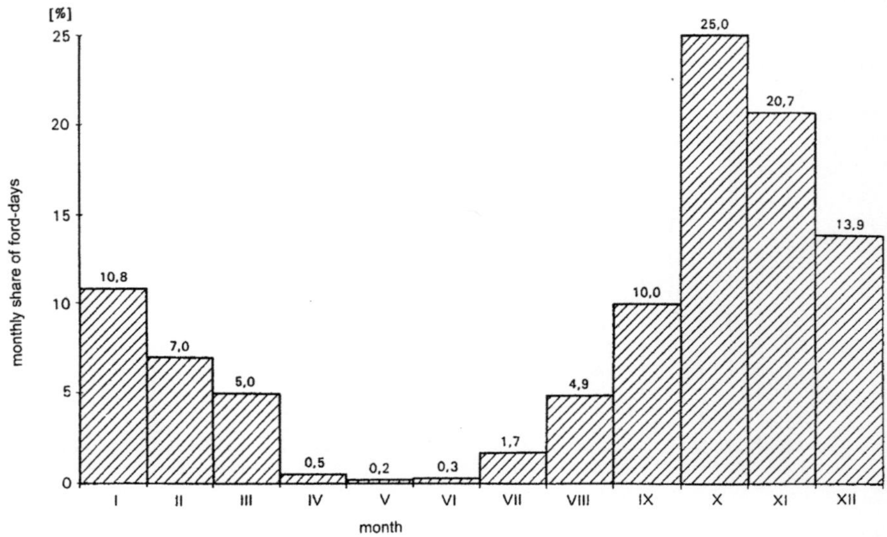

Fig. 11.10 Annual numbers of ford-days along the Danube reach between Regensburg and Sulina from 1961 to 1990

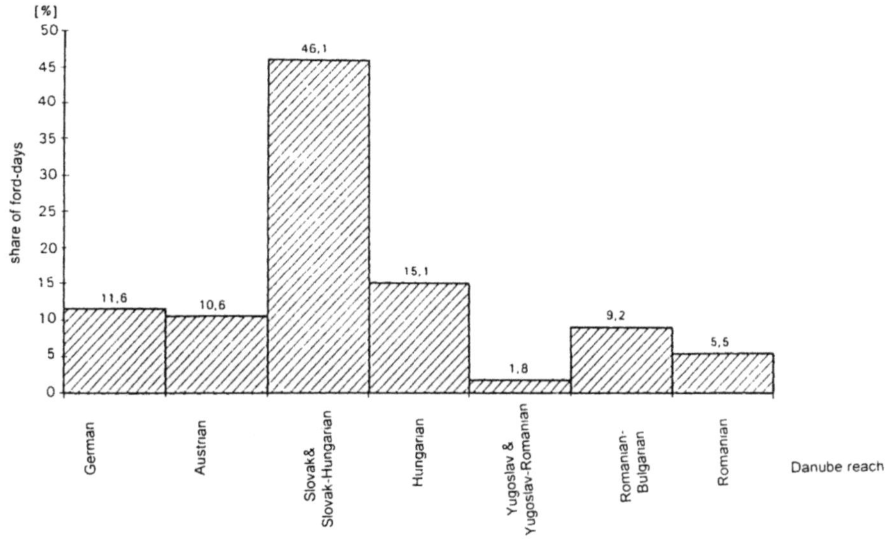

Fig. 11.11 Breakdown of ford-days by Danube stretches in 1961–1990

Along the Slovak and the joint Slovak–Hungarian Danube reach the share in the fall months increases further. In Hungary, the greatest number of ford-days is registered in November. Downstream, along the Yugoslavian, the Yugoslavian–Romanian and the Romanian–Bulgarian Danube reaches, the share of the late summer months increases, while that of January and February decreases. On the 236 fords of the Danube, a total number of 80,000 ford-days were registered. Almost half of them (46%) occurred on the fords of the Slovak and the joint Slovak–Hungarian Danube reach (Fig. 11.11). Relatively great numbers of ford-days were registered also on the Hungarian, German, Austrian and the joint Romanian–Bulgarian Danube reach.

The minimum ford depth values of the various years were mostly in the domain 8–14 dm or more, and only very seldom under 8 dm (down to 3 dm) (Fig. 11.8).

11.4.4 Changes in Ford Conditions Between 1961 and 1990

Ford conditions may considerably change from year to year. As already mentioned, during the 30 years of investigation, there wasn't a single year without some fords registered. Particularly great numbers of fords of long duration occurred in the years 1969, 1971, 1976, 1989 and 1990, in which at least one ford was observed in each month of the year. In these years, the number of fords was between 60 and 90, and the annual total number of ford-days between 220 and 290 (Fig. 11.8). The lowest numbers of fords were registered in the years 1966 and 1981, in each of which navigation was unhampered over as much as 8–9 months of the year.

As already mentioned, altogether 236 fords were registered between 1961 and 1990: in the first 15 years 180, and in the second 15 years 153 fords. There were

97 fords existing in both halves of the observation period. The change of ford conditions between 1961 and 1990 may be characterized with the following indices:

- Number of fords
- Total sum of ford-days
- Minimum ford depth

When considering the whole navigable reach of the Danube, the following average values can be obtained:

Index	1961–1975	1976–1990
Number of fords	50.4	48.5
Total number of ford-days	148.6	141.0
Minimum ford depth (dm)	11.2	10.1

One can see that both the number of fords and the total number of ford-days were lower by 4–5% in the second half of the investigation period than in the first one. This seems to indicate a favourable change. The minimum ford depth, however, is smaller by 1 dm in the second half, indicating an unfavourable tendency.

When considering, on the other hand, the changes in ford characteristics, instead of for the whole Danube, for its various partial reaches, one can see more marked differences (Table 11.1).

As for the number of fords, considerable decreases can be registered along the German, Austrian and Hungarian reaches, practically no changes along the Yugoslav and the joint Yugoslav–Romanian reaches, a slight increase on the Slovak and the joint Slovak–Hungarian reaches, and very strong increases along the joint Romanian–Bulgarian and the Romanian reaches.

The change in the index "total sum of ford-days" corresponds – save for a few exceptions – to that of the number of ford-days. On the Romanian Danube reach, the value of this index has trebled in the second half of the investigation period.

The average value of the annual minima of ford depth has favourably increased on the German, Austrian and Hungarian reaches. It has increased by almost 7 dm on the Romanian reach, and remained practically unchanged along the rest of the Danube.

Table 11.1 informs also about the impact of fords on navigation. As for the time duration of ford periods, their averages were rather unfavourable during the first half of the investigation period along the Upper and Central Danube, down to the Hungarian/Yugoslavian border, and in the second half also along the Romanian reach. As for the minimum ford depths, they were critical, during the first half period, on the German and Austrian reaches, while between 1976 and 1990 the most critical values were measured along the Romanian reach of the Danube.

Finally, also the impact of the fords of the various Danube reaches on the limitation of navigation on the whole Danube was investigated. It was shown that in 30% of cases the fords of the Slovak and Slovak–Hungarian Danube reach were critical,

Table 11.1 Mean ford characteristics of the various national stretches of the Danube

Parameter	Observation period	German Danube reach	Austrian Danube reach	Slovak & Slovak–Hungarian Danube reach	Hungarian Danube reach	Yugoslav & Yugosl.–Romanian Danube reach	Bulgarian–Romanian Danube reach	Romanian Danube reach
Average number of ford-days	1961–1975	14.0	8.2	15.3	9.5	2.5	5.4	2.8
	1976–1990	9.0	4.0	17.5	6.5	2.7	8.0	5.7
Total number of ford-days	1961–1975	83.4	116.8	115.6	100.5	53.3	54.7	23.9
	1976–1990	58.5	79.8	134.4	101.8	35.2	58.9	74.7
Average of annual minimum ford-depths (dm)	1961–1976	13.8	12.0	14.8	15.3	17.2	18.3	18.9
	1976–1990	15.1	14.6	13.9	17.2	17.3	17.2	12.1

while for Austria this value was 21%, for Germany 17%, for Romania 15% and for the Hungarian Danube reach 11%. The shares of the remaining national stretches were negligible.

11.5 Changes of the Riverbed of the Danube

The riverbed of the Danube is permanently changing. Whenever the drag force decreases, the solid material transported by the stream and its tributaries will be deposited, to be dragged again by the next flood wave. The river both erodes and builds its embankments. River training works generally achieve only local effects. They occasionally prevent local scouring, and in other cases increase the deposition of sediments. Dredging may also be expedient for certain purposes of river training, but its imprudent adoption may also lead to undesirable changes of the riverbed. River barrages modify basically and along longer reaches the processes of riverbed development.

As long as the sediment balance of a river reach is in equilibrium, i.e., the quantity of input and output sediment masses is approximately the same, no significant changes of the riverbed of the stretch in question are to be expected. When, however, this equilibrium tips over, changes of the riverbed begin immediately. These changes may result, depending on local factors, either in accretion or in scouring.

The sediment balance of the Danube underwent during the past 30 years– both in relation to in- and outputs – considerable changes.

The most important change on the input side was caused by the erection of a number of barrages on the German and Austrian Danube reaches and their respective tributaries.

The first follow-up volume to the Danube Monograph (RZD 1986) shows that the sediment transport of the Danube underwent considerable changes during the last decades. On the basis of the data collected between 1956 and 1985, one can see that while the annual sediment quantity supplied by the Danube into the Black Sea used to be 15–70 million t year^{-1} before the building of barrages, it decreased to 8–25 million t year^{-1} by the years following 1980. The major part of the thus missing quantity settled down partly in the impoundment spaces of barrages, and partly along the river reaches deepened by dredging. The average sediment transport of the different Danube reaches was estimated for the period 1956–1985 as follows (RZD 1986):

Danube reach	10^6 m^3 year^{-1}
Ingolstadt–Linz	0.35
Linz–Bezdan	1.8
Bezdan–Bogojevo	3.6
Bogojevo–Pančevo	7.2
Pančevo–Novo Selo	10.0
Novo Selo–Black Sea	22.0

Sediment transport was further modified by closing down of the riverbed at Čunovo. The determination of the actual sediment conditions can be carried out only after a certain time.

Before the establishment of barrages, the largest sediment quantities used to be produced along the German and Austrian reach of the Danube as well as in the surroundings of the Iron Gate. The Upper Danube had produced about 10–15% of the total sediment quantity of the whole river. The sediment mass arriving just downstream Bratislava was estimated to be 0.4–0.6 million m^3 $year^{-1}$ before the establishment of the upstream barrages. Presently the quantity reaching the close-down at Čunovo is still 0.2 million m^3 $year^{-1}$, but diminishes rapidly downstream dropping to 0.006–0.01 million m^3 $year^{-1}$ by rkm 1,800. Along the Danube reach downstream Orşova the order of magnitude of the sediment volume can hardly even be estimated.

After having considered the highly unreliable and uncertain input side of the sediment balance, let us have a look at its output side.

The quantity of sediment loss due to riverbed erosion is practically unknown. One can assume, however, that this quantity is not really significant, since the discharge rating curves of most gauging stations of the Danube hardly had changed before the start of intensive dredging of the riverbed.

The most significant component of the output side of the sediment balance was due to riverbed dredging during the period 1961–1990. The respective data published by the Danube Commission (Commission du Danube – b, Commission du Danube – c) do generally support this statement, while the exact resulting impacts of this dredging can be estimated only after several decades.

Along the Danube reach between Regensburg and Sulina, a volume of riverbed material as much as 420 million m^3 was dredged from 1961 to 1990, 350 million m^3 of which falls on the riverbed itself, including 83 million m^3 dredging for river training purposes. The remaining 70 million m^3 consist of harbour and other navigation dredging. In the river training works 11.3 million m^3 rubble was built in. Dividing these volumes by the length of the Danube reach considered (2,400 km), the specific sediment loss of 175 m^3 $myear^{-1}$ is obtained, which is of course the average of locally fluctuating values.

On the German Danube reach, the dredged volume used to be 0.6–0.7 million m^3 $year^{-1}$ around 1960, but this dropped gradually to 0.03 million m^3 $year^{-1}$ by 1990. A similar, but more moderate decrease was typical also on the Austrian reach. The annual data from Slovakia and the River Direction Rajka–Gönyű fluctuate in a wide domain, without any definite tendency. On the Hungarian Danube reach, an intensive dredging activity was started after 1970 for purposes of the construction industry, with strongly varying intensities among the various partial reaches (Fig. 11.12). Particularly conspicuous is the mass of more than 6 million m^3 dredged between rkm 1,720 and 1,710, causing a deepening of the river bottom by 1 m on a width of 600 m. From the 66 million m^3 dredged during 30 years from the Hungarian Danube reach, about 40 million m^3 served exclusively the construction industry (Laczay 1987, 1988, 1989).

The volume of gravel dredging, reported from Yugoslavia, concentrated onto a rather short Danube reach, surpasses all data described above (Fig. 11.13). Between

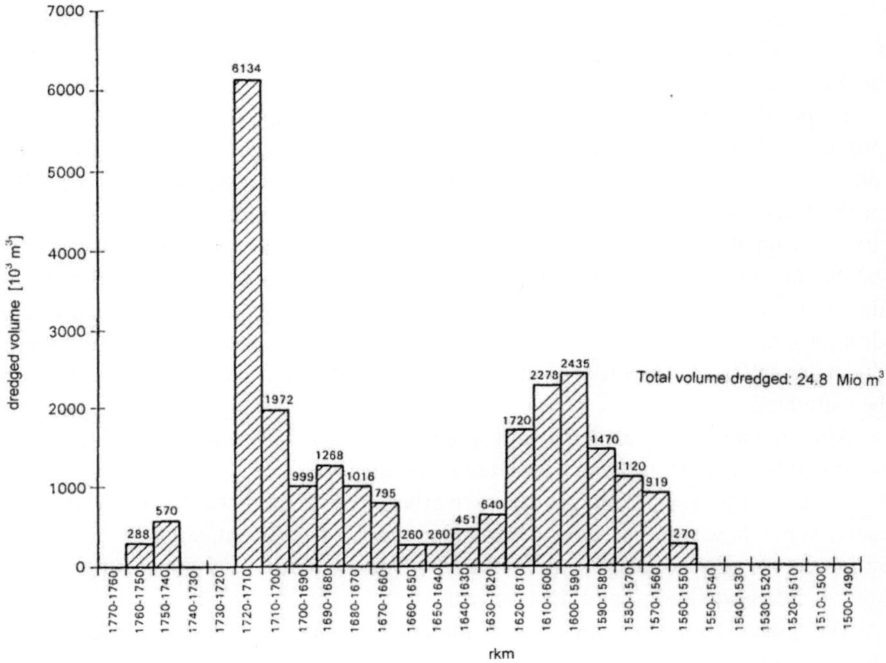

Fig. 11.12 Intensive dredging works on the Hungarian stretch of the Danube between 1975 and 1990

1968 and 1990, a sediment mass of 140 million m^3 was dredged, mostly from the impoundment space of the hydropower station Iron Gate I. Ninety percent of this mass was used by the construction industry.

Along the Danube reach downstream Orşova, the share of harbour dredging is gradually increasing, corresponding to the sediment transport conditions of the river, and equals locally even that of riverbed dredging. The "River Direction of the Lower Danube" reports about 1/6 of the total mass dredged from the whole Danube.

Summarizing the information communicated above, the overall breakdown of dredged material from the Danube may be the following:

Total volume 420 million m^3:
Including:
riverbed dredging 350 million m^3
harbour dredging: 70 million m^3
From riverbed dredging:
embouchure dredging 58 million m^3
for construction industry 176 million m^3
for river training and ford improvement 116 million m^3

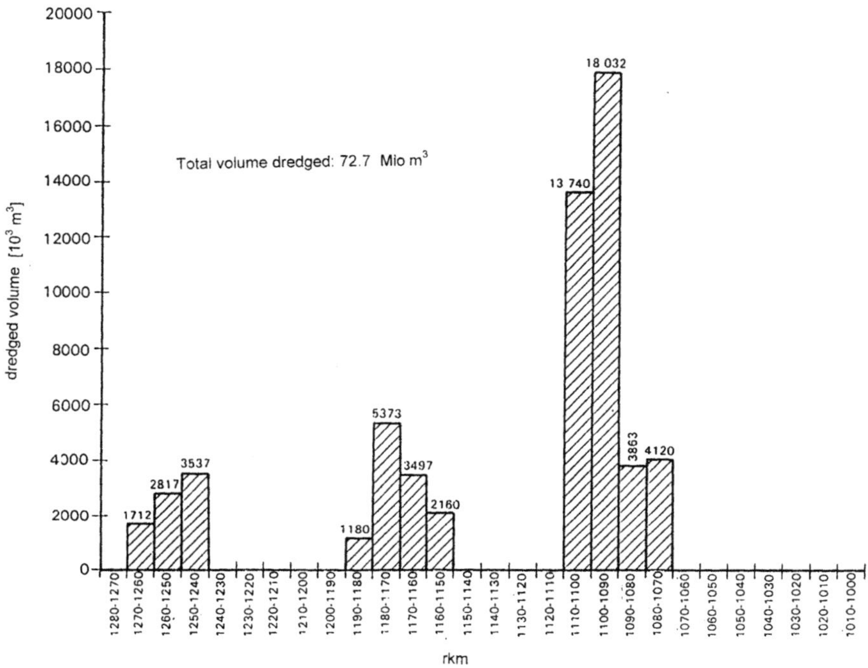

Fig. 11.13 Intensive dredging works on the Yugoslav stretch of the Danube between 1975 and 1990

11.6 Summary and Conclusions

Fords are important elements of the riverbed. Their formation, their changes and their cessation are the consequences of changes that have taken place in the discharge and sediment regime of the river due to natural and anthropogenic causes.

The present investigation contains an overlook on the fords of the Danube that was made possible by a database compiled from various publications of the Danube Commission. For the time being, such a processing of ford data (water depth, time duration) of a longer period (in this work for 1961–1990), is unique not only in the case of the Danube but also of other rivers of similar magnitude.

Along the Danube stretch between Regensburg (rkm 2,376) and the Danube Delta downstream Brăila (rkm 170) fords in non-uniform distribution in space were observed during the period investigated. The types of critical factors responsible for the genesis of fords vary along the Danube. On its German and Austrian stretch, submerged rock thresholds and the alluvial fans of tributaries were the most important factors. Along the Slovak and the joint Slovak–Hungarian stretch, the forks and entries of side-arm systems upstream the slope break of Gönyű had a major impact (44 fords only along the Danube stretch between rkm 1,900 and 1,800). On the lowland stretches of the Central Danube, the widening and curvatures of the riverbed acted as critical factors. Finally, along the Lower Danube, the great number of riverbed islands played a decisive role in ford formation.

The lowest water depths and the longest ford periods are observed, in accordance with the runoff regime, in the autumn and winter months. At many fords, the number of 200 ford-days per year was reached or surpassed, while the minimum water depth was less than 10 dm.

Ford conditions of the various stretches of the Danube have been changing in different ways between 1961 and 1990. On the German and Austrian stretch, the number of fords and their average time duration decreased by 30–50% due to the construction of river barrages. At the same time, the minimum water depths increased here by 1.5–2.5 dm. On the Slovak and the joint Slovak–Hungarian Danube stretch, the ford parameters did not noticeably improve in spite of large-scale measures of river training and dredging. Along the Lower Danube, rich in islands, ford characteristics grew perceptibly worse, partly due to a series of dry years (Gergov and Nenov 1979). Both the number and their time duration redoubled and even the minimum water depth decreased by 7 dm.

When establishing the impacts of fords on navigation, the integrated number of ford-days and the characteristics of peak fords have also to be taken into account. With regard to both characteristics mentioned, the less favourable was the Slovak and the joint Slovak–Hungarian Danube stretch, where the annual integrated sum of ford-days was as high as 125, while two thirds of these fords were critical at least once for the whole Danube.

The amount of suspended load, very important for ford formation, decreased between 1956 and 1983 by 65–75%. Due to the barrages built along the Upper Danube as well as to the closing of the riverbed at Čunovo, sediment no longer reaches the Danube stretch downstream Devín. The additional energy due to the sediment transport now missing was favourable for riverbed erosion. The main cause of riverbed deepening between 1961 and 1990 is however dredging. As a consequence, the Danube deepened considerably downstream Orşova, this deepening reaching locally the measure of 1 m. These deepenings are also expressed in the changes of the values of the so-called "navigation and regulation low-water stages", prescribed by the Danube Commission.

This knowledge about the ford conditions of the Danube enables the following conclusions (Goda 1971):

- During the second half of the twentieth century, very considerable changes took place in the riverbed of the Danube (due to the construction of barrages and the dredging of more than 400 million m^3 from the riverbed), whose magnitude reaches or even surpasses that of the river training activities carried out in the nineteenth century.
- Along the river stretches, inclined for ford formation, only the most unfavourable conditions of individual fords can be changed by adopting the traditional methods of river training and dredging. The overall ford characteristics of the given stretches remain however unchanged.

- An analysis of ford data allows a reliable evaluation of the efficiency of river training measures taken.
- Intensive dredging of the riverbed as well as natural erosion favour the coming into being of local erosion centres as well as the emergence of rock thresholds.
- Ford conditions of a given river stretch are influenced, in a given year, not only by its mean discharge, but also by its distribution within a year. Along the Central and the Lower Danube, the fords remain important factors of the processes forming the riverbed also in the future. Hence investigations should be carried out regarding the main characteristics of the fords and their relationships with the various triggering factors. On the realization of such investigations the following suggestions may be made:
- At selected river stretches, rich in fords, some selected fords should be observed continuously. At selected fords, undisturbed by dredging, relationships should be determined between the ford parameters and the morphological parameters (curve radius, width and depth of the riverbed), the hydraulic parameters (change in slope, flow velocity) and the sediment factors (sediment amount, particle diameter).
- Along selected stretches of the Lower Danube, rich in islands, systematic measurements should be carried out on selected island groups and their related fords, in order to improve the knowledge about these special processes of river morphology.

The database compiled in the framework of the project "The Fords of the Danube" can be considered as a relevant basis for comparisons. In the future it allows determination of tendencies and magnitudes of changes in ford conditions as well as improvements in the starting conditions of the research projects as suggested above.

Acknowledgments The present paper is an abridged version of the project report [17], compiled between 1994 and 1996 by the competent experts of the co-operating Danube Countries, under the co-ordination of the Author of this paper. The latter wishes to express his acknowledgement to all experts contributing, particularly to the Professors S. Prohaska (Yugoslavia), C. Bondar and V.-A. Stănescu (Romania) and G. Gergov (Bulgaria).

References

Commission du Danube (1966) Determinacion de l'étiage navigable et de regulation sur le Danube. Budapest

Commission du Danube (1988) Recomandations relatives á l'établissement des gabarits du chenal, des ouvrages hydrotechniques et autres sur le Danube. Budapest

Commission du Danube (1961–78a) Fiches des seuils sur le Danube, 1–18. Budapest

Commission du Danube (1961–78b) Information sur entretien du chenal du Danube de Regensburg á Sulina, vol. 1–18. Budapest

Commission du Danube (1979–90c) Information sur entretien du chenal et les seuils sur le Danube de Ulm á Sulina, vol. 1–12. Budapest

Commission du Danube (1970d) Recomandation relatives á la coordination du service hydrometeorologique sur le Danube, Budapest

Gergov G, Nenov Iv (1979) Izmeneniya visotnogo raspolozheniy ostrovov v bolgaro-ruminskom uchastke r. Dunay pri bitovom I narushennom rezhime [Changes in the altitudes of the island along the Bulgarian-Romanian Danube section under natural and influenced runoff regime]. In: Proceedings of the 10th Danube Conference, Vienna

Goda L (1971) A dunai gázlók és a hajózás [The fords of the Danube and the navigation], vol. 2. Vízügyi Közlemények, Budapest

Laczay I (1987) Folyószabályozás, ipari kotrás és partiszűrésű vízbázis [River training, industrial dredging and bank-filtered

Laczay I (1988) A folyószabályozás és az ipari kotrás hatása a Nagymaros-Budpaest közötti Duna-szakasz mederviszonyaira [Impact of river training and industrial dredging on the riverbed conditions of the Danube section between Nagymaros and Budapest], vol. 4. Vízügyi Közlemények, Budapest

Laczay I (1989) Ipari kotrások hatása a Komárom-Nagymaros közötti Duna-szakasz mederviszonyaira [Impact of industrial dredging on the riverbed conditions of the Danube section between Komárom and Nagymaros], vol. 3. Vízügyi Közlemények, Budapest

Nicolae T (1970) Prognoz minimal'nyh glubin na pekatah Dunaya na uchastke Orshova-Braila [Forecast of minimal water depths of the fords of the Danube on the stretch Orsova-Braila]. In: Proceedings of the 5th Danube Conference, Budapest

RZD (Regionale Zusammenarbeit der Donauländer) (1986) Die Donau und ihr Einzugsgebiet – Eine hydrologische Monographie [The hydrological Monograph of the Danube and its Catchment], vol. I–III. Bayerische Landesanstalt für Wasserwirtcshaft, München

RZD (Regionale Zusammenarbeit der Donauländer) (1993) Das Schwebstoff- und Geschieberegime der Donau [The suspended matter and bedload transport of the Danube]. Follow-up volume No. I to the Danube Manongraph. VITUKI, Budapest

RZD (Regionale Zusammenarbeit der Donauländer) (1996) Die Furten der Donau [The fords of the Danube]. Follow-up volume No. V/1 to the Danube Monograph. VITUKI, Budapest

RZD (Regionale Zusammenarbeit der Donauländer) (1999) Die Flussbettregulierung der Donau [River training of the Danube]. Follow-up volume No. V/3 to the Danube Monograph. VÚVH, Bratislava

Stănescu V-A (1967) Quelques problemes concernant les methodes emloyées pour établir la stabilité du lit inferior. IUGG, Colloque sur la morphologie des riviéres, Gentbrugge

Chapter 12
Forecast Uncertainties in the Operational Flood Forecasting of the Bavarian Danube Catchment

Stefan Laurent, Christine Hangen-Brodersen, Uwe Ehret, Inke Meyer, Katja Moritz, Alfons Vogelbacher, and Franz-Klemens Holle

Abstract Hydrological forecasts have become an important part of the flood information service, since they are calculated for all river catchments in the Bavarian Danube Catchment. Experiences with published forecasts during former flood events have shown the need for communicating the uncertainties associated with these forecasts to the civil protection authorities and the public. Therefore, methods for quantifying and representing these uncertainties have been developed and incorporated in the flood warning routine. A newly developed approach varies the dominant factors of uncertainty like the meteorological forecast in headwaters by including forecast ensembles. The remaining factors are represented by a static uncertainty measure derived from offline analysis and combined with the former. The total uncertainty is represented by the 10 and 90% exceedance probabilities published together with a single deterministic forecast via the internet.

Keywords Flood forecast · Uncertainty · Flood information system · Ensemble · Bavarian Danube Catchment

12.1 Introduction

In Germany, the responsibility for a flood information service is assigned to the federal states. Until 2000, the flood information service in Bavaria primarily served as a reporting service that collected water gauge records and forwarded these supplied with a trend comment. The amount of quantitative flood forecasts was limited. Simple methods were used, for example regression functions for gauges (Vogelbacher 2007). Triggered by the Whitsun flood of 1999, the Federal State of Bavaria initiated the innovation program "Quantitative Hydrology" that was part of

S. Laurent (✉)
State Office for Water Management, Kempten, Germany
e-mail: stefan.laurent@wwa-ke.bayern.de

M. Brilly (ed.), *Hydrological Processes of the Danube River Basin*,
DOI 10.1007/978-90-481-3423-6_12,

the "Action Program 2020" that aims at a sustainable flood protection. This program comprised the development of flood forecast models as well as the implementation of an automatic online rain gauge network and an optimisation of the existing river gauge network. The dissemination of flood information based on means of modern communication and its reliability was improved (Vogelbacher 2005).

Today, hydrological forecasts have become an important part of the flood information service in Bavaria, since hydrological models cover almost its total area. The hydrological forecasts are calculated daily on business days and more frequently during flooding periods in five regional flood forecast centres. The decision-makers within the Bavarian Water Management Authority can access the results, i.e., forecasts for around 600 gauge stations over the whole of Bavaria. Additionally, the forecasts of about 100 selected gauging stations are published in the web presence (www.hnd.bayern.de) for a horizon of 6–24 h depending on the catchment area.

Experiences with published forecasts during former flood events have shown the need for communicating the uncertainty associated with hydrological forecasts to the public and the responsible persons in civil protection. Expectations on the reliability of flood forecasts are high. Publishing only one forecast as a quasi-deterministic forecast with one value at a given time even raises these expectations by pretending to be exact. Therefore, one aim of computing uncertainty is to publish it along with the corresponding forecast. The resulting illustration should make the numeric forecast less absolute for the average user and communicate the probability of a certain water level to be reached.

For advanced users such as decision-makers in the water management authority, the published uncertainty should furthermore serve as a tool for better risk assessment. For example, the person in charge of operating a flood control basin can then base his decisions on probabilistic numbers in addition to the deterministic forecast instead of only on the latter. By judging, what probability threshold should be taken into account the responsibility for operators increases. Therefore, it is very important to teach users how to interpret the published forecasts, especially the probabilistic part. Understanding the sources of uncertainty and their meaning also helps improving the correct usage of published prognostic data. This task is especially important since, compared to the normal weather forecasts; flood forecasts rarely gain the same every-day importance for the average user. Therefore, users are mostly lacking personal experience in evaluating the reliability of hydrological forecasts.

12.2 Organisation of the Flood Information Service in Bavaria

Participants of the Bavarian flood information service are the flood information centre at the Bavarian Environment Agency in Munich, the regional flood forecast centres, the state offices for water management as the main reporting offices, offices

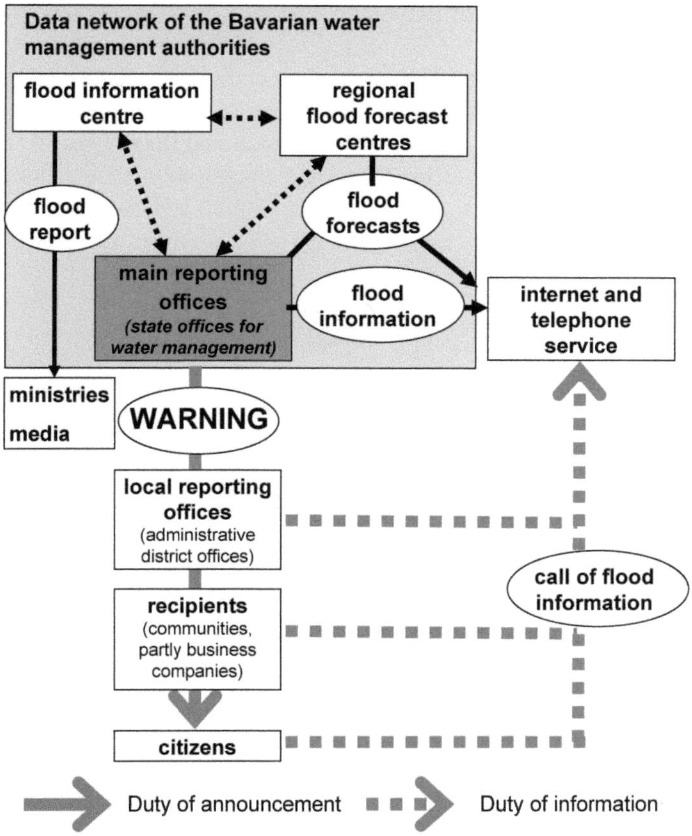

Fig. 12.1 Reporting and information scheme of the Bavarian flood information service

of the administrative districts as local reporting offices and the recipients in the reporting schemes (Fig. 12.1).

The Bavarian flood information service differentiates between warning limits and warning levels. The latter describe the impact of the flood in the area of a river gauge, while the first serve as criteria for the release of flood warnings.

The flood information service starts operations, when the warning limit at one main river gauge or at more than one river gauge is exceeded or is expected to be exceeded soon. The main reporting offices issue a flood warning to the local reporting offices by any communication means. The flood warnings have to be actively transmitted to the concerned recipients. Starting with the (signed) reception of a (pre-) warning, the recipient from then on bears the responsibility to inform himself about the threatening flood. For this purpose, the state offices for water management, the regional flood forecast centres and the flood information centre provide updated data, forecasts and flood reports via the website as well as flood news via a telephone service to the public and authorities.

12.3 Role and Operation of the Flood Forecast Centres

Five flood forecast centres corresponding to the main river catchments (Danube, Inn and Main) and Danube tributary catchments where large reservoirs have to be operated (Iller-Lech and Isar) are responsible for operational flood forecast (Fig. 12.2). The decentralised and river catchment-related organisation of the flood forecast allows the use of experiences with local peculiarities and the operation of large reservoirs in-situ.

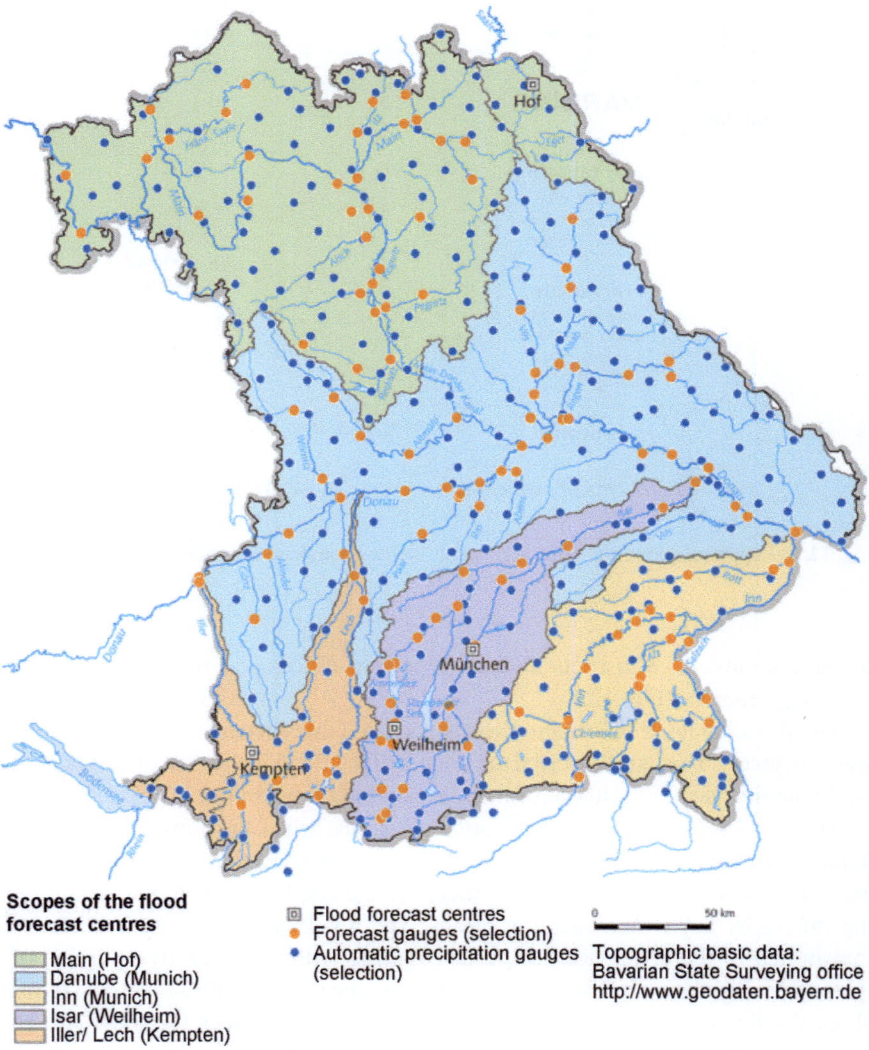

Fig. 12.2 Flood forecast centres in Bavaria (www.hnd.bayern.de/wir/hvz_karte.htm, accessed 25 April 2008, slightly modified)

The model-based flood forecast service supports the services of the regional water authorities and the flood information centre. In the case of the flood forecast centres Iller-Lech and Isar, respectively, the flood forecast service additionally assists in the management of reservoirs.

The state offices for water management are responsible for local flood forecasts and the dissemination of flood information in the regional flood information service, the application of models for reservoir operation and for the data service (collecting and providing data of the river gauge network).

Usually, the flood forecast centres produce water level and discharge forecasts once a day except on weekends and public holidays. The forecasts are published on the internet.

The regional flood forecast centre informs the main reporting offices within the forecast area and the flood information centre, if due to the flood forecast an extensive exceedance of the warning limits has to be expected. After receiving this information, the main reporting offices have to provide the river gauge data on an hourly basis.

In case of a flood event the flood forecasts are calculated three or four times per day at defined points of time. If necessary, the flood forecasts are updated every hour.

The forecast centres calculate forecasts for all river gauges implemented in the flood forecast models. The complete flood forecast simulation results are made available for the regional water authorities, regional governments and ministries by a Java-Client application within the data network of the Bavarian water management administration. In the intranet and internet, only forecasts at selected river gauges and for reduced forecast horizons are published. The regional water authorities have access to the flood forecasts for all river gauges and can use these results for their own forecasts.

12.4 Flood Forecast Model Systems

Until 2000, methods used in the flood information service were limited to empirical and empiric-synoptic approaches. Nowadays, forecast models are used that almost completely cover the area of Bavaria (Vogelbacher 2005) (Fig. 12.3).

The rainfall–runoff model predominantly used is LARSIM (Large Area Runoff Simulation Model, Ludwig and Bremicker 2006). Because of its robust and relatively simple components, the model is suitable for operational application (Gerlinger and Demuth 2000). At several German flood forecast centres, LARSIM is in operational use.

LARSIM can be applied as an event-based flood model as well as a water balance model for continuous simulations. For the time being, in Bavaria the operational LARSIM flood forecast models are implemented in event-based mode. The water balance model mode is currently being introduced. For the Iller catchment the water balance mode of LARSIM has been implemented, and is now in an operational pilot phase.

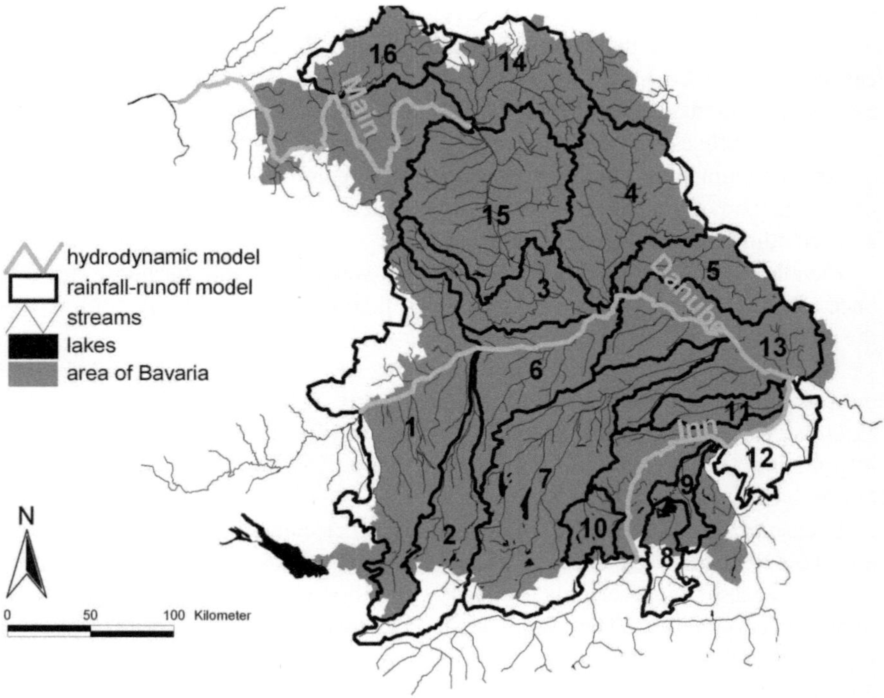

Fig. 12.3 Overview of the hydrodynamic and rainfall–runoff models used for operational flood forecasting in Bavaria. The numbers correspond to the models in Table 12.1

The rainfall–runoff models used are deterministic models, which calculate discharge rates as response to precipitation input. In the simplest case, the effective precipitation is determined with a runoff-coefficient depending on the antecedent moisture content of the area. According to non-linear relations between precipitation and resulting discharge, the runoff-producing precipitation is adjusted by superposition calculations.

The catchment parameters can be subdivided either in gridded or irregular sub-catchments. Each sub-catchment is characterised by coordinates, elevation, length of watercourses and schematic river cross-section with roughness coefficients. On a grid or sub-catchment basis, the relevant runoff processes are calculated using specific hydrological methods. Table 12.2 contains the methods applied in most of the LARSIM models used in Bavaria.

Snow accumulation and snowmelt can be considered as well as artificial influences (e.g. storage basins, diversions, water transfer between different catchments). The model calculations are based on hourly data of precipitation, discharge, precipitation forecasts, as well as on snowmelt calculations and forecasts.

The forecasts for the tributaries based on rainfall–runoff models are linked to hydrodynamic models.

Table 12.1 Rainfall–runoff models based on LARSIM used for operational flood forecasting in Bavaria (see Fig. 12.3)

Description	River catchment	Sub-catchment (sc)/grids (g)
Danube catchment		
(1) Danube upstream of Lech	Danube upstream of Lech (incl. Iller)	11,268 g
(2) Lech	Lech	177 sc
(3) Altmuehl	Altmuehl	2,638 g
(4) Naab	Naab	6,362 g
(5) Regen	Regen	3,091 g
(6) Danube incl. Paar	Danube downstream Donauwoerth up to Regensburg (without Lech, Altmuehl, Naab, Regen)	7,228 g
(7) Isar	Isar	1,055 sc
(8) Chiemsee	Chiemsee incl. Tiroler Achen	7,195 sc
(9) Traun	Alz downstream Lake Chiemsee	77 sc
(10) Mangfall	Mangfall	97 sc
(11) Rott	Rott	62 sc
(12) Inn	Inn in Bavaria (without Mangfall, Alz, Salzach, Rott)	2,682 sc
(13) Danube downstream Regensburg	Danube downstream Regensburg without Isar	7,931 g
Main catchment		
(14) Upper Main	Upstream of Regnitz	362 sc
(15) Regnitz	Regnitz	700 sc
(16) Fraenk. Saale	Fraenkische Saale	352 sc

Table 12.2 Methods applied in most of the LARSIM models used for operational flood forecasting in Bavaria (adapted from Ludwig and Bremicker 2006)

Module	Method
Areal precipitation for sub-catchments	Modified grid-point method
Snowmelt model	Snowmelt according to Knauf method
	Snow compaction according to Bertle
Effective precipitation	Runoff-coefficient (constant or nonconstant)
Base flow	Base flow yield (constant)
Runoff separation into overland flow and interflow	Interflow-index rate
Runoff concentration in sub-catchments	Modified Clark method for overland flow
	Linear storage for interflow
Channel flow	Translation-retention method
	Constant translation
	Williams method
	Volume/discharge relation

The basis for the flood routing model at the Main and Bavarian Danube rivers is the hydrodynamic model WAVOS (Wilke and Rademacher 2002).

The hydrodynamic model FLORIS 2000 (Reichel 2001) is run for the lower Bavarian Danube, Lech and for the Bavarian stretch of the Inn River from Kufstein to Passau. The standard operating regulations of the run-of-river power stations are included (Vogelbacher 2007).

The operational flood forecast model system works in a so-called served operation. First, the rainfall–runoff models have to be run to the connecting gauge for the hydrodynamic model. Besides forecasts of the rainfall–runoff models for the Bavarian region, external flood forecasts for tributaries are needed. For example, the Danube forecasts upstream of the Iller tributary are produced by the Federal State of Baden-Wuerttemberg. For the Inn catchment, flood forecasts are needed for the Inn upstream of Kufstein which are supplied by the federal state of Tyrol (Austria) and for the Inn tributary Salzach, supplied by the federal state of Salzburg (Austria). The Salzach river is within the responsibility of the federal government of Salzburg where they use HYDRIS (Hydrological information system for flood forecast) (Vogelbacher 2007).

Using rainfall–runoff models, the forecast period is limited by the weather and precipitation forecast. To adjust to the COSMO-EU model of the German Weather Service (see Table 12.3), 72 h forecasts are produced by default. However, forecasts for shorter periods are published on the internet. At present, the hydrodynamic models are run for 48 h forecasts (Vogelbacher 2007). Since December 2007, ensemble precipitation forecasts have been used operationally for the pilot catchments of

Table 12.3 Numerical weather forecasts used for operational flood forecasting in Bavaria

Model	Weather service/organisation	Grid step (km)	Interval (h)	Horizon (h)
GME	German Weather Service	40	12	174
COSMO-EU	German Weather Service	7	6	48–78
COSMO-DE	German Weather Service	2.8	3	21
COSMO-LEPS	German Weather Service	10	daily	132
SNOW 3	German Weather Service	1	6	42–72
GFS	NOAA (National Oceanic Atmospheric Administration)	~50	6	180
ALADIN Austria	Central Institute for Meteorology and Geodynamics (Austria)	9.6	12	48
MOSS	Meteomedia, Switzerland	Station based	daily	96

the Rivers Regen, Fraenkische Saale, Upper Main, Regnitz and Mangfall and are currently being evaluated.

12.5 Monitoring Network and Hydrometeorological Input Data

12.5.1 Water Level and Discharge Data

The river gauge network in Bavaria consists of about 600 stations, 560 of which are equipped with telemetric data transmission. 320 river gauges are so-called "report river gauges" for the flood information service. The extreme flood events in the Danube river catchment in May 1999 and August 2002 as well as in January 2003 in the Main river catchment revealed that equipment and data transfer were not sufficient. Failures and breakdowns of the measurement installation and data supply occurred. To ensure the data supply, in the meantime most of the river gauges have been equipped with redundant measurement devices and telecommunication channels. For telemetric data transfer, the conventional telephone network or mobile telephone systems GPRS resp. GSM are used. The remaining river gauges in the flood information service have at least redundant measurement devices. Furthermore, there were numerous river gauges where the extreme floodwater stages could not be assessed. Constructional measures will be taken within the next years.

For the flood forecast models, discharge is the observation, and thus reliable stage-discharge-relationships are needed. However, often discharge is not measured during flood situations and additionally it is associated with large uncertainties. To limit these uncertainties, alternative measurement methods, such as ultrasonic, acoustic Doppler current profiler (ADCP), tracer and radar measurements are being tested in a pilot phase. Partly, the implementation of the new methods has already been accomplished. For example, discharge tracer measurements are already in operation in alpine areas (Roth 2008).

To check and improve the rating curves especially in the high-flow extrapolation range, a project has been started at the Bavarian Environment Agency where hydraulic simulations for about half of the Bavarian river gauges shall be conducted.

There are also water level data available from the Hydrographical Services of Austria, from the Federal States of Baden-Wuerttemberg, Thuringia, Hesse as well as from the Federal Waterway and Navigation Administration for the national waterways. For the flood forecast models, discharge data at the power stations are required. These data are obtained in a data exchange with different operators of water power stations. The data of most of the external partners are imported per ftp- or http-request via the internet into the database of the flood information service.

12.5.2 Meteorological Data

Most of the precipitation data are provided by a joint automatic monitoring network, operated by the German Weather Service and the Bavarian Environment Agency with about 285 stations (107 of the Bavarian Environment Agency in co-operation with the State Offices for Water Management).

Additionally, data from the monitoring networks of the private company Meteomedia, the Bavarian Agency of Agriculture and the Bavarian Avalanche Warning Service are used.

Because of transboundary tributaries, also precipitation data of the Central Institute of Meteorology and Geodynamics, Vienna, and the Austrian hydrographical services as well as of the federal states of Baden-Wuerttemberg and Hesse are collected. In sum, precipitation data of 700 stations are available online (Vogelbacher 2007).

Besides precipitation, snow height, snow water equivalent, air temperature, wind speed and radiation are the most important meteorological observations for snow melt calculation. However, these quantities are measured in a lower resolution. The snow measurement network has been considerably increased within the last years. At present, the snow water equivalent is measured at approx. 120 stations in Bavaria in 1–3 day-intervals. Three automatic stations record snow water equivalent continuously.

The spatial assessment of precipitation can be improved by using weather radar measurements. The adjustment of radar signals to the measurement data of the ground precipitation network allows a spatially high-resolution assessment of rainfall events. The first operational products of the German Weather Service project RADOLAN (radar online-adjustment) were available in 2005 (DWD 2008). The composite data of the radar echoes and the products adjusted to the ground monitoring network are received each hour per FTP from the German Weather Service. Since 2008, the RADOLAN-data can be used directly as input data for the operational flood forecasts.

Several weather forecasts are available for operational flood forecasting in Bavaria (Table 12.3). Usually, the results of the numerical weather forecast models of the German Weather service are used. COSMO-EU is the main product of the German Weather service for the forecast horizon of 3 days (DWD 2008). The most recent product is COSMO-DE, which has been used for operational flood forecasting in Bavaria since 2008. The high horizontal resolution of 2.8 km allows direct simulation of thunderstorm cells of the size of only a few kilometres. Forecast runs are started every 3 h. Within the model runs, also updated radar measurements are integrated. It still has to be investigated whether this can help to improve the flood forecasts.

For Southern Bavaria, the results of the ALADIN-Austria model operated by the Central Institute of Meteorology and Geodynamics in Vienna are available (Vogelbacher 2007). Meteomedia provides a 4-day station-based precipitation forecast using a statistical correlation between the output of a “coarse-meshed” weather forecast model and the data of ground weather stations (Meteomedia 2008). For the purpose of comparison, forecast products such as the American GFS (Global Forecast System) model or of the ECWMF (European Centre for Medium-Range Weather Forecasts) in Reading (England) are used at times (Vogelbacher 2005).

To spatially assess the water release from snowmelt and rainfall, simulation results of the snowmelt model SNOW3 of the German Weather Service are used. Water yield and water equivalent of the snow cover are given in a spatial resolution of 1 × 1 km in an hourly time step (Vogelbacher 2005). The model is run from November to April and for the area of the federal states of Baden-Wuerttemberg, Rhineland-Palatinate, Saarland, parts of North Rhine-Westphalia, Hesse, Saxony and Bavaria as well as for Luxembourg and the French part of the Moselle river catchment and the river catchment Bregenzerach (Vorarlberg) (DWD 2008).

Weather and storm warnings of the German Weather Service are sent via FAX, e-mail or are made available via the internet. Further information can be obtained by telephone contact with the meteorologist in charge at the regional office of the German Weather Service in Munich (Bavarian Environment Agency 2008). For defined geographical regions in Bavaria, the meteorologist gives an evaluation for areal precipitation amounts.

Since December 2007, the model output of COSMO-LEPS (Consortium for Small-Scale Modelling – Limited-area Ensemble Prediction System) (DWD 2008) of 16 ensemble members has been available. The ensemble forecasts are in a test phase for operational use and applied to pilot areas.

12.6 Sources of Uncertainty Within Operational Flood Forecasting

12.6.1 Input Data

As mentioned above, hydrometeorological input data for runoff-simulation consist of hourly data of precipitation, snowmelt, runoff and meteorological forecasts. Each of these components bears a particular uncertainty, which affects the uncertainty of the simulation output.

In many cases, especially in headwaters, meteorological forecasts are the most dominant source of input data uncertainty. Often there are large differences in rainfall amounts between forecasts originating from different meteorological models and between the different model runs of the same model. A good example is the rainfall event in the upper catchment of the river Lech leading to major flooding in August 2005. One day before the event, the model COSMO-EU predicted a total amount of 215 mm precipitation for the whole rainfall event, the model GME 93 mm and the model GFS 164 mm. The measured precipitation was 169 mm. Figure 12.4 shows another example of the impact of different meteorological forecasts on the runoff forecast for a gauge at the tributary Regen in the Bavarian Forest.

Not only the amount, but also the spatial and temporal distribution of precipitation can vary between different forecasts and the measured precipitation affecting the output of the hydrological model, especially in smaller catchments. Therefore, ensemble forecasts should be included in the calculation of total output uncertainty to account for the dynamic uncertainty of the meteorological forecast.

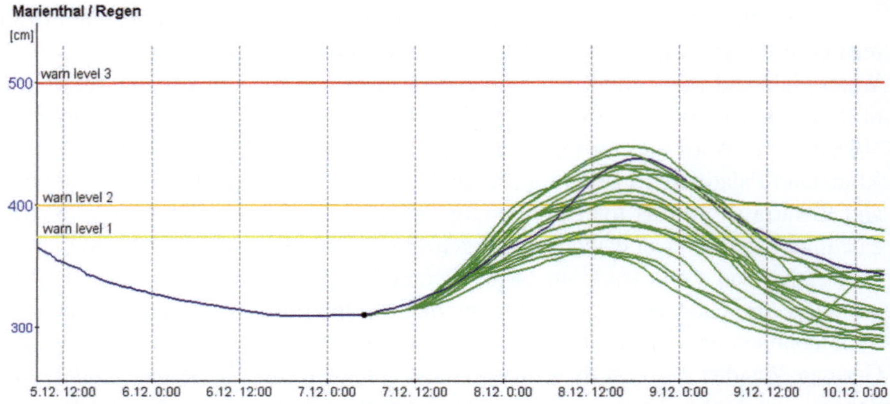

Fig. 12.4 Using all available meteorological forecasts as input into a rainfall–runoff model for a medium-sized catchment in the Bavarian Forest shows the broad variety in the resulting forecasts of water level (*green lines*). The observed water level is marked *blue*

Furthermore, observed hydrometeorological data contribute to input uncertainty. In Bavaria, about 500 precipitation stations provide the hydrological models with data. Nevertheless, these are point observations. The true spatial distribution has to be estimated. The measurement at the precipitation gauging station is inflicted with an uncertainty, too, mainly because of wind and evaporation. In addition, it is possible to use radar-derived precipitation observations for hydrological forecasts. However, especially in the process of transforming the observed radar reflectivity to rainfall rate, still large errors occur.

As observed runoff is an important input for hydrological models, its accurate determination is very important. Most of the gauging stations only measure water levels though. That way, runoff has to be calculated against water level using a discharge-rating curve based on measured runoff. The latter, however, is rarely measured during very low or high discharge. Therefore, especially during flood events, the uncertainty of observed discharge is rather high. Even if the discharge rating curves are regularly updated and improved, factors like flotsam, ice jam, erosion, dam failure and so forth may continuously change the relation between water level and discharge during flood. Improvements can be gained by directly measuring runoff for example by ultrasound or ADCP.

12.6.2 Model Simplification

If the forecast horizon lies within the travel time of a flood wave observed upstream, the runoff forecast is expected to be more accurate as it only involves the single process of flood routing. Although this applies to most cases, rather great uncertainties can appear in runoff forecasts even within these shorter forecast horizons. The main reason for this is the inadequate or missing reproduction of certain hydrological processes in the rainfall–runoff model. For example, if a meandering river overflows, this results in a runtime reduction of the channel line. The model LARSIM, however, cannot reproduce this effect. Furthermore, groundwater–river interactions, which can lead to significant reductions of the peak runoff in certain areas, cannot be accounted for in LARSIM.

12.6.2.1 Estimating Model Parameters

Model parameter calibration is conducted separately for each catchment based on three to five historic flood events. Very often, the extreme flood events in the past cannot be used, because of missing input data. Because of this, model calibration is often not optimal. It has to be improved and repeated especially after large floods.

In addition, model parameters can have different values for different hydrological conditions. Therefore, it is reasonable to use different sets of parameters for different hydrological situations. But again, often there are not enough observed flood events for each hydrological condition, which can be used for calibration.

12.6.2.2 Operational Practice

Another source of uncertainty is the operation of the hydrological models. Some parameter and model settings have to be adjusted by the forecaster according to the actual hydrological situation.

An example for this is choosing the right period for optimising the runoff coefficient in an operational run. Within this period, the simulated runoff is adapted to the observed runoff by automatically adjusting the runoff coefficient. Different periods may lead to different forecast results (Fig. 12.5). Another example is the often lacking information about discharge of smaller reservoirs. That way, the forecaster has to make an assumption, which also increases the uncertainty of the model output.

12.6.2.3 Relative Influence of the Different Sources of Uncertainty

The relative influence of the different sources of uncertainty depends on different factors like forecast horizon, meteorological conditions and up- or downstream

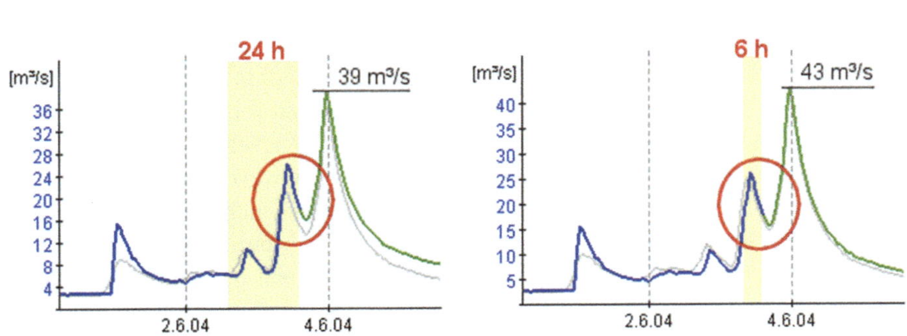

Fig. 12.5 Choosing different periods for optimising the runoff coefficient throughout the model may lead to different forecast results

location of the catchment. With increasing forecast horizon, the influence of the uncertainty of the meteorological forecast increases. However, during a period with almost no rainfall predicted, the influence of the uncertainty of the meteorological forecast should be very low. In headwater catchments, the uncertainty of the meteorological forecast is more dominant than in downstream catchments. If the forecast horizon lies within the runtime of the flood wave observed upstream, the uncertainty of the meteorological forecast has not as much influence as for example the uncertainty resulting from model simplification.

12.7 Estimating Uncertainty

To consider all the sources of uncertainty mentioned in the previous section, the optimal approach would be to do multiple forecast realisations by randomly varying all of the sources of uncertainty within their range (Monte Carlo simulation). The great drawback, however, is that if this is thoroughly done, the number of forecast realisations quickly amounts to quantities too large to handle within the time constraints of an operational environment.

Another operational constraint is that the river systems in Bavaria are tiled into a series of sub-catchments for operational forecasting (see Table 12.1). The Bavarian Danube Catchment, for example, consists of 13 individually simulated catchments. The hydrological forecasts from the headwater catchments are calculated first and then passed to the next downstream catchment and so forth. In such a series calculation, for the sake of coherence, the runoff forecast calculated in a catchment and the runoff forecast from the headwater catchment must be based on the same data (e.g. the same rainfall forecast). In operational forecasting, this also limits the number of Monte Carlo simulations that can be handled.

12.7.1 Operational Approaches in Bavaria

Because of operational constraints, a relatively straightforward approach for the consideration of forecast uncertainty has been applied in Bavaria so far. Long time series of former, archived flood forecasts have been compared to gauge observations calculating the relative error for each time step within the forecast horizon (Vogelbacher 2007). From the error distribution obtained for each gauge, the relative error on the 10 and 90% exceedance probability level is used to illustrate the uncertainty on each new forecast. This approach, however, does not take into consideration the dynamic nature of forecast uncertainty, which may vary with time.

Therefore, a new approach has been developed which is presented below. It varies the dominant factors of uncertainty and represents the rest by a static uncertainty measure derived from offline analysis. As this is work in progress, only preliminary results can be presented. First, the combination procedure is explained, which differs between headwater and downstream catchments. Then, the type and the calculation of the static uncertainty measure are explained in detail.

12.7.1.1 Headwater Catchments

In headwater catchments, experience shows that the dominating source of uncertainty is rainfall forecast. However, its magnitude may vary with the current weather situation; some weather patterns are easier to forecast than others. In the first case, the spread of the different rainfall forecasts available is much less than in the latter case. It is therefore reasonable to leave this source of uncertainty dynamic and represent the other sources with a static uncertainty value. The two types are combined as follows (Fig. 12.6):

With all available rainfall forecasts, calculate an ensemble of runoff forecasts. The spread varies with the spread of the rainfall forecasts.

Add the static uncertainty distribution to each time step of each runoff forecast by randomly sampling the empirical error distribution (e.g. 100 times). The uncertainty distribution is a function of forecast horizon, i.e. the uncertainty range widens with increasing forecast horizon.

For each forecast time step, calculate the empirical uncertainty distribution over the whole forecast ensemble and select the value on the desired uncertainty level (e.g. 10 and 90% exceedance probability). An example: If four runoff forecasts have been calculated and for each forecast time step, 100 samples are drawn from the static uncertainty distribution, 400 forecast values are available for the given forecast time step. They are ordered by magnitude, with the exceedance probability of the highest value being zero and the exceedance probability of the lowest being one. From the range of ordered runoff forecast values, pick the ones on the desired

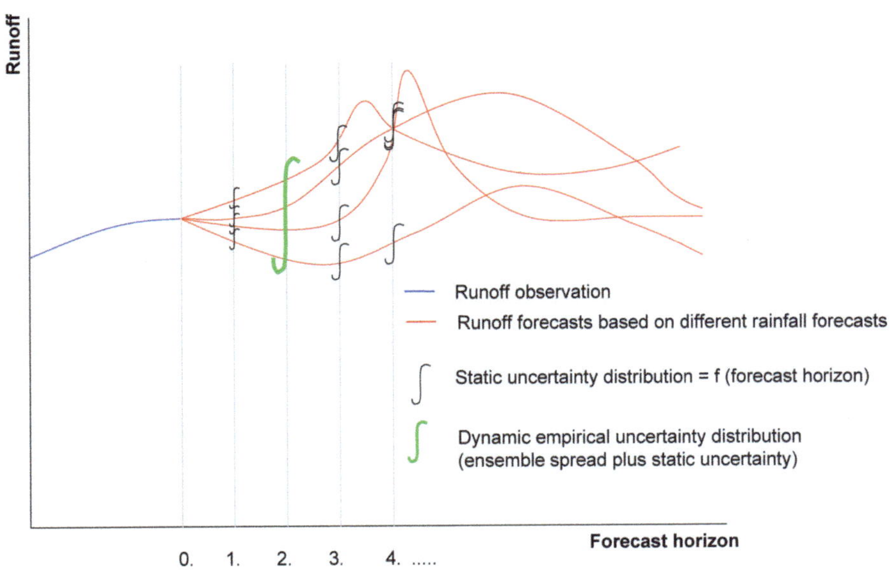

Fig. 12.6 Calculating the uncertainty distribution in a headwater catchment

exceedance probability levels. This is the uncertainty range for the current forecast time step.

12.7.1.2 Downstream Catchment

In downstream catchments, the relative influence of the rainfall forecast especially for a short forecast horizon is smaller. Here, the influence of uncertain gauge observations and the routing procedure in the hydrological model is usually more dominant and uncertainty ranges due to these factors may be much larger for a short forecast horizon than the spread from different rainfall forecasts. However, as mentioned above, operational constraints preclude the variation of gauge observations or the parameters of the routing algorithms. Therefore, a different approach is used in downstream catchments (Fig. 12.7):

With all available rainfall forecasts and the matching runoff forecasts from headwater catchments, calculate an ensemble of runoff forecasts. Without adding a static uncertainty measure, calculate the empirical uncertainty distribution and select the value on the desired uncertainty level (e.g. 10 and 90% exceedance probability). This is the dynamic uncertainty stemming only from the rainfall forecast uncertainty.

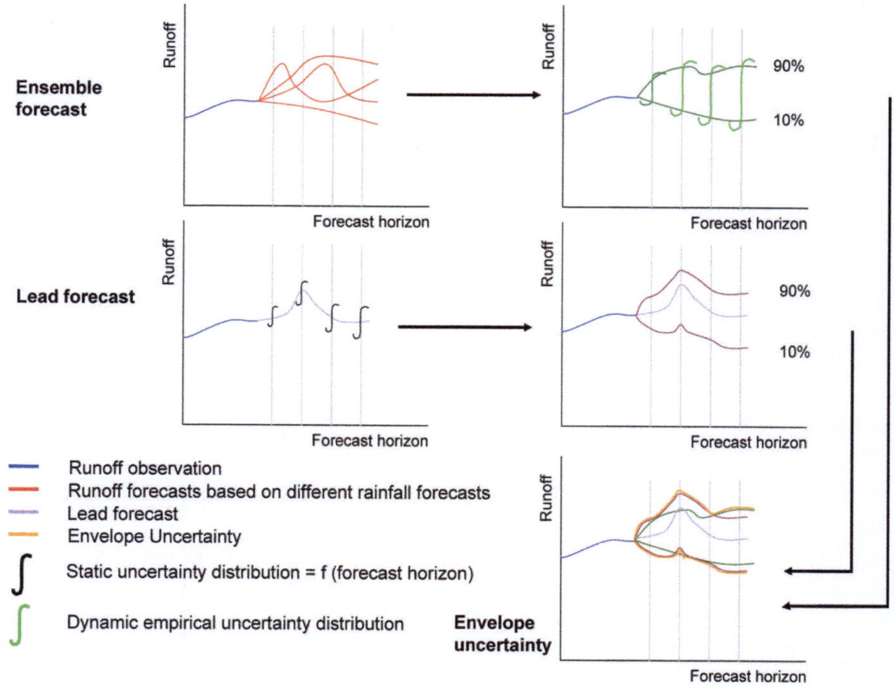

Fig. 12.7 Calculating the uncertainty distribution in a downstream catchment

Select the most likely rainfall forecast and the matching runoff forecast from upstream. The selection can be done on an objective base (e.g. long-term analysis of rainfall forecast quality) and/or a subjective base (e.g. the advice of a meteorologist). This forecast is regarded as the best single estimate of future runoff and termed "lead forecast".

To the lead forecast, add the static empirical uncertainty distribution and select the value on the desired uncertainty level (e.g. 10 and 90% exceedance probability). This is the static uncertainty stemming from all sources of uncertainty.

Combine the static and dynamic uncertainty by taking the enveloping curve of the two.

12.7.1.3 Determining the Empirical Uncertainty Distribution

One question has so far not been answered: How are the empirical uncertainty distributions determined? This also differs for headwater and downstream catchments as described in the following text.

In headwater catchments, the static uncertainty measure has to include all sources of uncertainty except the one stemming from the rainfall forecast (because this is dynamically accounted for by applying rainfall forecast ensembles). The uncertainty measure is therefore determined as follows (Fig. 12.8):

For a large number of historical floods, calculate runoff forecasts. Instead of taking real rainfall forecasts, rainfall observations are used. Thus, the uncertainty of rainfall forecasts is excluded.

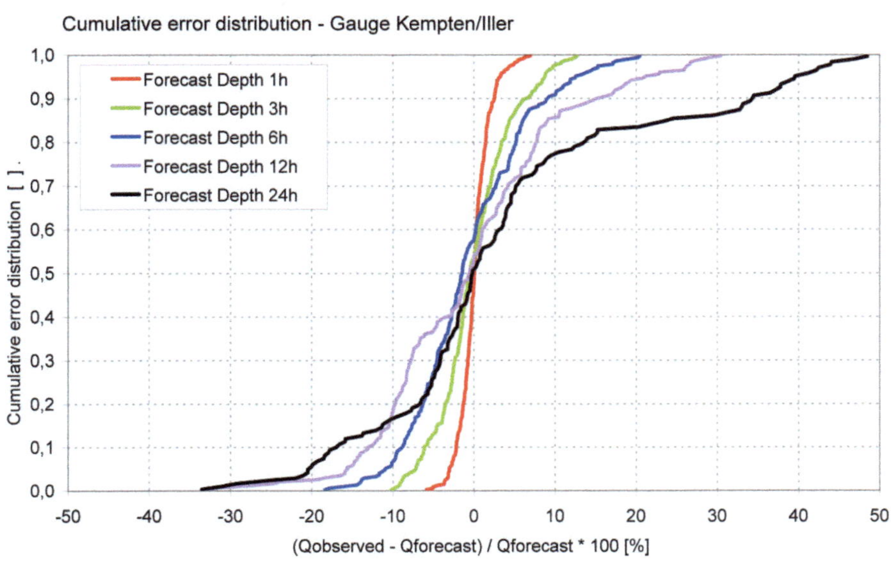

Fig. 12.8 Empirical error distribution for different forecast horizons in a headwater catchment, gauge Kempten at river Iller (Haag and Luce 2007)

Calculate a suitable error statistic for each forecast value. Here, the percent error $[(Q_{\text{observed}} - Q_{\text{forecast}})/Q_{\text{forecast}}] * 100$ was used.

From all runoff forecasts, collect the errors with the same forecast horizon and order them by magnitude. This is the empirical exceedance probability distribution of forecast errors (or, in other words, the forecast uncertainty).

From this distribution, errors on any desired level of exceedance probability can be drawn, for example on the 10 and 90% level.

In downstream catchments, the static uncertainty measure comprises all sources of uncertainty including the rainfall forecast uncertainty. The uncertainty measure can therefore be determined by analysing runoff forecasts produced by operational flood forecasting during the last years:

On the basis of the large number of archived runoff forecasts and matching gauge observations in Bavaria, calculate a suitable error statistic for each forecast value. Here, the percent error $[(Q_{\text{observed}} - Q_{\text{forecast}})/Q_{\text{forecast}}] * 100$ was used.

From all runoff forecasts, collect the errors with the same forecast horizon and order them by magnitude. This is the empirical exceedance probability distribution of forecast errors (or, in other words, the forecast uncertainty).

From this distribution, errors on any desired level of exceedance probability can be drawn, for example on the 10 and 90% level.

The so far developed approaches bear some weaknesses, which should be resolved step-by-step. Amongst others, the analysis of the static uncertainty based on old, archived runoff forecasts does not distinguish between errors in time and value. If, for example, the peak of a flood wave is forecast correctly in height, but not in time, the resulting differences, which are relatively high, influence the analysis in a questionable way. A possible solution might involve examining the minimal distance of the forecasted value to the observed as a vectorial deviation.

12.8 Communicating Uncertainty

Communicating uncertainty to the users of the hydrological forecast is a very important task. The focus of interest in this section is how to publish a graph for the public in the web presence. For this target group, the design should be simple and intuitive. As dealing with probabilistic numbers is not common to most users, descriptions and explanations should also be intelligible to all.

Up to now, publishing uncertainty of hydrological forecasts in the web presence is rarely done by operational hydrological services. Most figures only show one (deterministic) forecast without any ranges of uncertainty. Other services like the National Weather Service of the United States offer separate figures showing exceedance probabilities for water levels within a certain period (www.weather.gov/ahps). These are, however, calculated by statistic analysis rather than taking into account real-time forecasts.

The Lower Austrian and the Bavarian hydrological services show rather similar illustrations of their forecasts in the internet (www.noel.gv.at; www.hnd.bayern.de, Fig. 12.9): A distinct deterministic forecast ("lead forecast") is drawn as a single

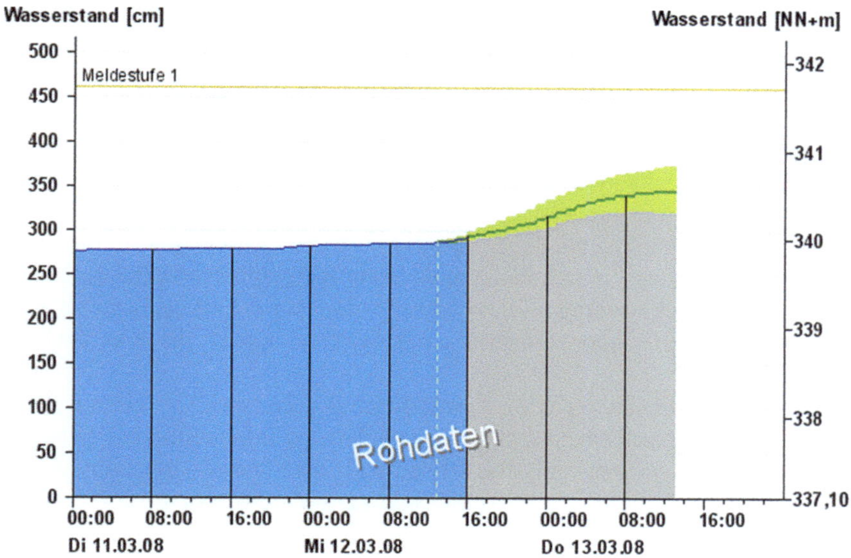

Fig. 12.9 The graph of the Bavarian flood information centre shows a deterministic lead forecast surrounded by an uncertainty range (www.hnd.bayern.de)

time series surrounded by an uncertainty range. In Lower Austria, the latter is quantified out of an ensemble of 50 meteorological forecast members marking the 10 and 90% exceedance probability (Komma et al. 2006).

For the Bavarian graph, the simplified approach described before has been used up to now; however it will be replaced by the newly developed approach in the future. Although, the method of illustration is probably going to be the same.

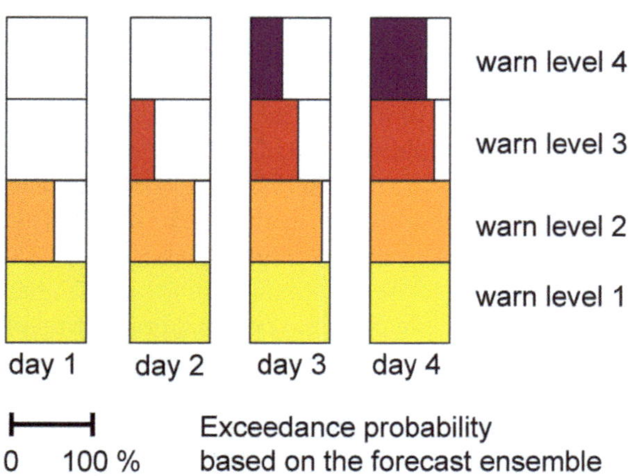

Fig. 12.10 Proposed illustration for a single gauging station showing the exceedance probabilities of the Bavarian warning levels for the following few days

The horizon of the published forecasts ranges from 6 to 24 h depending on the size of the catchment area. So far longer forecasts have not been published due to the increasing uncertainty. As information for a longer horizon is wanted, one possibility is to reduce the degree of accuracy of this information. Instead of publishing a distinct time series, the exceedance probability of certain water levels is shown for the following days (Fig. 12.10). In Bavaria, the flood warning system uses four warning levels, which are determined for each of the warning gauging stations, marking the flood impacts in this area. On the basis of the ensemble forecast including the calculated uncertainty the graph shows the exceedance probabilities of each warning level by filling its square proportionally with the corresponding colour. The more colourful a warning level appears the more likely it is reached.

In order to fulfil the needs of more advanced users such as decision-makers within the water management authority more detailed information has to be published. Instead of showing only the 10 and 90% exceedance probability a graph could also contain exceedance probabilities in steps of 10%.

12.9 Conclusion and Outlook

Estimating and communicating the uncertainty of hydrological forecasts has become an important task within operational forecasting today. The presented approach of varying the dynamic factor of uncertainty and combining it with a static part involves the use of meteorological ensemble forecasts. Especially in headwaters, where the precipitation forecast is the dominating source of uncertainty, this is seen as an advisable procedure. However, to date this approach has not operationally been fully implemented so findings cannot as yet be proven through operational results.

As input for the precipitation forecast a so-called poor man's ensemble, which consists of all available meteorological forecasts at that time, combined with the output of the model COSMO-LEPS as a "real ensemble" will be used. The latter, though, has the drawback that its results are based on older input due to the long computing time. Solutions also have to be found as to how to deal with the different forecast horizons of the poor man's ensemble.

Analysing the old, archived flood forecasts for each of the 100 gauges in Bavaria, for which forecasts are published, has to be automated in order to reduce time and effort. In doing so, different hydrological situations have to be distinguished in order to increase the significance of the analysis.

As mentioned before, communicating uncertainties associated with hydrological forecasts to the public and the civil authorities is an important task, which has to be enforced. Adequate illustrations and descriptions should be evaluated in relation to the normal and advanced user. Additionally, experiences gained during future flood events might help in further adaptation of methods of communication.

Overall, analysis and experiences with forecasts over the last years show that the accuracy achieved so far could be improved upon in many cases. Hence, calculating and communicating uncertainty is one goal. A major goal, though, remains reducing the existing uncertainty in the data, the model and its operation. Appropriate operations and projects remain a permanent task in modelling the Bavarian Danube Catchment.

References

Bavarian Enviornment Agency (2008) Gewässerkundlicher Bericht – Hochwasser August 2005. Bavarian Environment Agency, Augsburg

DWD (German Weather Service) (2008) www.dwd.de – Link to Special User – Water Management. http://www.dwd.de/bvbw/appmanager/bvbw/dwdwwwDesktop?_nfpb=true&_pageLabel=_dwdwww_spezielle_nutzer_wasserwirtschaft. Accessed 17 Mar 2008

Gerlinger K, Demuth N (2000) Operational flood forecasting for the Moselle River Basin. In: Bronstert A et al. (eds) European conference on advances in flood research – Proceedings (PIK-Report No. 65). Potsdam Institute for Climate Impact Research, Potsdam. http://www.pik-potsdam.de/research/publications/pikreports/.files/pr65.pdf. Accessed 25 March 2008

Haag I, Luce A (2007) Ermittlung geeigneter Verteilungsfunktionen zur Beschreibung statistischer Vorhersagefehler operationeller Wasserhaushaltsmodelle (Determination of suitable distribution functions for describing static forecast errors of operational water balance models). Report for the State Office for Water Management Kempten (unpublished). Karlsruhe

Komma J, Reszler Ch, Blöschl G, Haiden T (2006): Ensembleprognosen von Hochwasserabflüssen. Wiener Mitteilungen, Wasser, Abwasser, Gewässer 199:279–294

Ludwig K, Bremicker, M (eds) (2006) The Water Balance Model LARSIM – design, content and applications. Freiburger Schriften zur Hydrologie 22. Institute of Hydrology, Albert-Ludwig University Freiburg, Freiburg

Meteomedia (2008) MOS – Model output statistics. http://www.meteomedia.ch/index.php?id=162. Accessed 25 Mar 2008

Reichel G (2001) FluxDSS und FLORIS2000 – ein leistungsfähiges Paket zur Modellierung der Fließvorgänge in komplexen Systemen. Österreichische Wasser- und Abfallwirtschaft 53(5/6):170–171

Roth C (2008) Quality management of discharge measurements in Bavaria. IOP Conf. Ser.: Earth Environ Sci. doi: 10.1088/1755–1307/4/1/012014

Vogelbacher A (2005) Sonderthema – Stand der Hochwasservorhersage in Bayern (Special topic – actual state of flood forecast in Bavaria). In: Bavarian Environment Agency (ed) Gewässerkundlicher Jahresbericht für Bayern 2005. http://www.lfu.bayern.de/wasser/daten/gewaesserkundliche_jahresberichte/doc/jb2005.pdf. Accessed 25 Mar 2008

Vogelbacher A (2007) Unsicherheiten bei der Abflussvorhersage (Uncertainties of runoff forecasts). http://www.hnd.bayern.de/docs/unsicherheiten_abflussvorhersagen.pdf. Accessed 25 Mar 2008

Wilke K, Rademacher S (2002) Operationelle Wasserstand- und Durchflussvorhersagen im Rheingebiet. Österreichische Wasser- und Abfallwirtschaft 54(9/10):154–162

Chapter 13
SARIB

Lidija Globevnik, Matjaž Mikoš, Matej Padežnik, Sašo Petan, Ana Petkovšek, Andrej Vidmar, Radmila Milačič, Janez Ščančar, Ester Heath, Nives Ogrinc, and Mitja Brilly

Abstract An international team of 11 partners from eight countries was established to develop the SARIB project (http://www.sarib.net/). One part of this project is presented herein, including historical data, available GIS data from different sources and sediment quantity and quality data collected within the framework of the project. All data collected are available on http://www.ksh.fgg.uni-lj.si/sarib/. The Sava River is characterised by a torrential run-off and sediment regime upstream of the town of Sisak and by a lowland river regime downstream of it.

The environmental status of sediments of the Sava River is generally comparable to the Danube River. According to targeted organic analyses, it may be concluded that the pollution of the Sava River sediments is low and generally lower than the pollution of the Danube River sediments.

Keywords SARIB · The Sava River Basin · Chlorinated pesticides · Cation levels · The Sava watershed · Carbon and nitrogen dynamics

13.1 Introduction

The Sava River is an important tributary of the Danube River, with the second largest catchment area in the Danube River Basin. The Sava River provides 25% of the total Danubian discharge, more than any other tributary. The watershed area is 97,713 km^2 as evidented by the Sava Commission; it is not possible to estimate the exact area because the watershed contour line in the south-east part of the basin passes through the dinaric karst region in which the watershed divide cannot be exactly estimated.

The Sava River was the major national river of the Former Yugoslavia. Today it is an international river shared by seven countries: Slovenia, Croatia, Bosnia and

L. Globevnik (✉)
Faculty of Civil and Geodetic Engineering, University of Ljubljana, Ljubljana, Slovenia
e-mail: Lidija.Globevnik@guest.arnes.si

M. Brilly (ed.), *Hydrological Processes of the Danube River Basin*,
DOI 10.1007/978-90-481-3423-6_13, © Springer Science+Business Media B.V. 2010

Herzegovina, Montenegro, Kosovo, Albania, and Serbia. The Sava River springs in the Alpine Mountains in NW Slovenia, passes Croatia and Bosnia and Herzegovina, and after 945 km discharges into the Danube River in Belgrade.

The first integrated water management study of the river basin was carried out by UN experts 1972 [1]. The study was never accepted as a water management plan. In the 1970s, however, co-ordinated action was undertaken for a comprehensive study of the basin. By 1988 the major components of the study had been completed including an overview of monitoring and flood protection needs. After it began operating in 1994, the monitoring, laboratory and information management subgroup of the Environmental Programme of the Danube River Basin (Danube Programme) established five monitoring border and transborder stations on the Sava, one on the border between Slovenia and Croatia and four on the border between Croatia and BIH (Bosnia and Herzegovina) (http://www.icpdr.org/DANUBIS/, http://www.rec.org/danubepcu). On December 3, 2002, the riparian countries, Slovenia, Croatia, Bosnia and Herzegovina and Serbia and Montenegro, signed the Framework Agreement on the Sava River Basin and established the International Sava River Basin Commission (http://www.savacommission.org/). The Agreement regulates several aspects, among others comprehensive water management and environmental protection of the Sava River Basin.

In the development of the river basin management plan all countries are already collaborating under International Commission for the Protection of the Danube River (ICPDR) guidance. Up to 1991, the methodological bases for data collection had been reasonably unified over the catchment, but the data was lacking a lot of today's important aspects such as the ecological character of the river and its tributaries, inventory of pollution sources, dangerous substances, socio-economic parameters, cost and benefit implications etc. For the latter period, a great deal of data is missing due to insufficient monitoring (financing, recent warfare) and weak institutional and legal control over the use of water and land resources of the Sava River catchment. Many aspects of the river quality need scientific investigation. Furthermore, there is a need to link the knowledge of river quality state and environmental and health risk with pressures and their driving forces to propose efficient and beneficiary actions and measures for protection. In the project specific tools based on a combination of chemical analysis and biological effect methods will be developed and validated for the pollution of sediments and impact on water biota. These were the reasons for development of the SARIB (Sava River Basin) project in the Sixth Framework Program (Priority: Specific measures in support of international co-operation – Western Balkan countries).

An international team of 11 partners from eight countries was established to develop the project (http://www.sarib.net/). The project consists of six work packages: WP1 – Co-ordination, WP2 – Database and tools, WP3 – Development and Validation of Specific Tools divided into WP3.A – Pollution of sediments and water cycling processes and WP3.B – Availability and impact of pollutants on biota, WP 4 – Integrated system for the management of the Sava River quality, WP 5 – Social, economic and governance benefits and WP6 – Dissemination. In this paper, the main results of this research project achieved between 2003 and 2007 are briefly presented.

13.2 Database and Tools

Within the project, we developed two tools, the SARIB DataBase and the SARIB GIS. The first tool is a relational database. Here data on water chemistry and ecotoxicology obtained during SARIB research activities are integrated. It contains a list of sampling locations, results of laboratories analyses and information on hydrology that is related to a longitudinal profile. Information on water and related information needed for indicator evaluation are organised using the DPSIR (Driver-Pressure-State-Impact-Response) approach. All information is related.

13.2.1 The GIS Tool for the Sava River Basin

The main purpose of the raster image is positioning the Sava River Basin on the Balkan Peninsula, rapid orientation (administrative borders) and to provide basic information about relief data. The size of data cells is approximately 1,300 × 1,300 m. The raster image, mentioned above, was graphically re-designed, unnecessary data was deleted. The image was scaled, orientated, georeferenced and transformed to the WGS 1984 coordinate system, Lambert conformal conic projection.

The data source for administrative borders is the ESRI data DVD (Europe). Only administrative borders, which cover the Sava River Basin, were selected and transformed to the WGS 1984 coordinate system, Lambert conformal conic projection.

The Sava River Basin layer is the result of four participants' Sava River catchments: Slovenia, Croatia, Bosnia & Herzegovina and Serbia & Montenegro. Data was available from each country in DWG or SHP format, in different coordinate systems. Each piece of data was transformed into the WGS 1984 coordinate system, Lambert conformal conic projection (http://www.ksh.fgg.uni-lj.si/sarib/). The watershed area of the Sava River basin as reported in GIS format by national data sources and then integrated into the SARIB GIS Tool is 102112 km^2. This is 4399 km^2 more as evidented by the Sava Commision.

The Sava River springs in Slovenia and runs into the Danube River in Serbia. Almost all of the data was available in SHP or DWG format, provided by SARIB partners. The input data, provided by SARIB partners, was transformed into the Lambert conform conical projection, datum WGS84. The Sava river stream and its tributaries in Croatia were digitised, and all separated lines of the Sava River were connected to a single polyline, which represents the Sava River. Part of the Danube River, near the confluence of the Danube River and the Sava River, was digitised on the basis of a georeferenced map (1982), Lambert conform conical projection; scale 1:500,000 (http://www.ksh.fgg.uni-lj.si/sarib/).

Points where monitoring of water quality, water quantity and meteorological data occurred are properly georeferenced and incorporated into GIS. Locations of sampling points and quality data were provided by the Euro water net and include data along the Sava River and its tributaries (BOD5 years

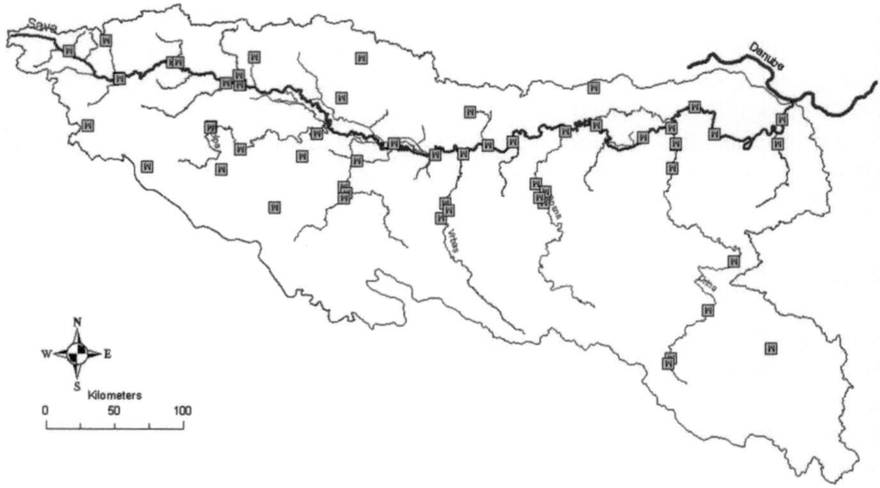

Fig. 13.1 The Sava River Basin, water quality sampling locations

1992–2005, N-tot years 1992–2005) (Fig. 13.1), http://dataservice.eea.europa.eu/dataservice/metadetails.asp?id=1081

Hydrological data were available in electronic form. Hydrological locations were available without geographic coordinates. The coordinates were found in the following publications/web sites: Hydrological Yearbook 1978, EIONET, Map of hydrological stations in Yugoslavia. Hydrological data from the hydrological stations of the Sava River and its tributaries was placed into one Excel file (http://www.ksh.fgg.uni-lj.si/sarib/).

Meteorological data were available in electronic and non-electronic forms (scanned pictures, tables). Most meteorological data were available without geographic coordinates. Coordinates were found in Hydrological Yearbook 1978 (Savezni hidrometeorološki zavod 1980). Additional meteorological stations were found to improve the frequency in the Sava River catchment, http://www.ksh.fgg.uni-lj.si/sarib/. The meteorological data of all (available) meteorological stations were placed into one Excel file, including precipitation, humidity and air temperature data.

Points where sampling data for the SARIB project were collected were also referenced and incorporated into GIS (http://www.ksh.fgg.uni-lj.si/sarib/). The distance from the confluence of the Sava and Danube Rivers is represented by points along the Sava stream. The source of these locations is the Hydrological Yearbook of Yugoslavia, 1976. The distance from the confluence is given in kilometres; precision is one decimal place.

Raster data were also collected and rearranged for SARIB GIS. Almost all of the CORINE land cover data were available in SHP format, provided by SARIB partners. Data was transformed from ETRS89 to WGS84 and then the Lambert conformal conical projection. During building of the CORINE land cover 2000 GIS system some problems occurred, relating to projection and geographical transformations. Because of these problems, some data was applied from the European Environment Agency (EEA). There is no vector data available for the SARIB partner Serbia & Montenegro. Dataset scale for CORINE land cover equals 1:100,000. CORINE land cover data was aggregated to the second level code (examples: polygons with the codes 111 and 112 were merged into polygons with the code 11; polygons with the codes 331, 332, 333, 334 and 335 were merged into polygons with the code 33).

From CORINE data set two GIS layers were developed, areas with significant point emissions and areas with significant diffuse emissions. For the first one, urban, industrial and commercial areas were merged (polygons with codes 111, 112, 121, 122 and 124). Arable (polygons with the codes 211 and 212) and complex agricultural areas (polygons with the code 242) were joined with areas covered by permanent crops (polygons with the codes 221 and 222) to represent data set of areas with significant diffuse emissions.

Geological data were available for three countries, all provided by SARIB partners. Available data was transformed into the Lambert conformal conical projection. Some problems relating to projection and geographical transformations occurred during transformation. The data source, provided by the Slovenian SARIB partner, includes a basic engineering geology map, at a scale of 1:200,000. The data source, provided by the SARIB partner Bosnia & Herzegovina, includes FAO classification. The geological classification layer of Serbia & Montenegro includes information on geological period and a basic geological foundation (http://www.ksh.fgg.uni-lj.si/sarib/).

13.2.2 Development of the SARIB Database

Data from sampling and laboratory testing during the SARIB project were put into one relational database with the use of replications. For each replicate a data quality assurance check was carried out prior to data input. Copies of replications were used independently at different locations, and the individual replicas were later synchronised with the "master" copy of the database. Basic relations of the SARIB database are shown in Fig. 13.2.

The relational database model was created in the MS ACCESS environment for the collation of all sampling data collected by all partners during the SARIB project. Instructions on adding data to the database were prepared and disseminated to all partners. The SARIB relational database contains a list of sampling locations and results of sampling campaign parameters (chemical analyses, biological parameters and hydromorphological data). SARIB partners sent data via their database replicate. The database contains data on measurements, laboratory tests and 2732 observations carried out and collected for the SARIB project. The SARIB database is available at http://www.ksh.fgg.uni-lj.si/sarib/.

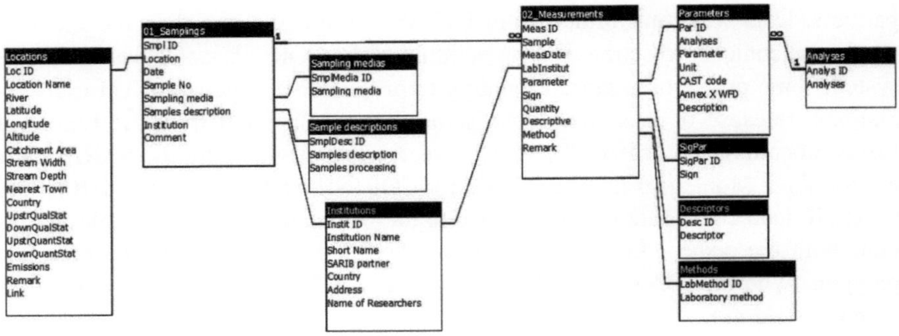

Fig. 13.2 The basic SARIB database relations

13.3 Characteristics of the Sava River Sediments

13.3.1 Granulometric Analysis of Gravel-Bar Sediments

13.3.1.1 Historical and SARIB Gravel-Bar Sampling

SARIB sampling was performed between October 24 and November 16, 2006 at 16 locations. The investigated reach of the Sava River extended from the river source in NW Slovenia to the boundary hydrological cross-section with Croatia and further to the confluence with the Una River in Croatia (Fig. 13.3). The identification of potentially suitable gravel bars for sampling was carried out through the examination of airborne orthophoto images from 2003, which are accessible in the Interactive Environmental Atlas of Slovenia (IEAS 2008), and of high-resolution space-born images from Google Maps (Google 2008). After the field survey approximately half of the previously identified gravel bars were sampled, some of them had already been exploited (mainly in the Slovenian lower Sava River reach) or they were inaccessible (in the Slovenian middle and upper Sava River reaches). The locations of the 16 sampled gravel bars are given in Table 13.1.

Two line transect samples per gravel bar were taken from 11 sites and one bulk sample per gravel bar from ten sites. The granulometric characteristics of the bed load were assumed to be similar to the grain size distribution of the riverbed, and were therefore determined by the previously mentioned volumetric (bulk) sampling of the riverbed subsurface and by transect (Wolman) sampling of the riverbed surface (Mikoš 2008). The bulk samples were air-dried, sieved and the fractions weighed. The line transect (Wolman) samples were not sieved but rather recalculated into the volumetric ones using empirical coefficients, and then rigidly combined with the Fuller grading curve into a volume sample (Anastasi 1984). The line transect sampling was very similar to the Wolman pebble count method, with minor differences. A rope, fixed with two wedges, was stretched along the gravel bar parallel to the stream direction and close to the ground surface (Fig. 13.4). Each pebble or cobble lying on the gravel-bar surface below the rope line was measured

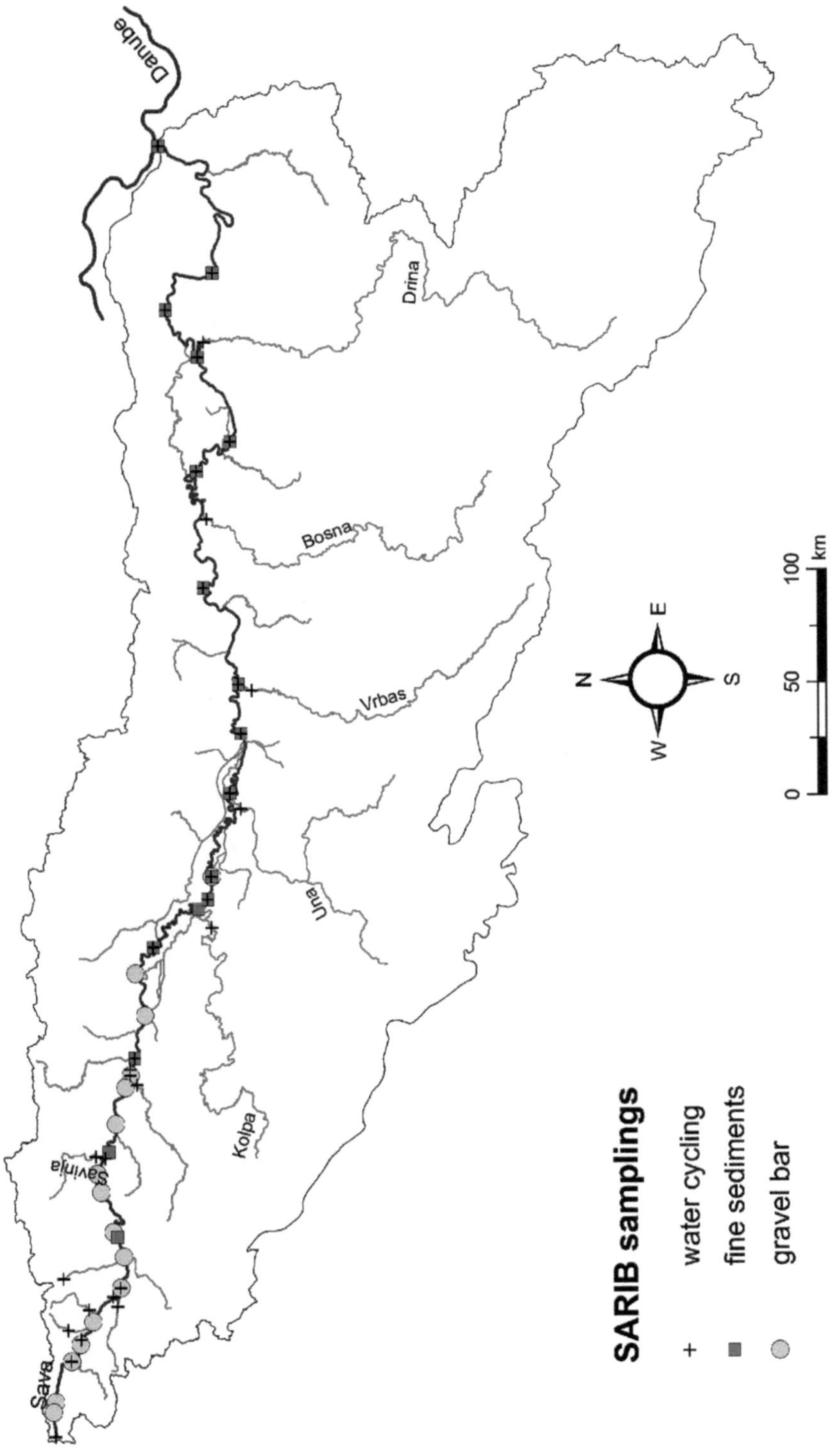

Fig. 13.3 The Sava River Basin with marked sampling locations

Table 13.1 Locations with obtained data on river bed granulometric characteristics

Place	River stationing (km)	Historical data (from the 1950s)	Historical data (from the late 1960s)	Historical data (from the late 1970s)	SARIB sampling (2006)
Čatež	737.5	1958	–	–	x
Vrbina	747.0	–	1968	–	x
Mrtvice	747.0	–	1968	–	x
Dolenje Brezovo	762.8	–	–	–	x
Radeče	784.0	–	1968	–	–
Hrastnik	792.9	–	–	–	x
Zagorje	801.0	–	–	–	x
Kresnice	828.8	–	–	–	x
Hotič	831.0	–	–	1976	–
Dol pri Ljubljani	841.4	–	–	–	–
Šentjakob	847.0	1958	–	–	–
Tacen	857.0	–	–	–	X
Zarica	875.0	–	1966	–	–
Okroglo	879.9	–	–	–	x
downstream Tržiška Bistrica	887.0	1952	–	–	–
Podnart	898.0	–	1968	–	–
Globoko	804.5	–	–	–	x
Blejski most	905.0	–	–	–	x
Mojstrana	928.0	–	1968	–	–
Na Belah	933.7	–	–	–	x
Belca	938.2	–	–	–	x
Kranjska Gora	940.4	–	1969	–	–
Jankomir	712.5	–	–	–	x
Hruščica	686.0	–	–	–	x
Lukavec	608.0	–	–	–	x

Fig. 13.4 The line transect sampling on the Vrbina (*left*; 747.0 km) and Dolenje Brezovo (*right*; 762.8 km) gravel bars

Table 13.2 Grain size classes applied in the pebble count method

Interval (Mm)	< 10	10–80	80–200	200–400	400–650	> 650
Number of classes	1	7 × 10 mm	6 × 20 mm	5 × 40 mm	5 × 50 mm	1

Fig. 13.5 Bulk sampling of the subsurface layer on the Na Belah gravel bar (933.7 km)

with the millimetre gauge. According to its measured middle axis (being approximated by an ellipsoid) it was sorted into the appropriate size class, which was chosen as shown in Table 13.2. Finally, the number of pebbles and cobbles in each of the classes was determined. Each line transect sample consisted of at least 100 pebbles, as is suggested for the Wolman pebble count method.

Because no additional field geodetic survey was performed during the SARIB project, many river morphological parameters, such as the longitudinal profile and some bed-load granulometric characteristics were obtained from the available historical data for ten locations (Table 13.1).

Prior to the bulk sampling, the gravel-bar surface layer with an area of approximately 1 m^2 and 20 cm thickness was removed (Fig. 13.5). The bulk samples were then taken uniformly from the uncovered area of the subsurface layer. The weight of bulk samples varied according to the grain size of the sampled gravel bar. As a rule, it ranged around 50 kg. Until mechanical sieving was carried out the bulk samples were stored in bags and then dried in air.

Analysis of the Line Transect Samples

Firstly, the numerical analysis was carried out in such a way that the pass-through fraction (grains finer than sieve size), was calculated from the average number of grains in each of the size classes from both line transect samples taken on one gravel bar using empirical coefficients (Anastasi 1984, Fehr 1987). Calculation of volumetric-weights via numerical analysis with rigid combination with the Fuller grading curve in the bulk sample was carried out using a simple spreadsheet program. After the parameter needed for the rigid combination with the Fuller grading curve was estimated (namely the overlapping interval of both grain size distributions), the grain size distribution of the volume sample was determined. In this way, the distribution of fines in the computed volume sample follows the Fuller curve, and the coarse grains are represented by the line transect sample.

Bulk Samples Analysis

The sieve analysis for coarse grains was carried out with a mechanical sieve. Dried bulk samples were sieved through the standard sieves: 2.8, 4, 8, 12.5, 16, 25, 31.5, 50, 63, 75, 90, 100 and 126 mm. The amount of sediments on sieves was precisely measured, with a weighing balance with a resolution of 0.1 g, and the grain size distribution determined. The size of the maximal grain from each bulk sample was also measured.

Characteristic Grain Analyses

Characteristic grains (such as d_{10}, d_{20} and so on until the 90% grain – d_{90}) were determined from the grain size distributions for both sampling methods, respectively, with implementation of linear interpolation. The arithmetic mean grain d_m was determined with the integration of the grain size distribution curve. Characteristic grains d_{16} and d_{84} were used for calculation of the geometric sorting index.

The determined characteristic grains were used for comparison of surface and subsurface layers. Because of the present riverbed armouring process along the Sava River due to restricted sediment supply from tributaries, the river is in many locations in the state of latent erosion. Therefore, during high flows, mainly fine grains are flushed over the coarse riverbed surface, and we expected the surface layer to be coarser than the subsurface layer.

The sieve analysis results were compared with those from the gravel-bar bulk sampling that was carried out at different locations between 1952 and 1976. The idea was to state the riverbed grain size distribution changes of the Sava River.

13.3.1.2 Results and Discussion

The results of the analyses of grain size distribution of the Sava River sediment samples are given on the project web page (http://www.ksh.fgg.uni-lj.si/sarib/.) and for the selected gravel bars are shown in Figs. 13.6, 13.7 and 13.8.

Each grain size diagram consists of up to three grain size distributions:

- grain size distribution based on numerical analysis of the line transect samples from the surface of a gravel bar;
- Grain size distribution based on calculation of volumetric-weights via numerical analysis with rigid combination with the Fuller grading curve, and
- grain size distribution based on weight (sieving) analysis of bulk samples taken from the subsurface of a gravel bar.

Fig. 13.6 Grain size distribution of samples taken at Na Belah (933.7 km)

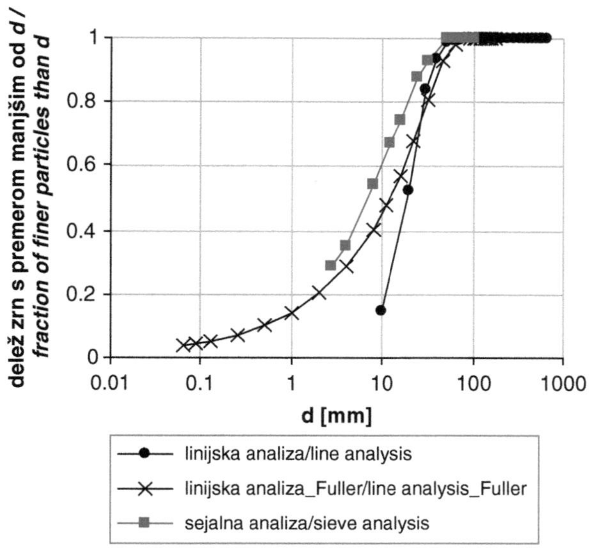

The size of the characteristic grains determined from the grain size distributions is shown longitudinally by sampling site in Figs. 13.9 and 13.10. Comparing the two diagrams (note that not all sampling sites are the same on both diagrams!) there is no clear decreasing trend for both layers, surface and subsurface of the riverbed. The diagrams show that grain size distribution of the surface layer is typically coarser than that of the subsurface.

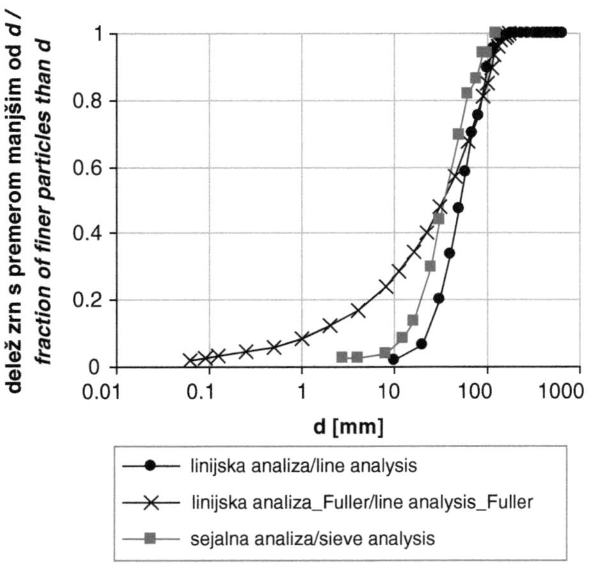

Fig. 13.7 Grain size distribution of samples taken at Dolenje Brezovo (762.8 km)

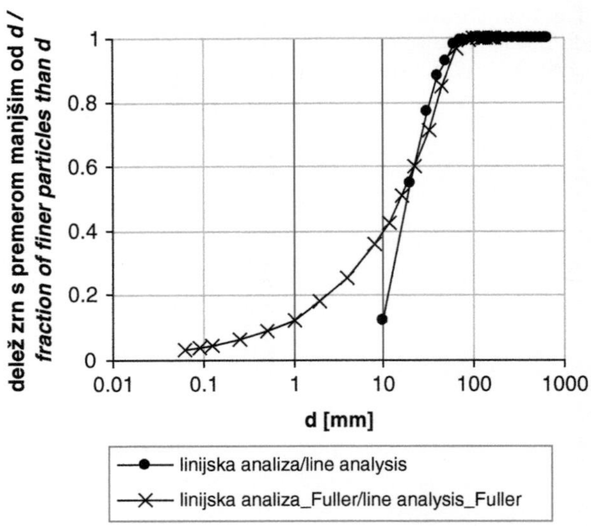

Fig. 13.8 Grain size distribution of samples taken at Vrbina (747.0 km)

Figure 13.11 shows the ratio between the size of the characteristic grain of the surface layer (determined with numerical analysis – count) and the subsurface layer (determined with weight analysis of bulk samples – sieving) at each sampling site. Because of the insensitivity of pebble count sampling towards finer grains, the ratios for finer characteristic grains (d_{10}, d_{20}, d_{30}) are as a rule several times higher than 1, however, for coarser characteristic grains (d_{95}, d_{80}, d_{70}) the ratio is closer to 1 by several tens of percent. There are two exceptions to be seen in Fig. 13.11. The first is

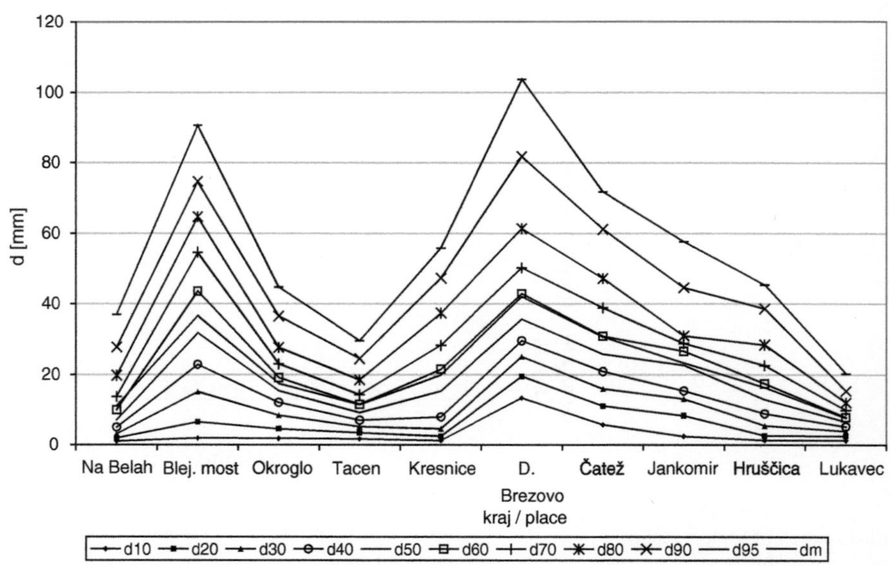

Fig. 13.9 Characteristic grains of the bulk samples

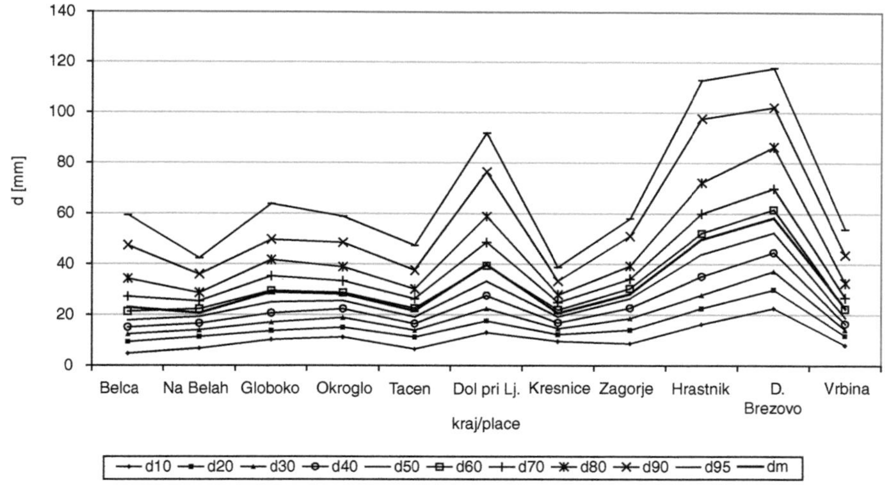

Fig. 13.10 Characteristic grains of the line transect samples

the grain size distribution of characteristic grains from Kresnice, which drops below the value of 1 for coarse characteristic grains. The sample was taken on a gravel bar 200 m upstream of a groundsill and should be therefore excluded from any analysis such as longitudinal trend.

Another exception is the grain size distribution of the characteristic grain at Dolenje Brezovo, progressing slowly from characteristic grain d_{10} to d_{95}, indicating great similarity between the samples of the surface and subsurface layers.

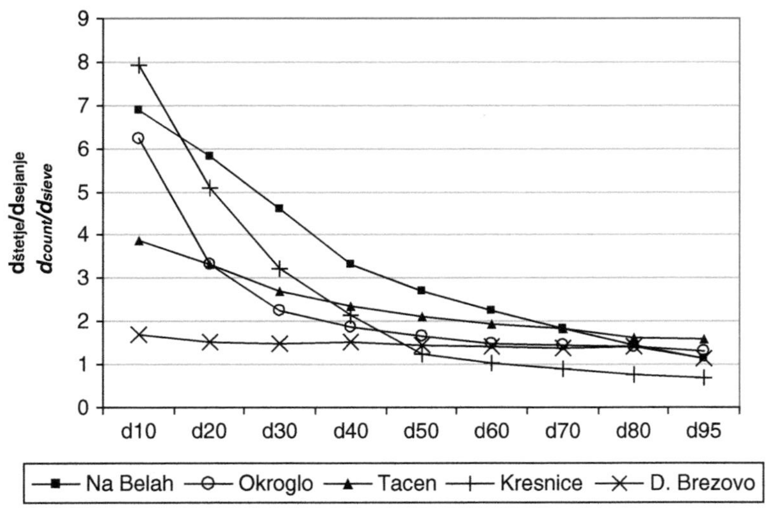

Fig. 13.11 Ratio between the size of the characteristic grain of the surface layer (determined with numerical analysis – count) and the subsurface layer (determined with weight analysis of bulk samples – sieving)

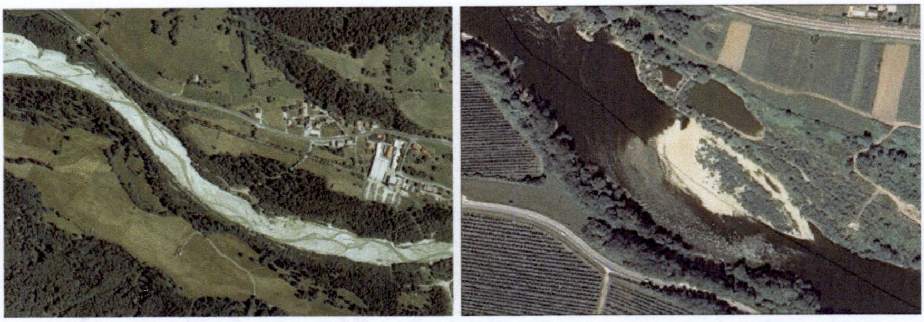

Fig. 13.12 Airborne orthophoto of the gravel bar at Na Belah (*left*; 933.7 km) and at Dolenje Brezovo (*right*; 762.8 km) (www.geopedia.si)

The pass-through (percent of finer grains) of the sieve analysis of the sample from Dolenje Brezovo (Fig. 13.7) shows that the occurrence of fine grains in the subsurface layer is very rare. This could be a consequence of the local inclination of the river channel (see Fig. 13.12, right) and the subsequent increase of flow velocity.

Similarly to Fig. 13.11, Fig. 13.13 shows the ratio between the size of the characteristic grain of the surface layer (determined from numerical analysis – with rigid composition of the line transect into volumetric samples with the Fuller grading curve) and the subsurface layer (determined with the weight analysis of bulk samples – sieving) per sampling site. It can be observed that the finer characteristic

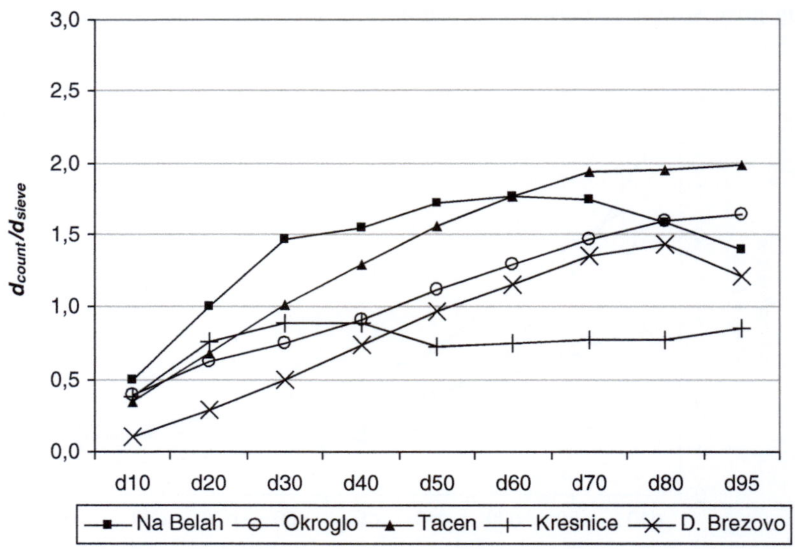

Fig. 13.13 Ratio between the size of the characteristic grain of the surface layer (determined from numerical analysis – with rigid composition of the line transect into volumetric samples with the Fuller grading curve) and the subsurface layer (determined with weight analysis of bulk samples – sieving)

Table 13.3 Geometric sorting index of the applied sample analyses

Location	Sieve analysis	Pebble count	Pebble count + Fuller
Belca	–	2.24	5.03
Na Belah	3.78	1.70	5.25
Blejski most	4.49	–	–
Globoko	–	1.91	5.28
Okroglo	3.18	1.76	5.27
Tacen	2.79	1.81	5.18
Dol pri Ljubljani	–	2.04	5.19
Kresnice	4.64	1.61	5.20
Zagorje	–	1.89	5.24
Hrastnik	–	2.00	5.20
Dolenje Brezovo	2.00	1.85	5.21
Vrbina	–	1.82	5.23
Čatež	2.40	–	–
Jankomir	2.41	–	–
Hruščica	3.89	–	–
Lukavec	2.55	–	–
Mean value	3.21	1.87	5.21
Standard deviation	0.93	0.16	0.06

grains are substantially smaller because of the rigid composition with the Fuller grading curve, and the coarse grains are somewhat larger.

The calculated values for the geometric sorting index are shown in Table 13.3. It is understandable that the maximum values for the geometric sorting index were found from the results of the numerical analysis with the fine grain distribution assumed according to the Fuller grading curve, and also the minimum values for the geometric sorting index from the results of the numerical analysis, which is insensitive to fine grains. The greatest standard deviation from the average value of the geometric sorting index was found when the sieve analysis was applied, while the results of the numerical analyses yield quite a uniform geometric sorting index. Thus, the sieve analysis offers the best insight into each of the individual sample characteristics. Weights of samples are given in Table 13.4.

Grain size distribution is strongly subjected to local flow conditions close to the gravel bar. This is also the reason why the comparison between the results of sampling in 2006 and sampling between 1952 and 1976 is a problematic one. Only for the Čatež gravel bar was sampling performed in both periods. From the longitudinally presented results in Fig. 13.14 only section changes in grain size distribution could be ascertained, originating from stream channel geomorphologic changes or newer flow disturbances due to recently built hydraulic structures. According to the results of other sampling series (VGI-VL 1980, Colarič 1983) it was possible to compare only d_{30}, d_{50}, d_{90} and d_m characteristic grains. Table 13.4 supplements Fig. 13.14 with site stations, year of sampling and weight of samples.

The results from the Čatež gravel bar show a slight decrease of the d_{50} characteristic grain, from 27.4 to 25.7 mm, while d_m decreased from 46.3 to 30.7 mm. The geometric sorting index has also decreased because of greater fine and smaller

Table 13.4 Sampling locations and weight of samples

Location	Year of sampling	Stationing (km)	Sample weight (kg)
Lukavec	2006	608.0	14.0
Hruščica	2006	686.0	18.1
Jankomir	2006	712.5	36.7
Čatež	1958/2006	737.5	125/88.9
Mrtvice	1968	747.0	125
Dolenje Brezovo	2006	762.8	58.2
Radeče	1968	784.0	300
Kresnice	2006	828.8	62.4
Hotič	1976	831.0	/
Šentjakob	1958	847.0	250
Tacen	2006	857.0	34.0
Zarica	1966	875.0	125
Okroglo	2006	879.9	44.1
Under Tržiška Bistrica	1952	887.0	125
Podnart	1968	898.0	125
Blejski most	2006	905.0	73.9
Mojstrana	1968	928.0	125
Na Belah	2006	933.7	38.4
Kranjska Gora	1969	940.4	125

coarse characteristic grains, as shown in Fig. 13.14. This also holds for the whole treated river section, as the ratio between d_{90} and d_{30} is considerably lower. At this point it should be noted that the sample weights of the older sampling series were much greater than those from the 2006 sampling (Table 13.4).

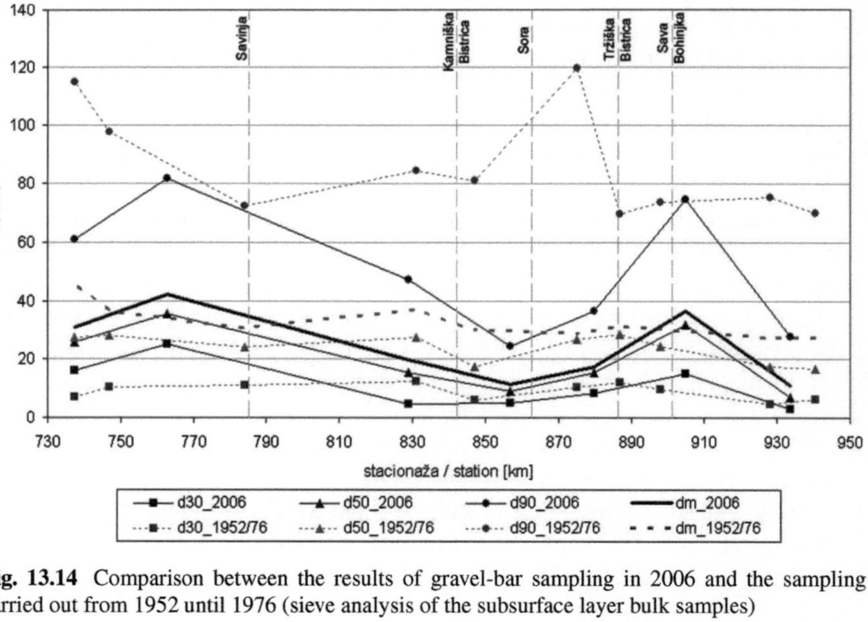

Fig. 13.14 Comparison between the results of gravel-bar sampling in 2006 and the sampling carried out from 1952 until 1976 (sieve analysis of the subsurface layer bulk samples)

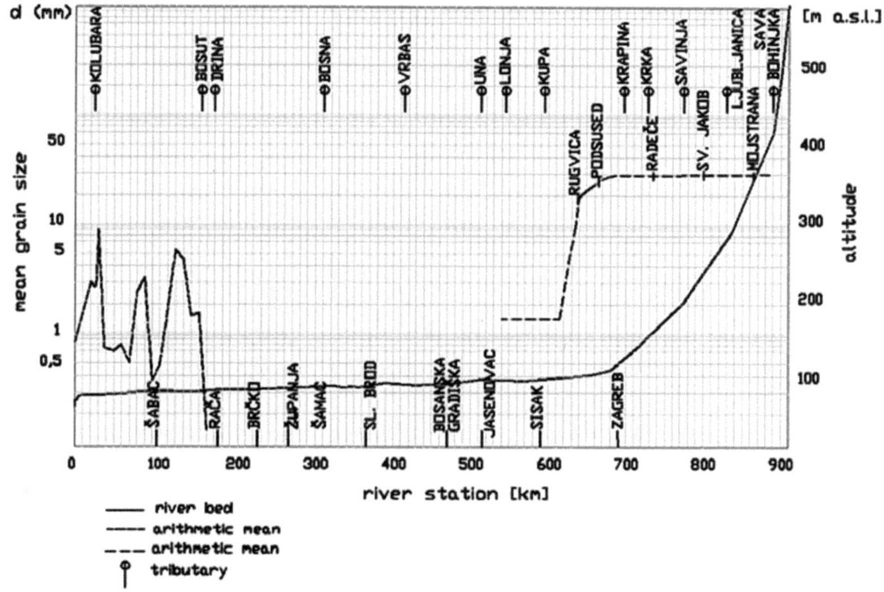

Fig. 13.15 The longitudinal profile of the Sava River showing the arithmetic mean size of river sediments

In general, the size of the arithmetic mean grain d_m along the Sava River flow in Slovenia has decreased. Above all this is noticeable on the river section between the confluence of Tržiška Bistrica and the HPP Boštanj dam in the lower flow of the river. The opposite, an increase of the arithmetic mean grain d_m, is observed at the Dolenje Brezovo gravel bar, which is strongly influenced by local flow conditions, and at the Blejski most gravel bar.

The Sava River in its upper reach in Slovenia is a typical gravel-bed river (mean grain size larger than 30 mm) with a sudden transition downstream of Zagreb into a sand-bed river (Fig. 13.15).

13.3.1.3 Conclusions

According to the results of sampling of the Sava River gravel-bar between 1952 and 1976 it was established that the bed-load arithmetic mean grain in the stretch between its source and Zagreb (Croatia) (Mikoš 2000) was of quite a uniform size of around 30 mm. The results of the 2006 gravel-bar sampling show a decrease of the arithmetic mean grain size, which is also more variable along the stream channel, while the grain sorting has become lower. It should be noted that the sample weights of the older sampling series were much greater than those from the 2006 sampling. The latest Sava River sediment sampling series also offers grain size distribution data for the gravel-bar surfaces, a detail that was not given enough attention during previous field campaigns on river sediments (Mikoš, 1999). The analysis has shown that the arithmetic mean grain of the surface layer is on average approx. 1.5-times larger than the arithmetic mean grain of the subsurface layer.

13.3.2 Granulometric Analysis of Fine Sediments

13.3.2.1 Method of Sampling

Sampling was performed in 2005 and 2006 at selected sampling sites along the Sava River from its spring to its confluence with the Danube River (Fig. 13.3; see Table 13.5 for details). Sediment cores were collected using a piston corer, which was released from a set distance above the riverbed, penetrated the sediment by free fall, and sucked the sediment into the core barrel via an upward moving piston as the core was retrieved. Plastic core liners were placed inside the core barrels to contain the sediment core sample, and to avoid extrusion and contamination. After the corer was retrieved, the liner was removed from the barrel, the top was capped, and the core set in a vertical position. Afterwards cores were extruded into 1,000-ml polyethylene containers and transported to the laboratory, where additional analyses were performed.

13.3.2.2 Results of Laboratory Work

Identifying soils in the laboratory was done by determining the gradation and plasticity characteristics of the materials. The gradation was determined by wet sieve

Table 13.5 Results of grain size analysis (*x*,*y*: Gauss-Krüger coordinate system in WGS84)

				d_{10}	d_{30}	d_{60}	<2 mm	<63 mm	<2 mm	Cu	Cc
LOCATION	x	y	USCS	mm	mm	mm	%	%	%	/	/
Jesenice	45.8617	15.6839	SP	0.126	0.178	0.244	96.05	0.60	0.00	1.90	1.03
Jevnica	46.0883	14.7475	SM	0.024	0.134	0.229	100.00	15.40	1.80	9.70	3.35
Vrhovo	46.0453	15.2153	M	0.003	0.007	0.023	100.00	78.90	8.60	8.60	0.85
Oborovo	45.6870	16.2468	M	0.003	0.011	0.102	99.41	50.40	8.30	40.50	0.50
Galdovo	45.4820	16.3859	SP-SM	0.094	0.163	0.211	100.00	7.80	1.20	2.20	1.34
Košutarica	45.2510	16.9526	M	0.002	0.011	0.071	100.00	58.00	10.40	27.20	0.86
Gradiška	45.0600	17.3000	M	0.001	0.008	0.059	100.00	62.50	12.40	39.50	0.79
Srbac	45.1088	17.5157	M	0.002	0.015	0.089	100.00	51.40	10.60	48.40	1.35
Slavonski Brod	45.1397	18.0740	M		0.002	0.01	100.00	90.80	29.00		
Županja	45.0398	18.6987	M		0.004	0.029	100.00	73.30	18.80		
Brčko	44.8822	18.8037	M		0.004	0.015	100.00	83.30	18.70		
Bosanska rača	44.9096	19.2955	M		0.003	0.008	100.00	96.80	24.10		
Sremska Mitrovica	44.9748	19.5932	SM	0.044	0.183	0.271	98.73	12.40	2.10	6.20	2.83
Šabac	44.7606	19.7075	M		0.002	0.011	100.00	88.20	27.30		
Beograd	44.8146	20.4465	M		0.012	0.108	100.00	55.60	15.80		
Crnac	45.4384	16.4253	SM	0.004	0.044	0.132	100.00	33.40	6.80	35.50	3.95
Lukovac	45.4015	16.5390	SM	0.002	0.012	0.13	100.00	45.80	9.70	62.70	0.58

Cu [–] ... coefficient of uniformity, expressed as d_{60}/d_{10}

Cc [–] ... coefficient of curvature, expressed as $d_{60}^2/$ (d10 x d_{60})

analysis. Grain sieve analysis is a determination of the proportions of particles lying within certain size ranges in a granular material by separation on sieves of different size openings. The result of this analysis is a grain-size curve, which is plotted as percent finer by weight against a logarithmic scale of grain size in millimetres.

Standards used for determination of the coefficients were standards for the general classification of soils for engineering purposes (Standard 1980), and the determination of particle size distribution (Standard 2004). Because of the very small particles in the silt/clay fraction, it is necessary to make this initial split of the sediment using a wet sieve analysis.

The general classification of soils for engineering purposes (USCS) is based on identifying soils according to their textural and plasticity qualities and on their grouping with respect to behaviour. Soils seldom exist in nature separately as sand, gravel, or any other single component. They are usually found as mixtures with varying proportions of particles of different sizes; each component part contributes its characteristics to the soil mixture. The USCS is based on those characteristics of the soil that indicate how it will behave as an engineering construction material:

- percentages of gravel, sand, and fines (silt and clay) (Fig. 13.16)
- shape of the grain-size distribution curve
- plasticity and compressibility characteristics

The comparison of the curves (Figs. 13.17 and 13.18) does not show any clear downstream trend for fine fractions in samples. The fine fraction is much more influenced by fresh sediment supply from the main Sava River tributaries such as the Kolpa, Vrbas, and Drina rivers. This again shows the importance of so-called sedimentary links for assessment of the sediment regime (Rice 1999).

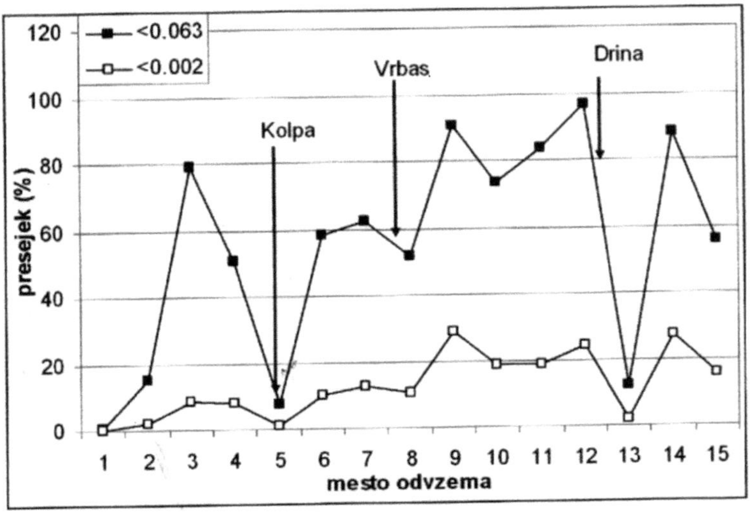

Fig. 13.16 The silt and clay (<0.063 mm) and clay fraction (<0.002 mm) at sampling sites

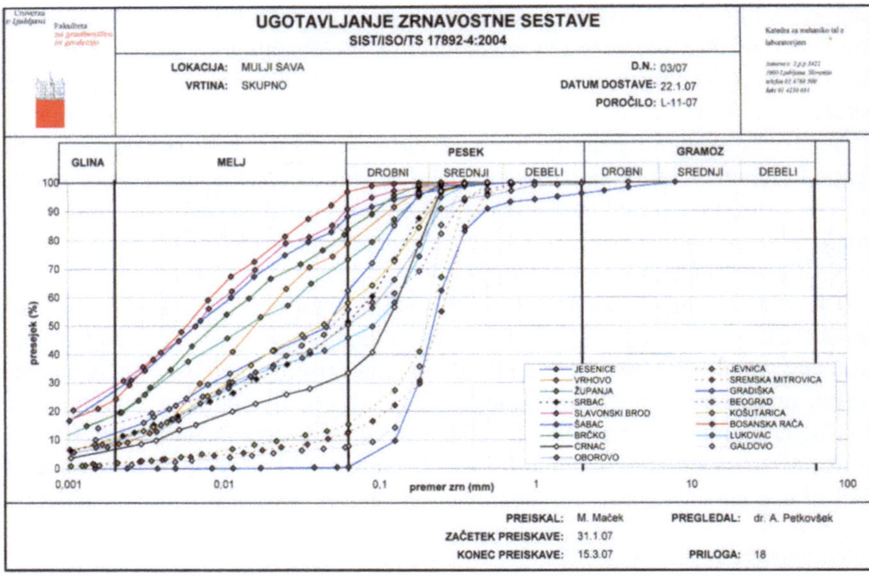

Fig. 13.17 Grain size distribution SIST/ISO/TS 17892-4:2004; all samples

Grain size distributions show that the samples taken are mostly silt (M) or clay (C). In all of the samples, except the Šabac sample, we noticed a uniform composition of sand particles (for example: Jesenice and Jevnica). There is also no clear correlation between the part of fine fractions and location in the Sava River Basin.

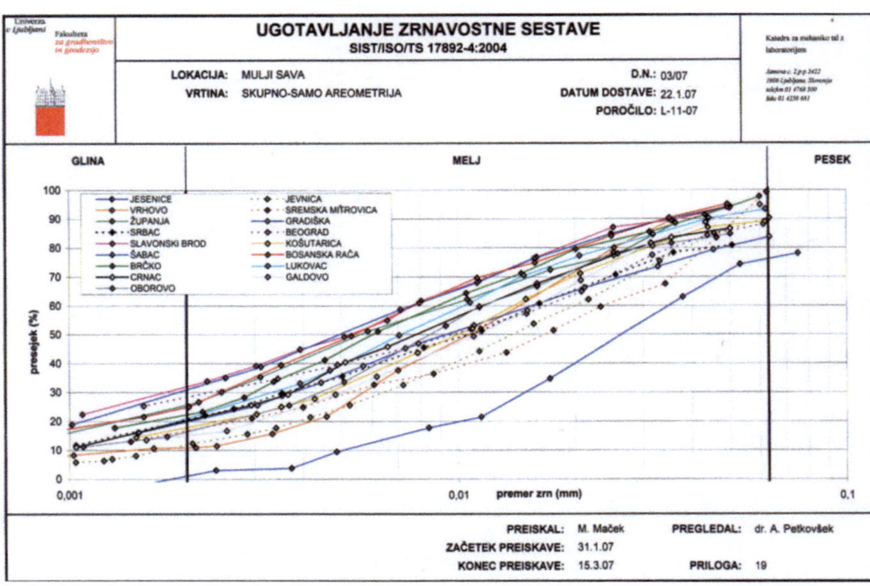

Fig. 13.18 Grain size distribution SIST/ISO/TS 17892-4:2004; all (areometry)

13.4 Pollution of Sediments

13.4.1 Introduction

Water is one of the most important natural resources. To protect water resources data on the presence of polluting substances in aqueous environments are necessary. For this purpose, monitoring of water quality is regularly carried out. In addition, chemical analyses of sediments are commonly applied since they reflect spatial and temporal variation of concentrations of elements (Loring and Rantala 1992, Škrbić and Čupić 2004) and organic pollutants (Škrbić et al. 2005, 2007). Among elements, Cd, Pb, Cu, Zn, Cr, Ni, Hg, As and P are often accumulated in sediments due to anthropogenic activities (Giusti and Taylor 2007, Grosbois et al. 2006, Hines et al. 2006, House and Denison 2002, Svete et al. 2001). The most frequently investigated organic pollutants that appear in sediments as a result of human activities are polycyclic aromatic hydrocarbons (PAH) (Škrbić et al. 2005, Barth et al. 2007), polychlorinated biphenyls (PCB) (Wang et al. 2007, Škrbić et al. 2007) and pesticides (Barth et al. 2007, Škrbić et al. 2007).

The national environmental agencies regularly monitor the quality of the Sava River water. However, data on the environmental status of the river basin are still lacking. Therefore, one of the aims of the SARIB project was to study the extent of pollution of selected elements and persistent organic pollutants in sediments along the Sava River. The choice of pollutants depended on the expected sources of pollution from the area investigated and followed also the recommendations of the Water Framework Directive. Total element concentrations and highly mobile element fractions (extraction in 0.11 mol l^{-1} acetic acid) were determined. For identification of the anthropogenic inputs of pollutants to sediments normalisation of total element concentrations to Al was performed. In addition, selected persistent organic pollutants were determined in sediments.

13.4.2 Materials and Methods

13.4.2.1 Sampling Sites, Sampling and Sample Preparation

Sampling was performed in April 2005, October 2005 and May 2006. Twenty sampling locations were selected along the Sava River with regard to sample representativeness and considering different anthropogenic sources of pollution. Samples were taken a few metres from the riverbank. From each location, about 3 kg of the top 15-cm sediment layer was collected using the piston plastic corers. Sediments from corers were extruded into 1,000-ml polyethylene containers, transported to the laboratory and homogenised. For comparability of analytical data to other river basins, wet sieving through a 63 μm sieve was applied (Borovec 2000, ICPDR 2002). After that samples were dried at 40°C for three days in the dark until constant weight, transferred into polypropylene containers and kept until the analysis at 4°C in the dark. The moisture content in sediments was determined based on the mass loss, by drying of samples to constant weight at 60°C. All the analyses were performed in two parallel determinations and the results expressed on a dry mass basis (details of sampling sites, sampling and sample preparation are presented in Milačič et al. 2009).

13.4.2.2 Analytical Procedures

Determination of total element concentrations in sediments: Cd, Pb, Zn, Cu, Ni, Cr, As, Al and P were determined by ICP-MS after microwave-assisted digestion using a mixture of nitric, hydrofluoric and hydrochloric acids (Ščančar et al. 2007).

Determination of the easily soluble element concentrations in 0.11 mol l^{-1} acetic acid: Easily soluble element concentrations were determined by shaking 2 g of sediments with 20 ml of 0.11 mol l^{-1} acetic acid on a mechanical shaker for 16 h (Svete et al. 2001). Element concentrations were determined by ICP-MS.

Normalisation to Al: In order to account for geochemical/geographical variations along the Sava River, normalisation by a conservative element Al was used. Significant deviations from the linear relationship may be used to differentiate between natural vs. anthropogenic inputs (Loring and Rantala 1992).

Determination of total Hg in sediments: Total Hg in sediments was determined by oxidative combustion using a DMA-80 Direct Mercury Analyser (www.milestonesci.com/merc-tech.php).

Determination of organotin compounds: For the analysis of organotin compounds (OTC) extraction in acetic acid was performed followed by derivatisation with sodium tetraethyl borate. Ethylated OTC species were then extracted into isooctane and their concentrations determined by GC-MS (Ščančar et al. 2007).

Determination of PAH: PAH were determined by accelerated solvent extraction – supercritical fluid extraction with methylene chloride and carbon dioxide (Heath et al. 2006, 2009).

Determination of PCB and chlorinated pesticides: PCB and selected chlorinated pesticides were extracted by Soxhlet extraction with hexane and determined after cleaning and fractionation on a Florisil column by GC-ECD (Heath et al. 2009).

13.4.3 Results and Discussion

13.4.3.1 Quality Control

The quality of data in element determinations was checked by the analysis of certified reference material CRM 320 (Trace Elements in River Sediment, Community Bureau of Reference, Geel, Belgium) and reference material IAEA 405 (Trace elements and methyl mercury in estuarine sediment, International Atomic Energy Agency, Vienna, Austria). To check the quality of data in OTC determinations reference material PACS 2 (Marine Sediment Reference Material for Metals and Other Constituents, National Research Council, Ottawa, Canada) was analysed. To verify the quality of data in organic pollutant determinations analysis of Reference Material IAEA-408: Organochlorine Compounds, Petroleum Hydrocarbons and

Sterols in a Sediment Sample from Mudflats of the Tagus Estuary (IAEA, Analytical Quality Control Services, Vienna, Austria) were performed.

Determined concentrations of elements and organic pollutants lay within the range of certified, reference and/or informative values, confirming the accuracy of the results reported in this study.

13.4.3.2 The Sava River Grain Size Distribution of Sediments

In Slovenia, the Sava River is a mountain river with a content of fine particles in sediment (<63 μm) between 40 and 60%. At the Slovenian–Croatian border the Sava River turns into a flat land river and the percentage of fine particles in sediments (<63 μm) is gradually increased, reaching up to 90% of the total sediment content (Milačič et al. 2009).

13.4.3.3 Analysis of the Total Element Concentrations in Sediments of the Sava River and Normalisation to Al

Total concentrations of selected elements in sediments of the Sava River are presented in Table 13.6, while normalisation of total element concentrations to Al in

Table 13.6 Determination of total concentrations of selected elements in sediments of the Sava River by ICP-MS and Hg by a DMA-80 Direct Mercury Analyser. Results represent average of two parallel samples ± standard deviation

Sampling site	Pb (mg kg^{-1})	Ni (mg kg^{-1})	Cr (mg kg^{-1})	Hg (mg kg^{-1})
Mojstrana	11.5 ± 0.1	12.6 ± 3.9	23.1 ± 1.3	0.035±0.001
Moste	40.9 ± 0.6	70.1 ± 1.1	184 ± 4	0.134±0.004
Jevnica	41.7 ± 3.7	27.3 ± 0.7	73 ± 1	0.388±0.043
Vrhovo	47.8 ± 3.4	36.9 ± 0.6	94 ± 2	0.397±0.011
Brežice	27.3 ± 2.7	17.6 ± 1.9	48.2 ± 0.6	0.247±0.006
Jesenice na Dol.	29.2 ± 6.6	12.3 ± 0.1	50.3 ± 0.8	0.235±0.042
Oborovo	41.1 ± 1.2	36.7 ± 1.1	88 ± 3	0.390±0.007
Galdovo	13.8 ± 0.4	16.3 ± 0.5	43 ± 1	0.091±0.005
Črnac	32.5 ± 1.0	30.8 ± 0.9	85 ± 3	0.366±0.004
Lukovec	32.7 ± 1.0	44.4 ± 1.3	99 ± 3	0.282±0.010
Košutarica	41.5 ±1.2	39.5 ± 1.2	127 ± 4	0.585±0.015
Gradiška	36.0 ± 1.1	41.9 ± 1.3	167 ± 5	0.629±0.025
Srbac	25.5 ± 0.5	79 ± 2	236 ± 5	0.376±0.009
Slavonski brod	25.5 ± 0.8	102 ± 3	186 ± 6	0.347±0.012
Županja	33.9 ±1.0	212 ± 7	381 ± 11	0.269±0.010
Brčko	52 ± 1	185 ± 4	312 ± 6	0.297±0.031
Bosanska Rača	122 ± 1	186 ± 4	273 ± 6	0.374±0.004
Sremska Mitr.	79 ± 1	177 ± 4	276 ± 6	0.444±0.091
Šabac	117 ± 1	163 ± 3	232 ± 5	0.624±0.012
Beograd	97 ± 1	82 ± 2	151 ± 3	0.275±0.030
ISQG	30.2	/	52.3	0.17
PEL	112	/	160	0.486

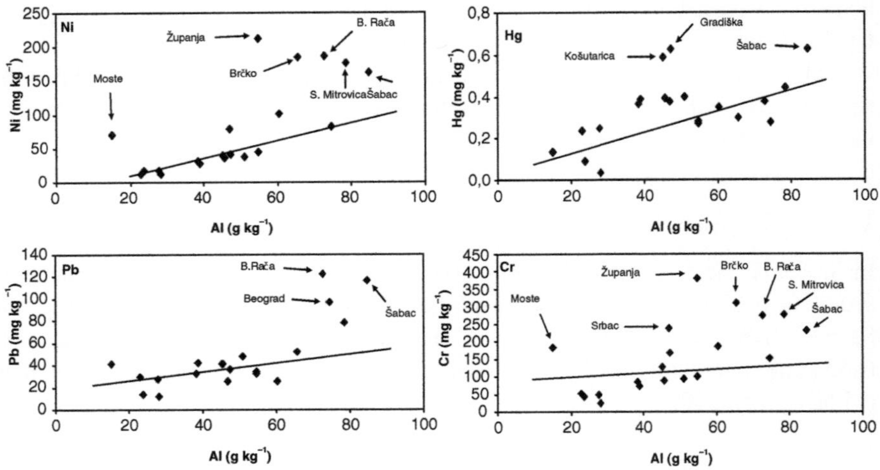

Fig. 13.19 Normalisation to Al of selected elements in sediments from the Sava River

sediments is presented in Fig. 13.19. To estimate the environmental status of sediments concentration of pollutants were compared to the Canadian Environmental Quality Guidelines (1999). In these guidelines concentrations of pollutants that support and maintain healthy aquatic life associated with bed sediments are obtained from available information on biological effects of sediment-associated chemicals. The Interim Sediment Quality Guidelines (ISQG) correspond to the threshold level effects below which adverse biological effects are not expected, while probable effects levels (PEL) characterise concentrations of pollutants that may affect the aquatic life. Data from Table 13.6 indicate that the lowest concentrations of elements are observed at sampling sites Mojstrana and Galdovo. Mojstrana is an unpolluted site close to the Sava Dolinka River spring (Ščančar et al. 2000), while Galdovo may not be considered as a representative site due to recent dredging of sediments at this sampling location.

Canadian Sediment Quality Guidelines for the Protection of Aquatic Life, 1999.

Interim Sediment Quality Guidelines (ISQG) corresponds to the threshold level effects below which adverse biological effects are not expected.

Probable effects levels (PEL) characterise concentrations of pollutants that may affect the aquatic life.

Data from Table 13.6 indicate that in general, the concentrations of elements in sediments of the Sava River gradually increase from the Sava River spring to its outflow to the Danube River. Among metals, Hg was found to be present in slightly elevated concentrations in the sediments of the Sava River. Concentrations in general ranged from 0.2 to 0.6 mg kg^{-1} Hg and at most sampling sites investigated exceeded the ISQG (0.17 mg kg–^{1} Hg) value. In Košutarica, Gradiška and Šabac Hg concentrations were around 0.6 mg kg^{-1} and exceeded the PEL value (0.486 mg kg^{-1} Hg). Normalisation to Al (Fig. 13.19) also indicated that higher values in Košutarica and Gradiška are most probably related to the oil refinery activities, while in Šabac the Hg input is most likely associated with the chemical industry. Slightly elevated Hg levels in the Slovenian part of the river were comparable to those reported previously (Kotnik et al. 2003), where Hg presence in the sediment was explained to be associated with former industrial pollution from a chemical plant in Hrastnik (until 1997 Hg cells were used in chlor-alkali production). Mercury concentrations in sediment are also comparable to the majority of sampling sites in the Danube River (around 0.4 mg kg^{-1} Hg) (ICPDR 2002). However, the concentrations of Hg in sediments of the Sava River are lower than in deposited sediment of the Seine River in France where Hg concentrations were around 1 mg kg^{-1} (Meybeck et al. 2007) and were much lower than those reported for contaminated sediments of the Soča River (10–20 mg kg^{-1} Hg) due to former mercury mining activities in Idrija, Slovenia (Hines et al. 2006).

Results from the data of Table 13.6 also indicate that the Sava River is moderately polluted with Cr and Ni at sampling site Moste in Slovenia (Acroni steel-works) and at sampling sites in Croatia, Bosnia and Herzegovina and Serbia, from Gradiška up to Šabac. At these sampling locations the concentrations of Cr are higher than those of PEL values (160 mg kg^{-1} Cr), and at most sampling sites investigated higher also than ISQG values (52.3 mg kg^{-1} Cr). For Ni there are no data on Canadian sediment quality standards. Normalisation data to Al indicated the same pattern of Cr and Ni inputs to sediments. At sampling site Moste the Cr and Ni contamination arises from the Acroni steel-works, while sampling sites from Gradiška up to Šabac indicate heavy industry and chemical industry activities along the Sava River in this area. The concentrations of Cr in industrial impacted sites range from 180 up to 380 mg kg^{-1} Cr and of Ni from 70 up to 200 mg kg^{-1} Ni. These concentrations are comparable to those determined in the sediments of the River Po in Italy (Vignati et al. 2003). However, the concentrations of Cr and Ni in the Sava River sediments are higher than most of the Cr and Ni values reported for the Danube River sediments, in which concentrations of both elements in general did not exceed 100 mg kg^{-1} (ICPDR 2002). Data from Table 13.6 further indicate that most of the Pb concentrations in the sediments of the Sava River exceeded the ISQG value (30.2 mg kg^{-1}) and at two sampling sites the PEL value (112 mg kg^{-1} Pb). From normalisation data to Al (Fig. 13.19) the anthropogenic input of Pb in Belgrade arises presumably due to heavy city traffic, while in Bosanska Rača high Pb concentrations in sediments are related to heavy traffic on the border between Bosnia and Herzegovina and Serbia. In Šabac, anthropogenic input of Pb in sediments is most likely associated with the chemical industry. Pb concentrations in the Sava River are in general comparable to concentrations of Pb in the Danube River (30–100 mg kg^{-1} Pb) (ICPDR 2002) and are much lower than those reported for the mining area (100–9,000 mg kg^{-1} Pb) (Svete et al. 2001). Cd concentrations in sediments of the Sava River in general did not exceed the ISQG value (0.7 mg kg^{-1}) (Milačič et al. 2009) and in general ranged from 0.22 to 1.4 mg kg^{-1}. These concentrations are in general lower than those determined in sediments of the Danube River (ICPDR 2002). Cd concentrations are also much lower than those determined at mining area sites (2–130 mg kg^{-1} Cd) (Svete et al. 2001). Zn concentrations in sediments of the Sava River ranged from 55 to 360 mg kg^{-1} and in general exceeded the ISQG value (124 mg kg^{-1} Zn) (Milačič et al. 2009). The determined Zn concentrations in sediments of the Sava River are comparable to most data reported for the Danube River (ICPDR 2002), but much lower than those determined in sediments at the mining areas of the Mežica valley, Slovenia (Svete et al. 2001). Concentrations of Cu in Sava River sediments in general ranged from 30 to 50 mg kg^{-1} and exceeded the ISQG value (18.7 mg kg^{-1} Cu), but were lower than the PEL value (108 mg kg^{-1} Cu) (Milačič et al. 2009). Concentrations of Cu in sediments of the Sava River do not represent anthropogenic inputs and are lower than reported for sediments of the Danube River (ICPDR 2002) and much lower than at contaminated sites (exploiting Cu) (Santos et al. 2003). Concentrations of As in the Sava River sediments range from 7 to 25 mg kg^{-1} and in general exceeded the ISQG value (7.24 mg kg^{-1}As) (Milačič et al. 2009). However, these As concentrations do not reflect anthropogenic inputs and are characterised by its natural background. Lastly, concentrations of total P in the sediments along the Sava River tend to increase from the spring toward the inflow into the Danube River. The highest concentrations (around 1,000 mg kg^{-1} of total P) were found before the accumulation dam of the hydroelectric power plant Vrhovo (Slovenia) and are impacted mostly by rural activities and downstream, in Croatia, Bosnia and Herzegovina and Serbia, where the use of P-containing fertilisers in agriculture are predominantly responsible for higher values found in the river sediments. In addition, the influence of the municipal sewage outflows (use of P-containing detergents in households) in big cities is also recorded by high P concentrations (around 1,000 mg kg^{-1} of total P) at sampling site Oborovo (outflow of the municipal sewage system in Zagreb, Croatia) and at the sampling site in the city of Belgrade before the Sava merges with the Danube River (Milačič et al. 2009). Similar P concentrations were found in sediments of the Danube River (ICPDR 2002).

13.4.3.4 Analysis of the Easily Soluble Element Fractions of Sediment of the Sava River

In order to estimate the extent of pollution of selected elements in sediments of the Sava River, extraction in 0.11 mol l^{-1} acetic acid was performed. In Fig. 13.20 the percentages of the easily soluble fractions of selected elements in sediments of the Sava River are presented. Data from Fig. 13.20 indicated that the percentages of the easily soluble metal fraction of Cr and Ni were low (below 0.3% of total Cr and below 16% of total Ni, respectively). Since these two elements exist primarily in the sparingly soluble forms, it is assumed that total Cr and Ni concentrations in sediments at industrially exposed sites do not represent a critical environmental burden. Analysis of the easily soluble concentrations of other elements was in general below 10% indicating their low mobility into the aquatic environment (Milačič et al. 2009). Exceptions were Cd (30–50%) and Zn (5–40%). Despite the high percentage of the easily soluble Cd content, these concentrations do not represent an environmental hazard, since the total Cd concentrations were low (Milačič et al. 2009).

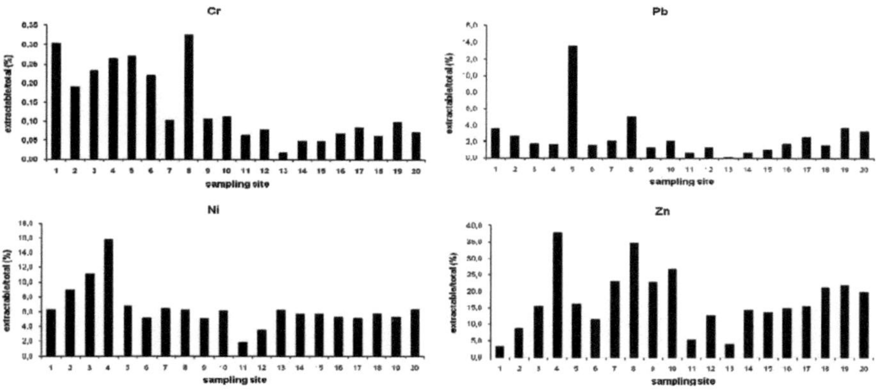

Fig. 13.20 Percentage of selected elements extracted from sediments of the Sava River with 0.11 mol l^{-1} acetic acid

13.4.3.5 Analysis of OTC in Sediments of the Sava River

Analysis of sediments from 20 locations in the Sava River indicated that concentrations of butyltins, octyltins and phenyltins were <2 ng Sn g^{-1} for butyltins, <3 ng Sn g^{-1} for octyltins and <10 ng Sn g^{-1} for phenyltins, respectively. So, sediments of the Sava River are not polluted with organotin compounds. Similar findings were reported for the Danube River sediments (ICPDR 2002) where, in general, TBT was not detected.

13.4.3.6 Analysis of PAH

Sediment samples collected downstream the Sava River were analysed and the content of 16 PAHs: Naphthalene, Anthracene, Phenanthrene, Fluoranthene, Benzo(a)anthracene, Chrysene, Benzo(a)pyrene, Benzo(g,h,i)perylene, Benzo-(k)fluoranthene, Indeno(1,2,3)pyrene, Acenaphtylene, Benzo(b)fluoranthene, Acenaphthene, Fluorene, Dibenzo(a,b)anthracene and Phenanthrene and their methylated analogues were determined (Heath et al. 2009). Table 13.7 shows the sum of 16 PAHs determined in sediment sample downstream Sava River.

Results revealed increasing PAH values downstream Črnac with four sites with significantly elevated PAH (the sum of 16 PAHs) concentrations, for example Županja, Brčko (up to 4,000 ng g^{-1}) and Bosanska Rača, Gradiška (about 2,000 ng g^{-1}) (Table 13.7). All four locations with elevated concentrations are situated downstream of oil fields (Črnac, Lukavec). The Canadian Environmental Quality Guidelines for separate PAHs in sediments quotes ISQG 6–111 ng g^{-1} and PEL 88–2,355 ng g^{-1}. Therefore, except for the above-mentioned locations, pollution with PAHs in sediments can be considered moderate along the Sava River.

When comparing PAH pollution in the Danube River sediments with the Sava River sediments, it was found that the Sava River sediment samples had generally lower PAH content determined. For instance, in the Danube River sediments

Table 13.7 Sum of 16 PAHs and 7 PCBs determined in the sediments of the Sava River (ng g^{-1})

	Sum 16 PAH	Sum 7 PCBs*
Mojstrana	55.46	0.308
Moste	467.01	1.912
Jevnica	178.35	1.221
Vrhovo	291.47	0.768
Brežice	51.30	0.545
Jesenice na Dol.	57.28	0.631
Oborovo	448.48	3.844
Galdovo	264.69	2.057
Črnac	1,185.90	2.242
Lukavec	1,044.25	0.255
Košutarica	1,254.62	5.482
Gradiška	1,900.37	2.310
Srbac	724.11	0.843
Slavonski brod	751.30	2.001
Županja	3,965.22	1.624
Brčko	3,550.37	2.123
Bosanska Rača	1,962.68	3.357
Sremska Mitrovica	966.77	3.273
Šabac	1,004.80	2.800
Beograd	334.43	3.410

*28: 2,4,4′-trichlorobiphenyl, 52: 2,2′5,5′-tetrachlorobiphelyl,101: 2,2′,4, 5,5′-pentachlorobiphenyl, 118: 2,3′,4,4′5-pentachlorobiphenyl, 138: 2,2,3, 4,4′,5′-hexachlorobiphenyl, 153: 2,2′,4,4′,5,5′-hexachlorobiphenyl, 180: 2, 2′,3,4,4′,5,5′-heptachlorobiphenyl

the PAH contamination profile (ICPDR 2002) is dominated by Phenanthrene and Anthracene in the majority of samples (up to 8 mg kg^{-1}), however the sum of the 16 PAHs rarely reached 2 mg kg^{-1}. Sava River sediments (Table 13.7) were more evenly polluted with regards to individual PAHs, where concentrations of individual PAHs (Heath et al. 2009) were in all cases below 1 mg kg^{-1}. In agreement with the above observations, the four most polluted sites have the highest contents of Phenantherene, Fluoranthene, Pyrene, Benzo (a) Anthracene and Chrysene (Heath et al. 2009).

A common method for estimating the source of PAH pollution derives from calculation of specific ratios of alkylated PAH and parent PAH (Methylphenanthrene/Phenanthrene and Methylpyrene/Pyrene) (Notar et al. 2001). In general, if the ratio between Methylphenanthrene/Phenanthrene (MePh/Ph) is between 0.5 and 1 and the ratio between Methylpyrene/Pyrene (MePy/Py) is smaller than 1, it can be assumed that the main sources of pollution are combustion processes. If the MePh/Ph ratio is between 2 and 6 and MePy/Py is greater than 2, there is a strong indication of fossil fuel pollution. Generally, the presence of Retene indicates forest fires as a source of PAH. Calculated MePh/Ph ratios (data not shown) show values between 0.5 and 1 at the following locations: Sava Moste, Vrhovo, Brežice and Jesenice, which was consistent with MePy/Py values below 1 (data not shown).

These data indicate that the main pollution in the Northern part of the Sava River (Slovenia) is derived from combustion processes which may result from local coal and wood heating. No data showed fossil fuels to be the source of PAH which was surprising since heavy petrochemical industry is located around Sisak (Croatia). Elevated Retene concentrations indicate five potential spots (data not shown) polluted with forest-fire PAHs (Vrhovo, Brežice, Jesenice, Srbac and Brčko), however no data were available to confirm these results.

13.4.3.7 Analysis of PCB

As an indicator for PCB pollution seven indicator PCBs were determined (seven congeners: 28: 2,4,4'-trichlorobiphenyl, 52: 2,2'5,5'-tetrachlorobiphelyl, 101: 2,2', 4,5,5'-pentachlorobiphenyl, 118: 2,3',4,4'5-pentachlorobiphenyl, 138: 2,2,3,4,4',5' -hexachlorobiphenyl, 153: 2,2',4,4',5,5'-hexachlorobiphenyl, 180: 2,2',3,4,4',5,5' -heptachlorobiphenyl) (Heath et al. 2009).

The sums of seven analysed PCBs downstream Sava River are presented in Table 13.7. Results showed no elevated concentrations in sampling sites downstream the Sava River (up to 6 ng g^{-1}). Even though Canadian Environmental Quality Guidelines quotes ISQG to be 34.1 ng g^{-1} and PEL for total PCBs 277 ng g^{-1}, it can be concluded that PCB pollution is not present downstream of the Sava River in significant levels (Table 13.7). However, slightly elevated values were shown at Košutarica sampling site which might be the influence of industrial activities.

Research in the Danube River (ICPDR 2002) showed that the sum of indicator PCBs never exceeded 0.005 mg kg^{-1}. In the Sava River, this number was exceeded only at the Košutarica sampling site. At all other locations the values were significantly below 5 mg kg^{-1}. Therefore, it may be concluded that pollution of both river basins is at comparable levels.

13.4.3.8 Analysis of Selected Chlorinated Pesticides

The presence of selected halogenated pesticides (Hexachlorobenzene, Heptachlor, Aldrine, p,p DDE, Lindane, p,p-DDD, p,p-DDT, Dieldrine and Endrine) was also evaluated in the Sava River sediments (Heath et al. 2009). Results of the three most abundant chlorinated pesticides (Hexachlorobenzene, p,p DDE and p,p-DDT) are presented in Table 13.8. Results revealed no elevated concentrations regarding selected pesticides except near Belgrade where hexachlorobenzene (HCB) was determined to be 90.8 ng g^{-1}. HCB was in the past widely used as a pesticide and also as a component for ammunition production. Even though no evidence exists, high HCB content in Belgrade sediment could be a result of the recent war conflict. However, repeated sampling in June 2007 at roughly the same location showed low HCB values (0.39 ng g^{-1}). To exclude analytical error, sediment collected in previous sampling was analysed again and high HCB content was confirmed (98.98 ng g^{-1}). These results indicated a point pollution of sediments of the Sava River with HCB near Belgrade. Among other chlorinated pesticides studied, most of the values were below 1 ng g^{-1}, except p,p-DDT at Galdovo (2.56 ng g^{-1}) and Košutarica

Table 13.8 Concentrations of most abundant chlorinated pesticides determined in the sediments of the Sava River

Pesticide (ng g^{-1})*	HCB	p,p DDE	p,p DDT
Mojstrana	0.007	0.004	0.002
Moste	0.076	0.095	0.171
Jevnica	0.195	0.095	0.005
Vrhovo	0.035	0.055	0.116
Brežice	0.096	0.062	0.281
Jesenice na Dol.	0.051	0.042	0.036
Oborovo	0.011	0.282	0.173
Galdovo	0.002	0.082	2.562
Črnac	0.008	0.131	0.163
Lukavec	0.130	0.350	0.145
Košutarica	0.859	0.277	1.288
Gradiška	0.476	0.092	0.248
Srbac	0.161	0.023	0.033
Slavonski Brod	0.341	0.187	0.063
Županja	0.096	0.283	0.203
Brčko	0.162	0.099	0.220
Rača	0.109	0.559	0.182
Sremska Mitrovica	0.612	0.090	0.259
Šabac	1.101	0.287	0.486
Beograd	90.823	0.582	0.649

* Nine halogenated pesticides were evaluated, here results of the three most abundant are shown.

(1.29 ng g^{-1}) and Endrine at Županja (0.98 ng g^{-1}) which may be a side-effect of intense agricultural activities in this area of the Sava River Basin (Heath et al. 2009). Canadian Environmental Quality Guidelines for separate pesticides are reported to be 1–4 ng g^{-1} (ISQG) and 4–65 ng g^{-1} (PEL) which confirms no significant organochlorine compound pollution levels in the Sava River Basin.

When comparing chlorinated pesticide values in sediments of the Danube and Sava River Basins it may be seen that HCB, which was elevated near the Belgrade (90.8 ng g^{-1}) Sava sampling site was also reported in higher values near the Budapest sampling site (23 ng g^{-1}) of the Danube River. Already the Budapest value had slightly exceeded the Canadian "Lowest effect Level" for HCB (ICPDR 2002) in sediments. However, repeated sampling showed significantly lower HCB values, indicating point pollution of the Sava River sediments. Among the rest of the chlorinated pesticides analysed in the Sava and Danube sediments, the Sava River generally contained lower values of all chlorinated pesticides analysed.

13.4.4 Conclusions

Data of the analysis of sediments sampled at 20 locations along the Sava River from its spring to its inflow into the Danube River indicate moderate elevation of

Hg in sediments (up to 0.6 mg kg^{-1}) and Cr and Ni (up to 400 and 210 mg kg^{-1}, respectively) in industrially impacted sites. However, the latter two elements exist primarily in sparingly soluble forms and therefore do not represent a heavy environmental burden. Pb, Zn, Cu, Cd and As in sediments of the Sava River were in general lower than PEL values and did not exhibit substantial pollution. P concentrations were slightly elevated (concentrations around 1,000 mg kg^{-1} of total P) in agricultural areas and close to big cities. Analysis of organic pollutants indicated that the Sava River is not polluted with butyltin, phenyltin or octyltin compounds. Among other organic pollutants PAH were present in sediments of the Sava River. Their concentrations increased downstream the Sava River. Concentrations of PCB in sediments of the Sava River were low. From selected pesticides p,p-DDT were found in sediments of Galdovo and Košutarica although the use of this persistent pesticide has been banned for many years. HCB pesticide was also determined in higher concentration at Šabac and in particularly high concentration at the Belgrade sampling site. However, in repeated sampling, it was confirmed that near Belgrade point pollution of HCB occurred.

The environmental status of sediments of the Sava River is in general comparable to the Danube River. However, Cd concentrations in sediments of the Sava River were in general lower, while Cr and Ni concentrations were in general higher than those determined in the Danube River. According to targeted organic analyses (organotin compounds, PAHs, PCBs and selected chlorinated pesticides) performed within the SARIB project it may be concluded that, with exception to some sampling sites (Županja, Brčko, Bosanska Rača, Gradiška and Belgrade), the pollution of the Sava River sediments is low and generally lower than the pollution of the Danube River sediments. The International Sava River Basin Commission, stakeholders, water management institutes and local authorities of the riparian countries may use the data of the investigation made within the SARIB project as a base for sustainable use, management and protection of the Sava River water resources.

13.5 Water Cycling Processes

13.5.1 Introduction

The primary objective of this study was to investigate hydrogeochemical dynamics using major elemental and stable isotopes to better understand transport and processes (natural and anthropogenic) within the Sava watershed. Special attention has been paid to carbon and nitrogen dynamics in the riverine system. For sustainable management of water supply, agriculture, flood–drought cycles and ecosystems, and human health, there is a basic need to improve the scientific understanding of water cycling processes in river basins. Additionally, based on the data obtained and experience, we are able to estimate the weakness and strengths of the studied catchments.

13.5.2 Sampling and Analytical Procedures

Sampling campaigns were carried out through three sampling seasons (autumn 2005, spring 2006 and fall 2006) at 33 selected sampling locations, 22 on the main river and 11 on the Sava river tributaries, from the source of the Sava river to Belgrade at its confluence with the Danube (Fig. 13.3). Samples were collected in October 2005, in May 2006, including three additional Bosnian sites (Srbac, Brčko, Bosanska Rača), and in October 2006. Sampling points were located at gauging stations on the Sava and at the mouth of tributaries before their confluence with the river. The following main tributaries of the river Sava were included in our study: Tržiška Bistrica, Kokra, Sora, Kamniška Bistrica, Savinja, Krka, (Slovenia) Kolpa/Kupa, Una, (Croatia/Bosnia) Vrbas, Bosna and Drina (Bosnia/ Serbia).

Samples of water and particulate material were subjected to stable isotope and chemical analysis. Temperature (T) and pH of samples were measured directly in the field. The field pH was determined on the NBS scale using two buffer calibrations with a reproducibility of ±0.02 pH unit. Samples for metals (Ca, Mg, K, Na, Al, Fe, SiO_2) were pre-treated with "suprapure" HNO_3. Aliquots for determination of total bicarbonate (HCO_3–), alkalinity, and concentrations of anions (Cl–, NO_3–, $SO_4{}^{2-}$) were stored in acid-washed HDPE bottles with no pre-treatment. In addition, samples for determination of stable isotopic composition of dissolved nitrogen ($\delta^{15}N_{NO3}$) were immediately filtered through a 0.2 μm filter. Samples for dissolved organic carbon (DOC) analyses were filtered, acidified, and stored at 4°C. Samples for carbon isotope analyses of dissolved inorganic carbon ($\delta^{13}C_{DIC}$) were preserved with $CuCl_2$ and capped in glass serum vials filled with no headspace. In addition, samples were taken for determining stable isotopic composition of sulphate ($\delta^{35}S_{SO4}$).

Particulate material for stable C and N isotopic analyses ($\delta^{13}C_{POC}$, $\delta^{15}N_{PN}$) was collected on pre-combusted Whatman glass-fibre filters (GF/F) by filtering 1–2.5 l of water using a pump connected to a Teflon filter holder. Filters were then wrapped in aluminium foil. Soil samples (n = 40) were collected from 13 locations along the riverbanks at different depths in the soil (usually 0, 5, 15, 20 and 25 cm), dried at 60°C, and ground and homogenised prior to isotope analysis.

Analyses for chemical and stable carbon, nitrogen and sulphur isotopes in Clark and Fritz 1997, Szramek et al. 2007, Kanduč et al. 2008, Ogrinc et al. 2008.

Stream discharge data from specific stations were obtained from the Environmental Agencies and related national institutions responsible for monitoring programmes in Slovenia, Croatia, Bosnia and Herzegovina and Serbia.

The partial pressure of CO_2 (pCO_2) and mineral saturation states for calcite and dolomite were calculated using the PHREEQC speciation program (Parkhurst and Appelo 1999). Principal component analysis (PCA) and factor analysis (FA) were applied to all data sets using data analysis software system STATISTICA (StatSoft inc. 2001).

13.5.3 Chemical Characteristics of Water in the Sava Watershed

Cation levels in the Sava take the following order: Ca^{2+} >> Mg^{2+} > Na^{+}> K^{+}, and the anion levels: HCO^{3-} >> $SO_4{}^{2-}$ ≥ Cl^{-} > NO^{3-}, which are the same as the distributions in global discharge-weighted rivers described by Meybeck (1987). The Sava waters are dominantly $HCO_3{}^{-}$ -Ca^{2+}-Mg^{2+} with only minor contributions from sulphate and chloride. In the upper Sava, dissolved Ca^{2+} and Mg^{2+} ions are largely supplied by the weathering of carbonate rocks, with smaller contributions from silicate weathering, as indicated by low Na^{+}, K^{+} and total dissolved Al concentrations. Carbonate mineral weathering dominates the geochemistry of the watershed. The Mg^{2+}/Ca^{2+} ratio <0.33 indicates that calcite mineral weathering dominates over dolomite mineral weathering. On the other hand, in the upper part of the Sava in Slovenia, there are nearly equal proportions of dolomite and calcite weathering (Szramek et al. 2007). $HCO_3{}^{-}$ and pCO_2 were observed to increase with distance from the source reflecting increased soil thickness and alluvium in the watershed. Silicate weathering is limited and difficult to separate from carbonate weathering and anthropogenic pollution. The data indicate that silicate weathering within the Sava watershed in Slovenia is limited and that less than 5% of $HCO_3{}^{-}$ from mineral weathering is from silicate mineral weathering (Szramek 2006). H_4SiO_4 (reported as SiO_2) does not show any clear trend downriver, with average values of 0.07 mmol l^{-1} found in autumn 2005 and 0.04 mmol l^{-1} in spring 2006. The latter lower concentrations could be the consequence of silica uptake by algae for formation of their frustules. The high contribution of chemical weathering is also evident when pollution is defined as the relative contribution of chloride, sulphate, and nitrate to the total anion concentration. It was found that over 80% of ions were derived from natural weathering processes in the main stream of the river, while the tributaries were more polluted, the % pollution reaching up to the 45% determined in the River Bosna (sampling location 26).

While the upper part of the Sava is controlled mainly by weathering of minerals, the lower part in Slovenia and partially in Croatia is subject to anthropogenic influence, mainly by agricultural activity. Nitrate inputs are controlled by land use in the Sava watershed, with the highest concentrations recorded in reaches dominated by hay fields and agricultural land in Croatia. The highest concentrations of $SO_4{}^{2-}$, Na^{+} and Cl^{-} were determined in the River Bosna, the main tributary of the Sava. The higher concentrations of Na^{+} and Cl^{-} were probably due to the influence of the salt mine at Tuzla, while higher $SO_4{}^{2-}$ concentrations found also in the River Savinja (sampling location 10) were the consequence of industrial activity.

Principal component analysis (PCA), based on all chemical variables and all locations and seasons, identifies the major components controlling the overall chemistry of the Sava and its tributaries. PCA on the entire data set identified five PCs with eigenvalues >1, explaining about 71% of the total variance in the entire data set. The dominant component 1 that explains 33.6% of the total variance has strong positive loading on conductivity, Na^{+} and Cl^{-} and moderate positive loadings on

Ca^{2+}, Mg^{2+} and SO_4^{2-}. This component represents predominantly the geogene processes of carbonate and feldspar weathering. On the other hand, strong positive loading of Na^+ and Cl^- cannot be attributed just to dissolution of halite, but to non-natural sources such as the Tuzla salt mine. The second component (11.0%) shows strong loadings with DOC and SiO_2 but no correlation with other parameters. The third component (10.0%) shows significant loading with HCO_3^- and NO_3^-. In the fourth component, strong loading of total dissolved Al and Fe was observed and, in the fifth, strong negative loading of T and concentration of dissolved oxygen. Components 2, 3 and 5 represents anthropogenic pollution sources and can be explained by the consumption of oxygen being a consequence of degradation of organic material, leading to decreased pH values. During this process, higher NO_3^- loadings, probably originating from fertilisers, promote the algal bloom that uses Si for formation of their frustules. On the other hand, the fourth component participated with total dissolved Al and Fe and indicated the origin being leachates from industrial wastes and/or manure piles.

Spatial variability shows that natural processes were more pronounced in the upper part of the Sava, including Upper Locations (sampling locations 1–7). Anthropogenic influence on the Sava watershed can further be seen from the higher loadings of K^+ and NO_3^- and DOC, attributed to greater agricultural activity in the Middle Locations (sampling locations 8–16), while at the Lower Locations (sampling locations 17–33) loads could be associated to leakage from municipal sewage systems. These are polluting sources, whose emissions are not related to seasonality but could be more pronounced during low water discharge. Seasonal variations allow the dilution process of the natural mineralisation during spring 2006 at higher discharge to be detected.

13.5.4 Sources and Transport of Carbon and Nitrogen

Carbon and nitrogen dynamics in the Sava watershed downstream to the Danube were studied by stable isotope and chemical approaches (Ogrinc et al. 2008). Riverine CO_2 concentrations were supersaturated up to forty-fold with respect to atmospheric equilibrium, resulting in large CO_2 emissions into the atmosphere. The total CO_2 efflux for the Sava at Belgrade sampling location 33 ranged between 4.97×10^7 and 3.1×10^8 mol day–^{1} in spring 2006 and between 3.18×10^7 mol day^{-1} and 1.98×10^8 mol day^{-1} in autumn 2006, representing between 6 and 19% of the river's dissolved inorganic carbon (DIC) transport. The overall annual DIC flux was estimated to be 2.1×10^{11} mol C. Thus, the Sava contributes ~0.7% of the global river carbon flux of 2.67×10^{13} mol C year^{-1} (Ludwig et al. 1996) and 23% of the annual DIC of the Danube. It was found that the flux of 2.2×10^6 mol C/(km^2 year), estimated on an areal basis, is up to 7-times higher in the Sava than in all other large rivers, including the Danube, due to the high weathering intensity (Szramek et al. 2007). Fluxes have been estimated, in mol C/(km^2 year), for the Danube, 1.0×10^6, St. Lawrence, 5.17×10^5, Yangtze, 8.76×10^5 and Amazon (3.12×10^5) rivers (Hélie et al. 2002, Probst et al. 1994, Szramek et al. 2007, Wu et al. 2007). As a result of the steady concentrations, the extremely high flux from the Sava is determined largely by the high discharge values. On the larger scale, the carbonate weathering intensity for the Danube Basin is inversely related to the catchment area; therefore, the Sava carbonate watershed has the highest weathering intensity within the basin. The Danube has a carbonate weathering intensity typical of that for all European rivers (Szramek et al. 2007). The annual organic carbon flux was lower and divided equally between DOC (2.1×10^{10} mol C) and POC (particulate organic carbon) (4.1×10^{9} mol C). The molar proportions of DIC:DOC:POC were 89:9:2, distinct from the mean proportions for world rivers draining to the oceans (DIC:DOC:POC = 45:37:18; Meybeck 1993) because of the greater contribution of DIC in the Sava.

According to the isotopic mass balance, tributaries of the Sava account for the major input of DIC flux (up to 60%) at its confluence with the Danube at Belgrade. Other processes influencing the production of DIC are carbonate dissolution, contributing between 32 and 42% of DIC, and respiratory CO_2 from the degradation of organic material, contributing 20–23% of DIC. These proportions are higher than those observed at the mouth of the Sava at the Croatian border, where carbonate dissolution contributes up to 26%, degradation of organic matter about 17% and exchange with atmospheric CO_2 up to 5% (Kanduč et al. 2007).

There was considerable temporal and spatial variation in the composition of particulate organic matter (POM) throughout the river system. On the basis of $\delta^{13}C_{POC}$ and $\delta^{15}N_{PN}$ and on C/N ratios, it was possible to identify four important sources of POM: phytoplankton, soil, fresh terrestrial derived material and aquatic vascular plants. The lowest $\delta^{13}C_{POC}$ values of ~30.7 and ~30.3‰, together with the corresponding C/N ratios of 6.7 and 7.9, indicate that phytoplankton was the major source of POM in the agricultural parts of the Sava. In these parts, the diatom alga *Stephanodiscus hantzschii*, an indicator of eutrophic conditions, was abundant. Periodic inputs of fresh terrestrial plant detritus are suggested by the high C/N ratio (>15), while the presence of macrophytes could be established by the $\delta^{15}N$ values. Overall it was found that, at ~59% of sampling sites, soil organic matter, which is largely refractory, was the major source of POM. Eighteen percent of the samples were dominated by plankton, 12% by periodic inputs of fresh terrestrial plant detritus, and about 11% by the contribution of aquatic vascular plants. $\delta^{15}N_{NO3}$ values ranged between +3.8 and +25.5‰ and did not correlate with NO_3^- concentrations, indicating that concentrations alone are insufficient to describe N sources. The sources of riverine N can be related to land use practices. In predominantly forested watersheds, nitrate was derived mainly from nitrification processes in soil, resulting in $\delta^{15}N_{NO3}$ values <6‰. Watersheds with a higher percentage of agricultural and/or urban land use typically showed $\delta^{15}N_{NO3}$ values between 6‰ and 9‰. Elevated $\delta^{15}N_{NO3}$ values up to +25.5‰ were caused predominantly by improperly treated sewage near large cities, such as Belgrade, and by manure-derived nitrate.

13.5.5 Sulphate and Isotopes of Sulphur

Sulphate concentrations vary seasonally from 0.02 in the spring sampling season to 0.34 mmol l^{-1} in the autumn sampling season, while a higher range was observed in tributaries, ranging from 0.08 mmol l^{-1} in spring to 0.59 mmol l^{-1} in autumn 2006. The highest sulphate concentrations were always observed in Bosna (sampling location 26) and Savinja (sampling location 10), arising from industry with sulphuric acid production, while in Tržiška Bistrica (sampling location 4) the higher concentrations were probably the consequence of the dissolution of Palaeozoic carbonates with evaporates forming the watershed. Elevated sulphate concentrations within the Sava watersheds can also be explained by anthropogenic input via acidic rain, which is commonly reported in Central European rivers. However, over 90% of measured sulphate concentrations in the Sava water samples were below 0.4 mmol l^{-1}. Upper stream locations showed lower concentrations, with mostly SO_4^{2-}/Cl^- molar ratios of 2:1, indicating atmospheric deposition as the source. The $\delta^{34}S_{SO4}$–depleted signature supports this assumption.

On the basis of the isotope data it is seen that two watersheds, the Tržiška Bistrica (sampling location 4) and Savinja (sampling location 10) with the highest sulphate concentrations (up to 0.6 mmol l^{-1}) have different sulphate sources. Kanduč and Ogrinc (2007) suggested that high SO_4^{2-} in the Savinja watershed was primarily associated with a productive agricultural region and a major industry. The Tržiška Bistrica is on the other hand a small watershed with the majority of the drainage area lying at high elevations. High $\delta^{34}S_{SO4}$ values suggested that SO_4^{2-} in this watershed is the result of gypsum and/or anhydrite weathering. The other two sampling locations, 23 and 26 (Rivers Drina and Bosna), also have higher $\delta^{34}S_{SO4}$ values of 9.5 and 11.7‰, respectively, indicating industrial influence on SO_4^{2-} concentrations.

From discharge and concentration measurements the annual sulphate flux was calculated to be 4.4×10^9 mol a^{-1} at Belgrade, sampling location 33. According to Ivanov (1983), the annual SO_4^{2-} flux from continents to oceans is 6.8×10^{12} mol a^{-1}, therefore the Sava accounts for 0.1% of the total sulphur flux. In order to determine different sources of sulphate in the Sava watershed, an isotope mass balance was performed at sampling location 33. Assuming that the sources of SO_4^{2-} in the Sava are the tributaries (F_{tri}), precipitation (F_p) and other sources (F_{other}), the contributions of these inputs were calculated to be 28, 8 and 64%, respectively. It was found that the four main tributaries contributing to the sulphate flux were the Savinja (11.8%), the Una (38.5%), the Vrbas (9.42%) and the Bosna (30.4%). The flux from other sources, that accounts for 64% of the SO_4^{2-}, may be derived from sources such as dissolution of evaporate minerals, the oxidation of sulphides, or from anthropogenic sources such as air pollution, smelting of sulphide ores and chemical industry. The fact that evaporates and sulphide minerals have a limited distribution in the bed rocks of the Sava channel, together with the calculated $\delta^{34}S_{other}$ value of 7.8‰, suggests that evaporates are not the missing contributor. Industrial pollution probably accounts for the bulk of the "other" flux. Unfortunately, the isotopic composition of sulphur from these industrial operations is not available and this precludes a more precise assessment of industrial impact on the Sava watershed. The research performed in the upper, Slovenian part of the Sava watershed indicated that the major source in the Sava watershed in Slovenia were tributaries (52%), other sources contributing only 40% (Kanduč and Ogrinc 2007).

13.6 Conclusions

The Sava River changes its own run-off regime from a torrential river in its upstream part in Slovenia to a slow lowland river in its downstream part. The run-off regime sharply changes over a short distance close to the town of Sisak, Croatia, Fig. 13.15. Changes in run-off regime cause also changes in sediment regime. The Sava River turns from a typical alpine gravel-bed river, that exhibits coarse sediment transport only a few times in a year during high flows into a sand-bed river with an important fraction of silt and clay that exhibits quite low bed-load transport. This known fact was again confirmed by an extensive field survey on gravel-bar sediments. The latest Sava River sediment sampling offers also grain size distribution data of the gravel-bar surfaces, a detail that was not given enough attention during previous field campaigns on river sediments (Mikoś, 1999). The analysis has shown that the arithmetic mean grain of the surface layer is on average approximately 1.5-times larger than the arithmetic mean grain of the subsurface layer.

Silty and clay riverbed sediments can accumulate higher amounts of solutes when compared to coarser sand or gravel sediments. A granulometric analysis of fine sediments from the Sava River (Figs. 13.17 and 13.18) does not show any clear downstream trend for fine fractions in samples. The fine fraction is much more influenced by fresh sediment supply from the main Sava River tributaries such as the Kolpa, Vrbas, and Drina rivers. This again shows the importance of so-called sedimentary links for the assessment of sediment regime (Rice 1999).

The environmental status of sediments of the Sava River is generally comparable to the Danube River sediment. However, Cd concentrations in sediments of the Sava River were in general lower, while Cr and Ni concentrations were in general higher than those determined for the Danube River. According to targeted organic analyses (organotin compounds, PAHs, PCBs and selected chlorinated pesticides) performed within the SARIB project it may be concluded that, with exception to some sampling sites mainly on the downstream part of the Sava River, the pollution of its sediments is low and generally lower than the pollution of the Danube River sediments.

A relational database model was developed in the MS ACCESS environment for the collation of sampling data collected by all partners during the SARIB project. The database contains data on measurement, laboratory tests and field observations carried out and collected for the SARIB project. It is available on (http://www.ksh.fgg.uni-lj.si/sarib/).

References

Anastasi G (1984) Geschiebeanalysen im Felde unter Berücksichtigung von Grobkomponenten. Mitteilungen VAW ETH Zürich 70:97

Barth JAC, Steidle D, Kuntz D, Gocht T, Movet C, von Tümpling W, Lobe I, Langenhoff A, Albrechtsen H-J, Janniche GS, Morasch B, Hunkeler D, Grathwohl P (2007) Depostion, persistence and turnover of pollutants: first results from the EU project Aqua Terra for selected river basins and aquifers. Sci Total Environ 376:40–50

Borovec Z (2000) Elements in size-fractionated bottom sediments of the Elbe River in its Czech part. Aquat Sci 62:232–251

Canadian Environmental Quality Guidelines (1999) Canadian Sediment Quality Guidelines for the Protection of Aquatic Life. Environment Canada Ottawa, Ontario. Canadian Council of Ministers of the Environment
Clark I, Fritz P (1997) Environmental isotopes in hydrology. Lewis publishers, New York
Colarič O (1983) Working notes – data on measurements, sampling and measures in the Sava River basin (Delovni zapiski – podatki o meritvah, vzorčevanjih in posegih na porečju Save). Water Management Institute, Ljubljana
Fehr R (1987) Geschiebeanalysen in Gebirgsflüssen. Mitteilungen VAW ETH Zürich 92:139
Giusti I, Taylor A (2007) Natural and anthropogenic contamination of the Fratta-Gorzone River (Veneto, Italy). Environ Monit Assess 134:211–231
Google (2008). Google maps. http://maps.google.com
Grosbois C, Meybeck M, Horowitz A, Ficht A (2006) The spatial and temporal trends of Cd, Cu, Hg, Pb and Zn in Seine River floodplain deposits (1994–2000). Sci Total Environ 356:22–37
Heath E, Ogrinc N, Faganeli J, Covelli S (2006) Sedimentary record of polycyclic aromatic hydrocarbons in the Gulf of Trieste (Northern Adriatic Sea). Water Air Soil pollut Focus 6:605–614
Heath E, Ščančar J, Zuliani T, Milačič R (2009) A complex investigation of the extent of pollution in sediments of the Sava River: Part 2: persistent organic pollutants. Environ Monit Assess. doi: 10.1007/s10661-009-0833-9
Hélie JF, Hillaire-Marcel C, Rondeau B (2002) Seasonal changes in the sources and fluxes of dissolved inorganic carbon through the St. Lawrence River – isotopic and chemical constraints. Chem Geol 186:117–138
Hines ME, Faganeli J, Adatto I, Horvat M (2006) Microbial mercury transformations in marine, estuarine and freshwater sediment downstream of the Idrija Mercury Mine, Slovenija. Appl Geochem 21:1924–1939
House WA, Denison FH (2002) Total phosphorus content of river sediments in relationship to calcium, iron and organic matter concentrations. Sci Total Environ 282–283: 341–351
ICPDR (International Commission for the Protection of the Danube River) Joint Danube Survey (2002) Technical Report of the International Commission for the protection of the Danube River
IEAS (2008) Interactive Environmental Atlas of Slovenia. http://kremen.arso.gov.si/NVatlas/
Ivanov MV (1983) The sulphur cycle in continental reservoirs. In: Ivanov MV, Freney RJ (eds) The Global biogeochemical sulphur cycle, SCOPE, vol 19. Wiley, Chichester, pp 297–356
Kanduč T, Ogrinc N (2007) Hydrogeochemical characteristics of the River Sava watershed in Slovenia. Geologija 50/1:157–177
Kanduč T, Szramek K, Ogrinc N, Walter LM (2007) Origin and cycling of riverine inorganic carbon in the Sava River watershed (Slovenia) inferred from major solutes and stable carbon isotopes. Biogeochemistry 86:137–154
Kanduč M, Trulsson M, Naji A, Burak Y, Forsam J, Podgornik R (2008) Weak- and strong-coupling electrostatic interactions between asymmetrically charged planar surfaces. Phys Rev E Stat Nonlin Soft Matter Phys 78: 061105-1–061105-14
Kotnik J, Horvat M, Milačič R, Ščančar J, Fajon V, Kryžanowski A (2003) Heavy metals in the sediments of the Sava River, Slovenia. Geologija 46:263–272
Loring DH, Rantala RRT (1992) Manual for the geochemical analysis of marine sediments and suspended particulate matter. Earth Sci Rev 32:325
Ludwig W, Amiotte Suchet P, Probst JL (1996) River discharges of carbon to the world's oceans; determining local inputs of alkalinity and of dissolved and particulate organic carbon. Comptes Rendus de l'Academie des Sciences, Serie II. Sciences de la Terre et des Planetes 323:1007–10014
Meybeck M (1987) Global chemical weathering of surficial rocks estimated from river dissolved loads. Am J Sci 287:401–428
Meybeck M (1993) Riverine transport of atmospheric carbon sources, global typology and budget. Water Air Soil Pollut 70:443–463

Meybeck M, Lestel L, Bonté P, Moilleron R, Colin J-L, Rousselot O, Hervé D, de Pontèves C, Grosbois C, Thevénot DR (2007) Historical perspective of heavy metals contamination (Cd, Cr, Cu, Hg, Pb, Zn) in the Seine River basin (France) following a DPISIR approach (1950–2005). Sci Total Environ 375:204–231

Mikoš M (1999) Problems with river sediments in the reservoirs of the hydro power plants on the Sava River – expertise (Problematika rečnih plavin v zajezitvenih prostorih hidroelektrarn na reki Savi – strokovno mnenje). University of Ljubljana (UL), Faculty of Civil and Geodetic Engineering, Ljubljana

Mikoš M (2000) Sediment budget of the Sava River from Jesenice to Mokrice (Prodna bilanca reke Save od Jesenic do Mokric). Gradbeni vestnik 49:208–219

Mikoš M (2008) River engineering – scriptum (Urejanje vodotokov – skripta). University of Ljubljana (UL), Faculty of Civil and Geodetic Engineering, Ljubljana

Milačič R, Ščančar J, Murko S, Kocman D, Horvat M (2009) A complex investigation of the extent of pollution in sediments of the Sava River: Part 1: selected elements. Environ Monit Assess. doi: 10.1007/s10661-009-0832-x

Notar M, Leskovšek H, Faganeli J (2001) Composition, distribution and sources of polycyclic aromatic hydrocarbons in sediments of the Gulf of Trieste, Northern Adriatic Sea. Mar Pollut Bull 42:36–44

Ogrinc N, Markovics R, Kanduč T, Walter LM, Hamilton SK (2008) Sources and transport of carbon and nitrogen in the River Sava watershed, a major tributary of the river Danube. Applied Geochemistry 23:3685–3698

Parkhurst DL, Appelo CAJ (1999) User's guide to PHREEQC (version 2)-a computer program for speciation, batch-reaction, one-dimensional transport, and inverse geochemical calculations. USGS Water-Resources Investigations Report 99-4259, Denver, CO

Probst JL, Mortatti J, Tardy Y (1994) Carbon river fluxes and weathering. CO_2 consumption in the Congo and Amazon River basin. Appl Geochem 9:1–13

Rice S (1999) The nature and controls on downstream fining within sedimentary links. J Sedim Res 69(1):32–39

Santos JC, Beltrán R, Gómez Ariza JL (2003) Spatial variations of heavy metals contamination in sediments from Odiel river (Southwest Spain). Environ Int 29:69–77

Savezni Hidrometeorološki Zavod (1980) Hidrološki godišnjak 1978 I. Savezni Hidrometeorološki Zavod, Beograd.

Ščančar J, Milačič R, Horvat M (2000) Comparison of various digestion and extraction procedures in analysis of heavy metals in sediments. Water Air Soil pollut 118:87–99

Ščančar J, Zuliani T, Turk T, Milačič R (2007) Organotin compounds and selected metals in the marine environment of Northern Adriatic Sea. Environ Monit Assess 127:271–282

Škrbić B, Čupić S (2004) Trace metal distribution in surface soils of Novi Sad and bank sediment of the Danube River. J Environ Sci Healt A 39:1547–1558

Škrbić B, Cvejanov J, Đurišić-Mladenović N (2005) Polycyclic aromatic hydrocarbons in surface soils of Novi Sad and bank sediment of the Danube River. J Environ Sci Healt A 40:29–42

Škrbić B, Cvejanov J, Đurišić-Mladenović N (2007) Organochloride pesticides and polychlorinated biphenyls in surface soils and bank sediment of the Danube River. J Environ Sci Healt B 42:311–319

Standard (1980) General classification of soils for engineering purposes (USCS). JUS U.B1.001

Standard (2004) Determination of particle size distribution. SIST/ISO/TS 17892-4:2004

StatSoft Inc. (2001) STATISTICA version 7.0. http://www.statsoftsa.com/

Svete P, Milačič R, Pihlar B (2001) Partitioning of Zn, Pb, and Cd in river sediments from a lead and zinc mining area using the BCR three-step extraction procedure. J Environ Monit 3:586–590

Szramek K (2006) Carbonate mineral weathering in mid-latitude watersheds: importance of calcite and dolomite dissolution on dissolved inorganic carbon acquisition and transport. PhD Thesis, University of Michigan

Szramek K, Williams E, Kanduč T, Ogrinc N, Walter Lynn M (2007) Relative weathering intensity of calcite vs. dolomite in carbonate-bearing temperate zone watershed: carbonate geochemistry and fluxes from catchments within the St. Lawrence and Danube River basins. Geochem Geophys Geosyst (G3) 8:1–26

VGI-VL (1980) Following sediment transport of main water courses in Slovenia – Phase I (Zasledovanje prodonosnosti glavnih vodotokov Slovenije vzdolž njihove dolžine, 1. faza). Water Management Institute – Hydraulic Laboratory (VGI-VL), Ljubljana

Vignati D, Pardos M, Diserens J, Ugazio G, Thomas R, Dominik J (2003) Characterisation of bed sediments and suspension of the river Po (Italy) during normal and high flow conditions. Water Res 37:2847–2864

Wang H, He M, Lin C, Quan X, Guo W, Yang Z (2007) Monitoring and assessment of persistent organochlorine residues in sediments from the Daliaohe River Watershed, Northeast of China. Environ Monit Assess 133:231–242

Wu Y, Zhang J, Liu SM, Zhang ZF, Yao QZ, Hong GH, Cooper L (2007) Sources and distribution of carbon within the Yangtze River system. Estuary Coastal and Shelf Science 71:13–25

www.milestonesci.com/merc-tech.php

Index

M. Brilly (ed.), *Hydrological Processes of the Danube River Basin*,
DOI 10.1007/978-90-481-3423-6, © Springer Science+Business Media B.V. 2010

D

MIX
Papier aus verantwortungsvollen Quellen
Paper from responsible sources
FSC® C105338

If you have any concerns about our products,
you can contact us on
ProductSafety@springernature.com

In case Publisher is established outside the EU,
the EU authorized representative is:
Springer Nature Customer Service Center GmbH
Europaplatz 3, 69115 Heidelberg, Germany

Printed by Libri Plureos GmbH
in Hamburg, Germany